HUMAN FACTORS IN ENGINEERING AND DESIGN

SIXTH EDITION

Mark S. Sanders, Ph.D.

California State University,
Northridge

Ernest J. McCormick, Ph.D.

Professor Emeritus of Psychological Sciences
Purdue University

McGRAW-HILL BOOK COMPANY

New York St. Louis San Francisco Auckland Bogotá
Hamburg Johannesburg London Madrid Mexico Milan Montreal New Delhi
Panama Paris São Paulo Singapore Sydney Tokyo Toronto

HUMAN FACTORS IN ENGINEERING AND DESIGN

1 2 3 4 5 6 7 8 9 0 DOCDOC 8 9 2 1 0 9 8 7

ISBN 0-07-044903-1

This book was set in Linotron by The Clarinda Company.
The editors were Jim Anker and John Morriss; the designer
was Carla Bauer; the production supervisor was Diane Renda.
Project Supervision was done by Chernow Editorial Services Inc.
R. R. Donnelley & Sons Company was printer and binder.

Library of Congress Cataloging-in-Publication Data

Sanders, Mark S.
    Human factors in engineering and design.

    McCormick's name appeared first on the 5th ed.
    Bibliography: p.
    Includes indexes.
    1. Human engineering. I. McCormick, Ernest J.
(Ernest James) II. McCormick, Ernest J. (Ernest James).
Human factors in engineering and design. III. Title.
TA166.S33 1987   620.8′2   86-10679
ISBN 0-07-044903-1

ABOUT
THE AUTHORS

Dr. Sanders received his M.S. and Ph.D. degrees in human factors from Purdue University. He is currently Professor of Psychology at California State University, Northridge, and Senior Staff Scientist at Essex Corporation. Also Dr. Sanders consults with various organizations and serves as an expert witness in cases involving human factors issues. He has executed or directed over 60 research and development contracts, subcontracts, and consulting activities. In addition, he has authored or coauthored over 70 technical reports, journal articles, and professional presentations. He received the Human Factors Society's Jack A. Kraft Award for his research on human factors issues in the mining industry. Dr. Sanders is a member of the Ergonomics Society, American Psychological Association, and Society of Automotive Engineers. He is a fellow in the Human Factors Society and has served that organization as secretary-treasurer and chair of the Education Committee and Educators Professional Group. Dr. Sanders is a member of the editorial board of *Human Factors* and *International Reviews of Ergonomics*.

Dr. Ernest J. McCormick is Professor Emeritus, Purdue University. His academic career as an industrial psychologist covered a span of 30 years at Purdue. His first edition of this text (then titled *Human Engineering*) was published in 1957. Dr. Sanders became a coauthor for the fifth (and now the sixth) edition. Dr. McCormick's other major publications include *Industrial and Organizational Psychology* (now in its eighth edition) and *Job Analysis: Methods and Applications*. He was responsible for development of the position analysis questionnaire (PAQ), a structured, computerized job analysis procedure being used by numerous organizations; he is president of PAQ Services, Inc. He has served on various advisory panels and committees, including the Army Scientific Advisory Panel, the Navy Advisory Board for Personnel Research, and the Committee on Occupational Classification and Analysis of the National Academy of Sciences. His awards include the Paul M. Fitts award of the Human Factors Society, the Franklin V. Taylor award of the Society of Engineering Psychologists, and the James McKeen Cattell award of the Society of Industrial and Organizational Psychology.

CONTENTS

PREFACE

This book deals with the field of *human factors,* or *ergonomics,* as it is also called. In simple terms, the term *human factors* refers to *designing for human use.* Five years ago, it would have been difficult to find very many people outside the human factors profession who could tell you what human factors or ergonomics was. Today, things are different. Human factors and ergonomics are in the news. Visual and somatic complaints of computer terminal users have been linked to poor human factors design. The incident at Three-Mile Island nuclear power station highlighted human factors deficiencies in the control room. The words *human factors* and especially *ergonomics* have also found their way into advertisements for automobiles, computer equipment, and even razors. The field is growing, as evidenced by the increase in the membership of human factors professional societies, in graduate programs in human factors, and in job opportunities.

We intended this book to be used as a textbook in upper-division and graduate-level human factors courses. We were also aware that this book has been an important resource for human factors professionals over the last five editions and 30 years. To balance these two purposes, we have emphasized the empirical research basis of human factors, we have stressed basic concepts and the human factors considerations involved in the topics covered, and we have supplied references for those who wish to delve into a particular area. We have tried to maintain a scholarly approach to the field, avoiding cutesy humor or inclusion of lots of pretty pictures that really do little to clarify the material. Unfortunately, there are times when our presentation may be a little technical or "dry," especially when we are presenting information that would be more appropriate for the practicing human factors specialist than for students. For this we apologize, but we hope that the book will be one students will want to keep as a valuable reference.

For students, we have written a workbook to accompany this text (published by Kendall-Hunt Publishing Co., Dubuque, Iowa). Included in the workbook, for each chapter, are a list of key terms and slef-contained projects that use concepts and information contained in this book.

There has been a virtual information explosion in the human factors field over the years. The first edition of this book, published in 1957, contained 16 chapters and 370 references. This edition contains 21 chapters and over 900 references. In 1972, the Human Factors Society (HFS) first published a proceedings of their annual meeting. It contained 106 papers and was 476 pages long. The proceedings for the 1985 HFS annual meeting contained over 300 papers and was 1164 pages long! In this book we have tried to cover both traditional and emerging areas of human factors, but it was impossible to include everything. The specific research material included in the text represents only a minute fraction of the vast amount that has been carried out in specific areas. It has been our intent to use as illustrative material examples of research that are relatively important or that adequately illustrate the central points in question. Although much of the specific material may not be forever remembered by the reader, we hope that the reader will at least develop a deep appreciation of the importance of considering human factors in the design of the features of the world in which we work and live. Appreciation is expressed to the many investigators whose research is cited. References to their work are included at the end of each chapter. To those investigators whose fine work we did not include, we apologize and trust they understand our predicament.

This edition represents some changes from the last edition, in addition to a general updating of the material (approximately one-third of the references are from the 1980s. Chapter 2 on information processing, for example, has been extensively revised to include current research and thinking. We have divided the visual displays chapter into two chapters (4 and 5) distinguishing between presentation of static and dynamic information. Human factors considerations related to computer systems are discussed in several chapters, including the chapters on visual displays (4 and 5), controls (10), illumination (14), and work station layout (12). We welcome comments and suggestions for making improvements in future editions.

Mark S. Sanders
Ernest J. McCormick

PART ONE

INTRODUCTION

CHAPTER 1

HUMAN FACTORS AND SYSTEMS

In the bygone millenia our ancestors lived in an essentially "natural" environment in which their existence virtually depended on what they could do directly with their hands (as in obtaining food) and with their feet (as in chasing prey, getting to food sources, and escaping from predators). Over the centuries they developed simple tools and utensils, and they constructed shelter for themselves to aid in the process of keeping alive and making life more tolerable.

The human race has come a long way from the days of primitive life to the present with our tremendous array of products and facilities that have been made possible with current technology, including physical accoutrements and facilities that simply could not have been imagined by our ancestors in their wildest dreams. In many civilizations of our present world, the majority of the "things" people use are made by people. Even those engaged in activities close to nature—fishing, farming, camping—use many such devices.

The current interest in human factors arises from the fact that technological developments have focused attention (in some cases dramatically) on the need to consider human beings in such developments. Have you ever used a tool, device, appliance, or machine and said to yourself, "What a dumb way to design this; it is so hard to use! If only they had done this or that, using it would be so much easier." If you have had such experiences, you have already begun to think in terms of human factors considerations in the design of things people use. In a sense, the goal of human factors is to guide the applications of technology in the direction of benefiting humanity. This text offers an overview of the human factors field; its various sections and chapters deal with some of the more important aspects of the field as they apply to such objectives.

HUMAN FACTORS DEFINED

Before attempting to define human factors, we should say a world about terms. *Human factors* is the term used in the United States and a few other countries. The term *ergonomics,* although used in the United States, is more prevalent in Europe and the rest of the world. Some people have tried to distinguish between the two, but we believe that any distinctions are arbitrary and that, for all practical purposes, the terms are synonymous. Another term that is occasionally seen (especially within the U.S. military) is *human engineering*. However, this term is less favored by the profession, and its use is waning. Finally, the term *engineering psychology* is used by some psychologists in the United States. Some have distinguished engineering psychology, as involving basic research on human capabilities and limitations, from human factors which is more concerned with the *application* of the information to the design of things. Suffice it to say, not everyone would agree with such a distinction.

We approach the definition of human factors in terms of its focus, objectives, and approach.

Focus of Human Factors

Human factors focuses on human beings and their interaction with products, equipment, facilities, procedures, and environments used in work and everyday living. The emphasis is on human beings (as opposed to engineering, where the emphasis is more on strictly technical engineering considerations) and how the design of things influences people. Human factors, then, seeks to change the things people use and the environments in which they use these things to better match the capabilities, limitations, and needs of people.

Objectives of Human Factors

Human factors has two major objectives. The first is to enhance the effectiveness and efficiency with which work and other activities are carried out. Included here would be such things as increased convenience of use, reduced errors, and increased productivity. The second objective is to enhance certain desirable human values, including improved safety, reduced fatigue and stress, increased comfort, greater user acceptance, increased job satisfaction, and improved quality of life.

It may seem like a tall order to enhance all these varied objectives, but as Chapanis (1983) points out, two things help us. First, only a subset of the objectives are generally of highest importance in a specific application. Second, the objectives are usually correlated. For example, a machine or product that is the result of human factors technology usually not only is safer, but also is easier to use, results in less fatigue, and is more satisfying to the user.

Approach of Human Factors

The approach of human factors is the systematic application of relevant information about human capabilities, limitations, characteristics, behavior, and motivation to the design of things and procedures people use and the environments in which they use them. This involves scientific investigations to discover relevant information about humans and their responses to things, environments, etc. This information serves as the basis for making design recommendations and for predicting the probable effects of various design alternatives. The human factors approach also involves the evaluation of the things we design to ensure that they satisfy their intended objectives.

Although no short catch phrase can adequately characterize the scope of the human factors field, such expressions as *designing for human use* and *optimizing working and living conditions* give a partial impression of what human factors is about. For those who would like a concise definition of human factors which combines the essential elements of focus, objectives, and approach discussed above, we present the following definition, modified slightly from Chapanis (1985): *Human factors discovers and applies information about human behavior, abilities, limitations, and other characteristics to the design of tools, machines, systems, tasks, jobs, and environments for productive, safe, comfortable, and effective human use.*

Discussion

There are several more or less established doctrines that characterize the human factors profession and that together distinguish it from other applied fields:

- Commitment to the idea that things, machines, etc. are built to serve humans and must be designed always with the user in mind
- Recognition of individual differences in human capabilities and limitations and an appreciation for their design implications
- Conviction that the design of things, procedures, etc. influences human behavior and well-being
- Emphasis on empirical data and evaluation in the design process
- Reliance on the scientific method and the use of objective data to test hypotheses and generate basic data about human behavior
- Commitment to a systems orientation and a recognition that things, procedures, environments, and people do not exist in isolation

We would be remiss if we did not at least mention what human factors is not. All too often, when people are asked what human factors is, they respond by saying what it is not. The following are three things human factors is not.

Human factors is not just applying checklists and guidelines. To be sure, human factors people develop and use checklists and guidelines; however, such aids are only part of the work of human factors. There is not a checklist or guideline in existence today that, if it were blindly applied, would ensure a good human factors product. Trade-offs, considerations of the specific application, and educated opinions are things that cannot be captured by a checklist or guideline but are all important in designing for human use.

Human factors is not using oneself as the model for designing things. Just because a set of instructions makes sense to an engineer, there is no guarantee others will understand them. Just because a designer can reach all the controls on a machine, that is no guarantee that everyone else will be able to do so. Human factors recognizes individual differences and the need to consider the unique characteristics of user populations in designing things for their use. Simply being a human being does not make a person a qualified human factors specialist.

Human factors is not just common sense. To some extent, use of common sense would improve a design, but human factors is more that just that. Knowing how large to make letters on a sign to be read at a specific distance, or selecting an audible warning that can be heard and distinguished from other alarms, is not determined by simple common sense. Knowing how long it will take pilots to respond to a warning light or buzzer is also not just common sense. Given the number of human factors deficiencies in the things we use, if human factors is based on just common sense, then we must conclude that common sense is not very common.

A HISTORY OF HUMAN FACTORS

To understand human factors, it is important to know from where the discipline came. It is not possible, however, to present more than just a brief overview of the major human factors developments. We have chosen to concentrate on developments in the United States, but several sources trace the history in other countries [see, for example, Edholm and Murrell (1973), Singleton (1982), and Welford (1976)].

Early History

It could be said that human factors started when early humans first fashioned simple tools and utensils. Such an assertion, however, might be a little presumptuous. The development of the human factors field has been inextricably intertwined with developments in technology and as such had its beginning in the industrial revolution of the late 1800s and early 1900s. It was during the early 1900s, for example, that Frank and Lillian Gilbreth began their work in motion study and shop management. The Gilbreths' work can be considered as one of the forerunners to what was later to be called human factors. Their work included the study of skilled performance and fatigue and the design of work stations and equipment for the handicapped. Their analysis of hospital surgical teams, for example, resulted in a procedure used today: a surgeon obtains an instrument by calling for it and extending his or her hand to a nurse who places the instrument in the proper orientation. Prior to the Gilbreths' work, surgeons picked up their own instruments from a tray. The Gilbreths found that with the old technique surgeons spent as much time looking for instruments as they did looking at the patient.

Despite the early contributions of people such as the Gilbreths, the idea of adapting equipment and procedures to people was not exploited. The major emphasis of behavioral scientists through World War II was on the use of tests for selecting the proper people for jobs and on the development of improved training procedures. The focus was clearly on fitting the person to the job. During World War II, however, it became clear that, even with the best selection and training, the operation of some of the complex equipment still exceeded the capabilities of the people who had to operate it. It was time to reconsider fitting the equipment to the person.

1945 to 1960: The Birth of a Profession

At the end of the war in 1945, engineering psychology laboratories were established by the U.S. Army Air Corp (later to become the U.S. Air Force) and U.S. Navy. At about the same time, the first civilian company was formed to do engineering psychology contract work (Dunlap & Associates). Parallel efforts were being undertaken in Britain, fostered by the Medical Research Council and the Department of Scientific and Industrial Research.

It was during the period after the war that the human factors profession was born. In 1949 the Ergonomics Research Society (now called simply the Ergonomics Society) was formed in Britain, and the first book on human factors was published, entitled *Applied Experimental Psychology: Human Factors in Engineering Design* (Chapanis, Garner, and Morgan, 1949). During the next few years conferences were held, human factors publications appeared, and additional human factors laboratories and consulting companies were established.

The year 1957 was an important year, especially for human factors in the United States. In that year the journal *Ergonomics* from the Ergonomics Research Society appeared, the Human Factors Society was formed, Division 21 (Society of Engineering Psychology) of the American Psychological Association was organized, the first edition of this book was published, and Russia launched *Sputnik* and the race for space was on. In 1959 the International Ergonomics Association was formed to link several human factors and ergonomics societies in various countries around the world.

1960 to 1980: A Period of Rapid Growth

The 20 years between 1960 and 1980 saw rapid growth and expansion of human factors. Until the 1960s, human factors in the United States was essentially concentrated in the military-industrial complex. With the race for space and staffed space flight, human factors quickly became an important part of the space program. As an indication of the growth of human factors during this period, consider that in 1960 the membership of the Human Factors Society was about 500; by 1980 it had grown to over 3000. More important, during this period, human factors in the United States expanded beyond military and space applications. Human factors groups could be found in many companies, including those dealing in pharmaceuticals, computers, automobiles, and other consumer products. Industry began to recognize the importance and contribution of human factors to the design of both workplaces and the products manufactured there. Also during this period another event occurred that would have a major impact on the public's awareness of human factors—the incident at Three-Mile Island nuclear power station. Despite all the rapid growth and recognition within industry, human factors was still relatively unknown to the average person in the street in 1980.

1980 and Beyond

Human factors continues to grow with membership in the Human Factors Society reaching over 4000 in 1986. The computer revolution propelled human factors into the public limelight. Talk of ergonomically designed computer equipment, user-friendly software, and human factors in the office seems to be part and parcel of virtually any newspaper or magazine article dealing with computers and people. Computer technology has provided new challenges for the human factors profession. New control devices, information presentation via computer screen, and the impact of new technology on people are all areas where the human factors profession is making contributions.

The role of human factors in the nuclear power industry came into sharp focus during the early 1980s. The U.S. Nuclear Regulatory Commission mandated that all nuclear power control rooms were to undergo a human factors review to identify and correct design deficiencies. Considerable contract work was generated by that mandate, and many human factors specialists turned their attention to the problems.

Another area that has seen a dramatic increase in human factors involvement is forensics and particularly product liability litigations. Courts have come to recognize the contribution of human factors expert witnesses for explaining human behavior and expectations, defining issues of defective design, and assessing the effectiveness of warnings and instructions. Approximately 15 percent of Human Factors Society members are involved in expert-witness work (Sanders, Bied, and Curran, 1986).

What does the future hold for human factors? We, of course, have no crystal ball, but it seems safe to predict continued growth in the areas established during the short history of human factors. Plans for building a permanent space station will undoubtedly mean a heavy involvement of human factors. Computers and the application of computer technology to just about everything will keep a lot of human factors people busy for a long time. We hope that in the future human factors will become more involved and recognized for its contribution to the quality of life and work, contributions that go beyond issues of productivity and safety and embrace more intangible criteria such as satisfaction, happiness, and dignity. Human factors, for example, should play a greater role in the future in improving the quality of life and work in underdeveloped countries.

HUMAN FACTORS PROFESSION

We offer here a thumbnail sketch of the human factors profession in the United States as depicted from surveys of the Human Factors Society membership and from an analysis of human factors graduate education programs in the United States.

Graduate Education in Human Factors

The *Directory of Human Factors Graduate Programs in the U.S.A.* (Sanders and Strother, 1985) lists 57 programs. The programs (with but a few exceptions) are housed in engineering departments (49 percent) or psychology departments (33 percent with an additional 9 percent being joint programs. Many of the programs listed, however, are just options or concentrations within another degree program (usually in industrial engineering or experimental psychology). This is especially true of engineering programs, where 50 percent are of this nature, compared to 26 percent of the psychology programs. From 1982 through 1984, engineering programs accounted

TABLE 1-1
ANALYSIS OF HUMAN FACTORS SOCIETY MEMBERSHIP BY ACADEMIC SPECIALTY
AND HIGHEST DEGREE HELD

Academic specialty	Highest degree held (%)			Total (%)
	Bachelor's	Master's	Doctoral	
Psychology	7.7	16.1	25.4	49.2
Engineering	3.7	6.3	4.5	14.5
Human factors/ergonomics	1.1	3.9	1.8	6.8
Industrial design	2.3	0.7	0.1	3.1
Education	0.3	0.8	1.5	2.6
Medicine, physiology, life sciences	0.8	0.8	0.9	2.5
Business administration	0.6	1.4	0.1	2.1
Computer science	0.3	0.7	0.3	1.3
Other	2.8	2.5	2.3	7.6
Total degrees	19.6	33.2	36.9	89.7
Students	—	—	—	10.2
Not specified	—	—	—	0.1
Total				100.0

Source: Knowles, 1986. Copyright by the Human Factors Society, Inc. and reproduced by permission.

for 67 percent of the master's degrees granted but only 54 percent of the doctoral degrees. In 1984 there were about 800 students (34 percent of which were female) in 55 programs that supplied such data (Sanders, 1985).

Who Are Human Factors People?

It is not entirely fair to assume that membership in the Human Factors Society defines a person as a human factors specialist. Nonetheless, an analysis of the backgrounds of its membership should give some insight into the profession as a whole. Table 1–1 shows the composition of the Human Factors Society membership by academic specialty and highest degree held. Psychologists comprise almost half of the membership and account for 69 percent of all doctoral degrees but only 48 percent of the master's degrees. More than half of the members do not possess doctoral degrees. The number of different academic specialties in Table 1–1 attests to the multidisciplinary nature of the profession.

Where Do Human Factors Specialists Work?

Sanders, Bied, and Curran (1986) conducted a survey of 564 members of the Human Factors Society to gather information regarding their work situations. Table 1–2 presents the percentages of those surveyed who reported working in various types of organizations. Of those working in industry, 28 percent worked in the aerospace industry, 31 percent in the computer and software industry, 18 percent in the communications industry, and 6 percent in other consumer product industries (18 percent worked in "other" industries). Among those working in business, 57 percent re-

TABLE 1-2
PERCENTAGE OF HUMAN FACTORS SOCIETY RESPONDENTS WORKING IN VARIOUS
TYPES OF ORGANIZATIONS

Type of organization	Percentage of respondents
Industry	39
Business	19
Government	20
Academia	18
Other	5

Source: Sanders, Bied, and Curran, 1986. Copyright by the Human Factors Society, Inc. and reproduced by permission.

ported working for a general research and development business. The others reported various businesses including health care, management consulting, and architectural services.

A majority (57 percent) of the respondents reported working for a "large" organization, yet 49 percent indicated that their immediate work groups consisted of 10 or less people. For most of the respondents, the number of human factors people in their work group was small. Some 24 percent reported no other human factors people in their work group besides themselves, and 25 percent reported only one or two others besides themselves. Such a situation can get a little lonely, but it can also be quite challenging.

What Do Human Factors People Do?

As part of the survey of Human Factors Society members, Sanders, Bied, and Curran provided a list of 63 activities and asked respondents to indicate for each activity whether they performed it rarely (if ever), occasionally, moderately often, or frequently. Table 1–3 lists all the activities for which 30 percent or more of the respondents indicated they performed moderately often or frequently in their work. The activities group nicely into four major areas: communication, management, system development, and research and evaluation.

How Do Human Factors People Feel about Their Jobs?

Sanders (1982), in another survey of the Human Factors Society membership, found that respondents rated their jobs especially high (compared to other professional and technical people) on skill variety (the degree to which the job requires a variety of different activities, skills, and talents) and autonomy (the degree to which the job provides freedom, independence, and discretion to the worker in scheduling and determining the procedures used to carry out the work). Sanders also assessed job satisfaction among his respondents and found them especially satisfied (compared to other professional and technical people) with their pay, security, and opportunities for personal growth. Their overall general level of satisfaction was also very high.

TABLE 1-3
ACTIVITIES PERFORMED MODERATELY OFTEN OR FREQUENTLY BY OVER 30 PERCENT OF HUMAN FACTORS SOCIETY RESPONDENTS

Activity	Percentage responding moderately often or frequently
Communication	
Write reports	80
Conduct formal briefing and presentations	59
Edit reports written by others	54
Write proposals	52
Evaluate relevance, worth, and quality of reports written by others	47
Review and summarize the literature	35
Management	
Schedule project activities	53
Manage and supervise others	49
Prepare budgets and monitor fiscal matters	38
System Development	
Determine system requirements	43
Verify system design meets human factors standards	43
Write system goals and objectives	40
Perform task analysis	37
Specify user requirements for hardware/software	31
Research and Evaluation	
Develop experimental designs to test theories or evaluate systems	44
Design data collection instruments and procedures	39
Determine proper statistical test for particular data set	38
Plan and conduct user-machine evaluations	36
Collect data in controlled laboratory setting	32
Develop criterion measures of human/system performance	31

Human Factors beyond the Human Factors Society

We have emphasized the Human Factors Society and its membership because it is the largest human factors professional group in the world and because a lot of interesting data are available about its membership. There are, however, other human factors groups in other countries, including Britain, West Germany, Japan, the Netherlands, Italy, France, Canada, Australia, Norway, Israel, Poland, Yugoslavia, Hungary, Austria, Brazil, and Mexico. In addition, other professional organizations have divisions or technical groups dealing with human factors. Such organizations include the American Industrial Hygiene Association, American Psychological Association, Institute of Electrical and Electronics Engineers, Society of Automotive Engineers, American Society of Mechanical Engineers, and the Association for Computing Machinery.

THE CASE FOR HUMAN FACTORS

Since humanity has somehow survived for these many thousands of years without people specializing in human factors, one might wonder why—at the present stage of history—it has become desirable to have human factors experts who specialize in worrying about these matters. As indicated before, the objectives of human factors are not new; history is filled with evidence of efforts, both successful and unsuccessful, to create tools and equipment which satisfactorily serve human purposes and to control more adequately the environment within which people live and work. But during most of history, the development of tools and equipment depended in large part on the process of evolution, of trial and error. Through the use of a particular device—an ax, an oar, a bow and arrow—it was possible to identify its deficiencies and to modify it accordingly, so that the next "generation" of the device would better serve its purpose.

The increased rate of technological development of recent decades has created the need to consider human factors early in the design phase, and in a systematic manner. Because of the complexity of many new and modified systems it frequently is impractical (or excessively costly) to make changes after they are actually produced. The cost of retrofitting frequently is exorbitant. Thus, the initial designs of many items must be as satisfactory as possible in terms of human factors considerations.

In effect, then, the increased complexities of the things people use (as the consequence of technology) place a premium on having assurance that the item in question will fulfill the two objectives of functional effectiveness and human welfare. The need for such assurance requires that human factors be taken into account early in the (usually long) design and development process.

SYSTEMS

A central and fundamental concept in human factors is the *system*. Various authors have proposed different definitions for the term; however, we adopt a very simple one here. A *system* is an entity that exists to carry out some purpose (Bailey, 1982). A system is composed of humans, machines, and other things that work together (interact) to accomplish some goal which these same components could not produce independently. Thinking in terms of systems serves to structure the approach to the development, analysis, and evaluation of complex collections of humans and machines. As Bailey (1982) states,

> The concept of a system implies that we recognize a purpose; we carefully analyze the purpose; we understand what is required to achieve the purpose; we design the system's parts to accomplish the requirements; and we fashion a well-coordinated system that effectively meets our purpose. (p. 192)

We discuss aspects of human-machine systems and then present a few characteristics of systems in general. Finally, we introduce the concept of system reliability.

Human-Machine Systems

We can consider a *human-machine system* as a combination of one or more human beings and one or more physical components interacting to bring about, from given inputs, some desired output. In this frame of reference, the common concept of machine is too restricted, and we should rather consider a "machine" to consist of virtually any type of physical object, device, equipment, facility, thing, or what have you that people use in carrying out some activity that is directed toward achieving some desired purpose or in performing some function. In a relatively simple form, a human-machine system (or what we sometimes refer to simply as a *system*) can be a person with a hoe, a hammer, or a hair curler. Going up the scale of complexity, we can regard as systems the family automobile, an office machine, a lawn mower, and a roulette wheel, each equipped with its operator. More complex systems include aircraft, bottling machines, telephone systems, and automated oil refineries, along with their personnel. Some systems are less delineated and more amorphous than these, such as the servicing systems of gasoline stations and hospitals and other health services, the operation of an amusement park or a highway and traffic system, and the rescue operations for locating an aircraft downed at sea.

The essential nature of people's involvement in a system is an active one, interacting with the system to fulfill the function for which the system is designed.

The typical type of interaction between a person and a machine is illustrated in Figure 1–1. This shows how the displays of a machine serve as stimuli for an operator, trigger some type of information processing on the part of the operator (including decision making), which in turn results in some action (as in the operation of a control mechanism) that controls the operation of the machine.

One way to characterize human-machine systems is by the degree of manual versus machine control. Although the distinctions between and among systems in terms of such control are far from clear-cut, we can generally consider systems in three broad classes: manual, mechanical, and automatic.

Manual Systems   A *manual system* consists of hand tools and other aids which are coupled by a human operator who controls the operation. Operators of such systems use their own physical energy as the power source.

Mechanical Systems   These systems (also referred to as *semiautomatic* systems) consist of well-integrated physical parts, such as various types of powered machine tools. They are generally designed to perform their functions with little variation. The power typically is provided by the machine, and the operator's function is essentially one of control, usually by the use of control devices.

Automated Systems   When a system is fully automated, it performs all operational functions with little or no human intervention. Robots are a good example of an automated system. Some people have the mistaken belief that since automated systems require no human intervention, they are not human-machine systems and involve no human factors considerations. Nothing could be further from the truth.

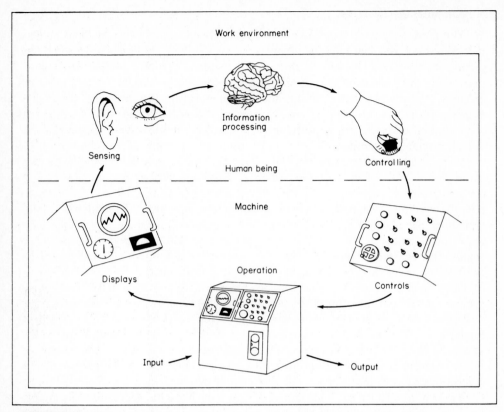

FIGURE 1-1
Schematic representation of a human-machine system. [*Source: Chapanis, 1976. From Chapanis, Alphonse. Engineering psychology. In Marvin D. Dunnette (Ed.),* Handbook of industrial and organizational psychology, *p. 701. Copyright © 1976 by Rand McNally College Publishing Company. Used by permission of Houghton Mifflin Company.*]

All automated systems require humans to install, program, reprogram, and maintain them. Automated systems must be designed with the same attention paid to human factors that would be given to any other type of human-machine system.

Characteristics of Systems

We briefly discuss a few fundamental characteristics of systems, especially as they relate to human-machine systems.

   Systems Are Purposive   In our definition of a system, we stressed that a system has a purpose. Every system must have a purpose, or else it is nothing more than a collection of odds and ends. The purpose of a system is the system goal, or objective, and systems can have more than one.

Systems Can Be Hierarchical   Some systems can be considered to be parts of larger systems. In such instances, a given system may be composed of more molecular systems (also called *subsystems*). When faced with the task of describing or analyzing a complex system, one often asks, ''Where does one start and where does one stop?'' The answer is, ''It depends.'' Two decisions must be made. First, one has to decide on the *boundary of the system,* that is, what is considered part of the system and what is considered outside the system. There is no right or wrong answer, but the choice must be logical and must result in a system that performs an identifiable function. The second decision is where to set the *limit of resolution* for the system. That is, how far down into the system is one to go? At the lowest level of analysis one finds *components*. A component in one analysis may be a subsystem in another analysis that sets a lower limit of resolution. As with setting system boundaries, there is no right or wrong limit of resolution. The proper limit depends on why one is describing or analyzing the situation.

Systems Operate in an Environment   The environment of a system is everything outside its boundaries. Depending on how the system's boundaries are drawn, the environment can range from the immediate environment (such as a work station, a lounge chair, or a typing desk) through the intermediate (such as a home, an office, a factory, a school, or a football stadium) to the general (such as a neighborhood, a community, a city, or a highway system). Note that some aspects of the physical environment in which we live and work are part of the natural environment and may not be amenable to modification (although one can provide protection from certain undesirable environmental conditions such as heat or cold). Although the nature of people's involvement with their physical environment is essentially passive, the environment tends to impose certain constraints on their behavior (such as limiting the range of their movements or restricting their field of view) or to predetermine certain aspects of behavior (such as stooping down to look into a file cabinet, wandering through a labyrinth in a supermarket to find the bread, or trying to see the edge of the road on a rainy night).

Components Serve Functions   Every component (the lowest level of analysis) in a system serves at least one function that is related to the fulfillment of one or more of the system's goals. One task of human factors specialists is to aid in making decisions as to whether humans or machines (including software) should carry out a particular system function. (We discuss this *allocation of function* process in more detail in Chapter 18.)

Components serve various functions in systems, but all typically involve a combination of four more basic functions: sensing (information receiving), information storage, information processing and decision, and action functions; they are depicted graphically in Figure 1–2. Since information storage interacts with all the other functions, it is shown above the others. The other three functions occur in sequence.

1 *Sensing (information receiving):* One of these functions is sensing, or information receiving. Some of the information entering a system is from outside the system, for example, airplanes entering the area of control of a control-tower operator, an

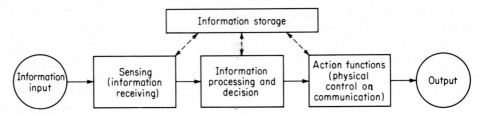

FIGURE 1-2
Types of basic functions performed by human or machine components of human-machine systems.

order for the production of a product, or the heat that sets off an automatic fire alarm. Some information, however, may originate inside the system itself. Such information can be feedback (such as the reading on the speedometer from an action of the accelerator), or it can be information that is stored in the system.

2 *Information storage:* For human beings, information storage is synonymous with memory of learned material. Information can be stored in physical components in many ways, as on punch cards, magnetic tapes and disks, templates, records, and tables of data. Most of the information that is stored for later use is in coded or symbolic form.

3 *Information processing and decision:* Information processing embraces various types of operations performed with information that is received (sensed) and information that is stored. When human beings are involved in information processing, this process, simple or complex, typically results in a decision to act (or, in some instances, a decision *not* to act). When mechanized or automated machine components are used, their information processing must be programmed in some way. Such programming is, of course, readily understood if a computer is used. Other methods of programming involve the use of various types of schemes, such as gears, cams, electric and electronic circuits, and levers.

4 *Action functions:* What we call the *action* functions of a system generally are those operations which occur as a consequence of the decisions that are made. These functions fall roughly into two classes. The first is some type of *physical control action* or process, such as the activation of certain control mechanisms or the handling, movement, modification, or alteration of materials or objects. The other is essentially a *communication action,* be it by voice (in human beings), signals, records, or other methods. Such functions also involve some physical actions, but these are in a sense incidental to the communication function.

Components Interact To say that components interact simply means that the components work together to achieve system goals. Each component has an effect, however small, on other components. One outcome of a system's analysis is the description and understanding of these component and subsystem relationships.

Systems, Subsystems, and Components Have Inputs and Outputs At all levels of a complex system there are inputs and outputs. The outputs of one subsystem or component are the inputs to another. A system receives inputs from the environment and makes outputs to the environment. It is through inputs and outputs that all the pieces interact and communicate. Inputs can be physical entities (such as materials and products), electric impulses, mechanical forces, or information.

It might be valuable at this time to distinguish between open-loop and closed-loop systems. A *closed-loop system* is continuous, performing some process which requires continuous control (such as in vehicular operation and the operation of certain continuously controlled chemical processes), and requires continuous feedback for its successful operation. The feedback provides information about any error that should be taken into account in the continuing control process. An *open-loop system,* when activated, needs no further control or at least cannot be further controlled. In this type of system the "die is cast" once the system has been put into operation; no further control can be exercised, such as in firing a rocket that has no guidance system. Although feedback with such systems obviously cannot serve continuous control, feedback can improve subsequent operations of the system.

In a system's analysis all the inputs and outputs required for each component and subsystem to perform its functions are specified. Human factors specialists are especially qualified to determine the inputs and outputs necessary for the human components of systems to successfully carry out their functions.

System Reliability

Unfortunately, nothing lasts forever. Things break or just fail to work, usually at the worst possible time. When we design systems, of course, we would like them to continue working. In this context, engineers speak of the *reliability* of a system or component to characterize its dependability of performance (including people) in carrying out an intended function. Reliability is usually expressed as the probability of successful performance (this is especially applicable when the performance consists of discrete events, such as starting a car). For example, if an automated teller machine gives out the correct amount of money 9999 times out of 10,0000 withdrawal transactions, we say that the reliability of the machine, to perform the function, is .9999. (Reliabilities for electronic and mechanical devices are often carried out to four or more decimal places; reliabilities for human performance, on the other hand, usually are carried no further than three decimal places.)

Another measure of reliability is *mean time to failure* (abbreviated MTF). There are several possible variations, but they all relate to the amount of time a system or individual performs successfully, either until failure or between failures; this index is most applicable to continuous types of activities. Other variations could also be mentioned. For our present discussion, however, let us consider reliability in terms of the probability of successful performance.

If a system includes two or more components (machine or human or both), the reliability of the composite system will depend on the reliability of the individual components and how they are combined within the system. Components can be combined within a system in *series,* in *parallel,* or in a combination of both.

Components in Series In many systems the components are arranged in series (or sequence) in such a manner that successful performance of the total system depends on successful performance of each and every component, person or machine. By taking some semantic liberties, we could assume components to be *in series* that may, in fact, be functioning concurrently and interdependently, such as a human operator using some type of equipment. In analyzing reliability data in such cases, two conditions must be fulfilled: (1) failure of any given component results in system failure, and (2) the component failures are independent of each other. When these assumptions are fulfilled, the reliability of the system for error-free operation is the product of the reliabilities of the several components. As more components are added in series, the reliability of the system *decreases*. If a system consisted of 100 components in series, each with a reliability of .9900, the reliability of the entire system would be only .365 (that is, 365 times out of 1000 the system would properly perform its function). The maximum possible reliability in a series system is equal to the reliability of the *least* reliable component, which often turns out to be the human component. In practice, however, the overall reliability of a series system is often much *less* than the reliability of the least reliable component.

Components in Parallel   The reliability of a system whose components are in parallel is entirely different from that whose components are in a series. With parallel components, two or more in some way are performing the same function. This is sometimes referred to as a *backup,* or *redundancy,* arrangement—one component backs up another so that if one fails, the other can successfully perform the function. In order for the entire system to fail, *all* the components in parallel must fail. Adding components in parallel *increases* the reliability of the system. For example, a system with four components in parallel, each with a reliability of .70, would have an overall system reliability of .992. Because humans are often the weak link in a system, it is common to see human-machine systems designed to provide parallel redundancy for some of the human functions.

Discussion   We have been discussing system reliability as if it were static and unchanging. As our own experience illustrates, however, reliability changes as a function of time (usually it gets worse). The probability that a 10-year-old car will start is probably lower than it was when the car was 1 year old. The same sort of time dependency applies to the reliability of humans, only over shorter periods. The probability of successful human performance often deteriorates over just a few hours of activity. Human reliability is discussed further in Chapter 2.

COVERAGE OF THIS TEXT

Since a comprehensive treatment of the entire scope of human factors would fill a small library, this text must be restricted to a rather modest segment of the total human factors domain. The central theme is the illustration of how the achievement of the two primary human factors objectives (i.e., functional effectiveness and human welfare) can be influenced by the extent to which relevant human considerations have been taken into account during the design of the object, facility, or environment in question. Further, this theme is followed as it relates to some of the more commonly recognized human factors content areas (such as the design of displays for presenting information to people, human control processes, and physical environment). Pursuing this theme across the several subareas would offer an overview of the content areas of human factors.

The implications of various perceptual, mental, and physical characteristics as they might affect or influence the objectives of human factors probably can best be reflected by the result of relevant research investigations and of documented operational experience. Therefore the theme of the text will generally be carried out by presenting and discussing the results of illustrative research and by bringing in generalizations or guidelines supported by research or experiences that have relevance to the design process in terms of human factors considerations. Thus, in the various subject or content areas much of the material in this text will consist of summaries of research that reflect the relationships between design variables, on the one hand, and criteria of functional effectiveness or human welfare, on the other hand.

We recognize that the illustrative material brought in to carry out this theme is in no way comprehensive, but we hope it will represent in most content areas some of the more important facets.

Although the central theme will, then, deal with the human factors aspects of the design of the many things people use, there will be some modest treatment of certain related topics, such as how human factors fits in with the other phases of design and development processes.

REFERENCES

Bailey, R. (1982). *Human performance engineering: A guide for systems designers.* Englewood Cliffs, NJ: Prentice-Hall.

Chapanis, A. (1976). Engineering psychology. In M. D. Dunnette (ed.), *Handbook of industrial and organizational psychology.* Chicago: Rand McNally.

Chapanis, A. (1983). Introduction to human factors considerations in system design. In C. M. Mitchell, P. Van Balen, and K. Moe (eds.), *Human factors considerations in systems design,* NASA Conference Publ. 2246. Washington: National Aeronautic and Space Administration.

Chapanis, A. (1985). Some reflections on progress. *Proceedings of the Human Factors Society 29th Annual Meeting.* Santa Monica, CA: Human Factors Society, pp. 1–8.

Chapanis, A., Garner, W., and Morgan, C. (1949). *Applied experimental psychology: Human factors in engineering design.* New York: Wiley.

Edholm, O., and Murrell, K. F. H. (1973). *The Ergonomics Research Society: A History 1949–1970.* Winchester, Hampshire: Warren & Sons, Ltd.

Knowles, M. (ed.) (1986). *Human Factors Society 1986 directory and yearbook.* Santa Monica, CA: Human Factors Society.

Sanders, M. (1982). HFS job description survey: Job dimensions, satisfaction, and motivation. *Proceedings of the Human Factors Society 25th Annual Meeting.* Santa Monica, CA: Human Factors Society.

Sanders, M. (1985). Human factors graduate education: An update. *Human Factors Society Bulletin,* 28(12), 1–3.

Sanders, M., Bied, B., and Curran, P. (1986). HFS membership job survey. *Human Factors Society Bulletin,* 29(3), 1–3.

Sanders, M., and Strother, L. (1985). *Directory of human factors graduate programs in the U.S.A.* Santa Monica, CA: Human Factors Society.

Singleton, W. T. (1982). *The body at work: Biological ergonomics.* London: Cambridge University Press.

Welford, A. T. (1976). Ergonomics: Where have we been and where are we going: I. *Ergonomics,* 19(3), 275–286.

CHAPTER 2

HUMAN FACTORS RESEARCH METHODOLOGIES

Human factors is in large part an empirical science. The central approach of human factors is the application of relevant information about human capabilities and behavior to the design of objects, facilities, procedures, and environments that people use. This body of relevant information is largely based on experimentation and observation. Research plays a central role in this regard, and the research basis of human factors is emphasized throughout this book.

In addition to gathering empirically based information and applying it to the design of things, human factors specialists also gather empirical data to evaluate the "goodness" of their designs and the designs of others. Thus, empirical data, and hence research, play a dual role in the development of systems: at the front end as a basis for the design and at the back end as a means of evaluating and improving the design. For this reason, in this chapter we deal with some basic concepts of human research as it relates to human factors. Our purpose is not to present a handbook of research methods, but rather to introduce some of the purposes, considerations, and trade-offs involved in the research process. For a more complete discussion of research methodologies relevant to human factors, refer to Meister (1985), Meister and Rabideau (1965), Kerlinger (1973), and Runkel and McGrath (1972).

AN OVERVIEW

As one would expect, most *human* factors research involves the use of human beings as subjects, so we focus our attention there. Not all human factors research, however, involves human subjects. Sanders and Krohn (1983), for example, surveyed underground mining equipment to assess the field of view from the operator's compartment. Aside from the data collectors, no humans were involved. Human factors research can usually be classified into one of three types: descriptive studies, experimental research, or evaluation research. Actually, not all human factors research fits neatly into only one category; often a particular study will involve elements of more than one category. Although each category has different goals and may involve the use of slightly different methods, all involve the same basic set of decisions: choosing a research setting, selecting variables, choosing a sample of subjects, deciding how the data will be collected, and deciding how the data will be analyzed.

We discuss all three types of research, organizing our discussion around these basic decisions. This will give us an opportunity to say a few things we want to say and introduce some concepts that will be popping up now and again in later chapters.

Before beginning our discussion of research methodologies, let us present some issues relevant to choosing one methodology over another (Meister, 1985):

- *Effectiveness*—the degree to which the method accomplishes its purpose. Very often the purpose of the research actually defines the methodology, although the researcher usually has considerable latitude in choosing the specific methods, variables, subjects, and data collection procedures.
- *Ease of use*—how easy it is to use the methodology in a particular application.
- *Cost*—how much it costs in terms of money, data requirements, equipment, personnel, and time.
- *Flexibility*—the degree to which the method can be used in various contexts and situations.
- *Range*—the number of phenomena, behaviors, and events that can be measured by using the method.
- *Validity*—the degree to which the method produces data that are like those occurring in real life (we have more to say about this later).
- *Reliability*—the degree to which the method produces data that are consistent over time and between applications (more on this later).
- *Objectivity*—the extent to which the method relies on data and procedures external to the one who applies the method.

As Meister points out, no method scores high on all these dimensions, and probably no method ever scores as high as we would like on even one. Trade-offs must be made, and the art lies in knowing what can be sacrificed and what cannot.

DESCRIPTIVE STUDIES

Generally speaking, descriptive studies seek to characterize a population (usually of people) in terms of certain attributes. Since we are usually interested in the behavior and characteristics of people, descriptive studies often seek such information. We present the results of many such studies throughout this book. Examples include surveys of the dimensions of people's bodies, hearing loss among people of different ages, people's expectations as to how a knob should be turned to increase the value on a display, and weights of boxes people are willing to lift.

Although descriptive studies are not very exciting, they are very important to the science of human factors. They represent the basic data upon which many design decisions are based. In addition, descriptive studies are often carried out to assess the magnitude and scope of a problem before solutions are suggested. A survey of operators to gather their opinions about design deficiencies and operational problems would be an example. In fact, the Nuclear Regulatory Commission (1981) requires such a survey as part of its mandated human factors control room review process.

Choosing a Research Setting

In choosing the research setting for a study, the fundamental decision is whether to conduct the study in the field, also referred to as the "real world," or in the laboratory. We present the pros and cons of using these two settings when we discuss experimental research. With descriptive studies, however, the choice is somewhat moot. The primary goal of descriptive studies is to generate data that describe a particular population of people, be they coal miners, computer operators, or the general civilian population. To poll such people, we must go to the real world. As we will see, however, the actual data collection may be done in a laboratory—often a mobile laboratory—which is, in essence, like bringing the mountain to Mohammad.

Selecting Variables

The selection of variables to be measured in research studies is such a fundamental and important question that we have devoted two later sections of this chapter to it. In descriptive studies two basic classes of variables are measured: *criterion variables* and *stratification* (or *predictor*) *variables*.

Criterion Variables Criterion variables describe those characteristics and behaviors of interest in the study. These variables can be grouped into the following classes according to the type of data being collected: *physical characteristics,* such as arm reach, stomach girth, and body weight; *performance data,* such as reaction time, visual acuity, hand grip strength, and memory span; *subjective data,* such as preferences, opinions, and ratings; and *physiological indices,* such as heart rate, body temperature, and pupil dilation.

Stratification Variables In some descriptive studies (such as surveys), it is the practice to select *stratified* samples that are proportionately representative of the population in terms of such characteristics as age, sex, education, etc. Even if a stratified sample is not used, however, information is often obtained on certain relevant personal characteristics of those in the sample. Thus, the resulting data can be analyzed in terms of the characteristics assessed, such as age, sex, etc. These characteristics are sometimes called *predictors*.

Choosing Subjects

Proper subject selection is critical to the validity of descriptive studies. Often, in such studies, the researcher expends more effort in developing a sampling plan and obtaining the subjects than in any other phase of the project.

Representative Sample   The goal in descriptive studies is to collect data from a sample of people representative of the population of interest. A sample is said to be *representative* of a population if the sample contains all the *relevant* aspects of the population in the same proportion as found in the real population. For example, if in the population of coal miners 30 percent are under 21 years of age, 40 percent are between 21 and 40 years, and 30 percent are over 40 years of age, then the sample—to be representative—should also contain the same percentages of each age group.

The key word in the definition of representative is *relevant*. A sample may differ from a population with respect to a nonrelevant variable and still be useful for descriptive purposes. For example, if we were to measure the reaction time of air traffic controllers to an auditory alarm, we would probably carry out the study in one or two cities rather than sampling controllers in every state or geographic region in the country. This is so because it is doubtful that the reaction time is different in different geographic regions; that is, geographic region is probably not a relevant variable in the study of reaction time. We would probably take great care, however, to include the proper proportions of controllers with respect to age and sex because these variables are likely to be relevant. A sample that is not representative is said to be *biased*.

Random Sampling   To obtain a representative sample, the sample should be selected randomly from the population. *Random selection* occurs when each member of the population has an equal chance of being included in the sample. In the real world, it is almost impossible to obtain a truly random sample. Often the researcher must settle for those who are easily obtained even though they were not selected according to a strict random procedure.

Even though a particular study may not sample randomly or include all possible types of people in the proportions in which they exist in the population, the study may still be useful if the bias in the sample is not relevant to the criterion measures of interest. How does one know whether the bias is relevant? Prior research, experience, and theories form the basis for an educated guess.

Sample Size   A key issue in carrying out a descriptive study is determining how many subjects will be used, i.e., the sample size. The larger the sample size, the more confidence one has in the results. Sampling costs money and takes time, so researchers do not want to collect more data than they need to make valid inferences about the population. Fortunately, there are formulas for determining the number of subjects required. [See, for example, Roebuck, Kroemer, and Thompson (1975).] Three main parameters influence the number of subjects required: *degree of accuracy desired* (the more accuracy desired, the larger the sample size required); *variance in the population* (the greater the degree of variability of the measure in the population, the larger the sample size needed to obtain the level of accuracy desired); and the *statistic being estimated,* e.g., mean, 5th percentile, etc. (some statistics require more subjects to estimate accurately than others; for example, more subjects are required to estimate the median than to estimate the mean with the same degree of accuracy).

How to Collect Data

Data in descriptive studies can be collected in the field or in a laboratory setting. Bobo et al. (1983), for example, measured energy expenditure of underground coal miners performing their work underground. Sanders (1981) measured the strength of truck and bus drivers turning a steering wheel in a mobile laboratory at various truck depots around the country.

Often, surveys and interviews are used to collect data. Survey questionnaires may be administered in the field or mailed to subjects. A major problem with mail surveys is that not everyone returns the questionnaires, and the possibility of bias increases. Return rates of less than 50 percent should probably be considered to have a high probability of bias.

How to Analyze Data

Once a study has been carried out and the data have been gathered, the experimenter must analyze the data. It is not our intention here to deal extensively with statistics or to discuss elaborate statistical methods. In the analysis of data from descriptive studies, however, usually fairly basic statistics are computed. Probably most readers are already familiar with most statistical methods and concepts touched on in later chapters, such as frequency distributions and measures of central tendency (mean, median, mode). For those readers unfamiliar with the concepts of standard deviation, correlation, and percentiles, we describe them briefly.

Standard Deviation *(S)* The standard deviation is a measure of the variability of a set of numbers around the mean. When, say, the reaction time of a group of subjects is measured, not everyone has the same reaction time. If the scores varied greatly from one another, the standard deviation would be large. If the scores were all close together, the standard deviation would be small. In a normal distribution (bell-shaped curve) approximately 68 percent of the cases will be within $\pm 1S$ of the mean, 95 percent within $\pm 2S$ of the mean, and 99 percent within $\pm 3S$ of the mean.

Correlation A correlation coefficient is a measure of the degree of relationship between two variables. Typically, we compute a linear correlation coefficient (e.g., Pearson product-moment correlation r) which indicates the degree to which two variables are *linearly* related, that is, related in a straight-line fashion. Correlations can range from $+1.00$, indicating a perfect positive relationship, through 0 (which is the absence of any relationship), to -1.00, a perfect negative relationship. A positive relationship between two variables indicates that high values on one variable tend to be associated with high values on the other variable, and low values on one are associated with low values on the other. An example would be height and weight because tall people tend to be heavier than short people. A negative relationship between two variables indicates that high values on one variable are associated with low values on the other. An example would be age and strength because older people tend to have less strength than younger people.

If the correlation coefficient is squared (r^2), it represents the *proportion of variance* (standard deviation squared) in one variable accounted for by the other variable.

This is somewhat esoteric, but it is important for understanding the strength of the relationship between two variables. For example, if the correlation coefficient between age and strength is 0.30 we would say that of the total variance in strength 9 percent ($.30^2$) was accounted for by the subjects' age. That means that 81 percent of the variance is due to factors other than age.

Percentiles   Percentiles correspond to the value of a variable below which a specific percentage of the group fall. For example, the 5th percentile standing height for males is 63.6 in (162 cm). This means that only 5 percent of males are smaller than 63.6 in (162 cm). The 50th percentile male height is 68.3 in (173 cm), which is the same as the median since 50 percent of males are shorter than this value and 50 percent are taller. The 95th percentile is 72.8 in (185 cm), meaning that 95 percent of males are shorter than this height. Some investigators report the *interquartile range*. This is simply the range from the 25th to the 75th percentiles, and thus it encompasses the middle 50 percent of the distribution. (Interquartile range is really a measure of variability.) The concept of percentile is especially important in using anthropometric (body dimension) data for designing objects, work stations, and facilities, as we will see in Chapter 12.

EXPERIMENTAL RESEARCH

The purpose of experimental research is to test the effects of some variable on behavior. The decisions as to what variables to investigate and what behaviors to measure are usually based on either a practical situation which presents a design problem or a theory that makes a prediction about variables and behaviors. Examples of the former include comparing how well people can edit manuscripts with partial-line, partial-page, and full-page computer displays (Neal and Darnell, 1984) and assessing the effect of seat belts and shoulder harnesses on functional arm reach (Garg, Bakken, and Saxena, 1982). Experimental research of a more theoretical nature would include a study by Hull, Gill, and Roscoe (1982) in which they varied the lower half of the visual field to investigate why the moon looks so much larger when it is near the horizon than when it is overhead.

Usually in experimental research the concern is whether a variable has an effect on behavior and the direction of that effect. Although the level of performance is of interest, usually only the relative *difference* in performance between conditions is of concern. For example, one might say that subjects missed, on the average, 15 more signals under high noise than under low noise. In contrast, descriptive studies are usually interested in describing a population parameter, such as the mean, rather than assessing the effect of a variable. When descriptive studies compare groups that differ on some variable (such as sex or age), the means, standard deviations, and percentiles of each group are of prime interest. This difference in goals between experimental and descriptive studies, as we will see, has implications for subject selection.

Choosing a Research Setting

The choice of research setting involves complex trade-offs. Research carried out in the field usually has the advantage of realism in terms of relevant task variables, environmental constraints, and subject characteristics including motivation. Thus, there is a better chance that the results obtained can be generalized to the real-world operational environment. The disadvantages, however, include cost (which can be prohibitive), safety hazards for subjects, and lack of experimental control. In field studies often there is no opportunity to replicate the experiment a sufficient number of times, many variables cannot be held constant, and often certain data cannot be collected because the process would be too disruptive.

The laboratory setting has the principal advantage of experimental control; extraneous variables can be controlled, the experiment can be replicated almost at will, and data collection can be made more precise. For this advantage, however, the research may sacrifice some realism and generalizability. Meister (1985) believes that this lack of realism makes laboratory research less than adequate as a source of applied human factors data. He believes that conclusions generated from laboratory research should be tested in the real world before they are used there.

For theoretical studies, the laboratory is the natural setting because of the need to isolate the subtle effects of one or more variables. Such precision probably could not be achieved in the uncontrolled real world. The real world, on the other hand, is the natural setting for answering practical research questions. Often, a variable that shows an effect in the highly controlled laboratory "washes out" when it is compared to all the other variables that are affecting performance in the real world.

In an attempt to combine the benefits of both laboratory and field research, researchers often use *simulations* of the real world in which to conduct research. A distinction should be made between *physical* simulations and *computer* simulations. Physical simulations are usually constructed of hardware and represent (i.e., look like, feel like, or act like) some system, procedure, or environment. Physical simulations can range from very simple items (such as a picture of a control panel) to extremely complex configurations (such as a moving-base jumbo jet flight simulator with elaborate out-of-cockpit visual display capabilities). Some simulators are small enough to fit on a desktop; others can be quite large, such as a 400-ft^2 underground coal mine simulator built by one of the authors.

Computer simulation involves modeling a process or series of events in a computer. By changing the parameters the model can be run and predicted results can be obtained. For example, workforce needs, periods of overload, and equipment downtime can be predicted from computer simulations of work processes. To develop an accurate computer model requires a thorough understanding of the system being modeled and usually requires the modeler to make some simplifying assumptions about how the real-world system operates.

In some cases, field research can be carried out with a good deal of control—although some people might say of the same situation that it is laboratory research that is being carried out with a good deal of realism. An example is a study in which Marras and Kroemer (1980) compared two distress signal designs (for flares). In one part of the study, subjects were taken by boat to an island in a large lake. They were told that their task was to sit in an inflatable rubber raft, offshore, and rate the visibility of a display which would appear on shore. They were given a distress signal to use in case of an emergency. While the subject was observing the display, the raft deflated automatically, prompting the subject to activate the distress signal. The time required for unpacking and successfully operating the device was recorded. The subject was then pulled back to shore.

Selecting Variables

In experimental research, the experimenter manipulates one or more variables to assess their effects on behaviors that are measured, while other variables are controlled. The variables being manipulated by the experimenter are called *independent variables* (IVs). The behaviors being measured to assess the effects of the IVs are called *dependent variables* (DVs). The variables that are controlled are called *extraneous, secondary,* or *relevant variables*. These are variables that can influence the DV; they are controlled so that their effect is not confused (confounded) with the effect of the IV.

Independent Variables   In human factors research, IVs usually can be classified into three types: (1) *task-related variables,* including equipment variables (such as length of control lever, size of boxes, and type of visual display) and procedural variables (such as work-rest cycles and instructions to stress accuracy or speed); (2) *environmental variables,* such as variations in illumination, noise, and vibration; and (3) *subject-related variables,* such as sex, height, age, and experience.

Most studies do not include more than a few IVs. Simon (1976), for example, reviewed 141 experimental papers published in *Human Factors* from 1958 to 1972 and found that 60 percent of the experiments investigated the effects of only one or two IVs and less than 3 percent investigated five or more IVs.

Dependent Variables   DVs are the same as the criterion variables discussed in reference to descriptive studies, except that physical characteristics are used less often. Most DVs in experimental research are performance, subjective, or physiological variables. We discuss criterion variables later in this chapter.

Choosing Subjects

The issue in choosing subjects for experimental research is to select subjects representative of those people to whom the results will be generalized. Subjects do not have to be representative of the target population to the same degree as in descriptive studies. The question in experimental studies is whether the subjects will be affected by the IV in the same way as the target population. For example, consider an experiment to investigate the effects of room illumination (high versus low) on reading text displayed on a computer screen. If we wish to use the data to design office environments, do we need to use subjects who have extensive experience reading text from computer screens? Probably not. Although highly experienced computer screen readers can read faster than novice readers (hence it would be important to include them in a descriptive study), the effect of the IV (illumination) is likely to be the same for both groups. That is, it is probably easier to read computer-generated text under low room illumination (less glare) than under high—no matter how much experience the reader has.

Sample Size   The issue in determining sample size is to collect enough data to reliably assess the effects of the IV with minimum cost in time and resources. There are techniques available to determine sample size requirements; however, they are beyond the scope of this book. The interested reader can consult Cohen (1969) for more details.

In Simon's (1976) review of research published in *Human Factors,* he found that 50 percent of the studies used less than 9 subjects per experimental condition, 25 percent used from 9 to 11, and 25 percent used more than 11. These values are much smaller than those typically used in descriptive studies. The danger of using too few subjects is that one will incorrectly conclude that an IV had no effect on a DV when, in fact, it did. The "danger," if you can call it that, of using too many subjects is that a tiny effect of an IV, which may have no practical importance whatsoever, will show up in the analysis.

How to Collect Data

The collection of data for experimental research is the same as for descriptive studies. Because experimental studies are often carried out in a controlled laboratory, often more sophisticated, computer-based methods are employed. As pointed out by McFarling and Ellingstad (1977), such methods provide the potential for including more IVs and DVs in a study and permit greater precision and higher sampling rates in the collection of performance data. Vreuls et al. (1973), for example, generated over 800 different measures to evaluate the performance of helicopter pilots doing a few common maneuvers. One, of course, has to be careful not to drown in a sea of data.

How to Analyze the Data

Data from experimental studies are usually analyzed through some type of inferential statistical technique such as analysis of variance (ANOVA) or multivariate analysis of variance (MANOVA). We do not discuss these techniques since several very good texts are available (Box, Hunter, and Hunter, 1978; Tabachnick and Fidell, 1983; Myers, 1979).

The outcome of virtually all such techniques is a statement concerning the statistical significance of the data. It is this concept that we discuss because it is central to all experimental research.

Statistical Significance   Researchers make statements such as "the IV had a significant effect on the DV" or "the difference between the means was significant." The term *significance* is short for *statistical significance*. To say that something is statistically significant simply means that there is a low probability that the observed effect, or difference between the means, was due to chance. And because it is unlikely that the effect was due to chance, it is concluded that the effect was due to the IV.

How unlikely must it be that chance was causing the effect to make us conclude chance was not responsible? By tradition, we say .05 or .01 is a low probability (by the way, this tradition started in agricultural research, earthy stuff). The experimenter selects one of these values which is called the *alpha level*. Actually any value can be selected, and you as a consumer of the research may disagree with the level chosen. Thus, if the .05 alpha level is selected, the researcher is saying that if the results obtained could have occurred 5 times or less out of 100 by chance alone, then it is unlikely that they are due to chance. The researcher says that the results are significant at the .05 level. Usually the researcher will then conclude that the IV was causing the effect. If, on the other hand, the results could have occurred more than 5 times out of 100 by chance, then the researcher concludes that it is likely that chance was the cause, and not the IV.

The statistical analysis performed on the data yields the probability that the results could have occurred by chance. This is compared to the alpha level, and that determines whether the results are significant. Keep in mind that the statistical analysis does not actually tell the researcher whether the results did or did not occur by chance; only the probability is given. No one knows for sure.

Here are a couple of things to remember when you are reading about "significant results" or results that "failed to reach significance": (1) results that are significant may still be due only to chance, although the probability of this is low. (2) IVs that are not significant may still be influencing the DV, and this can be very likely, especially when small sample sizes are used. (3) Statistical significance has nothing whatever to do with importance—very small, trivial effects can be statistically significant. (4) There is no way of knowing from the statistical analysis procedure whether the experimental design was faulty or uncontrolled variables confounded the results.

The upshot is that one study usually does not make a fact. Only when several studies, each using different methods and subjects, find the same effects should we be willing to say with confidence that an IV does affect a DV. Unfortunately, in most areas, we usually find conflicting results, and it takes insight and creativity to unravel the findings.

EVALUATION RESEARCH

Evaluation research is similar to experimental research in that its purpose is to assess the effect of "something." However, in evaluation research the something is usually a system or product. Evaluation research is also similar to descriptive research in that it seeks to describe the performance and behaviors of the people using the system or product.

Evaluation research is generally more global and comprehensive than experimental research. A system or product is evaluated by comparison with its goals; both intended consequences and unintended outcomes must be assessed. Often an evaluation research study will include a benefit-cost analysis. Examples of evaluation research include evaluating a new training program, a new software package for word processing, or an ergonomically designed life jacket. Evaluation research is the area where human factors specialists assess the "goodness" of designs, theirs and others, and make recommendations for improvement based on the information collected.

Evaluation research should be carried out under conditions representative of those in which the system or product is to be used and with subjects representative of the ultimate users of the system or product. This restriction virtually dictates that the research be carried out in the field. Evaluation is part of the overall systems design process, and we discuss it as such further in Chapter 18. Suffice it to say, evaluation research is probably one of the most challenging and frustrating types of research endeavors to undertake. The rewards can be great because the results are often used to improve the design of an actual system or product, but conducting the research can be a nightmare. Murphy's law seems to rule: "If anything can go wrong, it will." Extra attention, therefore, must be paid to designing data collection procedures and devices to perform in the often unpredictable and unaccommodating field setting. One of the authors recalls evaluating the use of a helicopter patrol for fighting crime against railroads. Murphy's law prevailed; the trained observers never seemed to be available when they were needed, the railroad coordination chairman had to resign, the field radios did not work, the railroads were reluctant to report incidences of crime, and finally a gang of criminals threatened to shoot down the helicopter if they saw it in the air.

CRITERION MEASURES IN RESEARCH

Criterion measures, as we discussed, are the characteristics and behaviors measured in descriptive studies, the dependent variables in experimental research, and the basis for judging the goodness of a design in an evaluation study. Any attempt to classify criterion measures inevitably leads to confusion and overlap. Since a little confusion is part and parcel of any technical book, we relate one simple scheme for classifying criterion measures to provide a little organization to our discussion.

In the human factors domain, three types of criteria can be distinguished (Meister, 1985): those describing the functioning of the *system,* those describing how the *task* is performed, and those describing how the *human* responds. The problem is that measures of how tasks are performed usually involve how the human responds. We briefly describe each type of criteria, keeping in mind the inevitable overlap among them.

System-Descriptive Criteria

System-descriptive criteria usually reflect essentially engineering aspects of the entire system. So they are often included in evaluation research but are used to a much lesser extent in descriptive and experimental studies. System-descriptive criteria include such aspects as equipment reliability (i.e., the probability that it will not break down), resistance to wear, cost of operation, maintainability, and other engineering specifications, such as maximum rpm, weight, radio interference, etc.

Task Performance Criteria

Task performance criteria usually reflect the outcome of a task in which a person may or may not be involved. Such criteria include (1) *quantity of output,* for example, number of messages decoded, tons of earth moved, or number of shots fired; (2) *quality of output,* for example, accidents, number of errors, accuracy of drilling holes, or deviations from a desired path; and (3) *performance time,* for example, time to isolate a fault in an electric circuit or amount of delay in beginning a task. Task performance criteria are more global than human performance criteria, although human performance is inextricably intertwined in task performance and in the engineering characteristics of the system or equipment being used.

Human Criteria

Human criteria deal with the behaviors and responses of humans during task performance. Human criteria are measured with performance measures, physiological indices, and subjective responses.

Performance Measures   Human performance measures are usually *frequency measures* (e.g., number of targets detected, number of keystrokes made, or number of times the "help" screen was used), *intensity measures* (e.g., torque produced on a steering wheel), *latency measures* (e.g., reaction time or delay in switching from one activity to another), or *duration measures* (e.g., time to log on to a computer system or time on target in a tracking task). Sometimes combinations of these basic types are used, such as number of missed targets per unit time. Another human performance measure is *reliability,* or the probability of errorless performance. There has been a good deal of work relating to human reliability, and we review some of it later in this chapter.

Physiological Indices   Physiological indices are often used to measure strain in humans resulting from physical or mental work and from environmental influences, such as heat, vibration, noise, and acceleration. Physiological indices can be classified by the major biological systems of the body: cardiovascular (e.g., heart rate or blood pressure), respiratory (e.g., respiration rate or oxygen consumption), nervous (e.g., electric brain potentials or muscle activity), sensory (e.g., visual acuity, blink rate, or hearing acuity), and blood chemistry (e.g., catecholamines).

Subjective Responses   Often we must rely on subjects' opinions, ratings, or judgments to measure criteria. Criteria such as comfort of a seat, ease of use of a computer system, or preferences for various lengths of tool handle are all examples of subjective measures. Subjective responses have also been used to measure perceived mental and physical workload. Extra care must be taken when subjective measures are designed because people have all sorts of built-in biases in the way they evaluate their likes, dislikes, and feelings. Subtle changes in the wording or order of questions, the format by which people make their response, or the instructions accompanying the measurement instrument can alter the responses of the subjects. Despite these shortcomings, subjective responses are a valuable data source and often represent the only reasonable method for measuring the criterion of interest.

Terminal versus Intermediate Criteria

Criterion measures may describe the terminal performance (the ultimate output of the action) or some intermediate performance that led up to the output. Generally, intermediate measures are more specific and detailed than terminal measures. Terminal measures are more valuable than intermediate ones because they describe the ultimate performance of interest. Intermediate measures, however, are useful for diagnosing and explaining performance inadequacies. For example, if we were interested in the effectiveness of a warning label that instructs people using a medicine to shake well (the bottle, that is) before using, the terminal performance would be whether the people shook the bottle. Intermediate criteria would include asking the people whether they recalled seeing the warning, whether they read the warning, and whether they could recall the warning message. The intermediate criterion data would be of value in explaining why people do not always follow warnings—is it because they do not see them, do not take the time to read them, or cannot recall them?

REQUIREMENTS FOR RESEARCH CRITERIA

Criterion measures used in research investigations generally should satisfy certain requirements. There are both practical and psychometric requirements. The psychometric requirements are those of reliability, validity, freedom from contamination, and sensitivity.

Practical Requirements

Meister (1985) lists six practical requirements for criterion measures, indicating that, when feasible, a criterion measure should (1) be objective, (2) be quantitative, (3) be unobtrusive, (4) be easy to collect, (5) require no special data collection techniques or instrumentation, and (6) cost as little as possible in terms of money and experimenter effort.

Reliability

In the context of measurement, reliability refers to the consistency or stability of the measures of a variable over time or across representative samples. This is different from the concept of human reliability, as we see later. Technically, reliability in the measurement sense is the degree to which a set of measurements is free from error (i.e., unsystematic or random influences).

Suppose that a human factors specialist in King Arthur's court were commanded to assess the combat skills of the Knights of the Roundtable. To do this, the specialist might have each knight shoot a single arrow at a target and record the distance off target as the measure of combat skill. If all the knights were measured one day and again the next, quite likely the scores would be quite different on the two days. The best archer on the first day could be the worst on the second day. We would say that the measure was unreliable. Much, however, could be done to improve the reliability of the measure, including having each knight shoot 10 arrows each day and using the average distance off target as the measure, being sure all the arrows were straight and the feathers set properly, and performing the archery inside the castle to reduce the variability in wind, lighting, and other conditions that could change a knight's performance from day to day.

Correlating the sets of scores from the two days would yield an estimate of the reliability of the measure. Generally speaking, test-retest reliability correlations around .80 or above are considered satisfactory, although with some measures we have to be satisfied with lower levels.

Validity

Several types of validity are relevant to human factors research. Although each is different, they all have in common the determination of the extent to which different variables actually measure what was intended. The types of validity relevant to our discussion are *face validity, content validity,* and *construct validity.* Another type of validity which is more relevant to determining the usefulness of tests as a basis for selecting people for a job is *criterion-related validity.* This refers to the extent to which a test predicts performance. We do not discuss this type of validity further; rather we focus on the other types.

Face Validity   Face validity refers to the extent to which a measure *looks as though* it measures what is intended. At first glance, this may not seem important; however, in some measurement situations (especially in evaluation research and experimental research carried out in the field), face validity can influence the motivation of the subjects participating in the research. Where possible, researchers should choose measures or construct tasks that appear relevant to the users. To test the legibility of a computer screen for office use, it might be better to use material that is likely to be associated with offices (such as invoices) rather than nonsense syllables.

Content Validity   Content validity refers to the extent to which a measure of some variable samples a domain, such as a field of knowledge or a set of job behaviors. In the field of testing, for example, content validity is typically used to evaluate achievement tests. In the human factors field, this type of validity would apply to such circumstances as measuring the performance of air traffic controllers. To have content validity, such a measure would have to include the various facets of the controllers' performance rather than just a single aspect.

Construct Validity   Construct validity refers to the extent to which a measure is really tapping the underlying "construct" of interest (such as the basic type of behavior or ability in question). In our Knights of the Roundtable example, accuracy in shooting arrows at stationary targets would have only slight construct validity as a measure of actual combat skill. This is depicted in Figure 2–1, which shows the overlap between the construct (combat skill) and the measure (shooting accuracy with stationary targets). The small overlap denotes low construct validity because the measure taps only a few aspects of the construct. Construct validity is based on a judgmental assessment of an accumulation of empirical evidence regarding the measurement of the variable in question.

Freedom from Contamination

A criterion measure should not be influenced by variables that are extraneous to the construct being measured. This can be seen in Figure 2–1. In our example of the

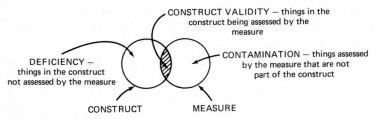

FIGURE 2-1
Illustration of the concept of construct validity and contamination in the measurement of research criteria.

knights, wind conditions, illumination, and quality of the arrows could be sources of contamination because they could affect accuracy yet are unrelated to the concept being measured, namely combat skill.

Sensitivity

A criterion measure should be measured in units that are commensurate with the anticipated differences one expects to find among subjects. To continue with our example of the knights, if the distance off target were measured to the nearest yard, it is possible that few, if any, differences between the knights' performance would have been found. The scale (to the nearest yard) would have been too gross to detect the subtle differences in skill between the archers.

Another example would be using a 3-point rating scale (uncomfortable-neutral-comfortable) to measure various chairs, all of which are basically comfortable. People may be able to discriminate between comfortable chairs; that is, some are more comfortable than others. With the 3-point scale given, however, all chairs would be rated the same—comfortable. If a 7-point scale were used with various levels of comfort (a more sensitive scale), we would probably see differences between the ratings of the various chairs. An overly sensitive scale, however, can sometimes decrease reliability.

HUMAN RELIABILITY

Human reliability is not the same as the reliability we discussed as a requirement for criterion measures. *Human reliability* is simply the probability of successful performance of a task or an element of a task by a human. As such, it is the same as systems reliability discussed in Chapter 1. Human reliability is expressed as a probability. For example, if the reliability of reading a particular display is .992, then out of 1000 readings we would expect 992 to be correct and 8 to be in error.

Historical Perspective

Interest in human reliability began in the 1950s when there was a desire to quantify the human element in a system in the same way that reliability engineers were quantifying the hardware element. Back then, human factors activities, especially in the aerospace industry, were often part of the reliability, or quality assurance, divisions of the companies. Thus it is understandable that human factors folks would want to impress their hosts and contribute information to already existing analytical methods used by their cohorts.

In 1962, the first prototype human reliability data bank (now we call them data bases) was developed and called, appropriately, the *Data Store*. Work has continued in the area on a relatively small scale by a handful of people. The Three-Mile Island nuclear power plant incident renewed interest in human reliability, and considerable work was done to develop data and methodologies for assessing human reliability in the nuclear industry.

We discuss some attempts at developing human reliability data bases and methodologies so that the flavor of the state of the art can be appreciated. We close with a discussion of some of the criticism leveled at the concept of human reliability.

Data Bases and Methodologies

Over the years, there have been several attempts to develop data bases of human reliability information. Topmiller, Eckel, and Kozinsky (1982) reviewed nine such attempts. We review two, Data Store and the Aerojet General Maintenance Performance Data Bank. In addition, two more elaborate methodologies for assessing human reliability are discussed, namely, the technique for human error rate prediction (THERP) and stochastic computer simulation models (Siegel and Wolf, 1969).

Data Store   Data Store was developed by the American Institute for Research (Munger, 1962; Munger, Smith, and Payne, 1962; Payne and Altman, 1962) particularly for tasks in the operation of electronic equipment. The data base consists of a compilation of performance data taken from 164 psychological studies culled from over 2000 examined. The data base consists of about 20 pages of tables indicating performance times and human reliability probabilities [the entire data base is reproduced in Topmiller, Eckel, and Kozinsky (1982)]. An example of part of one table is given in Table 2–1 for the operation of a joystick. Table 2–2 lists the components contained in Data Store; for each component several characteristics are listed that modify the response times and reliabilities. In using the data base, a series system approach is taken (see Chapter 1); that is, the reliabilities are multiplied to obtain the estimated reliability. For example, from Table 2–1, if a 15-in joystick had a range of movement of 50 degrees, the estimated reliability for using it would be .9927 (that is, .9967 × .9960). This series approach of multiplying reliabilities has been criticized as a flaw in the system that can result in grossly underestimating the human reliability of actual equipment operation (Meister, 1985).

Data Store has been incorporated into HECAD (Human Engineering Computer-Aided Design) by the Aerospace Medical Research Laboratory. This interactive computer graphics technique enables a crew station designer to lay out candidate configurations of a work station and automatically compute probabilities of successful task performance using Data Store information for the control and display features selected for the control panel.

TABLE 2-1
TIME AND RELIABILITY OF OPERATION OF JOYSTICKS OF VARIOUS LENGTHS WHEN THEY ARE MOVED VARIOUS DISTANCES

Dimension	Time to be added to base time,* s	Reliability
Stick length, in		
6–9	1.50	.9963
12–18	0.00	.9967
21–27	1.50	.9963
Stick movement, degrees		
5–20	0.00	.9981
30–40	0.20	.9975
40–60	0.50	.9960

*Base time, 1.93 s.
Note: Other data (not shown here) are given for variations in other dimensions, specifically, control resistance, presence or absence of arm support, and time lag between movement of control and corresponding movement of display.
Source: Munger, 1962.

TABLE 2-2
COMPONENTS CONTAINED IN DATA STORE HUMAN RELIABILITY DATA BANK

Inputs	Mediating processes	Outputs
Circular scales	Identification/recognition	Cable connections
Counters	Manipulation	Cranks
Labeling		Disconnecting
Lights		Joysticks
Linear scales		Knobs
Nonspeech		Levers
Scopes		Object positioning
Semicircular scales		Pushbuttons
Speech		Rotary selectors
		Speech
		Toggle switches
		Writing

Although Data Store is the most comprehensive, empirically based human reliability data base available, it does have its problems. In an attempt to validate the reliability data against performance of actual equipment operation, Goldbeck and Charlet (1974, 1975) found low, but positive, correlations (.33) between Data Store predictions (using HECAD) and actual reliabilities. Further, the molecular approach inherent in Data Store has also been criticized (Swain, 1967). Finally, the amount of data in the data base is small and is limited by the display and control technologies of more than a quarter century ago (pre-1960) when the data were collected.

Aerojet General Maintenance Performance Data Bank   This data bank (Irwin, Levitz, and Freed, 1964; also reproduced in Topmiller, Eckel, and Kozinsky, 1982) was developed specifically to predict reliabilities during checkout and maintenance activities performed on the *Titan II* rocket propulsion system. The data bank combined task analysis, expert judgment, and Data Store to estimate human reliability for selected maintenance functions. Reliabilities were estimated by using expert judges for approximately 60 task elements, such as read pressure gauge (reliability = .9952); install nuts, plugs, and bolts (.9962); verify switch position (.9966); or remove funnel from oilcan (.9980). These task elements were then combined to yield reliability estimates for about 60 different tasks, such as install gear box pressurization kit (.9198).

Technique for Human Error Rate Prediction   THERP has been developed in recent years mainly to assist in determining human reliability in nuclear power plants (Swain and Guttmann, 1983). THERP is a detailed procedure for analyzing a task and applying tables of human reliability estimates, such as the one shown in Table 2–3, to determine the overall reliability of the task. The actual procedure, although not overly complex, is more than we wish to deal with here. Suffice it to say, it does involve judgment calls on the part of the person doing the analysis.

Stochastic Simulation Models   These techniques are quite different from the previous approaches to determining human reliability. In essence, a computer is used to simulate the performance of a task or job. The computer performs the job over and over, sampling from probabilistic (i.e., stochastic) distributions of task element success and failure, and computes the number of times the entire task was success-

TABLE 2-3
EXAMPLES OF HUMAN RELIABILITY ESTIMATES FROM THERP

Description	Reliability
Select wrong control on a panel from an array of similar-appearing controls identified by labels only	.997
Omit an instruction when written procedures are available and should be used but are not	.950
Use a checklist properly	.500
Error in reading and recording quantitative information from:	
Analog meter	.997
Digital readout (4 or less digits)	.999
Chart recorder	.994

Source: Swain and Guttman (1983).

fully completed. Of course, the models and procedures are far more complex than depicted here. The advantage of using computer simulations is that one can vary aspects of the task and assess the effects on reliability. For example, Siegel, Leahy, and Wiesen (1977) in a study of a sonar system found that reliability for the simulated operators varied from .55 to .94 as time permitted the operators to do the task rose from 23 to 25 min. Here is an example in which a couple of minutes one way or the other could have profound effects on system performance. The results obtained from using such models, of course, are only as good as the models themselves and the data upon which the models are based.

Criticisms of Human Reliability

Despite all the effort at developing human reliability data banks and assessment methodologies, the majority of human factors specialists appear quite uninterested in the subject. Meister (1985) believes some of the problem is that the association of reliability with engineering is distasteful to some behavioral scientists and that some people feel behavior is so variable from day to day that it is ludicrous to assign a single point estimate of reliability to it.

Adams (1982) raises several practical problems with the concept of human reliability. Among them is that not all errors result in failure, and presently this problem is not handled well by the various techniques. Regulinski (1971) believes that point estimates of human reliability are inappropriate for continuous tasks, such as monitoring or tracking, and that tasks of a discrete nature may not be amenable to classical engineering reliability modeling. Meister (1985) counters that although our estimates may not be totally accurate, they are good approximations and hence have utility.

Some scientists object to the illusion of precision implicit in reliability estimates down to the fourth decimal (for example, .9993) when the data upon which they were based come nowhere near such levels of precision. There is also concern about the subjective components that are part of all human reliability techniques. Most of the reliability data used were originally developed by expert judgment—subjectively. In addition, the analyst must interject personal judgment into the reliability estimate to correct it for a particular application. And, of course, the biggest criticism is that the data are too scanty and do not cover all the particular applications we would like to see covered. Meister (1985), however, believes we cannot hibernate until some hypothetical time when there will be enough data—and as any human factors person will tell you, there are *never* enough data.

DISCUSSION

In this chapter we stressed the empirical data base of human factors. Research of all types plays a central role in human factors, and without it human factors would hardly be considered a science. A major continuing concern is the measurement of criteria to evaluate systems and the effects of independent variables. We outlined the major types of criteria used in human factors, with special reference to human reliability, and some of the requirements we would like to see in our measurement procedures. For some people, research methodology and statistics are not very exciting subjects. If these are not your favorite topics, think of them as a fine wine; reading about it cannot compare to trying the real thing.

REFERENCES

Adams, J. (1982). Issues in human reliability. *Human Factors,* 24, 1–10.

Bobo, M., Bethea, N., Ayoub, M., and Intaranont, K. (1983). Energy expenditure and aerobic fitness of male low seam coal miners. *Human Factors,* 25, 43–48.

Box, G., Hunter, W., and Hunter, J. (1978). *Statistics for experimenters: An introduction to design, data analysis, and model building.* New York: Wiley.

Cohen, J. (1969). *Statistical power analysis for the behavioral sciences.* New York: Academic.

Garg, A., Bakken, G., and Saxena, U. (1982). Effect of seat belts and shoulder harnesses on functional arm reach. *Human Factors,* 24, 367–372.

Goldbeck, R., and Charlet, J. (1974, June). *Task parameters for predicting panel layout design and operator performance* (WDL-TR-5480). Palo Alto, CA: Western Development Laboratories, Philco-Ford Corp.

Goldbeck, R., and Charlet, J. (1975, November). *Prediction of operator work station performance* (WDL-TR-7071). Palo Alto, CA: Western Development Laboratories, Philco-Ford Corp.

Hull, J., Gill, R., and Roscoe, S. (1982). Locus of the stimulus to visual accommodation: Where in the world, or where in the eye? *Human Factors,* 24, 311–320.

Irwin, I., Levitz, J., and Freed, A. (1964, August). Human reliability in the performance of maintenance. In *Proceedings of the Symposium on Quantification of Human Performance.* Albuquerque, NM: Electronic Industries Association.

Kerlinger, F. (1973). *Foundations of behavioral research* (2d ed.). New York: Holt.

McFarling, L., and Ellingstad, V. (1977). Research design and analysis: Striving for relevance and efficiency. *Human Factors,* 19, 211–219.

Marras, W., and Kroemer, K. (1980). A method to evaluate human factors/ergonomics design variables of distress signals. *Human Factors,* 22, 389–400.

Meister, D. (1985). *Behavioral analysis and measurement methods.* New York: Wiley.

Meister, D., and Rabideau, G. (1965). *Human factors evaluation in system development.* New York: Wiley.

Munger, S. (1962). *An index of electronic equipment operability: Valuation booklet.* Pittsburgh: American Institute for Research.

Munger, S., Smith, R., and Payne, D. (1962). *An index of electronic equipment operability: Data Store.* Pittsburgh: American Institute for Research.

Myers, J. (1979). *Fundamentals of experimental design* (3d ed.). Boston: Allyn and Bacon.

Neal, A., and Darnell, M. (1984). Text-editing performance with partial-line, partial-page, and full-page displays. *Human Factors,* 26, 431–442.

Nuclear Regulatory Commission (1981, September). *Guidelines for Control Room Design Process* (NUREG-0700). Washington.

Payne, D., and Altman, J. (1962). *An index of electronic equipment operability: Report of development.* Pittsburgh: American Institute for Research.

Regulinski, T. (1971, February). Quantification of human performance reliability research method rationale. In J. Jenkins (ed.), *Proceedings of U.S. Navy Human Reliability Workshop 22–23 July 1970* (Rep. NAVSHIPS 0967-412-4010). Washington: Naval Ship Systems Command.

Roebuck, J., Kroemer, K., and Thomson, W. (1975). *Engineering anthropometry methods.* New York: Wiley.

Runkel, P., and McGrath, J. (1972). *Research on human behavior*. New York: Holt.

Sanders, M. (1981). Peak and sustained isometric forces applied to a truck steering wheel. *Human Factors, 23,* 655–660.

Sanders, M. and Krohn, G. (1983, January). *Validation and extension of visibility requirements analysis for underground mining equipment*. Pittsburgh: Bureau of Mines.

Siegel, A., Leahy, W., and Wiesen, J. (1977, March). *Applications of human performance reliability—Evaluation Concepts and demonstration guidelines*. Wayne, PA: Applied Psychological Sciences.

Siegel, A., and Wolf, J. (1969). *Man-machine simulation models: Performance and psychological interactions*. New York: Wiley.

Simon, C. (1976, August). *Analysis of human factors engineering experiments: Characteristics, results, and applications*. Westlake Village, CA: Canyon Research Group.

Swain, A. (1967). Some limitations in using the simple multiplicative model in behavior quantification. In W. Askren (ed.), *Symposium on Reliability of Human Performance in Work*. (AMRL-TR-67-88). Wright-Patterson AFB, OH: Aerospace Medical Research Laboratories, pp. 17–31.

Swain, A., and Guttmann, H. (1983). *Handbook of human reliability analysis with emphasis on nuclear power plant applications: Final report* (NUREG/CR-1278). Washington: Nuclear Regulatory Commission.

Tabachnick, B., and Fidell, L. (1983). *Using multivariate statistics*. New York: Harper & Row.

Topmiller, D., Eckel, J., and Kozinsky, E. (1982). *Human reliability data bank for nuclear power plant operations:* vol. 1, *A review of existing human reliability data banks* (NUREG/CR-2744/1 of 2). Washington: Nuclear Regulatory Commission.

Vreuls, D., Obermayer, R., Goldstein, I., and Lauber, J. (1973, July). *Measurement of trainee performance in a captive rotary-wing device,* Rep. NAVTRAEQUIPCEN 71-C-0194-1. Orlando, FL: Naval Training Equipment Center.

PART TWO

INFORMATION INPUT

CHAPTER 3

INFORMATION INPUT AND PROCESSING

Today we hear a lot about information; in fact, we are reported to be in the throes of an information revolution. In industrialized countries the reception, processing, and storage of information are big business. Despite all the hoopla and advances in technology, still the most powerful, efficient, and flexible information processing and storage device on earth is located right between our ears—the human brain. On our jobs, in our homes, while driving in a car or just sitting on the beach, we are bombarded by all sorts of stimuli from our environment and from within our own bodies. These stimuli are received by our various sense organs (such as the eyes, ears, sensory receptors of the skin, and the proprioceptors in the muscles) and are processed by the brain.

The study of how humans process information about their environment and themselves has been an active area of research in psychology for over 100 years. Since about 1960, however, there has been an accelerated interest in the area, perhaps spurred by the information revolution and the associated advances in computer technology. At least one author (Anderson, 1980) gives some of the credit to the field of human factors and its early emphasis on practical problems of perception and attention. Various terms have been used to describe this growing area—*cognitive psychology, cognitive engineering,* and *engineering psychology,* to name a few. In this chapter we discuss the notion of information, including how it can be measured, coded, and displayed. A model of human information processing is presented and discussed along with the concept and methods of measuring mental workload. Finally, we close the chapter by looking at a few human factors implications of the information revolution itself.

INFORMATION THEORY

Our common notion of information is reflected in such everyday examples as what we read in the newspapers or hear on TV, the gossip over the backyard fence, or the instructions packed with every gadget we buy. This conception of information is fine for our everyday dealing; but when we investigate how people process information, it is important to have an operational definition of the term *information* and a method for measuring it. Information theory provides one such definition and a quantitative method for measuring it.

Historical Perspective

Before we discuss some of the rudiments of information within the context of information theory, it will be helpful to give a little historical perspective on the subject. During the late 1940s and early 1950s, information theory was developed within the context of communications engineering, not cognitive psychology. It quickly became, however, a source of ideas and experimentation within psychology.

Early investigators held out great promise that the theory would help unravel the mysteries of human information processing. Information theory was used to determine the information processing capacity of the different human sensory channels and was also employed extensively in problems of choice reaction time. Unfortunately, information theory has not lived up to its early promise in the field of human information processing, and its use has waned over the last 10 to 15 years until now its impact is quite negligible (Baird, 1984). The problems are (1) that most information concepts are descriptive rather than explanatory and (2) that they offer only the most rudimentary clues about the underlying psychological mechanisms of information processing (Baird, 1984). It is interesting, however, that in newer areas of cognitive psychology, where detailed models and theories have yet to be formulated, information theory continues to be applied.

Concept of Information

Information theory defines *information* as the reduction of uncertainty. The occurrences of highly certain events do not convey much information since they only confirm what was expected. The occurrences of highly unlikely events, however, convey more information. When the temperature warning light comes on in a car, for example, it conveys considerable information because it is an unlikely event. The "fasten seat belt" warning that comes on when the car is started conveys less information because it is expected. Note that the importance of the message is not directly considered in the definition of information; only the likelihood of its occurrence is considered. The "fasten seat belt" message is important; but because it comes on every time the car is started, it contains little (or no) information in the context of information theory.

Unit of Measure of Information

Information theory measures information in *bits* (symbolized by *H*). A *bit* is the amount of information required to decide between two equally likely alternatives. When the probabilities of the various alternatives are equal, the amount of information *H* in bits is equal to the logarithm, to the base 2, of the number *N* of such alternatives, or

$$H = \log_2 N$$

With only two alternatives, the information, in bits, is equal to 1.0 (because $\log_2 2 = 1$). The signal from the Old North Church to Paul Revere indicating whether the enemy was coming by land or by sea (two alternatives), for example, would convey 1.0 bit of information, assuming the two alternatives were equally probable. Four equally likely alternatives would convey 2.0 bits of information ($\log_2 4 = 2$). A randomly chosen digit from the set of numbers 0 to 9 would convey 3.322 bits ($\log_2 10 = 3.322$), and a randomly chosen letter from the set of letters A to Z would convey 4.7 bits ($\log_2 26 = 4.7$).

When the alternatives are not equally likely, the information conveyed by an event is determined by the following formula:

$$h_i = \log_2 \frac{1}{p_i}$$

where h_i is the information (in bits) associated with event *i*, and p_i is the probability of occurrence of that event. We are often more interested, however, in the average information conveyed by a series of events having different probabilities than in the information of just one such event. The average information H_{av} is computed as follows:

$$H_{av} = \sum_{i=1}^{N} p_i \left(log_2 \frac{1}{p_{i=1}} \right)$$

Let us go back to Paul Revere and the Old North Church. If the probability of a land attack were .9 and the probability of a sea attack were .1, then the average information would be 0.47 bit. The maximum possible information is always obtained when the alternatives are equally probable. In the case of Paul Revere, the maximum possible information in the Old North Church signal was 1.0 bit. The greater the departure from equal probability, the greater the reduction in information from the maximum. This leads to the concept of *redundancy,* which is the reduction in information from the maximum owing to unequal probabilities of occurrence. Percentage of redundancy is usually computed from the following formula:

$$\% \text{ Redundancy} = \left(1 - \frac{H_{av}}{H_{max}} \right) \times 100$$

For example, because not all letters of the English language occur equally often and certain letter combinations (e.g., th and qu) occur more frequently than others, the English language has a degree of redundancy of approximately 68 percent.

One other concept that may be encountered in the literature is bandwidth of a communication channel. The *bandwidth* is simply the rate of information transmission over the channel, measured in bits per second.

An Application of Information Theory

In choice reaction time experiments, the subject must make a discrete and separate response to different stimuli; for example, a person might be required to push one of four buttons depending on which of four lights comes on. Hick (1952) varied the number of stimuli in a choice reaction time task and found that reaction time to a stimulus increased as the number of equally likely alternatives increased. When he converted number of stimuli to bits and plotted the mean reaction time as a function of the information (bits) in the stimulus condition, the function was linear (that is, it was a straight line). (Scientists get very excited about linear relationships.) Reaction time appeared to be a linear function of stimulus information, but could it be just an artifact of the number of alternatives? That is, perhaps reaction time is just a logarithmic function of number of alternatives. Hyman (1953), however, realized that, according to information theory, he could alter stimulus information without changing the number of alternatives simply by changing the probability of occurrence of the alternatives. Lo and behold, using this technique, he found that reaction time was still a linear function of stimulus information when it was measured in bits. Thus, information theory and the definition of the bit made it possible to discover a basic principle: choice reaction time is a linear function of stimulus information. This is called the *Hick-Hyman law*.

DISPLAYING INFORMATION

Human information input and processing operations depend, of course, on the sensory reception of relevant external stimuli. It is these stimuli that contain the information we process. Figure 3-1 illustrates in a schematic way the various pathways by which information is received by an individual. Typically, the original source (i.e., the *distal* stimulus) is some object, event, or environmental condition. Information from these original sources may come to us *directly* (such as by *direct* observation of an airplane), or it may come to us *indirectly* through some intervening mechanism or device (such as radar or a telescope). In either case, the distal stimuli are sensed by the individual only through the energy that they generate (directly or indirectly) through *proximal* stimuli (light, sound, mechanical energy, etc.). In the case of *indirect* sensing, the *new* distal stimuli may be of two types. First, they may be *coded* stimuli, such as visual or auditory displays. Second, they may be *reproduced* stimuli, such as those presented by TV, radio, or photographs or through such devices as microscopes, microfilm viewers, binoculars, and hearing aids. In such cases the reproduction may be intentionally or unintentionally modified in some way,

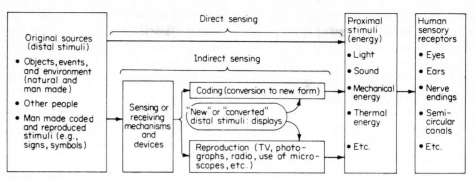

FIGURE 3-1
Schematic illustration of pathways of information from original sources (stimuli) to sensory receptors. Original sources can be directly sensed or indirectly sensed through coding or reproduction of the original stimuli.

as by enlargement, miniaturization, amplification, filtering, or enhancement. With either coded or reproduced stimuli, the new, or converted, stimuli become the actual distal stimuli to the human sensory receptors.

The human factors aspect of design enters into this process in those circumstances in which *indirect* sensing applies, for it is in these circumstances that the designer can design displays for presenting information to people. *Display* is a term that applies to virtually any indirect method of presenting information, such as a highway traffic sign, a family radio, or a page of braille print. We discuss in later chapters the human factors considerations in the design of visual displays (Chapters 4 and 5); auditory, tactual, and olfactory displays (Chapter 6); and speech displays (Chapter 7). It is appropriate in our present discussion of information input, however, to take an overview of the various circumstances and types of information that can be presented with displays.

When to Display Information

The question of when to display information is really a question of when to use indirect presentation of stimuli (see Figure 3-1). There are, of course, many original sources of information (i.e., distal stimuli) that people can sense directly without any problem. But in many circumstances direct sensing simply will not do, and some form of indirect display must be used. Some of these circumstances are listed here:

1 When distal stimuli are of the type that humans generally can sense, but cannot sense adequately under the circumstances because of such factors as
   a Stimuli at or below threshold values (e.g., too far, too small, or not sufficiently intense) that need to be amplified by electronic, optical, or other means
   b Stimuli that require reduction for adequate sensing (e.g., large land areas converted to maps)
   c Stimuli embedded in excessive noise that generally need to be filtered or amplified

    d Stimuli that are physically removed or obstructed and need to be reproduced or coded to be sensed

    e Stimuli that need to be sensed with greater precision than people can discriminate (e.g., temperature, weights and measures, and sound)

    f Stimuli that need to be stored for future reference (e.g., by photography and tape recorder)

    g Stimuli of one type that probably can be sensed better or more conveniently if converted to another type in either the same sensory modality (e.g., graphs to represent quantitative data) or a different modality (e.g., auditory warning devices)

    h Information about events or circumstances that by their nature virtually require some display presentations (such as emergencies, road signs, and hazardous conditions)

2 When distal stimuli are of the type that humans generally cannot sense or that are beyond the spectrum to which humans are sensitive, and so have to be sensed by sensing devices and converted into coded form for human reception (e.g., certain forms of electromagnetic energy and ultrasonic vibrations)

In these and other types of circumstances, it may be appropriate to transmit relevant information (stimuli) *indirectly* by some type of display. For our purposes, we consider a display to be any method of presenting information indirectly, in either *reproduced* or *coded* (symbolic) form. If a decision is made to use a display, there may be some option regarding the sensory modality and the specific type of display to use, since the method of presenting information can influence, for better or worse, the accuracy and speed with which information can be received. We discuss the issue of choosing a sensory modality a little later.

Types of Information Presented by Displays

In a general sense, information presented by displays can be considered *dynamic* or *static*. We often speak of the display as being dynamic or static, although it is really the information that has that quality. Dynamic information continually changes or is subject to change through time. Examples include traffic lights that change from red to green, speedometers, radar displays, and temperature gauges. In turn, static information remains fixed over time (or at least for a time). Examples include printed, written, and other forms of alphanumeric data; traffic signs; charts; graphs; and labels. With the advent of computer displays or visual display terminals (VDTs), the distinction between static and dynamic information is becoming blurred. Thus, much of the "static" information previously presented only in "fixed" format (as on the printed page) is now being presented on VDTs. For our purposes, we still consider such presentation as static information (covered in Chapter 4). Note, however, that even if such material is shown on a VDT, most of it, *as such,* still is static. That is, the specific information does not itself change but rather can be replaced by other information. This is in contrast with most dynamic information (such as speed) which *itself* is changeable.

Although information can be grossly categorized as static or dynamic, a more detailed classification is given below:

- *Quantitative information:* Display presentations that reflect the quantitative value of some variable, such as temperature or speed. Although in most instances the variable is dynamic, some such information may be static (such as that presented in nomographs and tables).
- *Qualitative information:* Display presentations that reflect the approximate value, trend, rate of change, direction of change, or other aspect of some changeable variable. Such information usually is predicated on some quantitative parameter, but the displayed presentation is used more as an indication of the change in the parameter than for obtaining a quantitative value as such.
- *Status information:* Display presentations that reflect the condition or status of a system, such as on-off indications; indications of one of a limited number of conditions, such as stop-caution-go lights; and indications of independent conditions of some class, such as a TV channel. (What is called *status information* here is sometimes referred to in other sources as *qualitative information*. But the terminology used in this text is considered more descriptive.)
- *Warning and signal information:* Display presentation used to indicate emergency or unsafe conditions, or to indicate the presence or absence of some object or condition (such as aircraft or lighthouse beacons). Displayed information of this type can be static or dynamic.
- *Representational information:* Pictorial or graphic representations of objects, areas, or other configurations. Certain displays may present dynamic images (such as TV or movies) or symbolic representations (such as heartbeats shown on an oscilloscope or blips on a cathode-ray tube). Others may present static information (such as photographs, maps, charts, diagrams and blueprints) and graphic representation (such as bar graphs and line graphs).
- *Identification information:* Display presentations used to identify some (usually) static condition, situation, or object, such as the identification of hazards, traffic lanes, and color-coded pipes. The identification usually is in coded form.
- *Alphanumeric and symbolic information:* Display presentations of verbal, numeric, and related coded information in many forms, such as signs, labels, placards, instructions, music notes, printed and typed material including braille, and computer printouts. Such information usually is static, but in certain circumstances it may be dynamic, as in news bulletins displayed by moving lights on a building.
- *Time-phased information:* Display presentations of pulsed or time-phased signals, e.g., signals that are controlled in terms of duration of the signals and of intersignal intervals, and of their combinations, such as the Morse code and blinker lights.

Note that some displays present more than one type of information, and users "use" the displayed information for more than one purpose. The kinds of displays that would be preferable for presenting certain types of information are virtually specified by the nature of the information in question. For presenting most types of information, however, there are options available regarding the kinds of displays and certainly about the specific features of the display.

Selection of Display Modality

In selecting or designing displays for transmission of information in some situations, the selection of the sensory modality (and even the stimulus dimension) is virtually a foregone conclusion, as in the use of vision for road signs and the use of audition for many various purposes. Where there is some option, however, the intrinsic advantages of one over the other may depend on a number of considerations. For example, audition tends to have an advantage over vision in vigilance types of tasks, because of its attention-getting qualities. A more extensive comparison of audition and vision is given in Table 3-1, indicating the kinds of circumstances in which each of these modalities tends to be more useful. These comparisons are based on considerations of substantial amounts of research and experience relating to these two sensory modalities.

The tactual sense is not used very extensively as a means of transmission of information, but does have relevance in certain specific circumstances, such as with blind persons, and in other special circumstances in which the visual and auditory sensory modalities are overloaded. Further reference to such displays are made in Chapter 6.

CODING OF INFORMATION

Many displays involve the coding of information (actually, of stimuli) as opposed to some form of direct representation or reproduction of the stimulus. Coding takes place when the original stimulus information is converted to a new form and displayed symbolically. Examples include radar screens that display aircraft as blips, maps that display population information about cities by using different sized letters to spell the names of cities, and electric resistors that use colored bands or stripes to indicate the resistance in ohms.

TABLE 3-1
WHEN TO USE THE AUDITORY OR VISUAL FORM OF PRESENTATION

Use auditory presentation if:	Use visual presentation if:
1 The message is simple.	1 The message is complex.
2 The message is short.	2 The message is long.
3 The message will not be referred to later.	3 The message will be referred to later.
4 The message deals with events in time.	4 The message deals with location in space.
5 The message calls for immediate action.	5 The message does not call for immediate action.
6 The visual system of the person is overburdened.	6 The auditory system of the person is overburdened.
7 The receiving location is too bright or dark-adaptation integrity is necessary.	7 The receiving location is too noisy.
8 The person's job requires moving about continually.	8 The person's job allows him or her to remain in one position.

Source: Deatherage, 1972, p. 124, table 4-1.

When information is coded, it is coded along various dimensions. For example, targets on a computer screen can be coded by varying size, brightness, color, shape, etc. An audio warning signal could be coded by varying frequency, intensity, or on-off pattern. Each of these variations constitutes a dimension of the displayed stimulus, or a *stimulus dimension*. The utility of any given stimulus dimension to convey information, however, depends on the ability of people (1) to identify a stimulus based on its position along the stimulus dimension (such as identifying a target as bright or dim, large or small) or (2) to distinguish between two or more stimuli which differ along a stimulus dimension (such as indicating which of two stimuli is brighter or larger).

Absolute versus Relative Judgments

The task of identifying a stimulus based on its position along a dimension and the task of distinguishing between two stimuli which differ along a dimension correspond to judgments made on an *absolute basis* and judgments made on a *relative basis,* respectively. A relative judgment is one which is made when there is an opportunity to compare two or more stimuli and to judge the relative position of one compared with the other along the dimension in question. One might compare two or more sounds and indicate which is the louder or compare two lights and indicate which is brighter. In some cases the judgment is merely whether the two stimuli are the same or different, as, for example, in judging the shape of two symbols.

In absolute judgments there is no opportunity to make comparisons; in essence, the comparison is held in memory. Examples of absolute judgments include identifying a given note on the piano (say, middle C) without being able to compare it with any others, grading the quality of wool into one of several quality categories without examples from each category to compare against, or identifying a symbol as belonging to one class of objects rather than another without other objects to compare it with.

Making Absolute Judgments along Single Dimensions

As one might expect, people are generally able to make fewer discriminations on an absolute basis than on a relative basis. For example, Mowbray and Gebhard (1961) found that people could discriminate on a relative basis about 1800 tone pairs differing only in pitch. On an absolute basis, however, people can identify only about 5 tones of different pitch.

In this connection Miller (1956) referred to the "magical number 7 ± 2" as the typical range of single-dimension identifications people can make on an absolute basis. This range of 7 ± 2 (that is, 5 to 9) seems to apply to many different dimensions, as shown in Table 3-2. With certain types of stimulus dimensions, however, people can identify more levels of stimuli. For example, most people can identify as many as 15 or more different geometric forms (some people might argue that geometric forms are multidimensional and hence not a valid test of the 7 ± 2 principle).

The limit to the number of absolute judgments that can be made appears to be a limitation of human memory rather than a sensory limitation. People have a hard

TABLE 3-2
AVERAGE NUMBER OF DISCRIMINATIONS AND CORRESPONDING NUMBER OF BITS
TRANSMITTED BY CERTAIN STIMULUS DIMENSIONS

Stimulus dimension	Average number of discriminations	Number of bits
Pure tones	5	2.3
Loudness	4–5	2–2.3
Size of viewed objects	5–7	2.3–2.8
Brightness	3–5	1.7–2.3

time remembering more than about five to nine different standards against which stimuli must be judged. Note that the number of different stimuli that can be identified can be increased with practice.

Making Absolute Judgments along Multiple Dimensions

Given that we are not very good at making absolute judgments of stimuli, how can we identify such a wide variety of stimuli in our everyday experience? The answer is that most stimuli differ on more than one dimension. Two sounds may differ in terms of both pitch and loudness; two symbols may differ in terms of shape, size, and color.

When multiple dimensions are used for coding, the dimensions can be orthogonal or redundant. When dimensions of a stimulus are *orthogonal*, the value on one dimension is independent of the value on the other dimension. That is, each dimension carries unique information, and all combinations of the two dimensions are possible. If shape (circle-square) and color (red-green) were orthogonal dimensions, then there would be red circles, green circles, red squares, and green squares and each would signify something different in the coding scheme.

When dimensions are *redundant*, knowing the value on one dimension helps predict the value on the other dimension. There are various degrees of redundancy between stimulus dimensions depending on how they are used. If shape and color were completely redundant dimensions, then, for example, all circles would be green and all squares would be red; in this scheme, knowing the value of one dimension completely determines the value of the other. If 80 percent of the squares were red and 80 percent of the circles were green, color and shape would be considered partially redundant; knowing the value of one dimension would help predict the value of the other, but the prediction would not always be correct.

Combining Orthogonal Dimensions   When dimensions are combined orthogonally, the number of stimuli that can be identified on an absolute basis increases. The total number, however, usually is less than the product of the number of stimuli that could be identified on each dimension separately. For example, if people could discriminate 7 different sizes of circles and 5 levels of brightness, we would expect that when size and brightness were combined, people would be able to identify more than 7 but less than 35 (7 × 5) different size-brightness combinations. As more orthogonal dimensions are added, the increase in the number of stimuli that can be identified becomes smaller and smaller.

Combining Redundant Dimensions   When dimensions are combined in a redundant fashion, more levels can be identified than if only a single dimension is used. For example, if size and brightness were combined in a completely redundant fashion, people would be able to identify more than the seven stimuli they could identify based on size alone.

Characteristics of a Good Coding System

In the next few chapters we discuss various aspects and uses of visual, auditory, tactual, and olfactory coding systems. The following characteristics, however, transcend the specific type of codes used and apply, across the board, to all types of coding systems.

Detectability of Codes   To begin, any stimulus used in coding information must be detectable; specifically, it has to be able to be detected by the human sensory mechanisms under the environmental conditions anticipated. Color-coded control knobs on underground mining equipment, for example, would likely not be detectable in the low levels of illumination typically found in underground mines.

Discriminability of Codes   Further, every code symbol, even though detectable, must be discriminable from other code symbols. Here the number of levels of coding becomes important, as we saw above. If 20 different levels of size are used to code information, it is highly likely that people will confuse one for another and hence fail to discriminate among them. In this connection, the degree of difference between adjacent stimuli along a stimulus dimension may also influence the ease of discrimination, but not as much as one might expect. Pollack (1952), for example, found that people could identify between 4 and 5 tones when all were close together in terms of frequency (pitch). When the tones were spread out (20 times the original range), the number of stimuli that could be identified only increased to 8. Mudd (1961) reported evidence that in the case of auditory intensity, the greater the interstimulus distance, the faster the subjects could identify the signals. No such effect was found for frequency differences, however.

Meaningfulness of Codes   A coding system should use codes meaningful to the user. Meaning can be inherent in the code, such as a bent arrow on a traffic sign to denote a curved road ahead, or meaning can be learned, such as using red to denote danger. The notion of meaningfulness is bound up in the idea of conceptual compatibility, which we discuss in the next section.

Standardization of Codes   When a coding system is to be used by different people in different situations, it is important that the codes be standardized and kept the same from situation to situation. If a new display is to be added to a factory already containing other displays, the coding system used should duplicate the existing coding schemes where the same information is being transmitted. Red, for example, should mean the same thing on all displays.

Use of Multidimensional Codes   The use of multidimensional codes can increase the number and discriminability of coding stimuli used. Depending on the specific demands of the task and the number of different codes needed, redundant or orthogonal coding should be considered.

COMPATIBILITY

One cannot discuss human information processing without reference to the concept of compatibility. *Compatibility* refers to the relationship of stimuli and responses to human expectations, and as such is a central concept in human factors designs. A major goal in any design is to make the system compatible with human expectations. The term *compatibility* was originally used by A. M. Small and was later refined by Fitts and Seeger (1953). The concept of compatibility implies a process of information transformation, or recoding. It is postulated that the greater the degree of compatibility, the less recoding must be done to process the information. This in turn should result in faster learning, faster response times, fewer errors, and a reduced mental workload. In addition, people like things that work as they expect them to work.

Types of Compatibility

Traditionally, three types of compatibility have been identified: *conceptual, movement,* and *spatial*. A fourth type of compatibility has emerged from studies of human information processing—*modality compatibility*. We discuss each in turn.

Conceptual Compatibility   Conceptual compatibility deals with the degree to which codes and symbols correspond to the conceptual associations people have. In essence, conceptual compatibility relates to how meaningful the codes and symbols are to the people who must use them. An aircraft symbol used to denote an airport on a map would have greater conceptual compatibility than would a green square. Another example of conceptual compatibility is the work being done on how to create meaningful abbreviations for use as command names in computer applications (Ehrenreich, 1985; Grudin and Barnard, 1985; Hirsh-Pasek, Nudelman and Schneider, 1982; Streeter, Ackroff, and Taylor, 1983). Some methods of generating abbreviations are easier to recall and use than others, but at present there is no conclusive evidence to recommend one technique over another when the goal is improved decoding performance (Ehrenreich, 1985).

Movement Compatibility   Movement compatibility relates to the relationship between the movement of displays and controls and the response of the system being displayed or controlled. An example of movement compatibility is where clockwise rotation of a knob is associated with an increase in the system parameter being controlled. That is, to increase the volume on a radio, we expect to turn the knob clockwise. Another example is a vertical scale in which upward movement of the pointer corresponds to an increase in the parameter being displayed. We discuss movement compatibility in more detail in Chapter 9.

Spatial Compatibility   Spatial compatibility refers to the physical arrangement in space of controls and their associated displays. For example, imagine five displays lined up in a horizontal row. Good spatial compatibility would be achieved by arranging the control knobs associated with those displays in a horizontal row directly below the corresponding displays. Poor spatial compatibility would result if the controls were arranged in a vertical column to one side of the displays. Spatial compatibility is also discussed further in Chapter 9.

Modality Compatibility   Modality compatibility refers to the fact that certain stimulus-response modality combinations are more compatible with some tasks than with others (Wickens, 1984). This is illustrated by a portion of a study carried out by Wickens, Sandry, and Vidulich (1983). The experiment was rather involved, but we need to discuss only a portion of it here. Subjects performed one of two tasks, a verbal task (respond to commands such as ''turn on radar-beacon tacan'') or a spatial task (bring a cursor over a designated target). Each task was presented through either the auditory modality (speech) or the visual modality (displayed on a screen). The subject responded either vocally (spoke the response) or manually (performed the command or used a controller to position the cursor). All combinations of presentation modality and response modality were used for the two types of tasks. The results are shown in Figure 3-2, which shows reaction times for the two tasks using the four presentation-response modality combinations. Note that in Figure 3-2 fast performance is shown at the top of the vertical axis. As can be seen, the most compatible combination for the verbal task is auditory presentation and vocal response. For the spatial task, the most compatible combination is visual presentation and manual response.

Origins of Compatibility Relationships

With the exception of modality compatibility, which probably has its origins in the way our brains are wired, compatibility relationships stem from two possible origins. First, certain compatible relationships are intrinsic in the situation, for example, turning a steering wheel to the right in order to turn to the right. In certain combinations of displays and controls, for example, the degree of compatibility is associated with the extent to which they are isomorphic or have similar spatial relationships. Second, other compatible relationships are culturally acquired, stemming from habits or associations characteristic of the culture in question. For example, in the United States a light switch is usually pushed up to turn it on, but in certain other countries it is pushed down. How such culturally acquired patterns develop is perhaps the consequence of fortuitous circumstances.

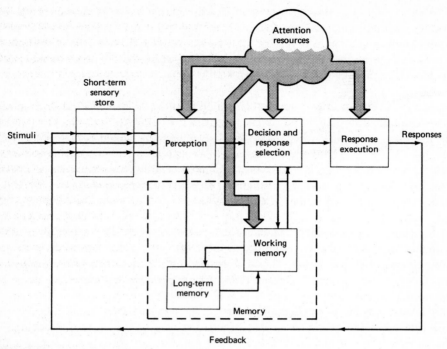

FIGURE 3-2
A model of human information processing showing the major processes, or stages, and
the interrelationships. (*Source: Wickens, 1984; fig. 1.1.*) Reprinted by permission of the
publisher.

Identification of Compatibility Relationships

To take advantage of compatible relationships in designing equipment or other items,
it is, of course, necessary to know what relationships are compatible. There generally
are two ways in which these can be ascertained or inferred. First, certain relationships are obvious, or manifest; this is particularly true in the case of spatial compatibility. A word of caution is in order here. What is obvious to you may not be
obvious to others, so it would be a good idea to check out the generality of the
assumed relationship. Second, when compatible relationships are not obvious, it is
necessary to identify them on the basis of empirical experiments. In later chapters
we present results from such studies that indicate the proportion of subjects who
choose specific relationships from several possibilities.

Discussion

Although different versions of compatibility involve the process of sensation and
perception and also response, the tie-in between these—the bridge between them—
is a mediation process. Where compatible relationships can be utilized, the probability of improved performance usually is increased. As with many aspects of human
performance, however, certain constraints need to be considered in connection with

compatibility relationships. For example, some such relationships are not self-evident; they need to be ascertained empirically. When this is done, it sometimes turns out that a given relationship is not universally perceived by people; in such instances it may be necessary to "figure the odds," that is, to determine the proportion of people with each possible association or response tendency and make a design determination on this basis. In addition, in some circumstances, trade-off considerations may require that one forgo the use of a given compatible relationship for some other benefit.

A MODEL OF INFORMATION PROCESSING

A model is an abstract representation of a system or process. Models are not judged to be correct or incorrect; the criterion for evaluating a model is *utility*. Thus, a *good model* is one that can account for the behavior of the actual system or process and can be used to generate testable hypotheses that are ultimately supported by the behavior of the actual system or process.

Models can be mathematical, physical, structural, or verbal. Information theory (which we discussed previously) is a mathematical model of information transfer in communication systems. Signal detection theory (which we discuss later) is also a mathematical model. We present here a structural model of human information processing which will organize our discussion of the process.

Remember, in the human information processing literature, numerous models have been put forth to "explain" how people process information (e.g., Broadbent, 1958; Haber and Hershenson, 1980; Sanders, 1983; Welford, 1976). Most of these models consist of hypothetical black boxes or processing stages that are assumed to be involved in the system or process in question. Figure 3-2 presents one such model (Wickens, 1984). The model depicts the major components, or stages, of human information processing and the hypothesized relationships between them. Most information processing models would agree with this basic formulation; however, some might add a box or two or divide one box into several. As we will see, the notion of attention resources and how they are allocated is still very much up in the air.

In the next several sections we discuss *perception*, with special emphasis on detection and signal detection theory; *memory*, including short-term sensory storage, working memory, and long-term memory; *decision making*, with special emphasis on the biases found in human decision making; and *attention*, including some of the controversy in the area and the notion of limited capacity and time sharing multiple tasks. The discussion of *response execution* and *feedback* is deferred to Chapters 7, 8, and 9.

PERCEPTION

There are various levels of *perception* that depend on the stimulus and the task confronting a person. The most basic form of perception is simple *detection;* that is, determining whether a signal or target is present. This can be made more complicated by requiring the person to indicate in which of several different classes the target belongs. This gets into the realm of *identification* and *recognition*. The simplest form of multilevel classification is making absolute judgments along stimulus dimensions, which we discussed previously.

The act of perception involves our prior experiences and learned associations. This is depicted in Figure 3-2 by the line connecting the long-term memory box to the perception box. Even the act of simple detection involves some complex information processing and decision making. This complexity is embodied in signal detection theory (Swets, 1964; Green and Swets, 1966).

Signal Detection Theory

Signal detection theory (SDT), also referred to as the Theory of Signal Detection (TSD), is applicable to situations in which there are two discrete states (signal and no signal) that cannot easily be discriminated (Wickens, 1984). Examples would include detecting a cavity on a tooth X-ray, detecting a blip on a radar screen, or detecting a warning buzzer in a noisy factory.

Concept of Noise   A central concept in SDT is that in any situation there is *noise* that can interfere with the detection of a signal. This noise can be generated externally from the person (e.g., noises in a factory other than the warning buzzer, or electronic static and false radar returns on a radar screen) and internally within the person (e.g., miscellaneous neural activity). This noise varies over time, thus forming a distribution of intensity from low to high. The shape of this distribution is assumed to be normal (bell-shaped). When a "signal" occurs, its intensity is added to that of the background noise. At any given time a person needs to decide if the sensory input (what the person senses) consists of only noise or noise plus the signal.

Possible Outcomes   Given that the person must say whether a signal occurred or did not occur, and given that there are only two possible states of reality (i.e., the signal did or did not occur), there are four possible outcomes:

- Hit: Saying there is a signal when there is a signal.
- *False alarm:* Saying there is a signal when there is no signal.
- *Miss:* Saying there is no signal when there is a signal.
- *Correct rejection:* Saying there is no signal when there is no signal.

Concept of Response Criterion   One of the major contributions of SDT to the understanding of the detection process was the recognition that people set a criterion along the hypothetical continuum of sensory activity and that this criterion is the basis upon which a person makes a decision. The position of the criterion along the continuum determines the probabilities of the four outcomes listed above. The best way to conceptualize this is shown in Figure 3-3. Shown are the hypothetical distributions of sensory activity when only noise is present and when a signal is added to the noise. Notice that the two distributions overlap. That is, sometimes conditions are such that the level of noise alone will exceed the level of noise plus signal.

SDT postulates that a person sets a criterion level such that whenever the level of sensory activity exceeds that criterion, the person will say there is a signal present.

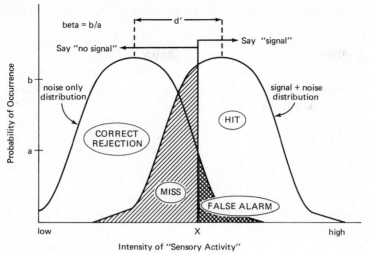

FIGURE 3-3
Illustration of the key concepts of signal detection theory. Shown are two hypothetical distributions of internal sensory activity, one generated by noise alone and the other generated by signal plus noise. The probabilities of four possible outcomes are depicted as the respective areas under the curves based on the setting of a criterion at X. Here d' is a measure of sensitivity, and beta is a measure of response bias. The letters a and b correspond to the height of the signal-plus-noise and noise-only distributions at the criterion.

When the activity level is below the criterion, the person will say there is no signal. Figure 3-3 also shows four areas corresponding to hits, false alarms, misses, and correct rejections based on the criterion shown in the figure.

Related to the position of the criterion is the quantity *beta*. Numerically, beta is the ratio (signal to noise) of the height of the two curves in Figure 3-3 at a given criterion point. If the criterion were placed where the two distributions cross, beta would equal 1.0. As the criterion is shifted to the right, beta increases and the person will say signal less often and hence will have fewer hits, but will also have fewer false alarms. We would say that such a person is being conservative. The criterion shown in Figure 3-3 is conservative and has a beta greater than 1.0. On the other hand, as the criterion is shifted to the left, beta decreases and the person will say signal more often and hence will have more hits, but also more false alarms. We would say such a person was being risky.

Influencing the Response Criterion   SDT postulates two variables that influence the setting of the criterion: (1) the likelihood of observing a signal and (2) the costs and benefits associated with the four possible outcomes. Consider the first. If you told your dentist that one of your teeth were hurting, thus increasing the likelihood that you had a cavity, the dentist would be more likely to call a suspicious spot on the X-ray of that tooth a cavity. That is, as the probability of a signal increases, the criterion is lowered (reducing beta), so that anything remotely suggesting a signal (cavity in our example) might be called a signal.

With regard to the second factor, costs and benefits, again consider the example of the dentist. What is the cost of a false alarm (saying there is a cavity when there is not)? You get your tooth drilled needlessly. What is the cost of a miss (saying there is no cavity when there is one)? The cavity may get worse, and you might lose your tooth. Under these circumstances the dentist might set a low criterion and be more willing to call a suspicious spot a cavity. But what if you go to the dentist every 6 months faithfully, so the dentist knows that if a cavity is missed this time, there will be other opportunities to detect it and correct it without the danger of loosing your tooth? In this case the dentist would set a more conservative criterion and would be less willing to call it a cavity—this time.

Concept of Sensitivity In addition to the concept of response criterion, SDT also includes a measure of a person's *sensitivity,* that is, the keenness or resolution of the sensory system. Further, SDT postulates that the response criterion and sensitivity are independent of each other. In SDT terms, sensitivity is measured by the degree of separation between the two distributions shown in Figure 3-3. The sensitivity measure is called d' and corresponds to the separation of the two distributions expressed in units of standard deviations (it is assumed that the standard deviations of the two distributions are equal). The greater the separation, the greater the sensitivity and the greater the d'. In most applications of SDT, d' ranges between 0.5 and 2.0.

Some signal generation systems may create more noise than others, and some people may have more internal noise than others. The greater the amount of noise, the smaller d' will be. Also the weaker and less distinct the signal, the smaller d' will be. Another factor that influences d' is the ability of a person to remember the physical characteristics of a signal. When memory aids are supplied, sensitivity increases (Wickens, 1984). In the case of the dentist, better X-ray equipment and film, or X-rays of actual cavities to compare with the X-ray under consideration, would increase sensitivity.

Applications of SDT SDT has been applied to a wide range of practical situations, including sonar target detection (Colquhoun, 1967), industrial inspection tasks (Drury, 1975), medical diagnosis (Swets and Pickett, 1982), eyewitness testimony (Ellison and Buckhout, 1981), and air traffic control (Bisseret, 1981). Long and Waag (1981), however, caution that serious difficulties may result from the uncritical acceptance and application of SDT to some of these areas. SDT was developed from controlled laboratory experiments where subjects were given many, many trials with precisely controlled signals and background noise levels. Long and Waag believe that many applied situations do not match these conditions and that erroneous conclusions may be drawn if SDT is applied without due concern for the differences.

MEMORY

The human memory system has been conceptualized as three subsystems or processes: *sensory storage, working memory,* and *long-term memory.* Working memory is the gateway to long-term memory. Information in the sensory memory subsystem must pass through working memory in order to enter long-term memory. Human memory is vast, but imperfect. We possess incredible amounts of information in long-term storage, yet we often have trouble retrieving it when needed. Despite our vast storage capacity, we often forget a name or number just seconds after we have heard or read it. In this section we follow the lead of Wickens (1984) and stress the way information is coded in the three memory subsystems and some of the practical implications of these coding schemes.

Sensory Storage

Each sensory channel appears to have a temporary storage mechanism that prolongs the stimulus representation for a short time after the stimulus has ceased. Although there is some evidence for tactual and olfactory sensory storage, the two sensory storage mechanisms we know most about are those associated with the visual system, called *iconic storage,* and the auditory system, called *echoic storage*.

If a visual stimulus is flashed very briefly on a screen, the iconic storage holds the image for a short time, allowing further processing of the image. The same sort of phenomenon occurs with auditory information in echoic storage. Iconic storage generally lasts less than about 1 s, while echoic storage can last a few seconds before the representation "fades" away. Information in the sensory stores is not coded but is held in its original sensory representation. Sensory storage does not require a person to attend to the information in order for it to be maintained there. In essence, sensory storage is relatively automatic, and there is little one can do to increase the length of the sensory representation. To retain information for a longer period, it must be encoded and transferred to working memory.

Working Memory[1]

To encode and transfer information from sensory storage to working memory and to hold information in working memory requires that the person direct attention to the process.

It is generally believed that information in working memory is coded with three types of codes: visual, phonetic, and semantic. Visual and phonetic codes are visual or auditory representations of stimuli. Each can be generated by stimuli of the opposite type or internally from long-term memory. For example, the visually presented word *DOG* is phonetically coded as the sound generated from reading the word. Further, if you hear the word *DOG,* you could generate a visual code (mental picture) of a dog. You can also form a visual image of an object from long-term memory without hearing or seeing it.

Semantic codes are abstract representations of the meaning of a stimulus rather than the sight or sound generated by the stimulus. Semantic codes are especially important in long-term memory. Although there is a natural progression of coding from the physical codes (visual and phonetic) to semantic codes, there is evidence

---

[1]Working memory is also called *short-term memory,* but the term *working memory* is currently preferred.

that all three codes for a particular stimulus can exist, at some level of strength, at the same time in working memory.

One practical consequence of the various codes in working memory is the finding that letters read from a page are automatically converted to a phonetic code. When people are presented a list of letters and asked to recall them, errors of recall tend to reflect acoustic rather than visual confusion (Conrad, 1964). Thus the letter E is more likely to be recalled as D (which sounds like E) than as F (which looks like E). This has implications for selecting letters to code information on, say, a computer display. If a subset of letters is to be used, they should be phonetically dissimilar.

Capacity of Working Memory   The only way information can be maintained in working memory is by rehearsal (convert speech), that is, the direction of attention resources to the process. This can be illustrated by presenting a person with four letters (e.g., J, T, N, and L) and having that person count backward by 3s from 187. After 15 s of counting backward (which prevents rehearsal of the letters), the person probably will not be able to recall the letters. Even with rehearsal, information in working memory can decay over time. This decay occurs more rapidly with more items in working memory. This is probably due to the fact that rehearsing a long list of items increases the delay between successive rehearsals of any given item.

Given that rehearsal is delayed as the number of items in working memory is increased, what is the maximum number of items that can be held in working memory? The answer is the magical number 7 ± 2 (that is, 5 to 9) which we discussed in connection with absolute judgments along stimulus dimensions (Miller, 1956). The key to understanding the 7 ± 2 limit of working memory is to know how an item is defined. Miller (1956) in his formulation of the 7 ± 2 principle recognized that people can "chunk" information into familiar units, regardless of size, and these can be recalled as an entity. Thus the limit is approximately 7 ± 2 *chunks*. For example, consider the letters C.A.T.D.O.G.R.A.T. Rather than consider this string as composed of 9 items, Miller would argue that it contains 3 chunks, CAT.DOG.RAT, and hence is well within the 7 ± 2 limit. In a similar manner, words could be combined into meaningful sentences and chunked, thus effectively increasing the capacity of working memory.

The manner in which the information is presented can aid in chunking. Digits can be recalled better if they are grouped into chunks of three to four items (for example, 458 321 691 is easier to recall than 458321691). Also, the more meaningful the chunks, the easier they are to recall (IB MJF KTV is more difficult to recall than IBM JFK TV). The practical implications of all this are as follows: (1) Avoid presenting more than five to nine chunks of information for people to remember. (2) Present information in meaningful and distinct chunks. (3) Provide training on how to recall information by chunking. Schneiderman (1980) indicates that if a menu of options is presented on a computer screen and all alternatives must be compared simultaneously with one another, the number of alternatives should be kept within the capacity limits of working memory. If more than seven alternatives are displayed on the screen, a person may tend to forget the first alternative before the last ones are read.

Long-Term Memory

Information in working memory is transferred to long-term memory by semantically coding it, that is, by supplying meaning to the information and relating it to information already stored in long-term memory. If you study for examinations by only repeatedly reading the textbook and course notes, you probably have difficulty recalling the information. One reason is that you have not taken the time to semantically encode the information. To recall more information, it must be analyzed, compared, and related to past knowledge. In addition, the more organized the information is initially, the easier it is to transfer to long-term memory; and the more organized the information is in long-term memory, the easier it is to retrieve. Retrieval is often the weak link in utilizing information stored in long-term memory. The use of *mnemonics* to organize information makes retrieval easier. Mnemonics that can be used to learn a list of items include using the first letters of the items on the list to make a word or sentence and forming bizarre images connecting the items on the list.

DECISION MAKING

Decision making is really at the heart of information processing. It is a complex process by which people evaluate alternatives and select a course of action. The process involves seeking information relevant to the decision at hand, estimating probabilities of various outcomes, and attaching values to the anticipated outcomes. Wickens (1984) distinguishes decision making from choice reaction time in that typical decision-making situations have a relatively long time frame in which to make a choice compared with the case of choice reaction time.

People, unfortunately, are not optimal decision makers and often do not act "rationally," meaning that they do not act according to objective probabilities of gain and loss. A number of biases are inherent in the way people seek information, estimate probabilities, and attach values to outcomes that produce this irrational behavior. The following is a short list of some of these biases (Wickens, 1984):

1 People give an undue amount of weight to early evidence or information. Subsequent information is considered less important.

2 Humans are generally conservative and do not extract as much information from sources as they optimally should.

3 The subjective odds in favor of one alternative or the other are not assessed to be as extreme or given as much confidence as optimally they should.

4 As more information is gathered, people become more confident in their decisions, but not necessarily more accurate. For example, people engaged in troubleshooting a mechanical malfunction are often unjustly confident that they entertained all possible diagnostic hypotheses.

5 Humans have a tendency to seek far more information than they can absorb adequately.

6 People often treat all information as if it were equally reliable, even though it is not.

7 Humans appear to have a limited ability to entertain a maximum of more than a few (three or four) hypotheses at a time.

8 People tend to focus on only a few critical attributes at a time and consider only about two to four possible choices that are ranked highest on those few critical attributes.

9 People tend to seek information that confirms the chosen course of action and to avoid information or tests whose outcome could disconfirm the choice.

10 A potential loss is viewed as having greater consequence and therefore exerts a greater influence over decision-making behavior than does a gain of the same amount.

11 People believe that mildly positive outcomes are more likely than mildly negative outcomes, but that highly positive outcomes are less likely than mildly positive outcomes.

12 People tend to believe that highly negative outcomes are less likely than mildly negative outcomes.

These biases explain, in part, why people do not always make the best decisions, given the information available. Maintenance workers troubleshooting a malfunctioning device, doctors diagnosing an illness, or you and I deciding which model of stereo or automobile to buy—all these are limited by the human capacity to process and evaluate information to arrive at an optimum decision. Understanding our capabilities and limitations can suggest methods for presenting and preprocessing information to improve the quality of our decisions. Computers can be used to aid the decision maker by aggregating the probabilities as new information is obtained; keeping track of several hypotheses, alternatives, or outcomes of sequential tests at one time; or selecting the best sources of information to test specific hypotheses. The area of computer-aided decision making will undoubtedly become more important in the future as new technologies combine with a better understanding of the capabilities and limitations of the human decision maker.

ATTENTION

We often talk about "paying attention." Children never seem to do it, and we occasionally do it to the wrong thing and miss something important. These sorts of everyday experiences suggest that attention can somehow be directed to objects or activities, and that things to which we are not paying attention are often not perceived (or at least recalled). Three types of situations or tasks involving the direction of attention are often encountered in work and at play. In the first, we are required to monitor several sources of information to determine whether a particular event has occurred. A pilot scanning the instruments, looking for a deviant reading, and a quarterback scanning the defensive line and backfield, looking for an opening, are examples of this type of task. Cognitive psychologists call this *selective attention*.

In the second type of task, a person must attend to one source of information and exclude other sources. This is called *focused attention*, and it is the sort of activity you engage in when you try to read a book while someone is talking on the telephone or you try to listen to a person talk in a crowded, noisy cocktail party.

In the third type of situation, called *divided attention*, two or more separate tasks must be performed simultaneously, and attention must be paid to both. Examples include driving a car while carrying on a conversation with a passenger and eating dinner while watching the evening news. We discuss selective, focused, and divided attention in a little more detail and indicate some of the human capabilities and limitations associated with them. A few guidelines for designing each type of situation are also given.

Selective Attention

Selective attention requires the monitoring of several *channels* (or sources) of information to perform a single task. Consider a person monitoring a dial for a deviate reading (the signal). Let us say that 25 such deviate readings occur per minute. These 25 signals could all be presented on the one dial, or they could be presented on five dials, each presenting only 5 signals per minute. Which would be better in terms of minimizing missed signals? The answer is one dial presenting 25 signals per minute. As the number of channels of information (in this case, dials) increases, even when the overall signal rate remains constant, performance declines. Conrad (1951) referred to this as *load stress,* that is, the stress imposed by increasing the number of channels over which information is presented. Conrad also discussed *speed stress,* which is related to the rate of signal presentation. Goldstein and Dorfman (1978) found that load stress was more important than speed stress in degrading performance. Subjects, for example, could process 72 signals per minute from one dial better than 24 signals per minute presented over three dials (that is 8 signals per minute per dial).

When people have to sample multiple channels of information, they tend to sample channels in which the signals occur very frequently rather than those in which the signals occur infrequently. Due to limitations in human memory, people often forget to sample a source when many sources are present, and people tend to sample sources more often than would be necessary if they remembered the status of the source when it was last sampled (Moray, 1981). Under conditions of high stress, fewer sources are sampled and the sources that are sampled tend to be those perceived as being the most important and salient (Wickens, 1984).

Guidelines for Selective-Attention Tasks   The following are a few guidelines gleaned from the literature that should improve selective-attention task performance:

1 Where multiple channels must be scanned for signals, use as few channels as possible, even if it means increasing the signal rate per channel.

2 Provide information to the person as to the relative importance of the various channels, so that attention can be directed more effectively.

3 Reduce the overall level of stress on the person so that more channels will be sampled.

4 Provide the person with some preview information as to where signals will likely occur in the future.

5 Train the person to effectively scan information channels and develop optimal scan patterns.

6 If multiple visual channels are to be scanned, put them close together to reduce scanning requirements.

7 If multiple auditory channels are to be scanned, be sure they do not mask one another.

8 Where possible, stimuli that require individual responses should be separated temporally and presented at such a rate that they can be responded to individually. Extremely short intervals (say, less than 0.5 or 0.25 s) should be avoided. Where possible, the person should be permitted to control the rate of stimulus input.

Focused Attention

With focused attention, the problem is to maintain attention on one or a few channels of information and not to be distracted by other channels of information. One factor that influences the ability of people to focus attention is the proximity, in physical space, of the sources. For example, it is nearly impossible to focus attention on one visual source of information and completely ignore another if the two are within 1 degree of visual angle from each other (Broadbent, 1982). Note that in a *selective-attention* task, close proximity would aid performance.

Anything that makes the information sources distinct from one another would also aid in focusing attention on one of them. For example, when more than one auditory message occurs at the same time, but only one must be attended to, performance can be improved if the messages are of different intensity (the louder one usually gets attended to), are spoken by different voices (e.g., male and female), or come from different directions.

Guidelines for Focused-Attention Tasks   The following are some guidelines that should aid in the performance of focused-attention tasks:

1 Make the competing channels as distinct as possible from the channel to which the person is to attend.

2 Separate, in physical space, the competing channels from the channel of interest.

3 Reduce the number of competing channels.

4 Make the channel of interest larger, brighter, louder, or more centrally located than the competing channels.

Divided Attention

When people are required to do more than one task at the same time, performance on at least one of the tasks often declines. This type of situation is also referred to as *timesharing*. It is generally agreed that humans have a limited capacity to process information, and when several tasks are performed at the same time, that capacity can be exceeded. Cognitive psychologists have put forth numerous models and theories to explain timesharing or divided-attention performance [for a review see Lane (1982) or Wickens (1984)]. Most of the recent formulations are based on the notion of a limited pool of resources, or multiple pools of limited resources, that can be directed toward mental processes required to perform a task. We briefly discuss the two major classes of such theories: single-resource theories and multiple-resource theories.

Single-Resource Theories of Divided Attention   Single-resource theories postulate one undifferentiated source of resources that are shared by all mental processes (e.g., Moray, 1967; Kahneman, 1973). These theories explain, very nicely, why performance in a timesharing situation declines as the difficulty of one of the tasks increases. The increase in difficulty demands more resources from the limited supply, thus leaving fewer resources for performing the other task.

The problem with single-resource theories is they have difficulty explaining (1) why tasks that require the same memory codes or processing modalities interfere more than tasks not sharing common codes and modalities; (2) why with some combinations of tasks increasing the difficulty of one task has no effect on performance of the others; and (3) why some tasks can be timeshared perfectly (Wickens, 1984). To explain these phenomena, multiple-resource theories were developed.

Multiple-Resource Theories of Divided Attention   Rather than postulating one central pool of resources, multiple-resource theories postulate several independent resource pools. Only when tasks share the same pools will performance be disrupted. Wickens (1984), for example, suggests three dimensions along which resources can be allocated:

1 *Stages:* Perceptual and central processing versus response selection and execution. Wickens and Kessel (1980) found, for example, that a tracking task that demanded response selection and allocation resources was disrupted by another tracking task, but not by a mental arithmetic task that demanded central processing resources.

2 *Modalities:* Auditory versus visual. Isreal (1980), for example, found that timesharing was better when the two tasks did not share the same modality (i.e., one was presented visually and the other auditorially) than when both tasks were presented visually or both were presented auditorially.

3 *Processing codes:* Spatial versus verbal. Two tasks that both require spatial memory coding or both require verbal coding will be timeshared more poorly than two tasks requiring different types of coding.

Discussion   Even with all the theory building and research in the area of divided attention, there is still much we do not understand. Even the notion of limited resources is not universally accepted by all (e.g., Navon, 1984, 1985). Some theories postulate competing influences, some of which should improve performance and others of which should degrade performance. How these factors come together to influence performance is still not entirely clear. Predictions on the outcome of timesharing real-world tasks, therefore, are still relatively primitive.

Guidelines for Divided-Attention Tasks   Despite the limitations referred to above, we can offer a few guidelines gleaned from the research evidence—some of which were cited above and some of which were not—that should improve performance in a timesharing or divided-attention situation.

1 Where possible, the number of potential sources of information should be minimized.

2 Where timesharing is likely to stress a person's capacity, the person should be provided with information about the relative priorities of the tasks so that an optimum strategy of dividing attention can be formulated.

3 Efforts should be made to keep the difficulty level of the tasks as low as possible.

4 The tasks should be made as dissimilar as possible in terms of demands on processing stages, input and output modalities, and memory codes.

5 Especially when manual tasks are timeshared with sensory or memory tasks, the greater the learning of the manual task, the less will be its effect on the sensory or memory tasks.

AGE AND INFORMATION PROCESSING

As medical knowledge and technology advance, the average life expectancy will tend to increase. The average age of people in the industrialized world is increasing. Despite the old adage that we grow older and wiser, there is evidence that age has some disruptive effects on our information processing capabilities.

Welford (1981) summarized the effects of age on information processing. Although the effects are probably progressive throughout adulthood, they usually do not become noticeable until after the age of 65 or so. There are, of course, large individual differences in the aging process, so it is not really possible to pinpoint the exact age for an individual at which these effects will become apparent. Welford discusses the following changes in information processing capability that occur with age:

1 A slowing of sensory-motor performance. This slowing appears to be due mainly to the difficulty in perception of incoming data and the choice of responses to the data. The actual execution of the motor response appears to be little affected by age. This slowing is most evident in tasks requiring continuous responses as opposed to tasks where an appreciable time lapses before the next response is required.

2 Increased disruption of working memory by a shift of attention during the time that the material is being held there.

3 Difficulty in searching for material in long-term memory. This is probably due to an increased difficulty in transferring information from working memory to long-term memory. It takes older people more trials to learn material; but once the material is learned, recall is little worse than that of younger subjects.

4 Difficulty in dealing with incompatibility, especially conceptual, spatial, and movement incompatibility.

Welford interprets these effects within a signal detection framework. He attributes them to a reduction in sensitivity d' due to weaker signals coming from the sense organs to the brain and to increased random neural activity (noise) in the brain. The response criterion (beta) is adjusted to compensate for the reduced d'; sometimes it is made more conservative, sometimes more risky, depending on the task.

Guidelines for Designing Information Processing Tasks for the Elderly

The above findings suggest the following considerations in designing a task for the elderly that requires information processing:

1 Strengthen signals displayed to the elderly. Make them louder, brighter, etc.

2 Design controls and displays to reduce irrelevant details that would act as noise.

3 Maintain a high level of conceptual, spatial, and movement compatibility.

4 Reduce timesharing demands.

5 Provide time between the execution of a response and the signal for the next response. Where possible, let the person set the pace of the task.

6 Allow more time and practice to initially learn material.

MENTAL WORKLOAD

In our everyday experiences we have encountered tasks that seem to tax our brains. In like manner, there are tasks that do not seem as demanding, such as balancing a checkbook (at least to some people it is not demanding). These experiences touch on the idea of mental workload as a measurable quantity of information processing demands placed on an individual by a task.

Since the middle to late 1970s, interest in defining and developing measures of mental workload has increased dramatically. A major area of focus of such work has been the assessment of the workload experienced by aircraft crew members, particularly pilots. As further indication of the growing concern, Wickens (1984) indicates that the Federal Aviation Administration will require certification of aircraft in terms of workload metrics and that the Air Force is imposing workload criteria on newly designed systems.

If a practical, valid method for measuring mental workload were available, it could be used for such purposes as (1) allocating functions and tasks between humans and machines based on predicted mental workload; (2) comparing alternative equipment and task designs in terms of the workloads imposed; (3) monitoring operators of complex equipment to adapt the task difficulty or allocation of function in response to increases and decreases in mental workload; and (4) choosing operators who have higher mental workload capacities for demanding tasks.

Concept of Mental Workload

The concept of mental workload is a logical extension of the resource models of divided attention discussed above. Resource models postulate a limited quantity of resources available to perform a task, and in order to perform a task, one must use some of or all these resources. Although there is no universally accepted definition of mental workload (McCloy, Derrick, and Wickens, 1983), the basic notion is related to the difference between the amount of resources available within a person and the amount of resources demanded by the task situation. This means that mental workload can be changed by altering either the amount of resources available within the person (e.g., keeping the person awake for 24 h) or the demands made by the task on the person (e.g., increasing the number of information channels). This concept of mental workload is similar to the concept of physical workload, which is discussed in Chapter 8.

Measurement of Mental Workload

Most of the research activity in the area of mental workload has been directed toward refining and exploring new measures of the concept. A useful measure of mental workload should meet the following criteria:

1 *Sensitivity*. The measure should distinguish task situations that intuitively appear to require different levels of mental workload.

2 *Selectivity*. The measure should not be affected by things generally considered not to be part of mental workload, such as physical load or emotional stress.

3 *Interference*. The measure should not interfere with, contaminate, or disrupt performance of the primary task whose workload is being assessed.

4 *Reliability*. The measure should be reliable; that is, the results should be repeatable over time (test-retest reliability).

5 *Acceptability*. The measuring technique should be acceptable to the person being measured.

The numerous measures of mental workload can be classified into four broad categories: primary task measures, secondary task measures, physiological measures, and subjective measures. Several reviews of mental workload measurement have appeared over the years (Meshkati, Hancock, and Robertson, 1984; Moray, 1982; Ogden, Levine, and Eisner, 1979; Wierwille, 1979; Williges and Wierwille, 1979) and should be consulted for a more complete discussion of the various methods under study. We briefly review each major category and give an example or two to illustrate the sort of work being done.

Primary Task Measures   Early attempts to quantify workload used a task analytic approach in which *workload* was defined simply as the time required to perform tasks divided by the time available to perform the tasks. Several computer-based modeling programs were developed to perform these calculations on a minute-by-minute basis throughout a scenario of tasks [one example is the statistical workload assessment model (SWAM) (Linton, 1975)]. The problem with this approach is that it does not take into account the fact that some tasks can be timeshared successfully while others cannot; nor do such methods take into account cognitive demands that are not reflected in increased task completion times.

Task performance itself would seem, at first glance, an obvious choice for a measure of mental workload. It is known, however, that if two versions of a task are compared, both of which do not demand resources in excess of those available, performance may be identical, yet one task could still be more demanding than the other. To overcome this problem, the *primary task workload margin* (Wickens, 1984) can be computed for various versions of a task. Basically, this technique involves changing a parameter of the task known to affect the demand for resources until performance on the task can no longer be maintained at some preset criterion level. How much that parameter could be changed before performance broke down is the measure of primary task workload margin. Performance on a difficult version of a task would break down with less of an increase in resource demands than would an easy version of the task.

A problem faced by all primary task measures of mental workload is that they are task-specific, and it is difficult, if not impossible, to compare the workloads imposed by dissimilar tasks.

Secondary Task Measures   The basic logic behind the use of secondary task measures is that spare capacity, not being directed to performance of the primary task, will be used by the secondary task. The greater the demand for resources made by the primary task, the less resources will be available for the secondary task and the poorer will be performance on the secondary task. In using this method, a person is usually instructed to maintain a level of performance on the primary task, so that differences in workload will be reflected in the performance of the secondary task. Sometimes this procedure is reversed; a person is instructed to devote all necessary resources to maintain performance on the secondary task, and performance on the primary task is taken as a measure of its difficulty. This is called a *loading task technique* (Ogden, Levine, and Eisner, 1979). The more demanding the primary task, the lower will be performance on it in the presence of a loading (secondary) task.

Some examples of secondary tasks that have been used include rhythmic tapping, reaction time, memory search, time estimation, production of random digits, and mental arithmetic. Casali and Wierwille (1984), in summarizing previous studies on the sensitivity of various measures of workload on perceptual, mediation, communication, and motor tasks, found that time estimation was the most sensitive secondary measure for all four types of tasks. Rhythmic tapping was sensitive only for the perceptual tasks, but it was not tested with the motor tasks.

There is a dilemma inherent in the secondary task methodology. In order to measure spare resource capacity, the secondary task should tap the same resources as those tapped by the primary task. If one accepts a multiple-resource model, then the tasks should share common modalities (visual, auditory, speech, manual) and common processing codes. Such tasks, however, by definition, interfere with the performance of the primary task. Thus, we cannot say whether we are measuring the workload imposed by just the primary task or that of the primary as interferred with by the secondary task.

Physiological Measures   Most physiological measures of mental workload are predicated on a single-resource model of information processing rather than a multiple-resource model. The logic is that information processing involves central nervous system activity, and this general activity or its manifestations can be measured. Some physiological measures that have been investigated include variability in heart rate (also called *sinus arhythmia),* evoked brain potential, pupillary response, respiration rate, and body fluid chemistry.

An example of the use of pupillary response as a measure of mental workload can be seen in the work done by Ahern (as cited in Beatty, 1982). Figure 3-4 shows pulpillary response (dilation) during the performance of mental multiplication tasks that differed in difficulty (multiplying pairs of one-digit numbers, to multiplying pairs of two-digit numbers). As can be seen, both the peak and duration of the pupillary response were greater as the difficulty of the mental operations involved increased.

Kramer, Wickens, and Donchin (1983) report on a specific component of event-

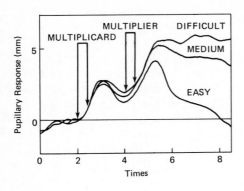

FIGURE 3-4

Pupillary responses during mental multiplication tasks that differ in cognitive difficulty. Both the amplitude and the duration of the dilation increase as a function of task difficulty. (*Source: Ahern as presented by Beatty, 1982; fig. 5.*) Copyright 1982 by the American Psychological Association. Reprinted by permission of the author.

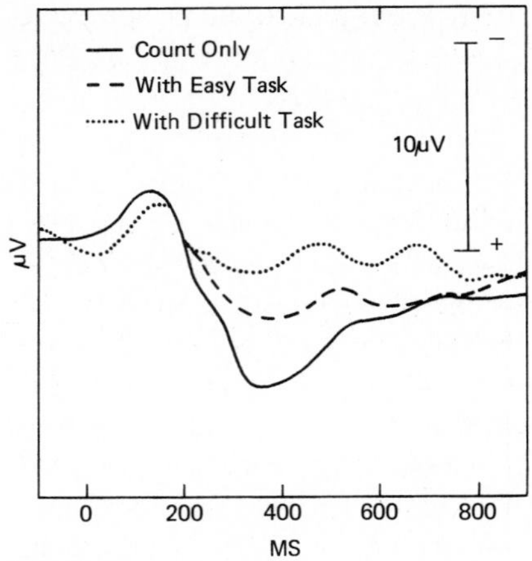

FIGURE 3-5

Effect of perceptual and cognitive task difficulty of the P300 event-related brain potential. As task difficulty increases, the magnitude of the P300 potential is decreased. (*Source: Adapted from Kramer, Wickens, and Donchin, 1983; fig. 3.*) Copyright by the Human Factors Society, Inc., and reproduced by permission.

related brain potentials called *P300* as a measure of mental workload. Event-related brain potentials are transient voltage oscillations in the brain ("brain waves") that can be recorded on the scalp in response to a discrete stimulus event. There is evidence to suggest that P300—a positive polarity voltage oscillation occurring approximately 300 ms after the occurrence of a task-relevant, low-probability event—is a manifestation of neuronal activity that is involved whenever a person is required to update an internal "model" of the environment or task structure. Typically, P300 responses are elicited by requiring subjects to covertly count occurrences of one of two events that are occurring serially. For example, two tones, high and low, are randomly presented one after another, and a subject is required to count the high tones. Each time a high tone occurs and the subject updates the count (or internal model), a P300 wave occurs. To the degree to which a primary task demands perceptual and cognitive resources, fewer resources will be available to update the count and the magnitude of the P300 wave will be reduced. This can be seen in Figure 3-5. The P300 response is shown when no additional primary task was present (count only) and when the response was paired with a primary tracking task that varied in perceptual and cognitive difficulty. As the tracking task increased in difficulty, the P300 waveform became flatter and less pronounced.

Physiological measures of mental workload offer some distinct advantages: they permit continuous data collection during task performance, they do not often interfere with primary task performance, and they usually require no additional activity or performance from the subject. The disadvantages are that bulky equipment often must be attached to the subject to get the measurements, and, except for perhaps the

P300 event-related brain potential, the measures do not isolate the specific stages of information processing being loaded by the primary task.

Subjective Measures   Some investigators have expressed the opinion that subjective ratings of mental workload came closest to tapping the essence of the concept (Sheridan, 1980). Rating scales are easy to administer and are widely accepted by the people who are asked to complete them. Perhaps the oldest and most extensively validated subjective measure of workload (although it was originally developed to assess handling characteristics of aircraft) is the *Cooper-Harper scale* (Cooper and Harper, 1969). Wierwille and Casali (1983) report that the scale can be used, with only minimal rewording, for a wide variety of motor or psychomotor tasks. In addition, Wierwille and Casali present a modified Cooper-Harper rating scale that can be used for perceptual, cognitive, and communications tasks. The scale combines a decision tree and a unidimensional 10-point rating scale. The rating scale goes from (1) very easy, highly desirable (operator mental effort is minimal and desired performance is easily attainable), through (5) moderately objectionable difficulty (high operator mental effort is required to attain adequate system performance), to (10) impossible (instructed task cannot be accomplished reliably).

Another approach to subjective measurement of mental workload treats the concept as a multidimensional construct. Sheridan and Simpson (1979) proposed three dimensions to define subjective mental workload: time load, mental effort load, and psychological stress. Reid, Shingledecker, and Eggemeier (1981) used these dimensions to develop the Subjective Workload Assessment Technique (SWAT). Subjects rate workload on each of the three dimensions (each with a 3-point scale), and the ratings are converted to a single interval scale of workload. Remember that the three underlying dimensions suggested by Sheridan and Simpson and used with SWAT were intuitively derived and have not been empirically validated. Further, Boyd (1983) found that when the dimensions were independently varied in a task, the ratings of the dimensions were not independent. That is, if only time load was increased in the task, subjects tended to increase their ratings on all three dimensions.

Discussion of Mental Workload Measures   Despite all the work that has been done and all the measures that have been proposed, there is still no real agreement on an adequate measure of mental workload. Thus, to measure mental workload, probably several measures will be needed to assess what is, in all likelihood, a multidimensional concept. In some cases where multiple measures of workload have been evaluated together, the different measures provide different results. When this occurs, the measures are said to *dissociate* (Yeh and Wickens, 1984; McCloy, Derrick, and Wickens, 1983). A common finding, for example, is that subjective measures dissociate from task performance. It appears that each may be measuring a different aspect of mental workload. Subjective measures appear to be more sensitive to the number of current tasks being performed, while task performance is more sensitive to the degree of competition for common resources among the various tasks being performed (Yeh and Wickens, 1984).

The ultimate success in developing a set of standardized mental workload assessment techniques will depend, in great part, on the development and refinement of our theories of human information processing.

HUMAN FACTORS IN THE INFORMATION REVOLUTION

Probably some of the most exciting developments in the information revolution are in the area of artificial intelligence (AI). We are not going to engage in philosophical discussion here as to whether computers will ever have the ability to think. The fact is that advances in computer technology and programming languages are endowing computers with abilities once considered strictly within the domain of humans. We have had to rethink the traditional allocation of functions between people and machines (see Chapter 18 for further discussion of allocation of function in system design). The new applications are presenting new and challenging human factors problems. We touch briefly on the two areas of expert systems and natural language interfaces, and we discuss a few general human factors areas of concern within each.

Expert Systems

An *expert system* is essentially a way to capture the knowledge and expertise of a subject-matter expert and transfer it to a computer program that, it is hoped, will emulate the problem-solving and decision-making performance of the expert (Hamill, 1984). Expert systems have been successfully developed for attacking very specific, well-defined problem areas, such as diagnosing and treating infectious blood diseases, making inferences about the structures of unknown chemical compounds, freeing oil well drill bits stuck in a drill shaft, and troubleshooting locomotive engines.

Kidd (1983) and Hamill (1984) discuss a few of the human factors issues involved with the development and use of expert systems. Human factors specialists can contribute to the methodologies used to extract the knowledge and inference rules from the experts. The interface between the system and user is an area also in need of human factors attention. Expert systems must communicate not only what the system is doing, but also why. This must be done in a manner that gives the user confidence in the system and permits intervention in the decision-making process by the user to redirect an obviously errant decision path (these systems are not perfect, and probably never will be). Cognitive psychologists should be able to contribute insights into human problem solving that could aid designers of future expert systems. Someday, maybe, an expert system will be created that emulates the knowledge and expertise of the human factors specialist—someday.

Natural Language Interfaces

A natural language interface allows a user of an interactive system to speak or write in ''plain English'' (or French, or Japanese, or whatever) and be ''understood'' by the machine. (Actually, no one really believes that the machine understands what is said, but rather only that it acts as if it did.) Natural language is incredibly complex; things we take for granted must be painstakingly ''explained'' to the computer. Consider the question "Which of our Washington staff lives in Virginia?" The computer has to figure out that Washington, in this question, refers to Washington, D.C., and not Washington state. Currently, natural language interfaces have been developed for very specific, well-defined domains, such as gathering facts from a specific data base of information. Human factors people have been involved in determining when and where natural language interfaces are likely to be more efficient than other types of interfaces (Small and Weldon, 1977; Thomas, 1978; Ehrenreich, 1981). Human factors people have also been helpful in determining the degree of structure that can be imposed in a natural language interface without destroying its appeal to the user (Gould, Lewis, and Becker, 1976).

Discussion

The prospects of artificial intelligence and its applications in expert systems and natural language interfaces have barely begun to crystallize. As these fields emerge, new problems and opportunities for integrating humans and machines into productive, efficient, and safe systems will emerge. Our understanding of the capabilities and limitations of human information processing will become more and more important in the design of future systems. Some people think that as computers take over more of the information processing and decision-making functions, human involvement in systems will become less and less. Do not believe it! The roles that humans play in systems of the future will undoubtedly change, but humans will be very much involved in such systems, doing tasks perhaps not even possible with today's technology. If we play our cards right, human factors will be a significant part of this exciting future.

REFERENCES

Anderson, J. (1980). *Cognitive psychology and its implications*. San Francisco: Freeman.

Baird, J. (1984). Information theory and information processing. *Information Processing and Management, 20*, 373–381.

Beatty, J. (1982). Task-evoked pupillary response, processing load, and the structure of processing resources. *Psychological Bulletin, 91*, 276–292.

Bisseret, A. (1981). Application of signal detection theory to decision making in supervisory control. *Ergonomics, 24*, 81–94.

Boyd, S. (1983). Assessing the validity of SWAT as a workload measurement instrument. In A. Pope and L. Haugh (eds.), *Proceedings of the Human Factors Society 27th Annual Meeting*. Santa Monica, CA: Human Factors Society.

Broadbent, D. (1958). *Perception and communication*. New York: Pergamon.

Broadbent, D. (1982). Task combination and selective intake of information. *Acta Psychologica, 50*, 253–290.

Casali, J., and Wierwille, W. (1984). On the measurement of pilot perceptual workload: A comparison of assessment techniques addressing sensitivity and intrusion issues. *Ergonomics, 27*(10), 1033–1050.

Colquhoun, W. (1967). Sonar target detection as a decision process. *Journal of Applied Psychology, 51*, 187–190.

Conrad, R. (1951). Speed and load stress in sensory-motor skill. *British Journal of Industrial Medicine, 8*, 1–7.

Conrad, R. (1964). Acoustic comparisons in immediate memory. *British Journal of Psychology,* 55, 75–84.

Cooper, G., and Harper, R. (1969, April). *The use of pilot ratings in the evaluation of aircraft handling qualities,* NASA Ames Tech. Rept., NASA TN-D-5153. Moffett Field, CA: NASA Ames Research Center.

Deatherage, B. (1972). Auditory and other sensory forms of information presentation. In H. Van Cott and R. Kinkade (eds.), *Human engineering guide to equipment design* (rev. ed.). Washington: Government Printing Office.

Drury, C. (1975). The inspection of sheet materials—Models and data. *Human Factors,* 17, 257–265.

Ehrenreich, S. (1981). Query languages: Design recommendations derived from the human factors literature. *Human Factors,* 23, 709–725.

Ehrenreich, S. (1985). Computer abbreviations: Evidence and synthesis. *Human Factors,* 27(2), 143–155.

Ellison, K., and Buckhout, R. (1981). *Psychology and criminal justice.* New York: Harper & Row.

Fitts, P., and Seeger, C. (1953). S-R compatibility: Spatial characteristics of stimulus and response codes. *Journal of Experimental Psychology,* 46, 199–210.

Goldstein, I., and Dorfman, P. (1978). Speed stress and load stress as determinants of performance in a time-sharing task. *Human Factors,* 20, 603–610.

Gould, J., Lewis, C., and Becker, C. (1976). *Writing and following procedural, descriptive and restrictive syntax language instructions.* IBM Research Rept. RC-5943.

Green, D., and Swets, J. (1966). *Signal detection theory and psychophysics.* New York: Wiley.

Grudin, J., and Barnard, P. (1985). When does an abbreviation become a word? and related questions. In *Proc. CHI '85 Human Factors in Computing Systems* (San Francisco, April 14–18, 1985). New York: Association of Computing Machinery, pp. 121–125.

Haber, R., and Hershenson, M. (1980). *The psychology of visual perception* (2d ed.). New York: Holt.

Hamill, B. (1984). Psychological issues in the design of expert systems. In M. Alluisi, S. de Groot, and E. Alluisi (eds.), *Proceedings of the Human Factors Society 28th Annual Meeting.* Santa Monica, CA: Human Factors Society.

Hick, W. (1952). On the rate of gain of information. *Quarterly Journal of Experimental Psychology.* 4, 11–26.

Hirsh-Pasek, K., Nudelman, S., and Schneider, M. (1982). An experimental evaluation of abbreviation schemes in limited lexicons. *Behavior and Information Technology,* 1, 359–370.

Hyman, R. (1953). Stimulus information as a determinant of reaction time. *Journal of Experimental Psychology,* 45, 423–432.

Isreal, J. (1980). Structural interference in dual task performance: Behavioral and electrophysiological data. Unpublished Ph.D. dissertation. Champaign: University of Illinois.

Kahneman, D. (1973). *Attention and effort.* Englewood Cliffs, NJ: Prentice-Hall.

Kidd, A. (1983). Human factors in expert systems. In K. Coombes (ed.). *Proceedings of the Ergonomics Society Conference.* London: Taylor & Francis.

Kramer, A., Wickens, C., and Donchin, E. (1983). An analysis of the processing requirements of a compex perceptual-motor task. *Human Factors,* 25(6), 597–621.

Lane, D. (1982). Limited capacity, attention allocation, and productivity. In W. Howell and E. Fleishman (eds.), *Human performance and productivity: Information processing and decision making,* vol. 2. Hillsdale, NJ: Lawrence Erlbaum Associates.

Linton, P. (1975, May). *VFA-STOL crew loading analysis,* NADC-57209-40. Warminster, PA: U.S. Naval Air Development Center.

Long, G., and Waag, W. (1981). Limitations on the practical applicability of d' and β measures. *Human Factors, 23*(3), 285–290.

McCloy, T., Derrick, W., and Wickens, C. (1983, November). Workload assessment metrics—What happens when they dissociate? In *Second Aerospace Behavioral Engineering Technology Conference Proceedings* (P-132). Warrendale, PA: Society of Automotive Engineers, pp. 37–42.

Meshkati, N., Hancock, P., and Robertson, M. (1984). The measurement of human mental workload in dynamic organizational systems: An effective guide for job design. In H. Hendrick and O. Brown, Jr. (eds.), *Human factors in organizational design and management.* Amsterdam: Elsevier.

Miller, G. (1956). The magical number seven, plus or minus two: Some limits on our capacity for processing information. *Psychological Review, 63,* 81–97.

Moray, N. (1967). Where is capacity limited? A survey and a model. *Acta Psychologica, 27,* 84–92.

Moray, N. (1981). The role of attention in the detection of errors and the diagnosis of errors in man-machine systems. In J. Rasmussen and W. Rouse (eds.), *Human detection and diagnosis of system failures.* New York: Plenum.

Moray, N. (1982). Subjective mental workload. *Human Factors, 24*(1), 25–40.

Mowbray, G., and Gebhard, J. (1961). Man's senses vs. information channels. In W. Sinaiko (ed.), *Selected papers on human factors in design and use of control systems.* New York: Dover.

Mudd, S. (1961, June). The scaling and experimental investigation of four dimensions of tone and their use in an audio-visual monitoring problem. Unpublished Ph.D. dissertation. Lafayette, IN: Purdue University.

Navon, D. (1984). Resources—A theoretical soup stone? *Psychological Review, 91,* 216–234.

Navon, D. (1985). *Attention division or attention sharing?* IPDM Rept. no. 20. Haifa, Israel: University of Haifa.

Ogden, G., Levine, J., and Eisner, E. (1979). Measurement of workload by secondary tasks. *Human Factors, 21*(5), 529–548.

Pollack, I. (1952). The information of elementary auditory displays. *Journal of the Acoustical Society of America, 24,* 745–749.

Reid, G., Shingledecker, C., and Eggemeier, T. (1981). Application of conjoint measurement to workload scale development. In R. Sugarman (ed.), *Proceedings of the Human Factors Society 25th Annual Meeting.* Santa Monica, CA: Human Factors Society.

Sanders, A. F. (1983). Towards a model of stress and human performance. *Acta Psychologica, 53,* 61–97.

Schneiderman, B. (1980). *Software psychology.* Cambridge, MA: Winthrop.

Sheridan, T. (1980). Mental workload: What is it? Why bother with it? *Human Factors Society Bulletin, 23,* 1–2.

Sheridan, T., and Simpson, R. (1979, January). *Toward the definition and measurement of the mental workload of transport pilots,* FTL Rept. R 79-4. Cambridge, MA: Massachusetts Institute of Technology, Flight Transportation Laboratory.

Small, D., and Weldon, L. (1977). *The efficiency of retrieving information from computers using natural and structured query languages,* Tech. Rept. SAI-78-655-WA. Arlington, VA: Science Applications, Inc.

Streeter, L., Ackroff, J., and Taylor, G. (1983). On abbreviating command names. *The Bell System Technical Journal, 62,* 1807–1826.

Swets, J. (ed.) (1964). *Signal detection and recognition by human observers: Contemporary readings*. New York: Wiley.

Swets, J., and Pickett, R. (1982). *The evaluation of diagnostic systems*. New York: Academic.

Thomas, J. (1978). A design-interpretation analysis of natural English with applications to man-computer interaction. *International Journal of Man-Machine Studies*. 10, 651–668.

Welford, A. T. (1981). Signal, noise, performance, and age. *Human Factors,* 23(1), 97–109.

Welford, H. (1976). *Skilled performance: Perceptual and motor skills*. Glenview, IL: Scott, Foresman.

Wickens, C. (1984). *Engineering psychology and human performance*. Columbus, OH: Merrill.

Wickens, C., and Kessel, C. (1980). The processing resource demands of failure detection in dynamic systems. *Journal of Experimental Psychology: Human Perception and Performance,* 6, 564–577.

Wickens, C., Sandry, D., and Vidulich, M. (1983). Compatibility and resource competition between modalities of input, central processing, and output. *Human Factors,* 25(2), 227–248.

Wierwille, W. (1979). Physiological measures of aircrew mental workload. *Human Factors,* 21(5), 575–593.

Wierwille, W., and Casali, J. (1983). A validated rating scale for global mental workload measurement applications. In A. Pope and L. Haugh (eds.), *Proceedings of the Human Factors Society 27th Annual Meeting*. Santa Monica, CA: Human Factors Society. pp. 129–133.

Williges, R., and Wierwille, W. (1979). Behavioral measures of aircrew mental workload. *Human Factors,* 21(5), 549–574.

Yeh, Y., and Wickens, C. (1984). Why do performance and subjective measures dissociate? In M. Alluisi, S. de Groot, and E. Alluisi (eds.), *Proceedings of the Human Factors Society 28th Annual Meeting*. Santa Monica, CA: Human Factors Society.

CHAPTER 4

VISUAL DISPLAYS OF STATIC INFORMATION

Visual displays are probably the most commonly used displays. This chapter deals primarily with displays that present static information, that is, displays of visual stimuli that are unchangeable or that remain in place for a reasonable time. Examples include displays to present alphanumeric, symbolic, identification, representational, status, and warning information. Chapter 5 deals primarily with displays that present stimuli representing dynamic, or changing, information regarding some specific parameter or condition, such as temperature, speed, pressure, altitude, and volume. Although the distinction between static and dynamic information is admittedly a thin one, it is used here somewhat as a matter of convenience for dividing the material between the two chapters.

In any event, since the use of visual displays (of whatever type) depends on the visual capabilities of people, we first discuss the process of seeing and the relevant features of the eye.

PROCESS OF SEEING

The eye is very much like a camera in that it has an adjustable lens through which light rays are transmitted and focused and a sensitive area (the retina) upon which the light falls. Figure 4–1 illustrates the principal features of the eye in cross section. Light rays that are reflected from an object enter the transparent *cornea* and pass through the clear fluid (*aqueous humor*) that fills the space between the cornea and the pupil and lens behind the cornea. The *pupil* is a circular, variable aperture whose size is changed through action of the muscles of the *iris,* becoming larger in dark surroundings and smaller in bright surroundings. The light rays that are transmitted through the opening of the pupil to the lens are refracted by the adjustable *lens,* and then they transverse the clear, jellylike *vitreous humor* that fills the eyeball back of the lens.

79

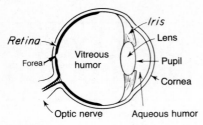

FIGURE 4-1
Principal features of the human eye in cross section. Light passes through the pupil, is refracted by the lens, and is brought to a focus on the retina. The retina receives the light stimulus and transmits an impulse to the brain through the optic nerve.

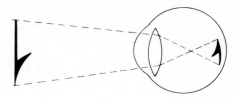

FIGURE 4-2
Illustration of the manner in which the image of an object is reproduced in inverted form on the retina of the eye.

In persons with normal or corrected vision, the light rays are brought to appropriate focus on the *retina*. The retina consists of about 6 or 7 million cones that tend to be concentrated near the center and about 130 million rods that tend to predominate in the outer areas of the retina around the sides of the eyeball. The cones are the dominant receptors for daytime vision. The rods are particularly important for use in dim light and at night. The *fovea,* in the dead center of the retina, is the area of greatest sensitivity. For an object to be seen clearly, the eye must be directed so that the image of the object is focused on the fovea. The image on the retina of what we see is actually inverted and reversed, as illustrated in Figure 4–2.

The cones and rods are connected to the optic nerve, which transmits neural impulses to the brain, which, in turn, integrates the many neural impulses to give us our visual impression of the outside world. In this process the inverted and reversed image on the retina is corrected, so we see things as they actually are.

Visual Abilities

The basic visual processes give rise to several visual abilities, a few of which are discussed below.

Visual Acuity   The ability of the eyes to differentiate between the detailed features of what we see, such as reading the fine print in an insurance contract or identifying a person across the street, is called *visual acuity*. Acuity depends very largely on the *accommodation* of the eyes, which is the adjustment of the lens of the eye to bring about proper focusing of the light rays on the retina. In normal accommodation, if one is looking at a far object, the lens flattens, and if one looks at a near object, the lens tends to bulge, in order to bring about proper focusing of the image on the retina.

In some individuals the accommodation of the eyes is inadequate. This causes the conditions that we sometimes call *nearsightedness* and *farsightedness*. When a person is nearsighted, the lens tends to remain in a bulged condition, so that while the

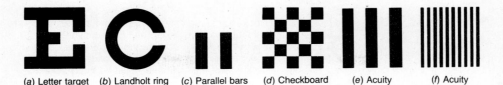

(a) Letter target   (b) Landholt ring   (c) Parallel bars   (d) Checkboard   (e) Acuity grating   (f) Acuity grating

FIGURE 4-3
Illustrations of various types of targets used in visual acuity tests and experiments. The features to be differentiated in targets a, b, c, d, and e are all the same size and would, therefore, subtend the same visual angle at the eye. With target a the subject is to identify each letter; with c, e, and f the subject is to identify the orientation (such as vertical or horizontal); and with b the subject is to identify any of four orientations. With target d the subject is to identify one checkerboard target from three others with smaller squares.

person may achieve a proper focus of near objects, a proper focus of far objects cannot be achieved. Farsightedness, in turn, is a condition in which the lens tends to remain too flat. A farsighted person may see clearly at far distance but encounters difficulty in seeing properly at near distance. Such conditions can sometimes be corrected by appropriate lenses which change the direction of the light rays before the rays reach the lens of the eye and thereby bring about proper focusing on the retina.

There really are different types of visual acuity. The most commonly used measure of acuity, *minimum separable acuity,* refers to the smallest feature, or the smallest space between the parts of a target, that the eye can detect. The visual targets used in measuring minimum separable acuity include letters and various geometric forms, such as those illustrated in Figure 4–3. Such targets can be varied in size and distance, the acuity of the subject being determined by the smallest target the person can properly identify. Such acuity usually is measured in terms of the reciprocal of the visual angle subtended at the eye by the smallest detail that can be discriminated (i.e., the angular subtense of that detail). The *visual angle* (VA) is measured in minutes of arc or seconds of arc; each of the 360° of a circle can be divided into 60 minutes of arc, each of which can be subdivided into 60 seconds of arc. The concept of the visual angle is illustrated in Figure 4–4. The formula for computing the visual angle is

$$VA \text{ (minutes)} = \frac{3438H}{D}$$

where H is the height of a visual stimulus and D is the distance from the eye. The measurements of H and D must be in the same units, such as inches, feet, centimeters, or meters.

The reciprocal of a visual angle of 1 minute of arc usually is used as a standard in scoring. This reciprocal, being unity, provides a base with which poorer or better levels of acuity can be compared. If, for example, one person can identify only a detail that subtends an arc of 1.5 minutes, the acuity score for that person is the reciprocal of 1.5 minutes, or 0.67. On the other hand, for a person who can identify a detail that subtends an arc of 0.8 minute, the score—the reciprocal of 0.8 minute— is 1.25.

Position of eye

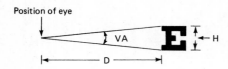

FIGURE 4-4
Illustration of the concept of visual angle; H =
height of visual stimulus, and D = distance from
the eye. In this illustration, the visual angle of the
specific elements of E could be derived (the
thickness of the elements would be the H value).

Vernier acuity refers to the ability to differentiate the lateral displacement, or slight offset, of one line from another that, if not so offset, would form a single continuous line (such as in lining up the "ends" of lines in certain optical devices). *Minimum perceptible acuity* is the ability to detect a spot (such as a round dot) from its background. In turn, *stereoscopic acuity* refers to the ability to differentiate the different images, or pictures, received by the retinas of the two eyes of a single object that has depth. (These two images differ most when the object is near the eyes and differ least when the object is far away.)

Convergence   As we direct our visual attention to a particular object, it is necessary that the two eyes converge on the object so that the images of the object on the two retinas are in corresponding positions; in this way we get an impression of a single object. The two images are said to be *fused* if they do so correspond. Convergence is controlled by muscles that surround the eyeball. Normally, as an individual looks at a particular object, these muscles operate automatically to bring about convergence. But some individuals tend to converge too much, and others tend not to converge enough. These conditions are called *phorias*. Since the double images that occur when convergence does not take place are visually uncomfortable, such people usually have compensated for this by learning to bring about convergence; however, muscular stresses and strains can occur in overcoming these muscular imbalances.

Color Discrimination   As indicated before, the cones of the retina are sensitive to differences in wavelength and thus are the basis for color discrimination. Various tests are available to measure the color discrimination of individuals, including the discrimination among certain colors. Some people, for example, have difficulty in discriminating between red and green or between blue and yellow. Complete color blindness is very rare.

Dark Adaptation   The adaptation of the eye to different levels of light and darkness is brought about by two functions. First, the pupil of the eye increases in size as we go into a darkened room, in order to admit more light to the eyes; it tends to contract in bright light, in order to limit the amount of light that enters the eye. A second function that affects how well we can see as we go from the light into darkness is a physiological process in the retina in which *visual purple* is built up. Under such circumstances the cones (which are color-sensitive) lose much of their sensitivity. Since in the dark our vision depends very largely on the rods, color discrimination is limited in the dark. The time required for complete dark adaptation is usually 30 min or more. The reverse adaptation, from darkness to light, takes place in some seconds, or a minute or two at most.

Conditions That Affect Visual Discriminations

The ability of individuals to make visual discriminations is, of course, dependent on their visual abilities, especially their visual acuity. Aside from individual differences, however, certain variables (conditions) external to the individual affect visual discriminations. Some of these variables are listed or discussed briefly below.

Luminance Contrast *Luminance contrast* (frequently called *brightness contrast,* or simply *contrast*) refers to the difference in luminance of the features of the object being viewed, in particular of the feature to be discriminated by contrast with its background (for example, an arrow on a direction sign against the background area of the sign). The luminance contrast is expressed by

$$\text{Contrast} = \frac{B_1 - B_2}{B_1} \times 100$$

where B_1 = percent of reflectance of brighter of two contrasting areas
B_2 = percent of reflectance of darker of two contrasting areas

The contrast between the print on this page and its white background is considerable. If we assume that the paper has a reflectance of 80 percent and that the print has a reflectance of 10 percent, the contrast is

$$\frac{80 - 10}{80} \times 100 = \frac{70}{80} \times 100 = 88\%$$

If the printing were on medium-gray paper rather than on white, the contrast would, of course, be much less.

When the contrast between a visual target and its background is low, the target must be larger for it to be equally discriminable to a target with greater contrast. This is illustrated, for example, by Figure 4–5, which shows, for various levels of contrast of parallel-bar targets, the relative differences in size required for equal discriminability.

FIGURE 4-5
Relative sizes of visual targets of parallel bars of various levels of contrast that would be required for equal discriminability. (*Source: Adapted from Heglin, 1973, fig. 1-10.*)

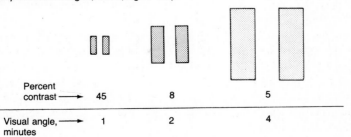

Amount of Illumination   This is discussed in Chapter 14.

Time   Within reasonable limits, the longer the viewing time, the greater the discriminability. (This is discussed further in Chapter 14, which deals with illumination.)

Luminance Ratio   The luminance ratio is the ratio between the luminance of any two areas in the visual field (usually the area of primary visual attention and the surrounding area). (This is also discussed further in Chapter 14.)

Glare   This is discussed in Chapter 14.

Combinations of Variables   Available evidence indicates that there are interaction effects on visual performance when various combinations of the above variables exist, such as the combined effects of contrast and motion as reported by Petersen and Dugas (1972). (Some of these effects are dicussed in Chapter 14.)

Movement   The movement of a target object or of the observer (or both) decreases the threshold of visual acuity. The ability to make visual discriminations under such circumstances is called *dynamic visual acuity* (DVA). It is usually expressed in degrees of movement per second. Acuity deteriorates rapidly as the rate of motion exceeds 60 degrees per second (60°/s) (Burg, 1966).

Age and Vision

Special mention should be made of the fact that visual skills, especially visual acuity, tend to deteriorate through age. The important implication for human factors is that if elderly people are to use visual displays, the displays should be designed accordingly (such as by providing alphanumeric characters that are large enough to see and by providing adequate illumination). We discuss this further in Chapter 14.

Perception

People certainly need to be able to ''see'' the relevant features of visual displays they use (such as traffic signs, electrocardiograph tapes, or pointers on pressure gauges). But the ability to see such features usually is not enough to make appropriate decisions based on the information in question. People must also understand the meaningfulness of what they see. Here we bump into the concept of perception as discussed in Chapter 3. Although perception does involve the sensing of stimuli, it concerns primarily the interpretation of that which is sensed, such as its identification or meaningfulness. This is essentially a cerebral rather than a sensory process. In many circumstances this interpretation process is, of course, fairly straightforward, but in the use of most visual displays it depends on previous learning (as by experience and training), such as learning shapes of road signs, color codes of electric wiring systems, abbreviations on a computer keyboard, implications of the zigs and zags of an electrocardiograph recording, or even the simple recognition of letters of the alphabet.

In considering visual displays, then, clearly the design should meet two objectives: the display must be able to be seen clearly and the design should help the viewer to correctly perceive the meaning of the display. As evidence of some increased awareness of the importance of perception in this process, we can cite the increased use of public signs with symbols to represent such concepts as "pedestrians," "no smoking," and "rest rooms." Although displays are designed for specific work situations, of course, adequate training of workers to interpret the meaningfulness of the specific display indications is needed; but appropriate design should capitalize on those display features that help people to understand the meaning, that is, to perceive correctly what they sense. We should add that certain types of visual stimuli play tricks on our perception. For example, a light of a given intensity usually is perceived as being brighter than it actually is when it is viewed under an ambiant light of lower intensity than when viewed under an ambient light of higher intensity. And a straight line usually seems distorted when it is viewed against a background of curved or radiating lines.

ALPHANUMERIC DISPLAYS

By all odds, the most commonly used displays are those with alphabetic and numeric characters. The form or technology of presenting such material has changed over the centuries, from the stone carvings in Egypt and elsewhere; to the Dead Sea scrolls; to the Rosetta Stone; to handwritten manuscripts; to books, magazines, and newspapers; to TV; to visual display terminals (VDTs); to cathode ray tubes (CRTs); and to several new technologies such as electroluminescence panels and electrochromic displays. This section of the chapter deals with some human factors aspects of designing alphanumeric displays—by whatever technology. In this regard, however, we should be familiar with the distinctions among certain human factors criteria:

- *Visibility:* The quality of a character or symbol that makes it separately visible from its surroundings. (This is essentially the same as the term *detectability* as used in Chapter 3.)
- *Legibility:* The attribute of alphanumeric characters that makes it possible for each one to be identifiable from others.[1] This depends on such features as stroke width, form of characters, contrast, and illumination. (This is essentially the same as the term *discriminability* as used in Chapter 3.)
- *Readability:* A quality that makes possible the recognition of the information content of material when it is represented by alphanumeric characters in meaningful groupings, such as words, sentences, or continuous text. (This depends more on the spacing of characters and groups of characters, on their combination into sentences or other forms, on the spacing between lines, and on margins than on the specific features of the individual characters.)

[1]For an excellent survey of legibility and related aspects of alphanumeric characters and related symbols, see Cornog and Rose (1967).

Typography

The term *typography* refers to the various features of alphanumeric characters, individually and collectively. For practical purposes, in everyday life most of the variations in typography adequately fulfill the human factors criteria mentioned above (visibility, legibility, and readability). However, there are at least four types of circumstances in which it may be important to use "preferred" forms of typography: (1) when *viewing conditions are unfavorable* (as with poor illumination or limited viewing time), (2) when the *information is important or critical* (as when emergency labels or instructions are to be read), (3) when viewing occurs at a *distance,* and (4) when people with *poor vision* may use the displays. The following discussion of typography applies primarily to these conditions. (Under most other conditions, variations in typography can have aesthetic values and still fulfill the criteria of visibility, legibility, and readability.)

Stroke Width   The stroke width of an alphanumeric character usually is expressed as the ratio of the thickness of the stroke to the height of the letter or numeral. Some examples are shown in Figure 4–6. The effects of stroke width are intertwined with the nature of the background (black on white versus white on black) and with illumination. A phenomenon called *irradiation* causes white features on a black background to appear to "spread" into adjacent dark areas, but the reverse is not true. Thus, in general, black-on-white letters should be thicker (have lower ratios) than white-on-black ones. Although these interactions are not yet fully understood, a few generalizations can be offered, adapted in part from Heglin (1973) (these are predicated on the assumption of good contrast):

• With reasonably good illumination, the following ratios are satisfactory for printed material: Black on white, 1:6 to 1:8; and white on black, 1:8 to 1:10.
• As illumination is reduced, thick letters become relatively more readable than

FIGURE 4-6
Illustrations of stroke width-to-height ratios of letters and numerals.

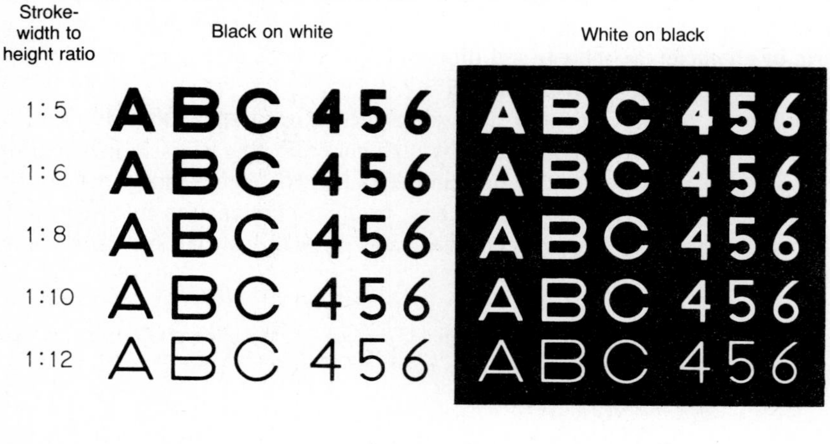

Stroke-width to height ratio	Black on white	White on black
1:5	A B C 4 5 6	A B C 4 5 6
1:6	A B C 4 5 6	A B C 4 5 6
1:8	A B C 4 5 6	A B C 4 5 6
1:10	A B C 4 5 6	A B C 4 5 6
1:12	A B C 4 5 6	A B C 4 5 6

thin ones (this is true for both black-on-white and white-on-black letters).

• With low levels of illumination or low contrast with background, printed letters preferably should be boldface type with a low stroke width-height ratio (such as 1:5).

• For highly luminous letters, ratios should be from 1:12 to 1:20.

• For black letters on a very highly luminous background, very thick strokes are needed.

Width-Height Ratio of Characters   The relationship between the width and height of a complete alphanumeric character is described as the *width-height ratio* and is expressed as a ratio (such as 3:5) or a percentage (such as 60 percent). A 3:5 ratio has its roots in the fact that some letters have five elements in height (such as the three horizontal strokes of the letter B plus the two spaces between these) and three elements in width (such as the two vertical strokes of the letter B plus the space between them). Because of these geometric relationships, a 3:5 ratio has come into fairly common use and, in general, is reasonably well supported by research. However, Heglin (1973) argues against the use of a fixed ratio for all letters, pointing out that the legibility of certain letters would be enhanced if their width were adjusted to their basic geometric forms (such as O being a perfect circle and A and V being essentially equilateral triangles).

Although a 3:5 width-height ratio is quite satisfactory for most purposes, wider letters are appropriate for certain circumstances, such as when the characters are to be transilluminated or are to be used for engraved legends (Heglin). Figure 4–7 shows letters with a 1:1 ratio that would be suitable for such purposes.

Styles of Type   There are over 30,000 *type styles* (or *typefaces,* or *fonts* of type) used in the printing trade. These fall into four major classes:

• *Roman:* This is the most common class. The letters have sherifs (little flourishes or embellishments).

• *Sans serif:* These styles (without serifs) tend to be modern in appearance and are used especially for headings, labels, etc.

• *Script:* These styles simulate modern handwriting.

• *Black letter:* These styles resemble the German manuscript handwriting of Gutenberg's time (fifteenth century).

Most of the type styles used for conventional text are Roman, and most of these are about equally legible. The other type styles tend to be used more for special purposes, such as for titles and headings, and for aesthetic effects. Script and black letter styles should never be used when visibility, legibility, and readability are critical.

In conventional text material, Roman types such as this is used, with uppercase (i.e., capital) and lowercase (small) letters (as illustrated in this sentence). *Italics such as this are used for emphasis, for titles of publications, to indicate special usage of words, etc.* **Boldface such as this is used for headings and labels, in some instances for special emphasis, and as a legibility aid under poor reading conditions.**

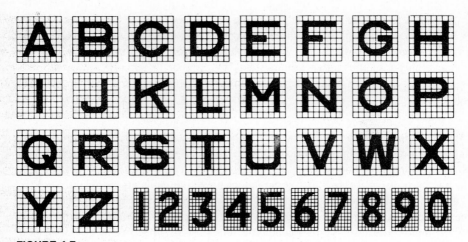

FIGURE 4-7
Letter and numeral font of United States Military Specification no. MIL-M-18012B (July 20, 1964); also referred to as NAMEL (Navy Aeronautical Medical Equipment Laboratory) or as AMEL. The letters as shown have a width-height ratio of 1:1 (except for I, J, L, and W). The numerals have a width-height ratio of 3:5 (except 1 and 4).

On the basis of research and experience in the military, certain type styles of uppercase letters and numerals have been developed and recommended (such as when words or abbreviations are to be used as labels). Two such sets are shown in Figures 4–7 and 4–8. These type styles are not specifically represented by standard type

TABLE 4–1
EXAMPLES OF GOTHIC STYLES

Very light styles	
Futura light	ABCDEFGHIJKLMNOP 1234567890
Sans serif light	ABCDEFGHIJKLM 1234567890
Vogue light	ABCDEFGHIJKLMNOPQR 1234567890
Light styles	
Futura book	ABCDEFGHIJKLMNOPQRST 1234567890
Sans serif medium	ABCDEFGHIJKLMNOPQ 1234567890
Tempo medium	ABCDEFGHIJKLMNOPQ 1234567890
Medium styles	
Futura medium	ABCDEFGHIHKLMNOPQRST 1234567890
Sans serif bold	**ABCDEFGHI 1234567890**

A B C D E F G H I J K L M
N O P Q R S T U V W X Y Z
O I 2 3 4 5 6 7 8 9

FIGURE 4-8
United States Military Standard letters and numerals MIL
Standard MS 33558 (ASG)(Dec. 17, 1957). The basic stroke
width-to-height ratio is 1:8, and the width is about 70 percent of
the height. These are sometimes referred to as AND (*Air Force-
Navy Drawing* 10400).

styles. However, certain commercial types have somewhat similar formats. A few
are illustrated in Table 4–1. These are examples of Gothic type styles.

Although many commercially available styles of type have been found through
experience and research to be quite legible under normal conditions, Heglin points
out that other designs could be developed that would be more legible, especially
under adverse reading conditions. He reports one experiment, for example, in which
some radically different numerals were used, with virtually no errors in legibility.
An example is given here:

$$| 2 \ \exists \ \supset \ \8$$

Illuminated Alphanumeric Characters Although most of the reading of alphanumeric material in the past has been done with printed or typed material (what
is now euphemistically called *hard copy*), the technological revolution has yielded
new methods of reproducing alphanumeric material. Two methods are of particular
interest, namely, those that result in *segmented characters* (especially numerals) and
in characters formed from a *dot matrix* (or grid).

Segmented numerals are formed from selected combinations of some of the seven
segments of the geometric form shown in Figure 4–9. Such numerals usually are
produced from some digital device such as digital clocks or counters. The numerals
often are represented by illumination of the specific line segments in question, these
being against a dark background.

In turn, dot-matrix characters are those generated with CRTs. The most common
types of CRTs are television (TV) and visual display terminals (VDTs). The complete image on a CRT is formed from combinations of many thousands of elements
of a matrix (or grid). This matrix consists of many horizontal *raster scan lines* (or
simply *scan lines*), each of which is made up of many separate elements called
pixels. (Scan lines are the continuous narrow strips of the picture area formed by one
horizontal "sweep" of the CRT; TVs in the United States, for example, have 525
scan lines.) The combination of scan lines (up to a few hundred) and pixels (up to a
few hundred) results in thousands of separate elements on a screen. These can be
thought of as the "dots" that form the complete matrix. The size of the matrix used
for individual characters can range from 5 × 7 (5 wide and 7 high) to 15 × 24. An
example of a letter formed from a 7 × 9 matrix is shown in Figure 4–9*b*. However,

a. Segmented Numeral

b. Dot Letter

FIGURE 4-9
Example of *(a)* a segmented numeral and *(b)* a dot letter. All numerals can be formed from the basic configuration of seven segments at the left. All letters and numerals can be formed from combinations of the dots at the right.

Type of numeral	Examples	Number of errors
NAMEL	**345**	187
Slanted segmented	345	391
Vertical segmented	345	388

FIGURE 4-10
Forms of numerals investigated by Plath, along with errors made in their identification under time-controlled conditions. (*Source: Adapted from Plath, 1970, fig. 1 and table 1.*)

the physical size of a character in inches or millimeters depends on the size of the screen and the size of the matrix of scan lines and pixels. Incidentally, the letters can be shown as light on a dark background or dark on a light background. (Most VDTs are light on dark background.)

The extensive use of segmented and dot-matrix characters naturally raises questions about their use in practice. To begin, it is probable that most such characters can be read adequately under good reading conditions (with good illumination, plenty of time, etc.). Their suitability should be tested under adverse circumstances. In the case of segmented numerals, Plath (1970), for example, compared the legibility of conventional numerals (the *NAMEL* numerals shown in Fig. 4–7) with two forms of segmented numerals slanted and vertical, as shown in Figure 4–10. Figure 4–10 also shows the errors in legibility under time-constrained conditions; the errors made with the conventional *NAMEL* numerals were many fewer than those for either of the segmented forms, thus giving the edge to conventional numerals.

In general, letters formed from a dot matrix are reasonably comparable to conventional printed letters in terms of visual performance criteria. Some such evidence comes from studies by Vartabedian (1971, 1973) in which he used letters such as those in Figure 4–11. These included uppercase and lowercase letters of both a dot-matrix design and a conventional stroke design. Controlling for letter size, Vartabedian found no appreciable difference in a timed task of searching for specific words for the two styles.

However, a word of caution is in order about the use of dot-matrix letters. If letters are formed from too small a matrix, the *resolution* of the image is poor.

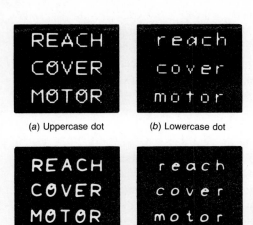

(a) Uppercase dot   (b) Lowercase dot

(c) Uppercase line   (d) Lowercase line

FIGURE 4-11
Examples of words used in cathode ray tube (CRT) displays for comparison of four types of generated letters. (*Source: Vartabedian, 1971, fig. 1.*)

Vartabedian (1973), for example, found that letters formed with a 7 × 9 matrix were more legible than those formed with a 5 × 9 matrix (even with the sizes controlled). And Pastoor, Schwarz, and Beldie (1983) found that those based on a 9 × 13 matrix were preferable to 7 × 9 matrix letters in terms of visual performance. Further, common practice with VDTs tends toward using larger matrices (8 × 16 or more) if high resolution is desirable (as in continuous office work).

Size of Alphanumeric Characters at Reading Distances   In the printing trade the size of type is measured in *points*. One point equals $\frac{1}{72}$ in. (0.35 mm). But this is the height of the *slug* on which the type is set. A close approximation of height in points is to consider 1 point (pt) as equivalent to $\frac{1}{100}$ in (0.25 mm) rather than $\frac{1}{72}$ in (0.35 mm) (Heglin, 1973). Some examples of different sizes are given below, with the sizes of the slugs and the approximate heights of the letters in inches.

This line is set in 4-pt type (slug = 0.55; letters = 0.04).

This line is set in 6-pt type (slug = 0.084; letters = 0.06).

This line is set in 8-pt type (slug = 0.111; letters = 0.08).

This line is set in 9-pt type (slug = 0.125; letters = 0.09).

This line is set in 10-pt type (slug = 0.139; letters = 0.10).

This line is set in 11-pt type (slug = 0.153; letters = 0.11).

This line is set in 12-pt type (slug = 0.167; letters = 0.12).

The type size in most printed material usually ranges from about 7 to 14 pt with the most common about 9 to 11 pt.

In selecting the size of alphanumeric characters, it is preferable to specify the height that would be optimum for the use in question. However, absolute standards of height are hard to come by. One fairly reasonable, general guide, however, is the type we read in newspapers and magazines. The wide use of this print size is indeed a reflection of its widespread suitability. The height of the commonly used 9- to 11-pt type ranges from 0.09 to 0.11 in (2.3 to 2.8 mm). Such sizes then might be accepted as a base for general use of alphanumeric material.

When the reading is critical or is performed under poor illumination or when the characters are subject to change (as on some instruments), the character heights should be increased. Certain sets of recommendations have been made for the size of alphanumeric characters for visual displays (such as on instrument panels) that take such factors into account. One such set is given in Table 4–2.

Note that the smallest height in that table is 0.05 in (1.27 mm). It is interesting to relate this height to conventional 20/20 vision as measured by the commonly used Snellen letter chart. That chart is based on the fact that a person with 20/20 vision can discriminate stroke widths (of letters) that subtend a visual angle of 1 minute of arc. Therefore, a letter 5 stroke widths high would subtend a visual angle of 5 minutes of arc. Translating that into dimensions at a reading distance of 28 in (71 cm) gives a letter height of 0.04 in (1.0 mm)—just a bit below the shortest height in Table 4–2. Somewhat higher characters would be needed by people who do not have 20/20 vision.

Size of Alphanumeric Characters at a Distance   In connection with the height of printed or lettered alphanumeric characters, it has been generally assumed that the legibility and readability are equal for various distances if the characters are increased for distance viewing so the visual angle subtended at the eye is the same. An illustration of such an extrapolation is shown in Figure 4–12 which is based substantially on the extrapolation of the guidelines for a 28 in viewing distance in Table 4–2.

Figure 4–12 includes variations in height for characters that are variable in position (versus fixed) and for low and high illumination. But it does not take into account variations for people with different levels of visual acuity. The National Bureau of Standards (NBS) has set forth a set of recommendations of stroke widths for persons with various Snellen vision test scores as shown in Figure 4–13 (Howett,

TABLE 4–2
ONE SET OF RECOMMENDED HEIGHTS OF ALPHANUMERIC CHARACTERS FOR CRITICAL AND NONCRITICAL USES UNDER LOW AND HIGH ILLUMINATION

	Height of numerals and letters	
	Low luminance (down to 0.03 fL)	High luminance (1.0 fL and above)
Critical use, position variable	0.20–0.30 in (5.1–7.6 mm)	0.12–0.20 in (3.0–5.1 mm)
Critical use, position fixed	0.15–0.30 in (3.8–7.5 mm)	0.10–0.20 in (2.5–5.1 mm)
Noncritical use	0.05–0.20 (1.27–5.1 mm)	0.05–0.20 (1.27–5.1 mm)

Note: For other viewing distances, in inches, multiply distance in inches by 28.)
Source: Adapted from Heglin (1973) and Woodson (1963).

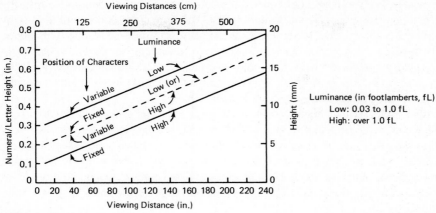

FIGURE 4-12
Heights of alphanumeric characters for various viewing distances for fixed and variable positions of characters and for low and high levels of luminance. (*Adopted from Heglin, 1973, fig. VIII-12 as based on Kubakawa et al., 1969.*)

1983). These were originally given in terms of stroke width, so letter heights would be 5 times as high (assuming a 3:5 width-to-height ratio). Howett's formulation is based on the ability of a person with 20/20 vision to read letters that subtend a visual angle of 5 minutes of arc. As indicated earlier, such a size would be slightly below the smallest height in Table 4–2, and corresponding sizes at different distances are as shown in Figure 4–12.

Size of Alphanumeric Characters on CRTs   Turning now to alphanumeric material presented in CRTs (especially VDTs), we see that certain investigators have recommended sizes that subtend a visual angle of at least 11 or 12 minutes of arc (Howell and Kraft, 1959; Shurtleff, 1967; and Snyder and Taylor, 1979). These

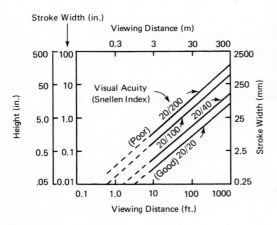

FIGURE 4-13
Minimum stroke widths (with letter heights) required for letter visibility as a function of viewing distance for persons with different levels of visual acuity. Note that the horizontal and vertical scales are expressed in log-log terms, not linear terms. (*Adapted from Howett, 1983, fig. 7.*)

values would result in heights of about 0.09 or 0.10 in (2.3 or 2.5 mm) at a conventional reading distance of about 28 in (71 cm), and they correspond quite well with conventional reading print. However, a couple of factors raise some questions about such guidelines. First, certain investigations have indicated that legibility and readability could be enhanced with larger sizes—up to 17 or 25 minutes of arc [0.14 to 0.20 in (3.56 to 5.08 mm) at a 28-in (71-cm) reading distance) (Hemingway and Erickson, 1969; Pastoor, Schwarz, and Beldie, 1983).

Another factor is that many VDTs are used at a shorter distance than the usual 28-in (71-cm) reading distance—frequently at about 18 in (about 46 cm). At such distances the sizes can be reduced considerably while an acceptable visual angle is maintained. As a guiding principle, the characters should be of a size to subtend a minimum visual angle of about 12 minutes of arc, but with larger configurations if any special considerations apply (such as time constraints, criticalness, poor viewing conditions, etc.).

Density of Alphanumeric Material   The *density* of alphanumeric material (sometimes called *local density*) refers to the compactness of the characters (how "tightly packed" they are). In a strict sense, density refers to the spacing between lines, although in some investigations the width of the letters has also been associated with density.

On the basis of review of relevant research Helander, Billingsley, and Schurick (1984) concluded that densely packed text (and narrower letters) usually is best for reading tasks in which meaningful text must be scanned and comprehended quickly. This advantage presumably is related to the eye fixations that occur during reading. Reading densely packed (and also smaller) characters requires less occular (and presumably less cognitive) work than reading less densely packed characters (Kolers, Duchnicky, and Ferguson, 1981). These investigators thus propose using VDTs with 80 characters per line in preference to smaller numbers such as 40.

Some confirming evidence about the use of high-density lines is reported by Moriarty and Scheiner (1984) who counted the words read by subjects reading copy from one sales brochure printed with regular spacing and one printed with close-set (high-density) type. Portions of the brochure are reproduced in Figure 4–14. The authors found that the close-set type was read more rapidly than the regularly spaced type.

Although the use of high-density characters thus seems to enhance reading performance, Tullis (1983) points out that one can go too far in squeezing the letters together. He then postulates the relevance of the inverted U-shaped hypothesis which applies to many aspects of human performance: For any given parameter (in this case, degree of density), there is some optimum range in terms of human performance, with both lower and higher levels resulting in poorer performance.

Sequences of Numerals and Letters   We are frequently required to store in long-term memory sequences of numerals and letters (such as license numbers, bank account numbers, and social security numbers) or to store in working memory such sequences as telephone numbers and numbers to be keypunched. The concept of chunking discussed in Chapter 3 would suggest that such sequences could be remembered most easily if they are broken into groups. This principle was confirmed by

Regular spacing of text type (regular density)

> The ESS Performance Series is both a choice and a statement. The choice is to continue ESS's long tradition of excellence by trimming costs without

Close-set text type (high density)

> The ESS Performance Series is both a choice and a statement. The choice is to continue ESS's long tradition of excellence by trimming costs without sacrificing performance and by omitting

FIGURE 4-14
Portions of an advertising brochure used in a study of reading speed of regularly spaced types and of close-set (high density) type. The close-set type was read more rapidly. (*Adapted from Moriarty and Scheiner, 1984, fig. 1.* Copyright 1984 by the American Psychological Association. Adapted by permission of the author.)

the study where Klemmer and Stocker (1974) had subjects enter a seven-digit number into a Touch-Tone telephone keyboard when the numbers were presented in three groupings:

3–4 (as in 123–4567)
No grouping (as in 1234567)
1–5–1 (as in 1–23456–7)

The results, shown in Fig. 4–15, show that the first grouping (3–4) was clearly superior in terms of both time and error (this applied to both forced-pacing and self-pacing conditions).

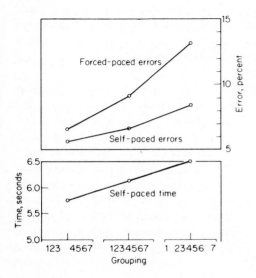

FIGURE 4-15
Time and errors in using a Touch-Tone telephone to enter seven digits presented in various groupings. (*Source: Klemmer and Stocker, 1974, fig. 2, p. 677.* Copyright © 1974 by the American Psychological Association. Reprinted by permission.)

The same general principle of grouping applies for letters used with digits, as demonstrated by Hull (1976). In his study, the subjects were given brief presentations of slides of six characters, and then they recorded what they recalled of the characters. In the phase of the study in which letters and digits were used in combination, the errors of recall were significantly fewer when the letters and the digits were in "patterns" of groups of 2 or 3 than when they were randomized. Some examples of the patterned and randomized materials are:

Patterned examples: WFC328; 769HRD; WF93RD; 76CH28
Random examples: Y1WJ15; CH9R2R; N3C6FD; HD7WC4

The separate grouping of the letters and of the numerals as shown in the patterned examples has the effect of separating them into chunks.

Human Factors Aspects of VDT Screens

We have referred to a few human factors aspects relating to VDT screens, particularly to certain features of alphanumeric characters shown on them. However, many other features regarding alphanumeric material presented on such screens have human factors implications. A partial inventory includes the format (various features of the overall "organization" of material presented on a VDT screen); paging versus scrolling (with *paging,* the entire page changes on the screen all at once; with *scrolling,* the lines of information move either continuously or one at a time); type of sequencing (the presentation on a screen of a series of discrete units of a complete message, as by separate words, lines, or chunks of information); rate of presentation of sequenced material; overall density (as well as local density, discussed above); grouping (the grouping on the screen of related information); the *menu,* or the nature of the list of options to which a VDT operator is to respond); color of characters and VDT screen; and direction of contrast (light characters on a dark background versus dark on light).

It is not feasible here to cover in detail the human factors implications of these and other features of VDTs. However, one aspect of display format is discussed as an illustration, in particular a comparison of narrative and structured display formats as related to the interpretation of the displayed information.

Narrative versus Structured VDT Format   A *narrative* format uses complete words and phrases, sometimes in the form of a list. A *structured* display presents the material in a systematic manner with relevant information being organized into groups with headings and sometimes using abbreviations. Examples are given in Figure 4–16 that deal with information regarding telephone lines; the same information is presented in both formats. The VDT operator uses such information in diagnosing problems in the lines. (The information in Figure 4–16 would be meaningful only to persons familiar with the data in question.)

A comparison of the two formats with subjects who were being trained in line-problem diagnosis indicated a distinct superiority of the structured format in terms of their speed of learning. It is reasonably clear that the organization of the structured format is an aid in interpreting displayed information, when such structure is feasible.

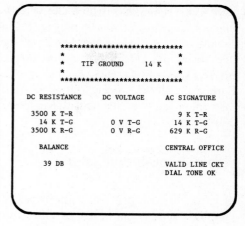

```
TEST RESULTS    SUMMARY: GROUND

 GROUND, FAULT T-G
 3 TERMINAL DC RESISTANCE
    >  3500.00 K OHMS T-R
    =    14.21 K OHMS T-G
    >  3500.00 K OHMS R-G
 3 TERMINAL DC VOLTAGE
    =     0.00 VOLTS  T-G
    =     0.00 VOLTS  R-G
 VALID AC SIGNATURE
 3 TERMINAL AC RESISTANCE
    =     8.82 K OHMS T-R
    =    14.17 K OHMS T-G
    =   628.52 K OHMS R-G
 LONGITUDINAL BALANCE POOR
    =      39   DB
 COULD NOT COUNT RINGERS DUE TO
    LOW RESISTANCE
 VALID LINE CKT CONFIGURATION
 CAN DRAW AND BREAK DIAL TONE
```

```
    ******************************
    *                            *
    *     TIP GROUND      14 K    *
    *                            *
    ******************************

 DC RESISTANCE       DC VOLTAGE        AC SIGNATURE

 3500 K T-R                              9 K T-R
   14 K T-G         0 V T-G             14 K T-G
 3500 K R-G         0 V R-G            629 K R-G


    BALANCE                          CENTRAL OFFICE

    39 DB                            VALID LINE CKT
                                     DIAL TONE OK
```

FIGURE 4-16
Examples of a narrative format and a structured format in presenting information on a VDT screen. (These examples both present the same information about the condition of telephone lines, for use in diagnosing line problems.) (*From Tullis, 1981, figs. 1 and 2.* Copyright by the Human Factors Society, Inc., and reproduced by permission)

Saying What Is Meant

Many members of the human factors clan play a critical role in getting information to people, such as by the use of labels, signs, directions, instructions, work aids, training materials, and other information displays. The development of such materials leads to a broad domain of communications that cannot be covered here, including principles of writing. However, a couple of horrible examples illustrate the importance of giving proper attention to such matters. One of these examples, reported by Chapanis (1967), is a notice beside an elevator that read as shown in the first box of Figure 4–17. That figure also indicates what the sign really meant and presents an

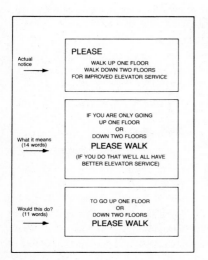

FIGURE 4-17
Examples of an elevator sign and of suggested revisions. (*Source: Adapted from Chapanis, 1967.*)

improved version that is both clearer and shorter. Broadbent (1977) reports another interesting example of a confusing instruction, on an electric razor: "This razor should only be used on the 220-V setting in the United Kingdom." This could be interpreted as meaning that outside the United Kingdom, this razor should be used on the 110-V setting. (Such an interpretation would, of course, ruin the razor in most European countries, where 220-V electricity is common.) The correct phrasing of the instruction is, "In the United Kingdom, this razor should be used only on the 220-V setting."

VISUAL CODING DIMENSIONS

As discussed in Chapter 3, sometimes there is a need to have a coding system that identifies various items of a given class. For example, various specific pipes in a plumbing system can be coded different colors, or various types of highway signs can be coded by different shapes. The items to be coded (such as different pipes or different types of highway signs) are called *referents*. Thus, a specific code such as an octagonal (eight-sided) road sign means "stop." The types of visual stimuli used in such circumstances are sometimes called *coding dimensions*. Some visual coding dimensions are color, geometric and other shapes, letters, numerals, flash rate (of a flashing light), visual angle (the positions of hands on a clock), size (variation in the size of circles), and even "chartjunk" [the term used by Tufte (1983) for the gaudy variations in shading and design of areas of graphs].

Single Coding Dimension

When various coding dimensions can be used, an experiment can be carried out to determine what dimension would be best. A couple of examples illustrate this approach.

FIGURE 4-18
Illustration of code symbols used in comparing coding of targets. (*Source: Hitt, 1961*).

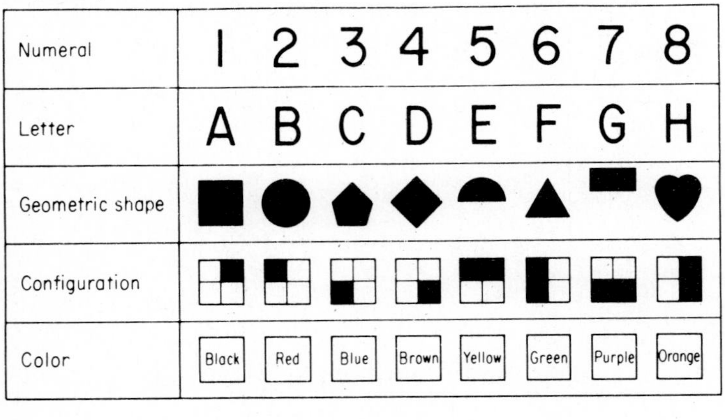

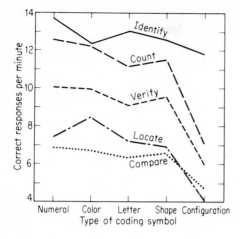

FIGURE 4-19
Relationship between coding method and performance in five tasks comparing targets on a display. (*Source: Adapted from Hitt, 1961.*)

One such study was done by Hitt (1961), who used the five codes shown in Figure 4–18 (numerals, letters, geometric shapes, configuration, and color) to represent various types of buildings or facilities in various sections of a maplike display. The subjects performed five different tasks in scanning the display (identification, location, counting, comparison, and verification). The results of the study, summarized in Figure 4–19, show that the numeric and color codes were best in terms of number of correct responses for most of the tasks, with the configuration codes being the worst.

In another study, Smith and Thomas (1964) used the four sets of codes shown in Figure 4–20 in a task of counting the number of items of a specific class (such as red, gun, circle, or B–52) in a large display with many other items of the same class.

FIGURE 4-20
Four sets of codes used in a study by Smith and Thomas (1964) (copyright © 1964 by the American Psychological Association and reproduced by permission). The notations under the color labels are the Munsell color matches of the colors used. (*Source: Munsell book of color, 1959.*)

	C-54	C-47	F-100	F-102	B-52
Aircraft shapes					
Geometric forms	Triangle	Diamond	Semicircle	Circle	Star
Military symbols	Radar	Gun	Aircraft	Missile	Ship
Colors (Munsell notation)	Green (2.5 G 5/8)	Blue (5 BG 4/5)	White (5 Y 8/4)	Red (5 R 4/9)	Yellow (10 YR 6/10)

(The density of items, that is, the total number of items in the display, was also varied.) The results, shown in Figure 4–21, show a clear superiority for the color codes for this task; this superiority was consistent for all levels of density.

Although these and other studies indicate that various visual coding dimensions differ in their relevance for various tasks and in various situations, definite guidelines regarding the use of such codes still cannot be laid down. Recognizing this, and realizing that good judgment must enter into the selection of visual codes for specific purposes, we see that a comparison such as that given in Table 4–3 can serve as at least partial guidance. In particular, Table 4–3 indicates the approximate number of levels of each of various visual codes that can be discriminated, along with some sideline comments about certain methods.

Multidimension Codes

Chapter 3 included a discussion of multidimension codes, indicating some of the variations in combining two or more dimensions. Heglin (1973) recommends, however, that no more than two dimensions be used together if rapid interpretation is required. In combining two visual codes, certain combinations do not "go well" together. Some simply cannot be used in combination. The combinations that are potentially useful are shown as Xs in Figure 4–22. Although combinations of codes can be useful in many circumstances, they are not always more effective than single-dimension codes.

Color Coding

Since color is a fairly common visual code, we discuss it somewhat further. An important question relating to color deals with the number of distinct colors persons

FIGURE 4-21
Mean time a and errors b in counting items of four classes of codes as a function of display density (copyright © 1964 by the American Psychological Association and reprinted by permission). The Xs indicate comparison data for displays of 100 items with color (or shape) held constant. (*Source: Smith and Thomas, 1964.*)

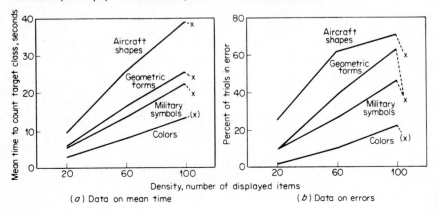

TABLE 4–3
SUMMARY OF CERTAIN VISUAL CODING METHODS

(Numbers refer to number of levels which can be discriminated on an absolute basis under optimum conditions.)

Alphanumeric	Single numerals, 10; single letters, 26; combinations, unlimited. Good; especially useful for identification; uses little space if there is good contrast. Certain items easily confused with each other.
Color (of surfaces)	Hues, 9; hue, saturation, and brightness combinations, 24 or more. Preferable limit, 9. Particularly good for searching and counting tasks. Affected by some lights; problem with color-defective individuals.*†
Color (of lights)	10. Preferable limit, 3. Limited space required. Good for qualitative reading.‡
Geometric shapes	15 or more. Preferable limit, 5. Generally useful coding system, particularly in symbolic representation; good for CRTs. Shapes used together need to be discriminable; some sets of shapes more difficult to discriminate than others.‡
Angle of inclination	24. Preferable limit, 12. Generally satisfactory for special purposes such as indicating direction, angle, or position on round instruments like clocks, CRTs, etc.§
Size of forms (such as squares)	5 or 6. Preferable limit, 3. Takes considerable space. Use only when specifically appropriate.
Visual number	6. Preferable limit, 4. Use only when specifically appropriate, such as to represent numbers of items. Takes considerable space; may be confused with other symbols.
Brightness of lights	3–4. Preferable limit, 2. Use only when specifically appropriate. Weaker signals may be masked.‡
Flash rate of lights	Preferable limit, 2. Limited applicability if receiver needs to differentiate flash rates. Flashing lights, however, have possible use in combination with controlled time intervals (as with lighthouse signals and naval communications) or to attract attention to specific areas.

*Feallock et al. (1966).
†M. R. Jones (1962).
‡Grether and Baker (1972).
§Muller et al. (1955).

with normal color vision can differentiate on an absolute basis. It has generally been presumed that the number was relatively moderate; for example, M. R. Jones (1962) indicated that the normal observer could identify about nine surface colors, varying primarily in hue. However, it appears that, with training, people can learn to make upward of a couple of dozen such discriminations when combinations of hue, saturation, and lightness are prepared in a nonredundant manner (Feallock et al., 1966). On the basis of this study, 24 colors were so chosen that only two of those within

	Color	Numeral and letter	Shape	Size	Brightness	Location	Flash rate	Line length	Angular orientation
Color		X	X	X	X	X	X	X	X
Numeral and letter	X		X			X	X		
Shape	X			X	X		X		
Size	X	X	X			X		X	
Brightness	X		X	X					
Location	X	X						X	X
Flash rate	X	X	X	X					X
Line length	X					X			X
Angular orientation	X					X	X	X	

FIGURE 4-22
Potential combinations of coding systems for use in compound coding. (*Source: Adapted from Heglin, 1973, table VI-6, VI-22.*)

the set were confused by subjects with others in the set under four experimental lighting conditions.[2]

When relatively untrained people are to use color codes, however, the better part of wisdom would argue for the use of smaller numbers. In this regard Conover and Kraft (1958) present four sets of colors for coding purposes when absolute recognition is required (barring the use of color-deficient people). These sets include, respectively, eight, seven, six, and five colors.[3]

In general, color coding has been found to be particularly effective for "searching" tasks—tasks that involve scanning an array of many different stimuli to "search-out" (or spot), or locate, or count those of a specific class. This advantage presumably is due to the fact that colors "catch the eye" more readily than other visual codes. Examples of searching tasks include the use of maps and navigational charts, searching for items in a file, and searching for color-coded electric wires of some specific color. Christ (1975) analyzed the results of 42 studies in which color codes

[2]The identifications of these colors in the new Federal Standard on Colors follow, with an (X) after those that were identifiable under all lighting conditions: 32648 (X), 31433, 30206 (X), 30219, 30257, 30111, 31136 (X), 31158, 32169, 32246, 32356, 33538 (X), 33434, 33695, 34552 (X), 34558, 34325 (X), 34258 (X), 34127, 34108 (X), 35189, 35231 (X), 35109, and 37144 (X).

[3]The Munsell (1959) notations for the colors in these four sets are given below:
  8-color: 1R; 9R; 1Y; 7GY; 9G; 5B; 1P; 3RP
  7-color: 5R; 3YR; 5Y; 1G; 7BG; 7PB; 3RP
  6-color: 1R; 3YR; 9Y; 5G; 5B; 9P
  5-color: 1R; 7YR; 7GY; 1B; 5P

were compared with other types of codes and found that color codes were generally better for searching tasks than most other codes (geometric shapes, letters, and digits).

Christ also found that color codes were better for "identification" tasks than were certain other codes, but color codes were generally not as good for such tasks as letters and numerals were. In this sense, an identification task involves the conceptual recognition of the meaningfulness of the code—not simple visual recognition of items of one color from others (such as identifying red labels from all other labels). Thus, we can see the possible basis for the advantage of letters and numerals, since they could aid in tying down the meaning of codes (such as grade A or size 12).

In general, then, color is a very useful coding dimension, but it is obviously not a universally useful system.

VISUAL CODES AND SYMBOLS

Civilization abounds with a wide assortment of visual codes, symbols, and signs that are intended to convey meaning to us, their use being part of almost all phases of human activity (travel, business, medicine, the sciences, religion, engineering, and recreation). Actually the use of graphic symbols as a means of communication goes back to early human history; it became part of the folklore of most cultures. An interesting collection of 20,000 such symbols from cultures all over the world was developed by Dreyfus (1972).

Comparison of Symbolic and Verbal Signs

Where there is some question as to whether to use a symbolic or verbal sign, a symbolic sign would probably be preferable *if* the symbol reliably depicted visually what it is intended to represent. One argument for this (supported by some research) is that symbols do not require the *recoding* that words or short statements do. For example, a road sign showing a deer conveys immediate meaning, whereas use of the words *Deer Crossing* requires the recoding from words to concept. Note, however, that some symbolic codes do not visually resemble the intended concepts very well and thus may require learning and recoding.

A study that supports the possible advantage of symbols was conducted by Ells and Dewar (1979) where subjects listened to a spoken traffic message (such as "two-way traffic") and were then shown a slide of a traffic sign (either a symbolic sign or a verbal sign) and asked to respond yes or no, depending on whether they perceived the sign as being the same as the spoken message. This was done under both visually "nondegraded" and "degraded" conditions. The mean reaction times, in Figure 4-23, show a systematic superiority for the symbolic signs, especially under the visually degraded conditions.

Objectives of Symbolic Coding Systems

When a coding system of symbols is developed, the objective is to use those symbols that best represent their referents (that is, the concepts or things the symbols are

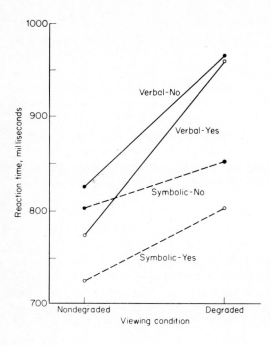

FIGURE 4-23
Mean reaction times of yes and no responses to symbolic and verbal traffic signs viewed under nondegraded and degraded viewing conditions. (*Source: Ellis and Deward, 1979, fig. 1, p. 168.*)

intended to represent). This basically depends on the strength of *association* of a *code symbol* with its *referent*. This association, in turn, depends on either of two factors: any already established association [Cairney and Siess (1982) call this *recognizability*] and the ease of learning such an association.

At a more specific level, in Chapter 3 we discussed certain guidelines for using coding systems: detectability, discriminability, compatibility, meaningfulness, standardization, and use of multidimensional codes.

Criteria for Selecting Coding Symbols

Some coding symbols already exist (or could be developed) that could be used with reasonable confidence. If there is any question about their suitability, however, they should be tested by the use of some experimental procedure. Various criteria have been used in studies with such symbols. A few criteria are discussed below:

Recognition On this criterion, subjects usually are presented with experimental symbols and asked to write down or to say what each represents.

Matching In some investigations, several symbols are presented to subjects along with a list of all the referents represented, and subjects are asked to match each symbol with its referent. This procedure usually yields a confusion matrix that indicates the number of times each symbol is confused with every other one (an example is given later). A variation consists of presenting the subjects with a complete array of symbols of many types and asking them to identify the specific symbol that represents each referent. Sometimes reaction time is measured as well as the number of correct and incorrect matches.

Preferences and Opinions   Still in other circumstances people are asked to express their preferences or opinions about experimental designs of symbols.

Examples of Code Symbol Studies

To illustrate the procedures used in code symbol studies, we give a few examples.

Mandatory-Action Symbols   The first example deals with a set of so-called mandatory-action messages and their respective symbols, illustrated in Figure 4–24 (concerning use of protection devices for the ear, eye, foot, etc.) (Cairney and Siess, 1982). These symbols were shown to a group of newly arrived Vietnamese in Australia, who were first told where the signs were to be used and then asked to state what they thought each sign meant. A week later the subjects were given a recall test in the same manner. Figure 4–24 shows, for each symbol, the percentage of correct responses for both the original test (O) and for the second test (R). The results of this particular study show rather dramatically how well a group of people with different cultural background could learn the intended meaning of those particular symbols. (Only the first symbol did not come through very well.)

This example illustrates the use of a recognition criterion and the desirable objective of using symbols that are easily learned, even if the symbols are not initially very recognizable.

Railway Symbols   As another example, Zwaga and Boersema (1983) carried out a study with the 29 information symbols used by the Netherlands railways. They

1: Must use ear protection	2: Must use eye protection	3: Must use foot protection	4: Must use hand protection	5: Must use head protection	6: Must use breathing protection
O → 10%	20%	30%	50%	37%	13%
R → 73	96	100	97	97	97

FIGURE 4-24
Symbols of mandatory-action messages used in a study of recognition and recall of such symbols. The percentages below the symbols are the percentages of correct recognition, as follows: O = original test; R = recall 1 wk later. (*Adapted from Cairney and Siess, 1982, fig. 1.*)

used a matching procedure in a study to test the strength of the association of the symbols with their referents. Each subject was shown a panel with the 29 symbols (in a random arrangement) and then was presented verbally with a particular referent and asked to report which symbol corresponded to that referent. Figure 4–25 shows 11 of the symbols, and Table 4–4 is a confusion matrix for these 11 symbols. This shows a wide range of accuracy for the 400 subjects from 77 (for symbol 9) to 395 (for symbol 7). Note that the confusion tends to be concentrated within three groups: 1 and 2 (entrance and exit symbols); 4, 5, and 6 (dealing with food); and 8, 9, and 10 (dealing with luggage). The results strongly indicated that certain symbols (those that were frequently confused with one another) should be modified to more clearly reflect their intended meanings.

Comparison of Exit Symbols for Visibility   The examples discussed above tell something about how well individual symbols are associated with their respective referents. When a specific symbol has a poor association with its referent, there can be a problem in selecting or designing a better symbol. One approach is to create alternative designs for the same referent and to test them.

Such a procedure was followed by Collins and Lerner (1983) with 18 alternate designs of exit signs, particularly under difficult viewing conditions simulating an emergency situation. (After all, you would like to be able to identify the exit if a

FIGURE 4-25
Eleven of 29 symbolic signs used by the Netherlands railway in an evaluation study of such signs. (*Adapted from Zwaga and Boersema, 1983, fig. 2.*)

TABLE 4-4
CONFUSION MATRIX SHOWING FREQUENCY WITH WHICH EACH OF 11 SYMBOLS WAS REPORTED BY SUBJECTS WHEN A CERTAIN REFERENT WAS PRESENTED (Number of subjects = 400)

Referent presented	Symbol reported by subject											
	1	2	3	4	5	6	7	8	9	10	11	12 to 29*
1 Entrance	219	93	6	2	2	1	2	4	3	2	8	58
2 Exit	87	264	6		2		1	2	1	2		35
3 Information office	2	8	337		1			2		1	3	44
4 Buffet	1		1	132	214	46					1	5
5 Restaurant		1		140	219	32					6	3
6 Food vending machines				60	54	256					7	20
7 Toilet			1				395	1	1		3	0
8 Left luggage	1	1			1		1	136	68	76	7	110
9 Luggage dispatch			6					80	77	20	10	207
10 Luggage lockers			1		2		1	107	20	185	6	78
11 Waiting room	8	8	3	73	37	9	1	1		3	240	17

*The table does not include specific data for the remaining symbols (nos. 12 to 29).
Source: Zwaga and Boersema (1983, table 3, p. 49).

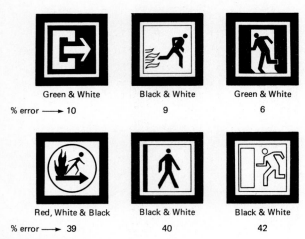

FIGURE 4-26
Examples of a few of the 18 exit signs used in a simulated
emergency experiment, with percentages of errors in
identifying them as exit signs. (*Adapted from Collins and
Lerner, 1983.*)

building were on fire!) The subjects were presented with 18 designs of exit symbols
under different levels of viewing difficulty (with very brief exposure time) and asked
to indicate whether each was, or was not, an exit sign. Certain of the designs are
shown in Figure 4–26, along with their percentages of error. (In reviewing the errors, it was found that certain symbols for ''No exit'' were confused with those for
''Exit.'')

Certain generalizations about the features of the best signs can be made: (1)
''Filled'' figures were clearly superior to ''outline'' figures, (2) circular figures were
less reliably identified than those with square or rectangular backgrounds, and (3)
simplified figures (as by reducing the number of symbol elements) seem beneficial.

FIGURE 4-27
Percentage of 330 pedestrians who responded that the symbol most clearly means
"Don't Walk." All symbols were shown in orange and red, but with the first two only
the color serves as the stimulus since there is no configuration involved. (*Source:
Adapted from Robertson, 1977, fig. 3 and table 4, p. 40.*)

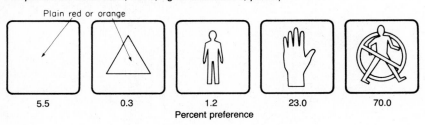

Percent preference

"Don't Walk" Signs   Still another procedure for evaluating various designs for a specific referent is to ask people for their preferences or opinions about the symbols in question. This procedure was used by Robertson (1977) in the evaluation of street crossing lights for pedestrians. In one phase of his investigation, 330 pedestrians in 10 cities were asked their opinions about the five designs of such signs shown in Figure 4–27. They were asked which of these symbols most clearly means "Don't Walk." The percentages of responses for the five symbols are also shown in Figure 4–27. Rather clearly, the fifth symbol, the circle with a slash, was preferred by most pedestrians. In connection with other questions regarding color, red was (as expected) clearly preferred over orange to represent "Don't Walk" and green over white to represent "Walk."

Perceptual Principles of Symbolic Design

Much of the research regarding the design of symbols for various uses has to be empirical, involving, for example, experimentation with proposed designs. However, experience and research have led to the crystallization of certain principles that can serve as guidelines in the design of symbols. Easterby (1967, 1970), for example, postulates certain principles that are rooted in perceptual research and generally would enhance the use of such displays. Certain of these principles are summarized briefly and illustrated. The illustrations are in Figure 4–28. Although these particular examples are specifically applicable to machine displays (Easterby, 1970), the basic principles would be applicable in other contexts.

Figure/Ground   Clear and stable figure-to-ground articulation is essential, as illustrated in Figure 4–28a.

FIGURE 4-28
Examples of certain perceptual principles relevant to the design of visual code symbols. These particular examples relate to codes used with machines. (*Source: Adapted from Easterby, 1970.*)

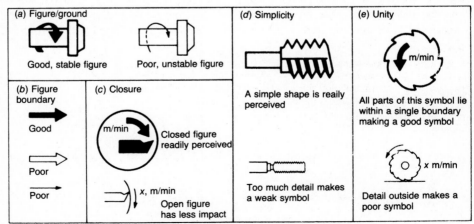

Figure Boundaries   A contrast boundary (essentially a solid shape) is preferable to a line boundary, as shown in Figure 4–28b. With different elements of a display to be depicted, the following practice should be followed: symbol dynamic—solid; moving or active part—outline; stationary or active part—solid.

Closure   A closed figure as illustrated in Figure 4–28c, enhances the perceptual process and should be used unless there is reason for the outline to be discontinuous.

Simplicity   The symbols should be as simple as possible, consistent with the inclusion of features that are necessary, as illustrated in Figure 4–28d.

Unity   Symbols should be as unified as possible. For example, when solid and outline figures occur together, the solid figure should be within the line outline figure, as shown in Figure 4–28e.

Standardization of Symbolic Displays

When symbol displays might be used in various circumstances by the same people, the displays should be standardized with a given symbol *always* being associated with the same referent. One example of such standardization is the system of international road signs. A few examples are shown in Figure 4–29. The National Park Service also has a standardized set of symbols to represent various services and concepts such as picnic areas, bicycle trails, and playgrounds.

PICTORIAL DISPLAYS

These displays are intended to present a visual representation of something, such as land areas, traffic routes, objects, and wiring or piping systems. In the development of such displays, the dominant guideline is that of simplicity. Obviously, the application of this principle needs to be within the constraints imposed by the operational requirements for "fidelity" of the configuration. The argument for simplicity arises from the fact that the perceptual processes of searching for relevant features take longer (and are subject to higher error rates) if an image is cluttered with irrelevant

FIGURE 4-29
Examples of a few international road signs. These are standardized across many countries, especially in Europe. Most of these signs are directly symbolic of their referents.

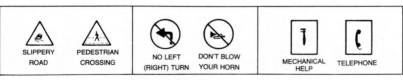

SLIPPERY ROAD	PEDESTRIAN CROSSING	NO LEFT (RIGHT) TURN	DON'T BLOW YOUR HORN	MECHANICAL HELP	TELEPHONE

(a) Danger signs   (b) Instruction signs   (c) Information signs

material. Within the constraints mentioned, there are two possible directions of simplification. One consists of removal of extraneous detail, as in the design of maps and navigational charts (the layout of maps and charts is a major human factors design problem). The other direction consists of the use of schematic representations. Conventional strip maps are examples. A more complex example is the London subway system, as shown in Figure 4–30.

In the design of pictorial displays, the designer needs first to answer two questions: (1) What information will the user need? (2) How can that information best be presented?

GRAPHIC REPRESENTATIONS

Many forms of graphic representations or graphs (bar charts, pie charts, line charts, etc.) find their way into printed form and onto TV and VDT screens. We hope that most readers have developed the wit to differentiate between those graphics that convey clearly the relevant information from those that muddy the waters or that convey misinterpretations (unintentional or intentional).

Objectives

As Tufte (1983) puts it, a graphic should convey a ''visual representation'' of data that is consistent with the numeric representation. People's visual representations are basically their perceptions, which can be misleading. For example, typically people perceive the areas of circles of increasing sizes to increase more slowly than the actual physical areas of the circles. Given such perceptual difficulties, Tufte proposes

FIGURE 4-30
Part of the London subway system given with a simplified schematic representation. The schematic form, used in the control center, is much easier for people to use. (*Photograph from London Transport.*)

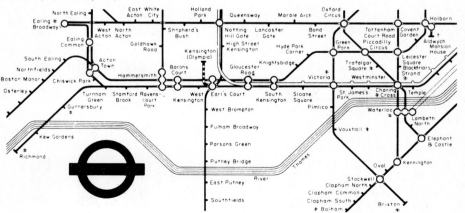

two principles of graphic displays that would give ". . . some assurance that per-
ceivers have a fair chance of getting the numbers right'':

1. The representation of numbers physically measured on the surface of the
graphic itself should be directly proportional to the quantities represented.

2. Clear, detailed, and thorough labeling should be used to defeat graphical distortion and ambiguity. Write out explanations of the data on the graphic itself. Label
important events in the data.

Graphic Features That Can Distort Perceptions

Certain features of graphics can distort the perceptions of people and thus reduce
what Tufte calls the *graphical integrity* of graphics. Examples of such features are
shown in Figure 4–31. Think of part 1 as showing curves for two groups of subjects,
A and B. The form of these curves can create the perception (actually an illusion)
that the *difference* between them increases with values of X. But this is not the case,
as shown by part 2. If one intended to focus on that difference, the format of part 2
would be better. Further, when part 1 is reviewed without reference to the values on
the Y axis, it could be perceived that both groups are lower on that variable than they
really are. This happens because the lowest Y value on the graph is 20; there is no
zero. Such a perception is corrected with part 2, which has the same curves but with
a zero base. The common perception of part 4 probably is that the value for condition
b is more than twice that for *a* and that *c* is more than 3 times *a*. Part 5 corrects
such a possible distortion by showing the three as 10, 20, and 30, since there is no

FIGURE 4-31
Examples of possible distortions in perceptions of data presented in graphics. Part 1 can
suggest that the difference between *A* and *B* increases; however, part 2 shows that this is not
the case. Part 1 also can imply that both *A* and *B* are lower than they really are; part 3 (with a *O*
base) corrects this impression. Part 4 can suggest disproportionate increases from condition *a* to
b to *c;* part 5 corrects for such an impression.

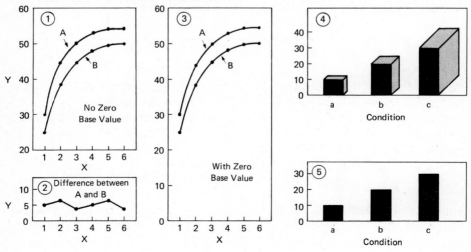

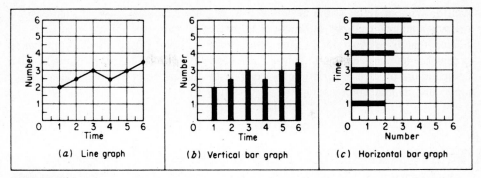

FIGURE 4-32
Illustrations of formats of multiple-trend charts that were compared on time and accuracy to
reading; some examples had 12 or 18 points instead of the 6 shown (line graphs generally were
superior). See text for discussion. (*Source: Adapted from Schutz, 1961.*)

confusion resulting from what Tufte calls *design* variations (as contrasted with *data*
variations).

Tufte's delightful and useful book (1983) includes many guidelines (with illustrations) that would enhance the correct perception of the data displayed. Although most
such guidelines are rooted in experience, there have been some modest research efforts dealing with graphic design (Macdonald-Ross, 1977).

One example of such research is reported by Schutz (1961). His study dealt with
a comparison of three formats for depicting trend data, as illustrated in Figure 4-
32*a, b,* and *c*—a line format, a vertical-bar format, and a horizontal-bar format
(Schutz, 1961). For each format, there were variations in the numbers of points
depicted (6, 12, or 18) and in the number of missing values. The subjects were
required to estimate the *trend* of the data and were scored on the time required to
make such estimates and on the accuracy of their estimates. On both criteria the line
graph proved to be preferable; presumably the line facilitated the perception of the
overall trend better than the bars (which, as such, might have added unnecessary
clutter).

Discussion

Properly designed graphics can present summarized statistical data in a fashion that
contributes to the integrity of the perception of the data. However, graphics will
never completely replace other types of presentations such as tabular material. For
some purposes tabular data may be preferable or may be used in combination with
graphics (Powers et al., 1984; Remus, 1984).

DISCUSSION

In our everyday lives we are bombarded with scads of visual displays such as newspapers, magazines, billboards, road signs, instructions, memoranda, warning signs,
maps, graphs, labels, directions, advertisements, TV, VDTs, ad infinitum. We can

well do without much of the information to which we are subjected; in such cases it matters not how well the information is presented. However, in the case of information that *is* relevant for us, the design of the displays used should conform to acceptable human factors principles. This chapter is not intended as a comprehensive compendium of such material, but we emphasize the importance of such design principles and include some examples.

REFERENCES

Broadbent, D. E. (1977). Language and ergonomics. *Ergonomics,* 8(1), 15–18.

Burg, A. (1966). Visual acuity as measured by static and dynamic tests: A comparative evaluation. *Journal of Applied Psychology,* 50(6), 460–466.

Cairney, P. T., and Siess, D. (1982). Communication effectiveness of symbolic safety signs with different user groups. *Applied Ergonomics,* 13(2), 91–97.

Chapanis, A. (1967, February). Words, words, words. *Human Factors,* 7(1), 1–17.

Christ, R. E. (1975). Review and analysis of color coding research for visual displays. *Human Factors,* 17(6), 542–570.

Collins, B. L., and Lerner, N. D. (1983). An evaluation of exit symbol visibility (NBSIR 82–2675). Washington: National Bureau of Standards.

Conover, D. W., and Kraft, C. L. (1958, October). *The use of color in coding displays* (TR 55–471). USAF, WADC.

Cornog, D. Y., and Rose, F. C. (1967, February). *Legibility of alphanumeric characters and other symbols: II. A reference handbook* (National Bureau of Standards, Miscellaneous 262–2). Washington: Government Printing Office.

Dreyfus, H. (1972). *Symbol sourcebook.* New York: McGraw-Hill.

Easterby, R. S. (1967). Perceptual organization in static displays for man/machine systems. *Ergonomics,* 10, 195–205.

Easterby, R. S. (1970). The perception of symbols for machine displays. *Ergonomics,* 13(1), 149–158.

Ellis, J. G., and Dewar, R. E. (1979). Rapid comprehension of verbal and symbolic traffic sign messages. *Human Factors,* 21 (2), 161–168.

Feallock, J. B., Southard, J. F., Kobayashi, M., and Howell, W. C. (1966). Absolute judgments of colors in the Federal Standards System. *Journal of Applied Psychology,* 50, 266–272.

Grether, W. F., and Baker, C. A. (1972). Visual presentation of information. In H. A. Van Cott and R. G. Kinkade (eds.), *Human engineering guide to equipment design* (rev. ed.). Washington: Government Printing Office.

Heglin, H. J. (1973, July). *NAVSHIPS display illumination design guide: II. Human factors* (NELC/TD223). San Diego: Naval Electronics Laboratory Center.

Helander, M. G., Billingsley, P. A., and Schurick, J. M. (1984). An evaluation of human factors research on visual display terminals in the workplace. Chapter 3 in *Human factors review.* Santa Monica, CA: Human Factors Society.

Hemingway, J. C., and Erickson, R. A. (1969). Relative effects of raster scan lines and image subtense on symbol legibility on television. *Human Factors,* 11(4), 331–338.

Hitt, W. D. (1961, July). An evaluation of five different coding methods. *Human Factors,* 3(2), 120–130.

Howell, W. C., and Kraft, C. L. (1959, September). *Size, blur, and contrast as variables affecting the legibility of alphanumeric symbols on radar-type displays,* Tech. Rept. 59–536. Wright-Patterson AFB, Ohio: Air Development Center.

Howett, G. L. (1983). *Size of letters required for visibility as a function of viewing distance and viewer acuity,* NBS Tech. Note 1180. Washington: National Bureau of Standards.

Hull, A. J. (1976). Human performance with homogeneous, patterned, and random alphanumeric displays. *Ergonomics,* 79(6), 741–750.

Jones, M. R. (1962). Color coding. *Human Factors,* 4, 355–365.

Klemmer, E. T., and Stocker, L. P. (1974). Effects of grouping of printed digits on forced-paced manual entry performance. *Journal of Applied Psychology,* 59(6), 675–678.

Kolers, P. A., Duchnicky, R. L., and Ferguson, D. (1981). *Eye movement measurement of readability of CRT displays. Human Factors,* 23(5), 517–527.

Kubakawa, C., et al. (eds.) (1969, November). *Databook for human engineering.* Prepared by Man Factors, Inc., San Diego, CA, for NASA.

Macdonald-Ross, M. (1977). How numbers are shown: A review of research on the presentation of quantitative data in texts. *Audio-Visual Communication Review,* 25, 359–409.

Moriarty, S. E., and Scheiner, E. C. (1984). A study of close-set type text. *Journal of Applied Psychology,* 69(4), 700–702.

Muller, P. F., Jr., Sidorsky, R. C., Slivinske, A. J., Alluisi, E. A., and Fitts, P. M. (1955, October). *The symbolic coding of information on cathode ray tubes and similar displays* (TR 55–375). USAF, WADC.

Munsell book of color (1959). Baltimore: Munsell Color Co.

Pastoor, S., Schwarz, E., and Beldie, I. P. (1983). The relative suitability of four dot-matrix sizes for text presentation on color television screens. *Human Factors,* 25(3), 265–272.

Petersen, H. E., and Dugas, D. J. (1972). The relative importance of contrast and motion in visual detection. *Human Factors,* 14(3), 207–216.

Plath, D. W. (1970). The readability of segmented and conventional numerals. *Human Factors,* 12(5), 493–497.

Powers, M., Lashley, C., Sanchez, P., and Schneiderman, B. (1984). An experimental comparison of tabular and graphic data. *International Journal of Man-Machine Studies,* 20, 545–566.

Remus, W. (1984). An empirical investigation of the impact of graphical and tabular data presentation on decision making. *Management Science,* 30(5), 533–541.

Robertson, H. D. (1977, June). Pedestrian preferences for symbolic signal displays. *Traffic Engineering,* 47(6), 38–42.

Schutz, H. G. (1961). An evaluation of formats for graphic trend displays—Experiment II. *Human Factors,* 3, 99–107.

Shurtleff, D. A. (1967). Studies in television legibility: A review of the literature. *Information Display,* 4, 40–45.

Smith, S. L., and Thomas, D. W. (1964). Color versus shape coding in information displays. *Journal of Applied Psychology,* 48, 137–146.

Snyder, H. L., and Taylor, G. B. (1979). The sensitivity of response measures of alphanumeric legibility to variations in dot matrix display parameters. *Human Factors,* 21(4), 457–471.

Tufte, E. R. (1983). *The visual display of quantitative information.* Cheshire, CT: Graphics Press.

Tullis, T. S. (1981). An evaluation of alphanumeric, graphic, and color information displays. *Human Factors,* 23(5), 541–550.

Tullis, T. S. (1983). The formatting of alphanumeric displays: A review and analysis. *Human Factors,* 25(6), 657–682.

Vartabedian, A. G. (1971). The effects of letter size, case, and generation method on CRT display search time. *Human Factors,* 13(4), 363–368.

Vartabedian, A. G. (1973). Developing a graphic set for developing cathode-ray-tube display using a 7 × 9 dot matrix. *Applied Ergonomics,* 4(1), 11–16.

Woodson, W. E. (1963, Oct. 21). *Human engineering design standards for spacecraft controls and displays.* General Dynamics Aeronautics Rept. GDS-63-0894-1, for NASA.

Zwaga, H.J., and Boersema, T. (1983). Evaluation of a set of graphic symbols. *Applied Ergonomics,* 14(1), 43–54.

CHAPTER 5

VISUAL DISPLAYS OF DYNAMIC INFORMATION

We live in a world in which things keep changing—or at least are subject to change. A few examples are natural phenomena (such as temperature and humidity), the speed of vehicles, the altitude of aircraft, our blood pressure and heart rate, traffic lights, the frequency and intensity of sounds, and the national debt. This chapter deals with some aspects of the design of displays used to present information about the parameters in our world that are subject to change, that is, that are dynamic. Such displays cover quite a spectrum, including those that present quantitative, qualitative, status, and representational information.

QUANTITATIVE VISUAL DISPLAYS

The objective of quantitative displays is to provide information about the quantitative value of some variable. In most cases, the variable changes or is subject to change (such as speed or temperature). But (with a bit of stretch) we also embrace the measurement of some variables that, in a strict sense, are more static (such as the length and weight of objects).

In the use of quantitative displays there is an explicit or implicit level of precision that is required or desired; that level of precision characterizes the concept of the *scale unit*. A medical thermometer, for example, usually is read to the nearest tenth of a degree, our conventional indoor and outdoor thermometer to the nearest whole degree, and some high-temperature industrial thermometer only to the nearest 10° or 100°.

Basic Design of Quantitative Displays

Conventional quantitative displays are mechanical devices of one of the following types:

1 Fixed scale with moving pointer
2 Moving scale with fixed pointer
3 Digital display

The first two are analog indicators in that the position of the pointer is analogous to the value it represents. Examples of these are shown in Figure 5–1.

Although conventional quantitative displays have moving mechanical parts (the pointer, the scale itself, or the numerals of a digital counter), modern technology makes it possible to present electronically generated features, thus eliminating the need for moving mechanical components. Examples of a few such designs are given in Figure 5–1*k, l,* and *m.*

Comparison of Different Designs   Over the years there have been a number of studies in which certain designs of conventional quantitative scales have been compared. Although the results of such studies are somewhat at odds with one another, certain general implications hold. For example, it is clear that for many uses digital displays (also called *counters*) are generally superior to analog displays (such as round, horizontal, and vertical scales) when the following conditions apply: (1) a precise numeric value is required, and (2) the values presented remain visible long enough to be read (are not continually changing). In one study, for example, Simmonds, Galer, and Baines (1981) compared digital speedometers with three dial displays and an experimental curvilinear display. (All the designs used were electronically generated; two of the dials and the curvilinear display used bars rather than pointers to represent numeric values.) The experimenters found the digital display consistently better in terms of both accuracy of reading and preference.

Although digital displays have a definite advantage for obtaining specific numeric values that tend to remain fixed long enough to be read, analog displays have advantages in other circumstances. Fixed-scale moving-pointer displays, for example, are particularly useful when the values are subject to frequent or continual change that would preclude the use of a digital display (because of limited time for reading any given value). In addition such analog displays have a positive advantage when it is important to observe the direction or rate of change of the values presented. By and large, analog displays with fixed scales and moving pointers are superior to those with moving scales and fixed pointers. In this regard Heglin (1973) offers the following list of factors to consider in the selection of analog displays (Heglin, p. II–30):

1 In general, a pointer moving against a fixed scale is preferred.

2 If numerical increase is typically related to some other natural interpretation, such as *more or less* or *up or down,* it is easier to interpret a straight-line or thermometer scale with a moving pointer (such as Figure 5–1*d* and *e*) because of the added cue of pointer position relative to the zero, or null, condition.

3 Normally, do not mix types of pointer-scale (moving-element) indicators when they are used for related functions—to avoid reversal errors in reading.

4 If a manual control over the moving element is expected, there is less ambiguity between the direction of motion of the control and the display if the control moves the pointer rather than the scale.

5 If slight, variable movements or changes in quantity are important to the observer, these will be more apparent if a moving pointer is used.

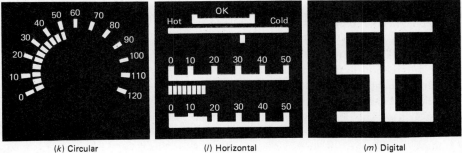

FIXED SCALE, MOVING POINTER

(*a*) Circular scales

(*b*) Circular scale with positive and negative values

(*c*) Semicircular or curved scale

(*d*) Vertical scale

(*e*) Horizontal scale

MOVING SCALE, FIXED POINTER

(*f*) Circular scale

(*g*) Open-window scales

(*h*) Vertical scale

(*i*) Horizontal scale

DIGITAL DISPLAY

(*j*) Digital display 2 7 9 4 3

ELECTRONIC DISPLAYS

(*k*) Circular

(*l*) Horizontal

(*m*) Digital

FIGURE 5-1
Examples of certain types of displays used in presenting quantitative information. The first three groups represent mechanical scales with moving pointers, scales, or counters (the digital display). The fourth group represents scales with electronically generated symbols of characters.

6 If you wish to have a numerical value readily available, however, a moving scale appearing in an open window can be read more quickly. (Such a scale is essentially a digital, or counter, display, as discussed.)

Although fixed scales with moving pointers are generally preferred to moving scales with fixed pointers, the former do have their limitations, especially when the range of values is too great to be shown on the face of a relatively small scale. In such a case, certain moving-scale fixed-pointer designs (such as rectangular open-window and horizontal and vertical scales) have the practical advantage of occupying a small panel space, since the scale can be wound around spools behind the panel face, with only the relevant portion of the scale exposed.

Further, research and experience generally tend to favor circular and semicircular scales (Figure 5–1*a, b,* and *c*) over vertical and horizontal scales (Figure 5–1*d* and *e*). However, in some circumstances vertical and horizontal scales would have advantages, as discussed in item 2 above.

Although many quantitative displays are used to obtain a specific quantitative value, in many circumstances a specific value is not really needed and some approximate value serves the purpose, such as the temperature of an automobile engine. In such a case, the underlying continuum (such as temperature) can be sliced into, say, three ranges (cold, normal, and hot). The use of a display for such a purpose represents a more qualitative than quantitative reading task. Qualitative scales are discussed later in this chapter.

Specific Features of Conventional Quantitative Displays

The ability of people to make visual discriminations (such as those required in the use of quantitative scales) is influenced in part by the specific features to be discriminated. Some of the relevant features of quantitative scales (aside from the basic types) are the numeric progressions used, length of the scale units, "interpolation" required, use of scale markers, width (thickness) of markers, design of pointers, and location of scale numbers. On the basis of a modest experiment that involved variations in these and other features, Whitehurst (1982) found the numeric progressions used and the length of scale units to be particularly important in terms of speed and accuracy of reading quantitative scales. A few features of such scales are discussed below. Although most of the research relating to such features has been carried out with conventional quantitative displays, it is probable that the implications would still be applicable to electronically generated displays.

Numeric Progressions of Scales   Every quantitative scale has some intrinsic numeric progression system that is characterized by the numeric difference between adjacent graduation markers on the scale and by the numbering of the major scale markers. In general, the garden variety of progression by 1s (0, 1, 2, 3, etc.) is the easiest to use. This lends itself readily to a scale with major markers at 0, 10, 20, etc., with intermediate markers at 5, 15, 25, etc., and with minor markers at individual numbers. Progression by 5s is also satisfactory and by 25 is moderately so. Some examples of scales with progressions by 1s and 5s are shown in Figure 5–2. Where large numeric values are used in the scale, the relative readabilities of the scales are

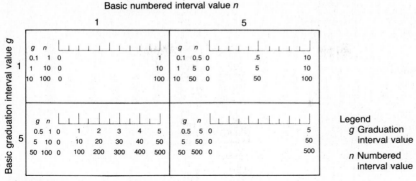

FIGURE 5-2
Examples of certain generally acceptable quantitative scales with different numeric progression systems (1s/5s). The values to the left in each case are, respectively, the graduation scale interval *g* (the difference between the minor markers) and the numbered scale interval *n* (the difference between numbered markers). For each scale there are variations of the basic values of the system, these being decimal multiples or multiples of 1 or 5.

the same if they are all multiplied by 10, 100, 1000, etc. Decimals, however, make scales more difficult to use, although for scales with decimals the same relative advantages and disadvantages hold for the various numeric progressions. The zero in front of the decimal point should be omitted when such scales are used.

Unusual progression systems (such as by 3s, 8s, etc.) should be avoided except under very special circumstances.

Length of Scale Unit   The length of the scale unit is the length on the scale that represents the numeric value that is the smallest unit to which the scale is to be read. For example, if a pressure gauge is to be read to the nearest 10 lb (4.5 kg) then 10 lb (4.5 kg) is the smallest unit of measurement; the scale is so constructed that a given length (in inches, millimeters, etc.) represents 10 lb (4.5 kg) of pressure. (Whether there is a marker for each such unit is another matter.)

The length of the scale unit should be such that the distinctions between the values can be made with optimum reliability in terms of human sensory and perceptual skills. Although certain investigators have reported acceptable accuracy in the reading of scales with scale units as low as 0.02 in (0.5 mm); most sets of recommendations provide for values ranging from about 0.05 to 0.07 in (1.3 to 1.8 mm), as shown in Figure 5-3. The larger values probably would be warranted when the use of instruments is under less than ideal conditions, such as when they are used by persons who have below-normal vision or when they are used under poor illumination or under pressure of time.

Design of Scale Markers   It is good practice to include a scale marker for each scale unit to be read. Figure 5-3 shows the generally accepted design features of scale markers for normal viewing conditions and for low-illumination conditions. This illustration is for the conventional progression scheme with major markers rep-

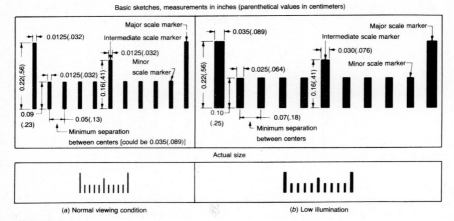

FIGURE 5-3
Recommended format of quantitative scales, given length of scale unit and gradation markers. Format *a* is proposed for normal illumination conditions under normal viewing conditions and *b* for low illumination. (*Source: Adapted from Grether and Baker, 1972, p. 88.*)

resenting intervals of 1, 10, 100, etc., and minor markers at the scale units (the smallest values to be read), such as 0.1, 1, 10.

Scale Markers and Interpolation The concept of interpolation (relative to quantitative scales) usually applies to the estimation of specific values between markers when not all scale units have markers. However, usually it is desirable to have a scale marker for each unit to be read. In such instances, pointers between markers are rounded to the nearest marker. (Although such "rounding" has also been called interpolation, we do not use the term in this sense.) If scales were to be much more compressed than those in Figure 5–3, the scale markers would be crowded together, and this could affect the reading accuracy (especially under time constraints and low illumination). In such circumstances, it is better to use a scale that does require interpolation. Actually, people are moderately accurate in interpolation; Cohen and Follert (1970) report that interpolation of fifths, and even of tenths, may yield satisfactory accuracy in many situations. Even so, where high accuracy is required (as with certain test instruments and fine measuring devices), a marker should be placed at every scale unit, even though this requires a larger scale or a closer viewing distance.

Scales are read most quickly when the pointers are precisely on the markers (Whitehurst, 1982). In other instances a person needs to round to the nearest scale unit, but such rounding must, of course, be tolerated.

Design of Pointers The few studies that have dealt with pointer design leave some unanswered questions, but some of the common recommendations follow: Use pointed pointers (with a tip angle of about 20°); have the tip of the pointer meet, but not overlap, the smallest scale markers; have the color of the pointer extend from the tip to the center of the scale (in the case of circular scales); and have the pointer close to the surface of the scale (to avoid parallax).

Combining Scale Features Several of the features of quantitative scales discussed above have been integrated into relatively standard formats for designing scales and their markers, as shown in Figure 5–3. Although these formats are shown in a horizontal scale, the features can, of course, be incorporated in circular or semicircular scales. Also, the design features shown in this figure should be considered as general guidelines rather than as rigid requirements, and the advantages of certain features may, in practical situations, have to be traded off for other advantages.

One does not have to look very hard to find examples of poor instrument design. The scale in Figure 5–4 (assuming that the amperes should be read to the nearest 10) suffers from shortness of scale units and inadequate intermediate markers. The original meter design in Figure 5–5a suffers from various ailments, which are corrected in the redesigned version (Figure 5–5b).

Scale Size and Viewing Distance The above discussion of the detailed features of scales is predicated on a normal viewing distance of 28 in (71 cm). If a display is to be viewed at a greater distance, the features have to be enlarged in order to maintain, at the eye, the same visual angle of the detailed features. To maintain that same

FIGURE 5-4
Example of a poorly designed ampere scale.

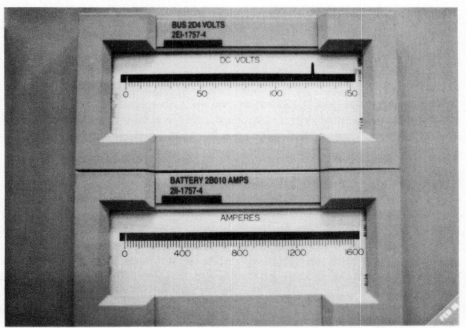

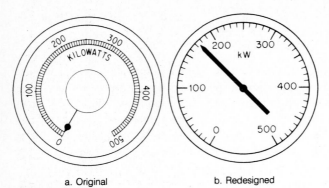

a. Original   b. Redesigned

FIGURE 5-5
Illustration of two designs of a meter. The one at the right
would be easier to read because it is bolder and less cluttered
than the one at the left. It has fewer gradation markers, and the
double arc line has been eliminated. The scale length is
increased by placing the markers closer to the perimeter;
although this requires that the numerals be placed inside the
scale, the clear design and the fact that the numerals are
upright probably would partially offset this disadvantage.
(*Source: Adapted from* Applied ergonomics handbook, *1974,
fig. 3-1.*)

visual angle, the following formula can be applied for any other viewing distance in
inches (*x*):

$$\text{Dimension at } x \text{ in} = \text{dimension at 28 in} \times \frac{x \text{ in}}{28}$$

Specific Features of Electronic Quantitative Displays

The previous section dealt largely with conventional, mechanical quantitative displays. Although much of that discussion presumably applies to corresponding features of electronic displays, certain features of electronic displays create special design problems.

One such problem, discussed by Green (1984), deals with the bars of bar-type
displays (such as Figure 5–1*l*). In his study, various combinations of several automobile displays were used, including variations of bar-type designs for several purposes. He concluded that, for bar-type displays, only one of the segments of the bar
should be shown. (An example of this feature is the top display of those in Figure
5–1*l*.) The extension of the bar to the zero position (as shown in the middle example)
was confusing to some people. However a bar-type design (whether with only a
single segment or with an extended bar) would be particularly inappropriate for a
variable such as a fuel gauge since there would be *no* indication for an empty tank.
The complete bar-type display, however, probably would be more appropriate for a
variable such as speed, since it would grow longer or shorter with changing values.

Design of Altimeters

Aircraft altimeters represent a type of quantitative display of special interest. In fact, the initial interest in quantitative displays was triggered because of numerous instances in which aircraft accidents had been attributed to misreading of an earlier altimeter model. The model consisted of a dial with three pointers representing, respectively, 100, 1000, and 10,000 ft (30.5, 305, and 3050 m), like the second, minute, and hour hands of a clock. That model required the reader to combine the three pieces of information, and this combining was the basis for the errors made in reading the altimeter.

In connection with the design of altimeters, Roscoe (1968) reports a study that dealt directly with certain features of altimeters but that has implications for quantitative displays in certain other circumstances. In particular, the study compared three design variables: (1) vertical versus circular scales, (2) integrated presentations of three altitude values (present altitude, predicted altitude 1 minute away, and command altitude), and (3) circular (analog) versus counter (digital) presentation. The four designs that embody the combinations of these three variables are shown in Figure 5–6.

Without going into details, the results of the study (given in Figure 5–6) show that design *a* (the integrated vertical scale) was clearly the best in terms of time and errors. The explanation for this given by Rocoe is primarily its pictorial realism in representing relative positions in vertical space by a display in which *up* means *up* and *down* means *down*. Design *b*, which represents vertical space in a distorted manner (around a circle), did not fare as well as *a* but was generally superior to *c* and *d*, both of which consisted of *separate* displays of the three altitude values rather than an *integrated* display. The results can be interpreted as supporting the advantage of compatible designs (as illustrated by the vertical display representing altitude) and the advantage of integrated displays (where they are appropriate) as contrasted with separate indications for various related values. [Note that, although we have referred to the advantage of digital displays or counters, in this study they were not as good as a vertical (analog) display.]

QUALITATIVE VISUAL DISPLAYS

In using displays for obtaining qualitative information, the user is primarily interested in the approximate value of some continuously changeable variable (such as temperature, pressure, or speed) or in its trend, or rate of change. The basic underlying date used for such purposes usually are quantitative.

Quantitative Basis for Qualitative Reading

Quantitative data may be used as the basis for qualitative reading in at least three ways: (1) for determining the status or condition of the variable in terms of each of a limited number of predetermined ranges (such as determining if the temperature gauge of an automobile is cold, normal, or hot); (2) for maintaining approximately some desirable range of values [such as maintaining a driving speed between 50 and 55 mi/h (80 and 88 km/h)]; and (3) for observing trends, rates of change, etc. (such

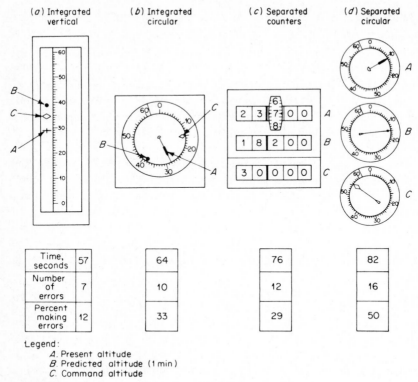

Legend:
 A. Present altitude
 B. Predicted altitude (1 min)
 C. Command altitude

FIGURE 5-6
Four display designs for presenting (A) present altitude, (B) predicted altitude (in 1 min), and (C) command altitude, and three criteria (mean time for 10 trials, number of errors, and percent of 24 subjects making errors). The displays are shown in overly simplified form. (*Source: Adapted from Roscoe, 1968*).

FIGURE 5-7
Illustration of color coding of sections of instruments that are to be read qualitatively.

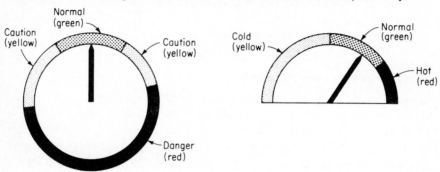

as noting the rate of change in altitude of an airplane). In the qualitative use of quantitative data, however, evidence suggests that a display that is best for a quantitative reading is not necessarily best for a qualitative reading task.

Some evidence for support of this contention comes from a study in which open-window, circular, and vertical designs are compared (Elkin, 1959). In one phase of this study, subjects made qualitative readings of high, OK, or low for three ranges of numerical values, using a quantitative scale, and employed the same three scales to make strictly quantitative readings. The average times taken (shown in Table 5–1) show that, although the open-window (digital) design took the shortest time for quantitative reading, it took the longest time for qualitative reading. In turn, the vertical scale was best for qualitative reading. Thus different types of scales vary in effectiveness for qualitative versus quantitative reading.

Design of Qualitative Scales

As indicated above, many qualitative scales represent a continuum of values that are sliced into a limited number of ranges (such as cold, normal, and hot); in other instances, specific ranges have particular importance to the user (such as representing a danger zone). In such cases the perception of the correct reading is aided by some method of coding the separate ranges. One way to do this is to use color codes, as illustrated in Figure 5–7.

Another method is to use some form of *shape coding* to represent specific ranges of values. It is sometimes possible, in the design of such coded areas, to take advantage of any natural "compatible" associations people may have between coding features and the intended meanings. One study taking such an approach is reported by Sabeh, Jorve, and Vanderplas (1958) in which they were interested in identifying the shapes with the strongest associations with the intended meanings of certain military aircraft instrument readings. Having solicited a large number of designs initially, the investigators selected the seven shown in Figure 5–8. These were presented to 140 subjects, along with the seven following "meanings": caution, undesirable, mixture lean, mixture rich, danger—upper limit, danger—lower limit, and dangerous vibration. Figure 5–8 shows the number of subjects out of 140 who selected the indicated meaning to a statistically significant level.

TABLE 5-1
TIMES FOR QUALITATIVE AND
QUANTITATIVE READINGS WITH THREE
TYPES OF SCALES

Type of scale	Average reading time, s	
	Qualitative	Quantitative
Open-window	115	102
Circular	107	113
Vertical	101	118

Source: Elkin, 1959.

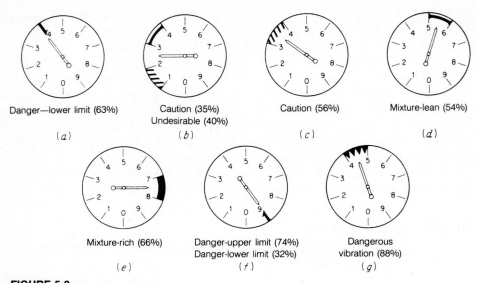

FIGURE 5-8
Association of coded zone markings with the intended meanings of specific scale values of military aircraft instruments. The numbers in parentheses are the percentages of individuals (out of 140) who reported significant associations with the meanings listed. (*Adapted from Sabeh, Jorve, and Vanderplas, 1958.*)

The argument for the use of precoded displays for qualitative reading (when this is feasible) is rooted in the nature of the human perceptual and cognitive processes. To use a strictly quantitative display to determine if a given value is within one specific range or another involves an additional cognitive process of allocating the value that is read to one of the possible ranges of values that represent the categories that have operational meaning. The initial perception of a precoded display immediately conveys the meaning of the display indicator.

Note that quantitative displays with coded zones also can be used to reflect trends, directions, and rates of change. Further, they can be used for quantitative reading if the scale values are included. If in the use of quantitative date for qualitative reading it is not appropriate to precode certain zones, some conventional form of quantitative display should be used. The particular choice, however, depends on the relative importance of the qualitative versus quantitative readings.

In the use of electronically generated displays, segments of a total scale can be identified in various ways, as illustrated in Figure 5–1*l* as the top display for automobile temperature.

Check Reading

The term *check reading* refers to the use of an instrument to ascertain whether the reading is normal. Usually this is done with a quantitative scale, but the normal condition is represented by a specific value or narrow range of values. In effect, check reading is a special case of qualitative reading. As such, the normal reading value should be clearly coded, as discussed above.

If several or many instruments for check reading are used together in panels, their configuration should be such that any deviant reading stands out from the others. Research relating to various possible configurations has indicated that (with round instruments) the normal position preferably should be aligned at the 9 o'clock (or possibly 12 o'clock) positions. A 9 o'clock alignment is illustrated in Figure 5–9. The advantage of such a systematic alignment is based on human perceptual processes, in particular what is referred to as the *gestalt,* that is, the human tendency to perceive complex configurations as complete entities, with the result that any feature that is "at odds" with the configuration is immediately apparent. Thus, a dial that deviates from a systematic pattern stands out from the others. In the case of panels of such dials, the addition of extended lines *between* the dials can add to the gestalt, thus helping to make any deviant dial stand out more clearly (Dashevsky, 1964). This is illustrated in the top row of dials of Figure 5–9.

STATUS INDICATORS

In a sense, some *qualitative* information approximates an indication of the *status* of a system or a component, such as the use of some displays for check reading to determine if a condition is normal or abnormal or the qualitative reading of an automobile thermometer to determine if the condition is hot, normal, or cold. However, more strictly status indications reflect separate, discrete conditions such as on and off or (in the case of traffic lights) stop, caution, and go. If a qualitative instrument is to be used *strictly* for check reading or for identifying a particular status (and *not* for some other purpose such as observing trends), the instrument could be converted to a status indicator.

The most straightforward, and commonly used, status indicators are lights, such as traffic lights. In some cases redundant codes are used, as with most traffic lights that are coded both by color (red, yellow, and green) and by location (top, middle,

FIGURE 5-9
A panel of dials used for check reading. When all the "normal" readings are aligned at the 9 o'clock (or 12 o'clock) position, any deviant reading can be perceived at a glance. In some instances an extended line is shown between the dials, as illustrated in the top row; this can aid in making the deviant dial more distinct.

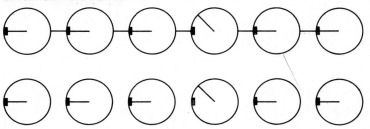

and bottom). Although many status indicators are lights, other coding systems can be used, such as the controls of some stoves that are marked to identify the off position.

SIGNAL AND WARNING LIGHTS

Flashing or steady-state lights are used for various purposes, such as indications of warning (as on highways); identification of aircraft at night; navigation aids and beacons; and to attract attention, such as to certain locations on an instrument panel. There apparently has been little research relating to such signals, but we can infer some general principles from our knowledge of human sensory and perceptual processes.

Detectability of Signal and Warning Lights

Of course, various factors influence the detectability of lights. Certain such factors are discussed below.

Size, Luminance, and Exposure Time   The absolute threshold for the detection of a flash of light depends in part on a combination of size, luminance, and exposure time. Drawing in part on some previous research, Teichner and Krebs (1972) depicted the minimum sizes of lights (in terms of visual angle of the diameter in minutes of arc) that can be detected 50 percent of the time under various combinations of exposure times (in seconds) and luminance (in millilamberts), these relationships being shown in somewhat simplified form in Figure 5–10. From this it is clear that the luminance threshold decreases linearly as a function of exposure time up to a particular value, but the exposure time at which the luminance threshold levels off decreases systematically with target size (actually the *area* of the light, as implied

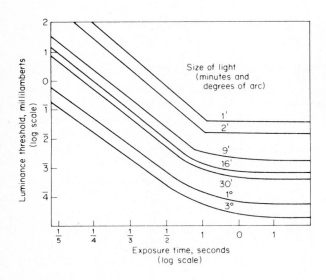

FIGURE 5-10
Minimum sizes of lights (in minutes and degrees of arc) that can be detected 50 percent of the time under varying combinations of exposure time and luminance. (*Source: Adapted from Tichner and Krebs, 1972.*)

by its diameter). Thus, the very minimal, lower-bound detectability of lights (50 percent accuracy) is seen to be a function of size, luminance, and exposure time. (Operational values should well exceed those in Figure 5–10.)

Color of Lights   Another factor related to the effectiveness of signal lights is color. Using response time as an indication of the effectiveness of four different colors, Reynolds, White, and Hilgendorf (1972) report the following order (from fastest to slowest): red, green, yellow, and white. However, the background color and ambient illumination can interact to influence the ability of people to detect and respond to lights of different colors. In general, the researchers found that if a signal has good brightness contrast against a dark background, and if the absolute level of brightness of the signal is high, the color of the signal is of minimal importance in attracting attention. But with low signal-to-background brightness contrast, a red signal has a marked advantage, followed by green, yellow, and white in that order.

Flash Rate of Lights   In the case of flashing lights, the flash rate should be *well below* that at which a flashing light appears as a steady light (the flicker-fusion frequency), which is approximately 30 times per second. In this regard, rates of about 3 to 10 per second (with duration of at least 0.05 *s*) have been recommended for attracting attention (Woodson and Conover, 1964, pp. 2–26). Markowitz (1971) makes the point that the range of 60 to 120 flashes per minute (1 to 2 per second), as used on highways and in flyways, appears to be compatible with human discrimination capabilities and available hardware constraints.

In discussing the range of rates for use as an automobile deceleration signal, Mortimer and Kupec (1983) suggest a range from 1 to 9 per second. A rate below 1 per second could involve too long a wait for a driver of a following car to apply the brake, and the flashing character of tungsten-filament lamps is generally degraded above rates above 9 per second. For most practical purposes, flash rates from about 3 to 9 per second probably are preferable, with 4 per second being suggested as an optimum (Heglin, 1973).

In most circumstances, a single flashing light is used. However, in some situations lights with different flash rates might be used to signify different things. A very obvious possibility would be with automobile tail lights, where different flash rates could signify different deceleration rates. In this regard clearly the perceptual skills of people are such that no more then three different rates should be used (Mortimer and Kupec, 1983; Tolin, 1984). With more than three different rates, people could not differentiate clearly one rate from the others. [This is based on the concept of the *just noticeable difference* (JND), or the difference in the magnitude of a given type of stimulus that can be just noticeable to people.]

Context of Signal Lights.   Various situational and environmental variables can influence people's awareness of signal lights. A couple of examples will illustrate the effects of such variables. The first example deals with the perception of various types of roof-mounted, rotating-beam emergency lights, such as those used on ambulances and police cars (Berkhout, 1979). Subjects viewed the stimulus vehicle at night on an unused gravel road. The vehicle either was not moving or was moving toward or away from the subjects (who were in a parked car). The subjects viewed the stimulus vehicle for 7 s and were required to indicate if the vehicle was moving and, if so, to indicate the direction and whether it was moving fast or slow. The results were somewhat complex and are not discussed in detail here, but we do mention a couple of results.

Table 5–2 presents some of the results for six of the light systems viewed while the vehicle was standing still. First, in the majority of trials many of the subjects misperceived the situation and thought the stationary vehicle was moving either toward or away from them. Second, there was a tendency for the blue lights to be seen as moving toward the observer more often than the red lights, while the red ones were seen more often as moving away. Berkhout points out that the red Twin Beacon (with a side-to-side flash pattern) is especially prone to produce a dangerous illusion of receding motion when actually the vehicle is standing still. An emergency vehicle parked on the shoulder of a road and displaying this light could, therefore, run an increased risk of a rear-end accident. (However, this same red Twin Beacon performed relatively well for indicating direction of motion when the vehicle was actually moving.) Given the responses of observers from other aspects of the study, it should be added that no single light system was consistently superior to the others. Rather, the results indicated the disturbing fact that a light that was most suitable for one particular dynamic situation did not necessarily perform well in another.

The second example deals with the problems of discriminating signal lights with other lights in the background. (Traffic lights in areas with neon and Christmas lights represent serious examples of such a problem.) In an interesting investigation Crawford (1963) used both steady and flashing signal lights against a background of (irrelevant) steady lights, flashing lights, and admixtures of both, using a criterion of time to identify the signal lights. His conclusions are summarized as follows:

Signal light	Background lights	Comment
Flashing	Steady	Best
Steady	Steady	OK
Steady	Flashing	OK
Flashing	Flashing	Worst

The presence of even one flashing background light (out of many) seriously affected the identification of a flashing signal light.

Recommendations Regarding Signal and Warning Lights

Here are some recommendations about signal and warning lights (based largely on Heglin, 1973):

- *When should they be used?* To warn of an actual or a potential dangerous condition.
- *How many warning lights?* Ordinarily only one. (If several warning lights are required, use a master warning or caution light and a word panel to indicate specific danger condition.)

TABLE 5-2
SUBJECTS' RESPONSES TO VARIOUS EMERGENCY LIGHT SYSTEMS WHEN VEHICLE
WAS STATIONARY

Light system		Subjects' response of movement, %		
		Toward	Still*	Away
Single dome	Red	17	47	36
	Blue	26	44	30
Twin sonic	Red	16	46	38
	Blue	31	42	27
Twin Beacon	Red	9	36	55
	Blue	20	36	44
Blue lights (all three)		26	40	34
Red lights (all three)		14	43	43

*Still is the correct response.
Source: Adapted from Berkhout, 1979, table 4. Copyright by the Human Factors Society, Inc. and
reproduced by permission.

• *Steady state or flashing?* If the light is to represent a continuous, ongoing condition, use a steady state light unless the conditon is especially hazardous; continuous flashing lights can be distracting. To represent occasional emergencies or new conditions, use a flashing light.

• *Flash rate:* If flashing lights are used, flash rates should be from about 3 to 10 per second (4 is best) with equal intervals of light and dark. If different flash rates are to represent different levels of some variable, use no more than three different rates.

• *Warning-light intensity:* The light should be at least twice as bright as the immediate background.

• *Location:* The warning light should be within 30° of the operator's normal line of sight.

• *Color:* Warning lights are normally red because red means danger to most people. (Other signal lights in the area should be other colors.)

REPRESENTATIONAL DISPLAYS

Most representational displays that depict changeable conditions consist of a background with elements superimposed that tend to change positions or configurations. Many such displays represent the position and movement of vehicles such as aircraft or ships. We use aircraft-position displays to illustrate the nature of such displays and of certain human factors aspects.

Aircraft-Position Displays

The problem of representing the position and movement of aircraft has haunted designers—and pilots—for years. One major facet of this problem relates to the basic movement relationships to be depicted by the display, there being two such relation-ships, as shown in Figure 5–11 and described as follows:

- *Moving aircraft:* The earth (specifically, the horizon) is fixed, with the aircraft moving in relation to it (moving-aircraft or outside-in display).
- *Moving horizon:* The aircraft is fixed, with the horizon moving in relation to it (moving-horizon or inside-out display). Most aircraft displays are of this type.

This is basically a problem of visual perception, specifically with respect to what is referred to as the *figure and ground phenomenon.* Psychologically, the part of a dynamic field of view that is perceived as stationary is called the *background,* or *ground,* and any object that is moving in relation to the ground is called the *figure.* Since we have grown up in a world in which the earth and buildings are fixed, the moving-aircraft display (with a fixed horizon) usually is the most compatible. How-ever, many pilots have become accustomed to the moving-horizon type of display (with the aircraft fixed).

Aircraft displays also can be characterized as to whether they are pursuit or com-pensatory displays. This distinction actually applies to any tracking task in which there are two moving symbols, sometimes called the *target* and the *follower.* In aircraft, the target can be the heading (the intended direction) or the command alti-tude (the intended altitude). The aircraft is the follower. With a pursuit display the positions of both symbols are shown relative to a fixed background (such as direction or altitude relative to the earth). The difference between the two is error. With a compensatory display, one of the symbols is fixed and the other moves relative to it, with the difference being the error. Research generally has shown that pursuit dis-plays are superior, although there are certain exceptions. (See Chapter 9 for more discussion of tracking displays.)

FIGURE 5-11
The two basic movement relationships for depicting aircraft attitude, namely, the moving-aircraft (outside-in) and the moving-horizon (inside-out) relationship.

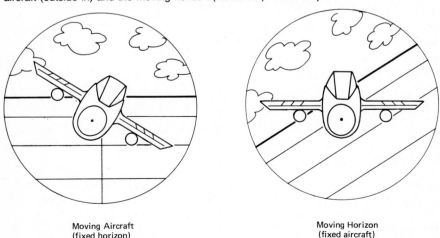

Moving Aircraft (fixed horizon)	Moving Horizon (fixed aircraft)

Principles of Aircraft-Position Displays   Although the evidence regarding aircraft-position displays is still not entirely definitive, Roscoe, Corl, and Jensen (1981) have crystallized a few principles that seem to have general validity. Four of these principles are discussed here. (Certain other principles are not discussed because they deal with some of the complex aspects of the control dynamics of aircraft.)

1 *Principle of pictorial realism*. This principle is an assertion that a display should present a spatial analog, or image, of the real world, in which the position of the object (the aircraft) is seen in depth as well as in up-down and left-right frames of reference.

2 *Principle of integration*. Following this principle, *related* information should be presented in integrated displays, as discussed before and illustrated in Figure 5–6a.

3 *Principle of compatible motion*. In the use of aircraft displays, this principle is in conformity with the use of moving-aircraft (outside-in) displays with fixed horizons.

4 *Principle of pursuit presentation*. This refers to the preferable use of pursuit displays as contrasted with compensation displays.

Discussion of Representational Displays

Although the previous discussion dealt with aircraft-position displays, the same basic problems apply to most displays in which other moving objects (such as vehicles) change positions relative to a background. The same human factors principles usually would apply to other circumstances, albeit with situational variations (the displays in aircraft control towers, for example, have their own special features and problems).

Aside from displays of vehicles and their backgrounds, there are other displays that represent changing conditions and relationships, such as CRT displays used in medical and laboratory settings and in automobile repair shops. So far relatively little research has been done on the use of such displays in these settings.

CONSPICUITY OF PEOPLE AND VEHICLES

There are many circumstances in which it is important to identify the presence or movement of people and vehicles, such as aircraft, ships, boats, motorcycles, road and other construction workers, school crossing guards, miners, pedestrians, etc. The most common method for increasing such conspicuity includes the use of lights, paint (as on vehicles), and special clothing (for personnel). The use of bright orange apparel for school guards and flaggers on road projects is an example. For purposes of illustration a couple of examples are discussed.

Conspicuity of Motorcycle and Motorcyclist

In an interesting study of the conspicuity of motorcycles and motorcyclists in actual traffic situations, Olson, Halstead-Nussloch, and Sivak (1981) experimented with several methods of enhancing conspicuity, especially at intersections. Some of the methods used were as follows:

• *For the vehicles:* Equipping vehicles with a fairing (a surface to increase the frontal area ahead of the handlebars), this being covered with orange or green fluorescent fabric or (at night) retroreflective fabric (that shows up at night); low-beam headlamp; modulating headlamp (with changing brightness); reduced-brightness headlamp; and running lights (turn-signal lights on all the time but not flashing).

• *For the rider:* Orange or green fluorescent apparel (complete outfit, or vest, or helmet).

The study was carried out in actual traffic, with the motorcycle rider (one of the experimenters) manuevering in traffic. Other experimenters (in a following car) measured the gap in traffic that unsuspecting car drivers left in front of the motorcycle before moving in ahead of it. Virtually all the methods were better than a control condition in preventing the unsuspecting drivers from squeezing in ahead of the motorcycle. In general, however, the investigators proposed the following methods for improving conspicuity:

• *For daytime:* Drive with headlights on at all times (modulating headlights are a bit better, but more expensive), and use high-visibility materials (they are better when they are worn by the rider than when they are fitted to the motorcycle).

• *For nighttime:* Wear retroreflective garments and use running lights. (Combinations of fluorescent and retroreflective fabrics are available which can provide both daytime and nighttime conspicuity. It is also possible to treat ordinary fabrics with retroreflective beads to make them retroreflective without changing their appearance.)

Conspicuity of Coal Miners

Coal miners are subject to a variety of hazards, including being struck by mobile equipment. Thus, methods for improving the conspicuity of miners could reduce accident rates. Retroreflective materials would seem to be useful in this situation. In a simulated experiment with such materials, Beith, Sanders, and Peay (1982) used various configurations of such material (some of which are shown in Figure 5–12) as well as the presently mandated (in the United States) 6 in^2 (39 cm^2) of retroreflective material on the helmet.

The results indicated that all the experimental configurations (full suit, armbands, and zebra shirt) increased detection accuracy when compared with the minimum mandated condition. However, given the cost benefits (since retroreflective fabric is expensive) the armband and zebra shirt configurations were recommended; these

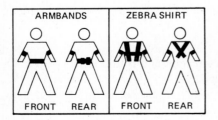

FIGURE 5-12
Configurations of armbands and zebra shirts of retroreflective material used in a study of conspicuousness of coal miners. (*From Beith, Sanders, and Peay, 1982.* Copyright by the Human Factors Society Inc. and reproduced by permission.)

TABLE 5–3
GENERAL GUIDE TO VISUAL DISPLAY SELECTION

To display	Select	Because	Example
Go, no go, start, stop on, off	Light	Normally easy to tell if it is on or off.	
Identification	Light	Easy to see (may be coded by spacing, color, location, or flashing rate; may also have label for panel applications).	
Warning or caution	Light	Attracts attention and can be seen at great distance if bright enough (may flash intermittently to increase conspicuity)	
Verbal instruction (operating sequence)	Enunciator light	Simple "action instruction" reduces time required for decision making.	RELEASE EJECT
Exact quantity	Digital counter	Only one number can be seen, thus reducing chance of reading error.	5 2 9 0 0
Approximate quantity	Moving pointer against fixed scale	General position of pointer gives rapid clue to the quantity plus relative rate of change.	
Set-in quantity	Moving pointer against fixed scale	Natural relationship between control and display motions.	
Tracking	Single pointer or cross pointers against fixed index	Provides error information for easy correction.	
Vehicle attitude	Either mechanical or electronic display of position of vehicle against established reference (may be graphic or pictorial)	Provides direct comparison of own position against known reference or base line.	

Source: Heglin, 1973, table II–4.

would increase detectability in the periphery by 100 percent over the minimum mandated conditions.

DISCUSSION

There are many varieties of visual displays that represent changeable conditions, and a complete set of guidelines for their variations is far beyond the scope of this text. However, certain guidelines offered by Heglin (1973) are given in Table 5–3. Certain of Heglin's guidelines have been covered in the text, although a few have not been discussed.

REFERENCES

Applied ergonomics handbook. (1974). Guilford, Surrey, England: IPC Science and Technology Press.

Beith, B. H., Sanders, M. S., and Peay, J. M. (1982). Using retroreflective material to enhance the conspicuity of coal miners. *Human Factors,* 24 (6), 727–735.

Berkhout, J. (1979). Information transfer characteristics of moving light signals. *Human Factors,* 21(4), 445–455.

Cohen, E., and Follert, R. L. (1970). Accuracy of interpolation between scale graduations. *Human Factors,* 12(5), 481–483.

Crawford, A. (1963). The perception of light signals: The effect of mixing flashing and steady irrelevant lights. *Ergonomics,* 6, 287–294.

Dashevsky, S. G. (1964). Check-reading accuracy as a function of pointer alignment, patterning, and viewing angle. *Journal of Applied Psychology,* 48, 344–347.

Elkin, E. H. (1959, February). *Effect of scale shape, exposure time and display complexity on scale reading efficiency* (TR 58–472). USAF, WADC, Wright-Patterson Air Force Base, Ohio.

Green, P. (1984). *Driver understanding of fuel and engine gauges,* Tech. Paper Series 840314. Warrendale, PA: Society of Automotive Engineers.

Grether, W. F., and Baker, C. A. (1972). Visual presentation of information. In H. A. Van Cott and R. G. Kinkade (eds.), *Human engineering guide to equipment design* (rev. ed.). Washington: Government Printing Office.

Heglin, H. J. (1973, July). *NAVSHIPS display illumination design guide: II. Human factors* (NELC/TD223). San Diego: Naval Electronics Laboratory Center.

Markowitz, J. (1971). Optimal flash rate and duty cycle for flashing visual indicators. *Human Factors,* 13 (5), 427–433.

Mortimer, R. G., and Kupec, J. D. (1983). Scaling of flash rate for a deceleration signal. *Human Factors,* 25 (3), 313–318.

Olson, P. L., Halstead-Nussloch, R, & Sivah, M. (1981). The effect of improvements in motorcycle/motorcyclist conspicuity on driver behavior. *Human Factors,* 23 (2), 237–248.

Reynolds, R. F., White, R. M., Jr., and Hilgendorf, R. I. (1972). Detection and recognition of colored signal lights. *Human Factors,* 14 (3), 227–236.

Roscoe, S. N. (1968). Airborne displays for flight and navigation. *Human Factors,* 10 (4), 321–332.

Roscoe, S. N., Corl, L., and Jensen, R. S. (1981). Flight display dynamics revisited. *Human Factors,* 23 (3), 341–353.

Sabeh, R., Jorve, W. R., and Vanderplas, J. M. (1958, March). *Shape coding of aircraft instrument zone markings,* Tech. Note 57–260. USAF, WADC, Wright-Patterson Air Force Base, Ohio.

Simmonds, G. R. W., Galer, M., and Baines, A. (1981). *Ergonomics of electronic displays,* Tech. Paper Series 810826. Warrendale, PA: Society of Automotive Engineers.

Teichner, W. H., and Krebs, M. J. (1972). Estimating the detectability of target luminances. *Human Factors,* 14 (6), 511–519.

Tolin, P. (1984). An information transmission of signal flash rate discriminability. *Human Factors,* 26 (4), 489–493.

Whitehurst, H. O. (1982). Screening designs used to estimate the relative effects of display factors on dial reading. *Human Factors,* 24 (3), 301–310.

Woodson, W. E., and Conover, D. W. (1964). *Human engineering guide for equipment designers* (2d ed.). Berkeley: University of California Press.

CHAPTER 6

AUDITORY, TACTUAL, AND OLFACTORY DISPLAYS

We all depend on our auditory, tactual, and olfactory senses in many aspects of our lives, including hearing our children cry or the doorbell ring, feeling the smooth finish on fine furniture, or smelling a cantaloupe to determine if it is ripe. As discussed in Chapter 3, information can come to us directly or indirectly. A child's natural cry is an example of direct information, while a doorbell's ring would be indirect information that someone is at the door. It is becoming increasingly possible to convert stimuli that are intrinsically, or directly, associated with one sensory modality into stimuli associated (indirectly) with another modality. Such technological developments have resulted in increased use of the auditory, tactual, and—to a lesser extent—olfactory senses. Using buzzers to warn of fire or to warn blind people of physical objects in their path are examples. Often the indirect stimulus can be more effective than the actual direct stimulus. For example, Kahn (1983) reported that sleeping subjects responded faster and more often to an auditory fire alarm, of sufficient intensity, than to the presence of either heat or the smell of smoke. Heat or smoke awoke the subjects only 75 percent of the time, while the alarm was effective virtually 100 percent of the time.

In this chapter we discuss the use of auditory, tactual, and olfactory senses as means of communication. Our focus is on indirect information sources rather than direct, and the topic of speech is deferred until Chapter 7.

HEARING

In discussing the hearing process, first we describe the physical stimuli to which the ear is sensitive, namely, sound vibrations, and then we discuss the anatomy of the ear.

Nature and Measurement of Sound

Sound is created by vibrations from some source. Although such vibrations can be transmitted through various media, our primary concern is with those transmitted through the atmosphere to the ear. Two primary attributes of sound are *frequency* and *intensity* (or amplitude).

Frequency of Sound Waves   The frequency of sound waves can be visualized if we think of a simple sound-generating source such as a tuning fork. When it is struck, the tuning fork is made to vibrate at its *natural* frequency. In so doing, the fork causes the air molecules to be moved back and forth. This alternation creates corresponding increases and decreases in the air pressure.

The vibrations of a simple sound-generating source, such as a tuning fork, form *sinusoidal* (or *sine*) *waves,* as shown in Figure 6–1. The height of the wave above the midline, at any given time, represents the amount of above-normal air pressure at that point. Positions below the midline in turn, represent the reduction in air pressure below normal.

One feature of simple sine waves is that the waveform above the midline is the mirror image of the waveform below the midline. Another feature is that the waveform pattern repeats itself again and again. One complete cycle is shown in Figure 6–1. The number of cycles per second is called the *frequency* of the sound. Frequency is expressed in hertz (abbreviated Hz), which is equivalent to cycles per second. On the musical scale, middle C has a frequency of 256 Hz. Any given octave has double the frequency of the one below it, so one octave above middle C has a frequency of 512 Hz. In general terms, the human ear is sensitive to frequencies in the range of 20 to 20,000 Hz, but it is not equally sensitive to all frequencies. In addition, there are marked differences among people in their relative sensitivites to various frequencies.

The frequency of a physical sound is associated with the human sensation of pitch. *Pitch* is the name given to the highness or lowness of a tone. Since high frequencies yield high-pitched tones and low frequencies yield low-pitched tones, we tend to think of pitch and frequency as synonymous. Actually there are other factors that influence our perception of pitch besides frequency. For example, the intensity of a

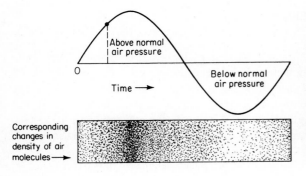

FIGURE 6-1
Sinusoidal wave created by a simple sound-generating source. The magnitude of the alternating changes in air pressure can be represented by a sine wave, shown in the upper part of the figure. The lower part of the figure depicts the changes in the density of the air molecules caused by the vibrating source and corresponding to the sine wave changes in pressure.

tone can influence our perception of pitch. When they are increased in intensity, low-frequency tones (less than about 1000 Hz) become lower in pitch, while high-frequency tones (greater than about 3000 Hz) become higher in pitch. Tones between 1000 and 3000 Hz, on the other hand, are relatively insensitive to such intensity-induced pitch changes. Incidentally, complex sounds such as those produced by musical instruments are very stable in perceived pitch regardless of whether the instruments are played loudly or softly—happily for music lovers!

Intensity of Sound   Sound intensity is associated with the human sensation of *loudness*. As in the case of frequency and pitch, several factors besides intensity can influence our perception of loudness. We, however, have deferred the discussion of loudness to Chapter 16.

Sound intensity is defined in terms of power per unit area, for example, watts per square meter (W/m^2). Because the range of power values for common sounds is so tremendous it is convenient to use a logarithmic scale to characterize sound intensity. The *Bel* (B), named after Alexander Graham Bell, is the basic unit of measurement used. The number of bels is the logarithm (to the base 10) of the ratio of two sound intensities. Actually, a more convenient and commonly used measure of sound intensity is the *decibel* (dB), where 1 dB = 0.1 B.

Unfortunately, there are no instruments available for directly measuring the sound power of a source. However, since sounds are pressure waves that vary above and below normal air pressure, these variations in air pressure can be directly measured. Luckily, sound power is directly proportional to the *square* of the sound pressure. The sound-pressure level (SPL), in decibels, is therefore defined as

$$\text{SPL (dB)} = 10 \log \frac{P_1{}^2}{P_0{}^2}$$

where $P_1{}^2$ is the sound pressure squared of the sound one wishes to measure and $P_0{}^2$ is the reference sound pressure squared that we choose to represent 0 dB. With a little algebraic magic, we can simplify the above equation so that we deal with sound pressure rather than sound pressure squared.

The equation for SPL becomes

$$\text{SPL (dB)} = 20 \log \frac{P_1}{P_0}$$

Note that P_0 cannot equal zero, because the ratio P_1/P_0 would be infinity. Therefore, a sound pressure greater than zero must be used to represent 0 dB. Then, when P_1 equals P_0, the ratio will equal 1.0, and the log of 1.0 is equal to zero; i.e., SPL will equal 0 dB. The most common sound-pressure reference value used to represent 0 dB is 20 micronewtons per square meter (20 $\mu N/m^2$). The μN stands for 0.000001 newton (0.000001 N).[1] This sound pressure is roughly equivalent to the lowest inten-

---

[1] This can also be expressed as 0.0002 dyne per square centimeter (0.0002 dyn/cm^2) or 0.0002 microbar (0.0002 μbar).

sity, 1000 Hz pure tone, which a healthy adult can just barely hear under ideal conditions. It is possible, therefore, to have sound-pressure levels that are actually less than 0 dB; that is, they have sound pressures less than 20 $\mu N/m^2$.

The decibel scale is a logarithmic scale, so an increase of 10 dB represents a 10-fold increase in sound power and a 100-fold increase in sound pressure (remember that power is related to the square of pressure). In like manner, it can be shown that doubling the sound power will raise the SPL by 3 dB. Another consequence of using a logarithmic scale is that the ratio of two sounds is computed by subtracting (rather than dividing) one decibel level from the other. So when we speak of the signal-to-noise ratio, we are actually referring to the difference between a meaningful signal and the background noise in decibels. For example, if a signal is 90 dB and the noise is 70 dB, the signal-to-noise ratio is +20 dB.

FIGURE 6-2
Decibel levels for various sounds. Decibel levels are *A*-weighted sound levels measured with a sound-level meter. (*Source: Peterson and Gross, 1972, fig. 2-1, p. 4.*)

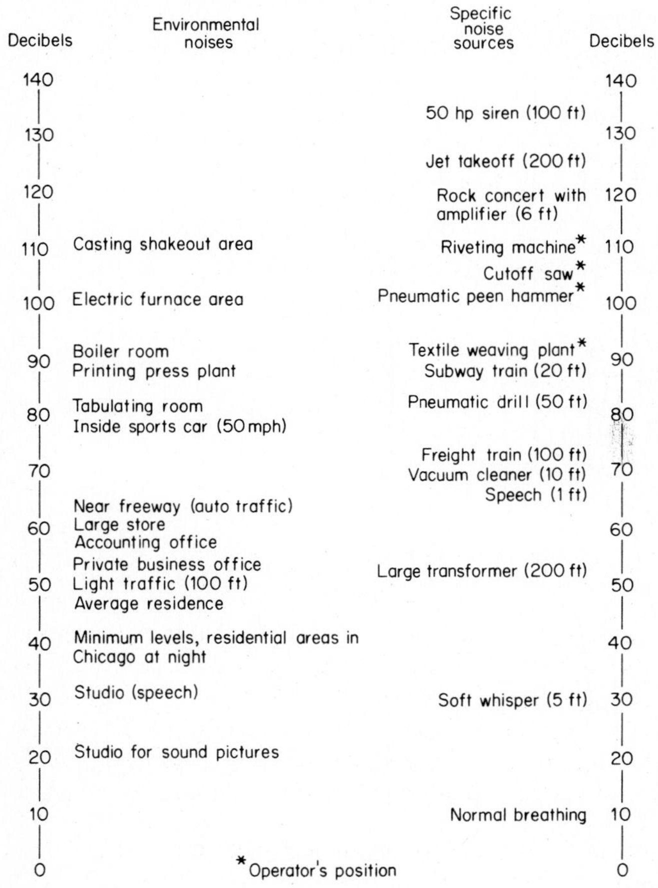

Decibels	Environmental noises	Specific noise sources	Decibels
140			140
		50 hp siren (100 ft)	
130			130
		Jet takeoff (200 ft)	
120			120
		Rock concert with amplifier (6 ft)	
110	Casting shakeout area	Riveting machine*	110
		Cutoff saw*	
100	Electric furnace area	Pneumatic peen hammer*	100
90	Boiler room / Printing press plant	Textile weaving plant* / Subway train (20 ft)	90
80	Tabulating room / Inside sports car (50 mph)	Pneumatic drill (50 ft)	80
70		Freight train (100 ft) / Vacuum cleaner (10 ft) / Speech (1 ft)	70
60	Near freeway (auto traffic) / Large store / Accounting office		60
50	Private business office / Light traffic (100 ft) / Average residence	Large transformer (200 ft)	50
40	Minimum levels, residential areas in Chicago at night		40
30	Studio (speech)	Soft whisper (5 ft)	30
20	Studio for sound pictures		20
10		Normal breathing	10
0	*Operator's position		0

Sound pressure is measured by sound-level meters. The American National Standards Institute (ANSI) and the American Standards Association have established a standard to which sound-level meters should conform. This standard requires that three different *weighting* networks (designated A, B, and C) be built into such instruments. Each network responds differently to low or high frequencies according to standard *frequency-response* curves. We have more to say about this in Chapter 16 when we discuss noise. For now, we need to be aware that when a sound intensity is reported, the specific weighting network used for the measure must be specified, for example, the A-weighted sound level is 45 dB,'' ''sound level (A) = 45 dB,'' ''SLA = 45 dB.'' or ''45 dBA.'' Figure 6–2 shows the decibel scale with examples of several sounds that fall at varying positions.

FIGURE 6-3
Waveform of a complex sound formed from three individual sine waves. (*Source: Minnix, 1978, fig. 1-11.*)

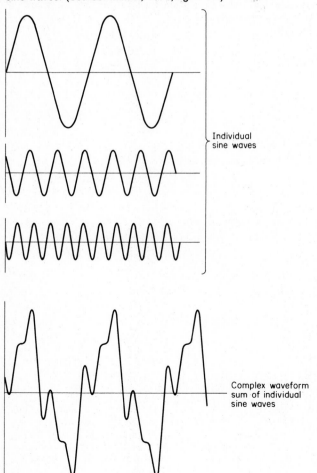

Individual sine waves

Complex waveform sum of individual sine waves

Complex Sounds Very few sounds are pure tones. Even tones from musical instruments are not pure, but rather consist of a fundamental frequency in combination with others (especially harmonic frequencies that are multiples of the fundamental). Most complex sounds, however, are nonharmonic. Complex sounds can be depicted in two ways. One is a waveform which is the composite of the waveforms of its individual component sounds. Figure 6–3 shows three component waveforms and the resultant composite waveform.

The other method of depicting complex sounds is by use of a sound spectrum which divides the sound into frequency bands and measures the intensity of the sound in each band. A frequency-band analyzer is used for this purpose. The four curves of Figure 6–4 illustrate spectral analyses of the noise from a rope-closing machine. Each curve was generated with a different sized frequency band (*bandwidth*), namely, an octave, a half octave, one-third of an octave, and a thirty-fifth of an octave. The narrower the bandwidth, the greater the detail of the spectrum and the lower the sound level within each bandwidth. There have been various practices in the division of the sound spectrum into octaves, but the current preferred practice as set forth by ANSI is to divide the audible range into 10 bands, with the *center* frequencies being 31.5, 63, 125, 250, 500, 1000, 2000, 4000, 8000, and 16,000 Hz. (Many existing sets of sound and hearing data are presented that use the previous practice of defining octaves in terms of the *ends* of the class intervals instead of their *centers.*)

Anatomy of the Ear

The ear has three primary anatomical divisions: the outer ear, the middle ear, and the inner ear. These are shown schematically in Figure 6–5.

Outer Ear The outer ear, which collects sound energy, consists of the external part (called the *pinna,* or *choncha*); the auditory canal (the *meatus*), which is a bayonet-shaped tube about 1 in long that leads inward from the external part; and the eardrum (the *tympanic membrane*) at the end of the auditory canal. The resonant properties of the auditory canal contribute to the sensitivity of the ear in the frequency range 2000 to 5000 Hz, enhancing sound-pressure levels by as much as 12 dB (Davies and Jones, 1982).

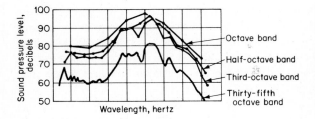

FIGURE 6-4
Spectral analyses of noise from a rope-closing machine. Analyzers used varying bandwidths. The narrower the bandwidth, the greater the detail and the lower the sound-pressure level within any single bandwidth. (*Source: Adapted from* Industrial noise manual, *1966, p. 25.*)

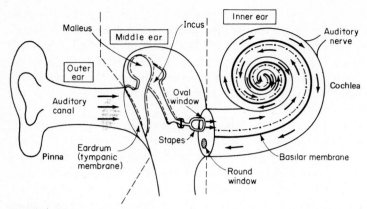

FIGURE 6-5
Schematic drawing of the ear. Shown are the outer, middle, and inner
ear structures.

Middle Ear The middle ear is separated from the outer ear by the tympanic membrane. The middle ear includes a chain of three small bones called *ossicles* (the *malleus,* the *incus,* and the *stapes*). These three ossicles, by their interconnections, transmit vibrations from the eardrum to the oval window of the inner ear. The stapes acts something like a piston on the oval window, its action transmitting the changes in sound pressure to the fluid of the inner ear, on the other side of the oval-window membrane. Due to the large surface area of the tympanic membrane and the lever action of the ossicles, the pressure of the foot of the stapes against the oval window is amplified to about 22 times that which could be effected by applying sound waves directly to the oval window (Guyton, 1969, p. 300).

The middle ear also contains two muscles attached to the ossicles. The *tensor tympani muscle* attaches to the malleus, and the *stapedius muscle* attaches to the stapes. These muscles represent a protection for the inner ear against intense sounds. When the muscles tighten, they reduce sound transmitted to the inner ear. This reaction is called the *acoustic,* or *aural, reflex.* The reflex occurs when the ear is exposed to a sound that is about 80 dB above threshold level; the reflex appears to be more responsive to broadband sounds than to pure tones and to lower frequencies than to higher frequencies (Kryter, 1985). The reflex can provide as much as 20-dB attenuation. The muscles will remain flexed for up to 15 min in the presence of intense steady state noise. When the ear is exposed to intense impulse noise, such as gunfire, there is a delay (or latency) of about 35 to 150 milliseconds (ms) before the muscles contract. Because of this delay there is not much protection provided against the initial impulse; however, the relaxation time following an impulse can be as long as 2 to 3 s, with most of it occurring within about 0.5 s. Protection, therefore, is provided against subsequent impulses when the time between impulses is less than about 1 s.

Inner Ear The inner ear, or *cochlea,* is a spiral-shaped affair that resembles a snail. If uncoiled, it would be about 30 mm long, with its widest section (near the oval window) being about 5 or 6 mm. The inner ear is filled with a fluid. The stapes of the middle ear acts on this fluid like a piston, driving it back and forth in response to changes in the sound pressure. These movements of the fluid force into vibration a thin membrane called the *basilar membrane,* which in turn transmits the vibrations to the *organ of Corti.* The organ of Corti contains hair cells and nerve endings that are sensitive to very slight changes in pressure. The neural impulses picked up by these nerve endings are transmitted to the brain via the *auditory nerve.*

Conversion of Sound Waves to Sensations

Although the mechanical processes involved in the ear have been known for some time, the procedures by which sound vibrations are ''heard'' and differentiated are still not entirely known. In this regard, the numerous theories of hearing fall generally into two classes. The *place* (or resonance) theories are postulated on the notion that the fibers at various positions (or places) along the basilar membrane act, as Geldard (1972) puts it, as harp or piano strings. Since these fibers vary in length, they are differentially sensitive to different frequencies and thus give rise to sensations of pitch. On the other hand, the *temporal* theories are based on the tenet that pitch is related to the time pattern, or interval, of the neural impulses emitted by the fibers.

As Geldard (1972) comments, it is not now in the cards to be able to accept one basic theory to the exclusion of any others as explaining all auditory phenomena. Actually, Geldard indicates that the current state of knowledge would prejudice one toward putting reliance in a place theory as related to high tones and in a temporal theory as related to low tones, one principle thus giving way to the other in the middle range of frequencies. As with theories in various areas of life, this notion of the two theories being complementary still needs to be regarded as tentative pending further support or rejection from additional research. Moore (1982), for example, offers a theory or model based on both place and temporal information.

One consequence of all this is that the ear is not equally sensitive to all frequencies of sound. Although we discuss this more fully in Chapter 16, it is important to know that, in general, the ear is less sensitive to low-frequency sounds (20 to approximately 500 Hz) and more sensitive to higher frequencies (1000 to approximately 5000 Hz). That is, a 4000-Hz tone, at a given sound-pressure level, will seem to be louder than, say, a 200-Hz tone at the same sound-pressure level.

Masking

Masking is a condition in which one component of the sound environment reduces the sensitivity of the ear to another component. Operationally defined, *masking* is the amount that the threshold of audibility of a sound (the masked sound) is raised by the presence of another (masking) sound. In studying the effects of masking, an experimenter typically measures the absolute threshold (the minimum audible level) of a sound (the sound to be masked) when presented by itself and then measures its threshold in the presence of the masking sound. The difference is attributed to the

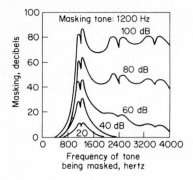

FIGURE 6-6
The effects of masking pure tones by a pure tone of
1200 Hz (100, 80, 60, 40, and 20 dB). (*Source: Wagel
and Lane, 1924.*)

masking effect. The concept of masking is central to discussions of auditory displays.
In selecting a particular auditory signal for use in a particular environment, we must
consider the masking effect of any noise on the reception of that signal.

The effects of masking vary with the type of masking sound and with the masked
sound itself—whether pure tones, complex sound, white noise, speech, etc. Figure
6–6 illustrates the masking of a pure tone by another pure tone. A few general
principles can be gleaned from this. The greatest masking effect occurs near the
frequency of the masking tone and its harmonic overtones. As we would expect, the
higher the intensity of the masking tone, the greater the masking effect. Notice,
however, that with low-intensity masking tones (20 to 40 dB) the masking effect is
somewhat confined to the frequencies around that of the masking tone; but with
higher-intensity masking tones (60 to 100 dB), the masking effect spreads to higher
frequencies. The masking of pure tones by narrowband noise (i.e., noise concentrated in a narrow band of frequencies) is very similar to that shown in Figure 6–6.

In the masking of pure tones by wideband noise, the primary concern is with the
intensity of the masking noise in a "critical band" around the frequency of the
masked tone. The size of the critical band is a function of the center frequency, with
larger critical bands being associated with higher frequencies.

The nature of masking effects depends very much upon the nature of the two
sounds in question, as discussed by Geldard (1972, pp. 215–220) and by Deatherage
(1972). Although these complex effects are not discussed here, the effects of masking
in our everyday lives are of considerable consequence, as, for example, the noise of
a hair dryer drowning out the sound of the telephone ringing.

AUDITORY DISPLAYS

The nature of the auditory sensory modality offers certain unique advantages for
presenting information as contrasted with the visual modality. One set of comparisons of these two senses was given in Table 3–1. On the basis of such comparisons
and of other cues, it is possible to identify certain types of circumstances in which
auditory displays would be preferable to visual displays:

- When the origin of the signal is itself a sound
- When the message is simple and short
- When the message will not be referred to later
- When the message deals with events in time
- When warnings are sent or when the message calls for immediate action
- When continuously changing information of some type is presented, such as aircraft, radio range, or flight-path information
- When the visual system is overburdened
- When speech channels are fully employed (in which case auditory signals such as tones should be clearly detectable from the speech)
- When illumination limits use of vision
- When the receiver moves from one place to another
- When a verbal response is required

Obviously the application of these guidelines should be tempered with judgment rather than being followed rigidly. There are, of course, other circumstances in which auditory displays would be preferable. In the above guidelines particular mention should be made of the desirability of restricting auditory messages to those that are short and simple (except in the case of speech), since people do not recall complex messages from short-term memory very well.

We can consider four types of human functions, or tasks, involved in the reception of auditory signals: (1) *detection* (determining whether a given signal is present, such as a warning signal), (2) *relative discrimination* (differentiating between two or more signals presented close together), (3) *absolute identification* (identifying a particular signal of some class when only one is presented), and (4) *localization* (determining the direction from which the signal is coming). Relative discrimination and absolute identification can be made on the basis of any of several stimulus dimensions, such as intensity, frequency, and duration.

Detection of Signals

The detection of auditory signals can be viewed within the framework of signal detection theory (as discussed in Chapter 3). Signals can occur in "peaceful" surroundings or in environments that are permeated by ambient noise. As indicated in the discussion of signal detection theory, if at all possible, the signal plus noise (SN) should be distinct from the noise (N) itself. When this is not the case, the signal cannot always be detected in the presence of the noise; this confusion is, in part, a function of masking. When the signal (the masked sound) occurs in the presence of noise (i.e., the masking sound), the threshold of detectability of the signal is elevated—and it is this elevated threshold that should be exceeded by the signal if it is to be detected accurately. Failure to consider this forced New York City in 1974 to abandon legislation aimed at regulating the decibel output of newly manufactured automobile horns—the quieter horns could not be heard over the noise of traffic (Garfield, 1983).

In quiet surroundings (when one does not have to worry about clacking typewriters, whining machines, or screeching tires), a tonal signal about 40 to 50 dB above absolute threshold normally would be sufficient to be detected. However, such detectability would vary somewhat with the frequency of the signal (Pollack, 1952) and its duration. With respect to duration, the ear does not respond instantaneously to sound. For pure tones it takes about 200 to 300 ms to "build up" (Munson, 1947) and about 140 ms to decay (Stevens and Davis, 1938); wideband sounds build up and decay more rapidly. Because of these lags, signals less than about 200 to 500 ms in duration do not sound as loud as those of longer duration. Thus, auditory signals (especially pure tones) should be at least 500 ms in duration; if they have to be shorter than that, their intensity should be increased to compensate for reduced audibility. Although detectability increases with signal duration, there is not much additional effect beyond a few seconds in duration.

In rather noisy conditions, the signal intensity has to be set at a level far enough above that of the noise to ensure detectability. In this regard Deatherage (1972) has suggested a rule of thumb for specifying the optimum signal level: The signal intensity (at the entrance of the ear) should be about midway between the masked threshold of the signal in the presence of noise and 110 dB.

Fidell, Pearsons, and Bennett (1974) developed a method for predicting the detectability of complex signals (complex in terms of frequency) in noisy backgrounds. Although we do not present the details of the procedure, it is based on the observation that human performance is a function of the signal-to-noise ratio in the single one-third octave band to which human sensitivity is highest. The basic data employed in the proposed method are therefore the one-third octave band spectra of both the signal and background noise.

Use of Filters   Under some circumstances it is possible to enhance the detectability of signals by filtering out some of the noise. This is most feasible when the predominant frequencies of the noise are different from those of the signal. In such a case, some of the noise can be filtered out and the intensity of the remaining sound raised (signal plus the nonfiltered noise). This has the effect of increasing the signal-to-noise ratio, thus making the signal more audible.

The central nervous system is also capable of filtering out noise if the circumstances are right. Presenting a signal to one ear and noise to both ears will increase detectability compared to the situation where both the noise and signal are presented to both ears. This appears to be due to filtering of the noise somewhere in the central nervous system. The exact mechanism is not known, but the brain has the capacity to compare the signals received by both ears and, in effect, cancel some of the noise.

Increasing Detectability of Signals   Mulligan, McBride, and Goodman (1984) list several procedures to increase the detectability of a signal in noise, several of which we have already mentioned: (1) reduce the noise intensity in the region of frequencies (i.e., critical bandwidth) around the signal's frequency; (2) increase the intensity of the signal; (3) present the signal for at least 0.5 to 1 s; (4) change the frequency of the signal to correspond to a region where the noise intensity is low; (5) phase-shift the signal and present the unshifted signal to one ear and the shifted signal to the other; and (6) present the noise to both ears and the signal to only one ear.

Relative Discrimination of Auditory Signals

The relative discrimination of signals on the basis of intensity and frequency (which are the most commonly used dimensions) depends in part on interactions between these two dimensions. In many real-life situations, however, discriminations depend on much more than just frequency and intensity differences. Talamo (1982), for example, discusses the difficulties faced by equipment operators monitoring machine sounds for indications of impending mechanical failure. Discriminations in such cases may involve relative strengths of harmonics or even the relative phases of sounds.

Discriminations of Intensity Differences   Figure 6–7 (Deatherage, 1972) reflects the just-noticeable differences (JNDs) for certain pure tones and for wideband noise of various sound-pressure levels. It is clear that the smallest differences can be detected with signals of higher intensities, at least 60 dB above the absolute threshold. The JNDs of signals above 60 dB are smallest for the intermediate frequencies (1000 and 4000 Hz). These differences in JNDs, of course, have implications for selecting signals if the signals are to be discriminated by intensity.

Discrimination of Frequency Differences   Some indication of the ability of people to tell the difference between pure tones of different frequencies is shown by the results of an early study by Shower and Biddulph (1931) as depicted in Figure 6–8. This figure shows the JNDs for pure tones of various frequencies at various

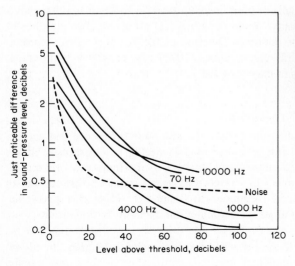

FIGURE 6-7
The just-noticeable differences (JNDs) in sound intensity for pure tones of selected frequencies and for wideband noise. (*Source: Deatherage, 1972, p. 147, as based on data from Riesz, 1928, and Miller, 1947.*)

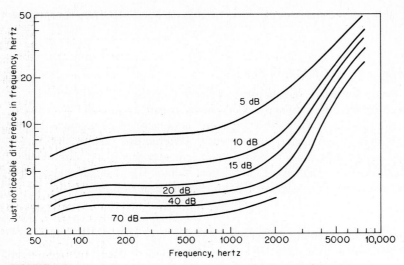

FIGURE 6-8
Just-noticeable differences in frequency for pure tones at various levels above threshold. (*Source: Shower and Biddulph, 1931.*)

levels above threshold. The JNDs are smaller for frequencies below about 1000 Hz (especially for high intensities) but increase rather sharply for frequencies above that. Thus, if signals are to be discriminated on the basis of frequency, it usually would be desirable to use signals of lower frequencies. This practice may run into a snag, however, if there is much ambient noise (which usually tends to consist of lower frequencies) that might mask the signals. A possible compromise is to use signals in the 500- to 1000-Hz range (Deatherage, 1972). It is also obvious from Figure 6–8 that the JNDs are smaller for signals of high intensity than for those of low intensity, thus suggesting that signals probably should be at least 30 dB above absolute threshold if frequency discrimination is required. Duration of the signal is also important in discrimination of frequency differences. Gales (1979) reports that discrimination is best when stimulus duration is in excess of 0.1 s.

Absolute Identification of Auditory Signals

The JND for a given stimulus dimension reflects the minimum difference that can be discriminated on a relative basis. But in many circumstances it is necessary to make an absolute identification of an individual stimulus (such as the frequency of a single tone) presented by itself. The number of "levels" along a continuum that can be so identified usually is quite small, as mentioned in Chapter 3. Table 6–1 gives the number of levels that can be identified for certain common auditory dimensions.

Multidimensional Coding   If the amount of information to be transmitted by auditory codes is substantial (meaning that absolute identification of numerous signals

TABLE 6–1
LEVELS OF AUDITORY DIMENSIONS
IDENTIFIABLE ON AN ABSOLUTE
BASIS

Dimension	Level(s)
Intensity (pure tones)	4–5
Frequency	4–7
Duration	2–3
Intensity and frequency	9

Source: Deatherage, 1972; Van Cott and
Warrick, 1972.

needs to be made), it is possible to use a multidimensional code system. For example, Pollack and Ficks (1954) used various combinations of several dimensions, such as direction (right ear versus left ear), frequency, intensity, repetition rate, on-and-off time fraction, and duration. Such a system obviously imposes the requirement for training the receiver regarding the ''meaning'' of each unique combination. In using such multidimensional codes, however, it is generally better to use more dimensions with fewer steps or levels of each (such as eight dimensions with two steps of each dimension) than to use fewer dimensions and more levels of each (such as four dimensions with four steps of each). It is on the basis of the many different facets (dimensions) of sounds that we can identify the voices of individuals, a dripping water faucet, or a squeaking hinge.

Sound Localization

The ability to localize the direction from which sound waves are emanating is called *stereophony*. The primary cues used by people to determine the direction of a sound source are differences in both intensity and phase of the sounds. If a sound source is directly to one side of the head, the sound reaches the nearer ear approximately 0.8 ms before it reaches the other ear. This results in the sound in one ear being out of phase with the sound in the other ear, and this is an effective cue for localizing a sound source when the frequency of the sound is below 1500 Hz. The problem with using phase differences as the localization cue is that one cannot tell whether the sound is forward of or behind the head without turning the head from side to side. At low frequencies, sound waves go around the head easily, and even if the sound is on one side, there is little or no intensity difference at the two ears. At high frequencies (above 3000 Hz), however, the head effectively shadows the ear and intensity differences can be marked. The effectiveness of intensity differences as a cue to sound localization can be demonstrated with a home stereo recording system. If speakers are placed at each end of a room and the volume of one speaker is increased while the volume of the other is decreased, the sound will appear to move across the room. Listeners will be able to ''follow'' the sound by pointing to it as it ''moves.''

Localizing sound in the midrange frequencies (1500 to 3000 Hz) is relatively difficult because such frequencies produce neither effective phase nor intensity difference cues. The ability to turn the head from side to side greatly increases accuracy of localization under almost all circumstances. Without moving (tilting) the head, localization in the vertical (up-down) direction is rather poor.

Caelli and Porter (1980) provide a practical example of research on sound localization, i.e., determining the direction of an ambulance siren. Subjects sat in a car and heard a 2-s burst of a siren (120 dBA at the ambulance) 100 m from the car. They were required to estimate its direction and distance. Subjects tended to overestimate the distance of the siren, often by a factor of 2 or more. That is, the subjects thought the ambulance was farther away than it really was, a dangerous type of error in a real-life situation. With respect to localization, subjects were often 180° off in their perceptions of direction. One-third of all the test trials were classified as reversals of perceived direction. Particularly poor were perceptions when the siren was behind the driver. One-half to two-thirds of the time the subjects thought the siren was in front of them! It was also found that localization was poorer when the driver-side window was rolled down. With the window down, the sound intensity to the window-side ear was higher than to the passenger-side ear, even when the sound was coming from the passenger side. This gave false localization cues to the subjects and resulted in poor localization performance.

Principles of Auditory Display

As in other areas of human factors engineering, most guidelines and principles have to be accepted with a few grains of salt since specific circumstances, trade-off values, etc., may argue for their violation. With such reservations in mind, a few guidelines for the use of auditory displays are given. These generally stem from research and experience. Some are drawn in part from Mudd (1961) and Licklider (1961).

I General principles
    A *Compatibility:* Where feasible, the selection of signal dimensions and their encoding should exploit learned or natural relationships of the users, such as high frequencies associated with up or high and wailing signals with emergency.
    B *Approximation:* Two-stage signals should be considered when complex information is to be presented. The signal stages would consist of:
        1 Attention-demanding signal: to attract attention and identify a general category of information.
        2 Designation signal: to follow the attention-demanding signal and designate the precise information within the general class indicated by the first signal.
    C *Dissociability:* Auditory signals should be easily discernible from any ongoing audio input (be it meaningful input or noise). For example, if a person is to listen concurrently to two or more channels, the frequencies of the channels should be different if it is possible to make them so.
    D *Parsimony:* Input signals to the operator should not provide more information than is necessary.

E *Invariance:* The same signal should designate the same information at all times.

II Principles of presentation

  A *Avoid extremes of auditory dimensions:* High-intensity signals, for example, can cause a startle response and actually disrupt performance.

  B *Establish intensity relative to ambient noise level:* The intensity level should be set so that it is not masked by the ambient noise level.

  C *Use interrupted or variable signals:* Where feasible, avoid steady state signals and, rather, use interrupted or variable signals. This will tend to minimize perceptual adaptation.

  D *Do not overload the auditory channel:* Only a few signals should be used in any given situation. Too many signals can be confusing and will overload the operator. (For example, during the Three-Mile Island nuclear crisis, over 60 different auditory warning signals were activated.)

III Principles of installation of auditory displays

  A *Test signals to be used:* Such tests should be made with a representative sample of the potential user population to be sure the signals can be detected and discriminated by them.

  B *Avoid conflict with previously used signals:* Any newly installed signals should not be contradictory in meaning to any somewhat similar signals used in existing or earlier systems.

  C *Facilitate changeover from previous display:* Where auditory signals replace some other mode of presentation (e.g., visual), preferably continue both modes for a while to help people become accustomed to the new auditory signals.

Auditory Displays for Specific Purposes

For illustration, a few examples of special-purpose auditory displays are discussed.

Warning and Alarm Signals The unique features of the auditory system make auditory displays especially useful for signaling warnings and alarms. Various types of available devices each have their individual characteristics and corresponding advantages and limitations. A summary of such characteristics and features is given in Table 6–2 (Deatherage, 1972, p. 126).

As one example of research in this area, Adams and Trucks (1976) tested reaction time to eight different warning signals (''wail,'' ''yelp,'' ''whoop,'' ''yeow,'' intermittent horns, etc.) in five different ambient noise environments. Across all five environments, the most effective signals were the ''yeow'' (descending change in frequency from 800 to 100 Hz every 1.4 s) and the ''beep'' (intermittent horn, 425 Hz, on for 0.7 s, off for 0.6 s). The least effective signal was the ''wail'' (slow increase in frequency from 400 to 925 Hz over 3.8 s and then a slow decrease back to 400 Hz). Figure 6–9 shows the mean reaction time to these three signals as a function of overall signal intensity in one of the ambient noise environments tested. As can be seen, reaction time decreased with increased signal intensity. The differ-

TABLE 6–2
CHARACTERISTICS AND FEATURES OF CERTAIN TYPES OF AUDIO ALARMS

Alarm	Intensity	Frequency	Attention-getting ability	Noise-penetration ability
Diaphone (foghorn)	Very high	Very low	Good	Poor in low-frequency noise
Horn	High	Low to high	Good	Good
Whistle	High	Low to high	Good if intermittent	Good if frequency is properly chosen
Siren	High	Low to high	Very good if pitch rises and falls	Very good with rising and falling frequency
Bell	Medium	Medium to high	Good	Good in low-frequency noise
Buzzer	Low to medium	Low to medium	Good	Fair if spectrum is suited to background noise
Chimes and gong	Low to medium	Low to medium	Fair	Fair if spectrum is suited to background noise
Oscillator	Low to high	Medium to high	Good if intermittent	Good if frequency is properly chosen

Source: Deatherage (1972, table 4–2).

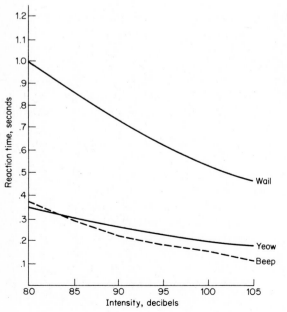

FIGURE 6-9
Mean reaction time to three auditory warning signals in an ambient noise environment. Reaction time decreases with increased signal intensity. (*Source: Adapted from Adams and Trucks, 1976, fig. 4.*)

ence in reaction time between the most effective and the least effective signals also decreased somewhat as intensity increased.

This relationship between reaction time and signal intensity may only hold for simple reactions, that is, where the same response is to be made for all signals. Where a different response is to be made to different signals (choice reaction time), the relationship between reaction time and signal intensity is not so simple. Van der Molen and Keuss (1979) used two signals (1000 and 3000 Hz) and required the subjects to make either the same response to both (simple reaction time) or a different response to each (choice reaction time). Figure 6–10 illustrates their results. For simple reaction time, increases in signal intensity resulted in faster reaction times. For choice reaction time, however, the fastest reaction times were for moderate-intensity signals, rather than high-intensity signals. What appears to be happening is that the high-intensity signals elicit a startle reflex, which may be helpful for efficient task performance only if the same response is demanded by both signals. But when one of two responses must be provided, the startle reflex acts to degrade performance. The intensity level of warning devices should, therefore, be chosen with concern for the response requirements placed on the operator.

In the selection or design of warning and alarm signals, the following general design recommendations have been proposed by Deatherage (1972, pp. 125, 126) and by Mudd (1961), these being in slightly modified form.

• Use frequencies between 200 and 5000 Hz, and preferably between 500 and 3000 Hz, because the ear is most sensitive to this middle range.

• Use frequencies below 1000 Hz when signals have to travel long distances (over 1000 ft), because high frequencies do not travel as far.

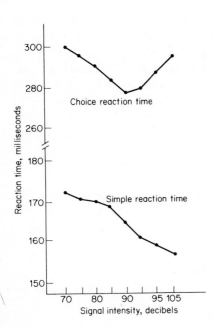

FIGURE 6-10
Relationship between reaction time and signal intensity for choice reaction time and simple reaction time tasks. (*Source: Adapted from Van der Molen and Keuss, 1979, fig. 1.*)

• Use frequencies below 500 Hz when signals have to "bend around" major obstacles or pass through partitions.

• Use a modulated signal (1 to 8 beeps per second, or warbling sounds varying from 1 to 3 times per second), since it is different enough from normal sounds to demand attention.

• Use signals with frequencies different from those that dominate any background noise, to minimize masking.

• If different warning signals are used to represent different conditions requiring different responses, each should be discriminable from the others, and moderate-intensity signals should be used.

• Where feasible, use a separate communication system for warnings, such as loudspeakers, horns, or other devices not used for other purposes.

Weiss and Kershner (1984) report that the auditory alarms depicted in Figure 6–11 have been found to be easily discernible and discriminable in numerous tests at nuclear power plants. Although 12 can be discriminated on a relative basis, if an absolute identification is required, it would probably be best to limit the number to 5.

Aids for the Blind With advances in technology, several devices have been built which use auditory information to aid the blind in moving about in their environment. Kay et al. (1977) distinguish two types of mobility aids—clear-path indicators (where the aim is to provide essentially a go–no-go signal indicating whether it is safe to proceed along the line of travel) and environmental sensors (which attempt to allow users to build up a picture of their environment). Examples of clear-path indicators would include *Pathsounder* (Russell, 1969), which emits an audible buzzing sound when there is something 6 ft ahead and a distinctively different high-pitched beeping sound when an object is within 3 ft of the user; *Single-Object Sensor* (Kay et al., 1977), which emits repetitive clicks, coded in terms of repetition rate to the distance of objects; and *Sonic Pathfinder* (Heyes, 1984) which emits eight musical notes to indicate the range of an object located within a 120° arc in eight 30-cm zones in front of the user. Dodds, Clark-Carter, and Howarth (1984) evaluated the Sonic Pathfinder with six totally blind subjects. All subjects used a long cane, with and without the device, to navigate a course. The Sonic Pathfinder was designed to be used with a long cane or guide dog because it cannot detect holes and curbs. The results clearly showed that fewer contacts occurred and that the subjects were more aware of obstacles when they used the device. Not surprisingly, from our discussion of resource competition in Chapter 3, the auditory display did not interfere with the subjects' reception and processing of tactile changes in the ground's surface, but it did reduce their reception and processing of environmental sounds. It was also noted that the musical information could not be processed at a rate commensurate with a comfortable walking speed and that the subjects tended to walk more slowly when using the device.

An example of an environmental sensor is the *Sonicguide* (Kay, 1973), which is built into eyeglass frames (as are both the Single-Object Sensor and the Sonic Pathfinder). Sonicguide is one of the more complex mobility aids. It gives the user three

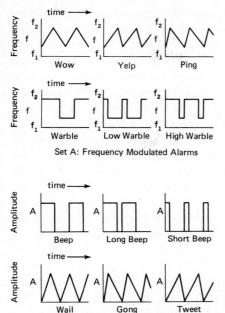

FIGURE 6-11
Twelve auditory alarms that are easily discernible and discriminable on a relative basis as determined in numerous tests at nuclear power plants. (*Source: Weiss and Kershner, 1984; figs. 6 and 7.*)

types of information about an object: frequency differences for determining distance, directional information, and information about the surface characteristics of an object. The complex information also tends to overload users, and they tend to walk more slowly when using the device.

In another application of using sound to aid the blind, Mansur (1984) reported on the use of a continuously varying pitch to represent two-dimensional graphical data. The pitch corresponded to the Y axis (ordinate) of the graph, and time was used to represent the X axis (abscissa). The method was compared with tactual graphs and was found to convey concepts (such as symmetry, monotonicity, and slopes) more quickly and with the same level of accuracy as the tactual presentations.

CUTANEOUS SENSES

In everyday life people depend on their cutaneous (somesthetic, or skin) senses much more than they realize. There is, however, an ongoing question as to how many cutaneous senses there are; this confusion exists in part because of the bases on which the senses can be classified. As Geldard (1972, pp. 258, 259) points out, we can classify them *qualitatively* (on the basis of their observed similarity—that is, the sensations generated), in terms of the *stimulus* (i.e., the form of energy that triggers the sensation, as thermal, mechanical, chemical, or electrical), or *anatomically* (in accordance with the nature of the sense organs or tissues involved). With respect to the anatomic structures involved, it is still not clear how many distinct types of nerve endings there are, but for convenience Geldard considers the skin as housing three more or less separate systems of sensitivity, one for pressure reception, one for pain, and one responsive to temperature changes. Although it has been suggested that any particular sensation is triggered by the stimulation of a specific corresponding type of nerve ending or receptor, this notion seems not to be warranted. Rather, it is now generally believed that the various receptors are specialized in their functions through the operation of what Geldard refers to as the principle of *patterned response*. Some of the cutaneous receptors are responsive to more than one form of energy (such as mechanical pressure and thermal changes) or to certain ranges of energy. Through complex interactions among the various types of nerve endings, being stimulated by various forms and amounts of energy, we experience a wide "variety" of specific sensations that we endow with such labels as *touch, contact, tickle,* and *pressure* (Geldard, 1972).

For the most part, tactual displays have utilized the hand and fingers as the principal receptors of information. Craft workers, for centuries, have used their sense of touch to detect irregularities and surface roughness in their work. Interestingly, Lederman (1978) reports that surface irregularities can be detected more accurately when the person moves an intermediate piece of paper or thin cloth over the surface than when the bare fingers are used alone. Inspection of the coachwork of Aston Martin sports cars, for example, is done by rubbing over the surface with a cotton glove on the hand.

Not all parts of the hand are equally sensitive to touch. One common measure of touch sensitivity is the *two-point threshold,* the smallest distance between two pressure points at which the points are perceived as separate. Figure 6–12 shows the median two-point threshold for fingertip, finger, and palm (Vallbo and Johansson, 1978). The sensitivity increases (i.e., the two-point threshold becomes smaller) from the palm to the fingertips. Thus, tactual displays that require fine discriminations are best designed for fingertip reception. Tactual sensitivity is also degraded by low skin temperatures; therefore, tactual displays should be used with extreme caution in low-temperature environments. Stevens (1982), however, found that cooling or warming the object being felt caused a marked improvement (averaging 41 percent) in the skin's acuity relative to thermally neutral stimulation. With this bit of background information, let us look at or, more appropriately, feel our way through some specific types of tactual displays.

TACTUAL DISPLAYS

Although we depend very much upon our cutaneous senses in everyday living, these senses have been used only to a limited degree as the basis for the intentional transmission of information to people by the use of tactual displays. The primary uses of tactual displays to date have been as substitutes for hearing, especially as aids to the deaf, and as substitutes for seeing, especially as aids to the blind.

The most frequent types of stimuli used for tactual displays have been mechanical vibration or electric impulses. Information can be transmitted by mechanical vibration based on such physical parameters as location of vibrators, frequency, intensity, or duration. Electric impulses can be coded in terms of electrode location, pulse rate, duration, intensity, or phase. Electric energy has some advantages over mechanical vibrators, as noted by Hawkes (1962). Electrodes are easily mounted on the body surface and have less bulk and a lower power requirement than mechanical vibrators. Sensitivity to mechanical vibration is dependent on skin temperature, whereas the amount of electric current required to reach threshold apparently is independent of skin temperature. A disadvantage of electric energy is that it can elicit pain.

Substitutes for Hearing

Tactual displays have generally found three applications as substitutes for hearing: reception of coded messages, perception of speech, and localization of sound.

Reception of Coded Messages   Both mechanical and electric energy forms have been used to transmit coded messages. In an early example, using mechanical vibrators, Geldard (1957) developed a coding system using 5 chest locations, 3 levels of intensity, and 3 durations, which provided 45 unique patterns (5 $\times$ 3 $\times$ 3), which in turn were used to code 26 letters, 10 numerals, and 4 frequently used words. One subject could receive messages at a rate of 38 words per minute, somewhat above the military requirements for Morse code reception of 24 words per minute.

Perception of Speech   Kirman (1973) reviews numerous attempts to transmit speech to the skin through mechanical and electrical stimulation. All attempts to build successful tactile displays of speech have been disappointing at best. One reason given for the poor performance of such devices is that the resolving power of

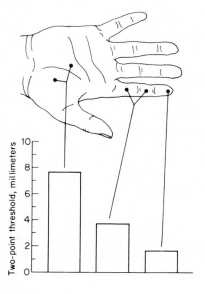

FIGURE 6-12
Median two-point threshold for fingertip, finger, and palm. (*Source: Vallbo and Johansson, 1978, fig. 6.*)

the skin, both temporally and spatially, is too limited to deal with the complexities of speech (Loomis, 1981).

Localization of Sounds   Richardson and his colleagues (Frost and Richardson, 1976; Richardson and Frost, 1979; Richardson, Wuillemin, and Saunders, 1978) report that subjects are able to localize a sound source by use of a simple tactual display with an accuracy level comparable to normal audition. The sound intensity is picked up by microphones placed over each ear. The outputs of these microphones are amplified and fed to two vibrators, upon which the subjects rest their index fingers. The intensity of vibration on each finger is then proportional to the intensity of the sound reaching the ear. It is this difference in intensity, as discussed earlier, that allows localization of sound with normal audition.

Substitutes for Seeing

Tactual displays have been most extensively used as substitutes for seeing—as an extension of our eyes. The most exciting possibilities in the use of tactual displays lie in this realm, especially with the advances being made in low-cost microelectronics technology. As an example of possible future applications, Dobelle (1977) has developed and tested an experimental unit that uses direct electrical stimulation of the brain. It produces genuine sensations of vision—indeed, volunteers who have been blind for years complain of eyestrain. A television camera picks up a black-and-white picture which is converted to a matrix of 60 or so electrodes implanted in the primary visual cortex of the brain. Each electrode, when stimulated, produces a single point of light in the visual field. Several subjects have had the implant and have been able to recognize white horizontal and vertical lines on black backgrounds. There are, of course, problems with such a technique, and the best that can be expected in the foreseeable future, according to Dobelle, is low-resolution black-and-white slow-scan images.

We now review some applications of tactual displays as substitutes for seeing, ranging from the relatively mundane to the more sophisticated.

Identification of Controls   A very practical use of the tactual sense is with respect to the design of control knobs and related devices. Although these are not *displays* in the conventional sense of the term, the need to correctly identify such devices may be viewed within the general framework of displays. The coding of such devices for tactual identification includes their shape, texture, and size. Figure 6–13, for example, shows two sets of shape-coded knobs such that the knobs within each group are rarely confused with each other (Jenkins, 1947). In Chapter 10, we discuss in more detail methods of tactually coding controls.

Reading Printed Material   Probably one of the most widely known tactual displays for reading printed material is braille printing. Braille print for the blind consists of raised dots formed by the use of all the possible combinations of six dots numbered and arranged thus:

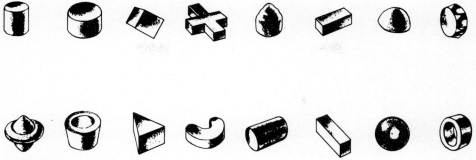

FIGURE 6-13
Two sets of knobs for levers that are distinguishable by touch alone. The shapes in each set are rarely confused with each other. (*Source: Jenkins, 1947.*)

$$1 \quad . \quad . \quad 4$$
$$2 \quad . \quad . \quad 5$$
$$3 \quad . \quad . \quad 6$$

A particular combination of these represents each letter, numeral, or common word. The critical features of these dots are position, distance between dots, and dimension (diameter and height), all of which have to be discriminable to the touch.

Because of the cost of braille printing there are only limited amounts of braille reading material available to blind persons. The combination of research relating to the sensitivity of the skin and new technological developments, however, now permit us to convert optical images of letters and numbers to tactile vibrations that can be interpreted by blind persons. The Optacon (for *op*tical-to-*ta*ctile *con*verter), one such device developed at the Stanford University Electronics Laboratories, consists of essentially two components. One is an optical pickup "camera," about the size of a pocketknife, and a "slave" vibrating tactile stimulator that reproduces images that are scanned by the camera (Bliss et al., 1970; Linvill, 1973). A photograph of the Optacon is shown in Figure 6–14. The 144 phototransistors of the camera, in a 24 × 6 array, control the activation of 144 corresponding vibrating pins that "reproduce" the visual image. The vibrating pins can be sensed by one fingertip. The particular features of the vibrating pins and of their vibrations were based on extensive psychophysical research relating to the sensitivity of the skin to vibrations. [For what interest they may be, the specifications of the vibrating pins are: size of pin, 10 mils; space between 6 rows, 100 mils; space between 24 pins in each row, 50 mils; frequency of vibrations, 230 Hz; depth of skin indentation, 2.6 mils (Bliss et al., 1970)].

The letter O, for example, produces a circular pattern of vibration that moves from right to left across the user's fingertip as the camera is moved across the letter.

Learning to "read" with the Optacon takes time. For example, one subject took about 160 hours (160 h) of practice and training to achieve a reading rate of about

FIGURE 6-14
Photograph of Optacon for use in
converting visual images to tactile
vibrations. (*Photograph, Stanford
Electronics Laboratories. The Optacon is
produced by Telesensory Systems, Inc.*)

50 words per minute. Most users, however, only achieve rates of about half that.
This rate is well below that of most readers of braille, some of whom achieve a rate
of 200 words per minute. At the same time, such a device does offer the possibility
to blind persons of being able to read conventional printing at least at a moderate
rate.

Navigation Aids To navigate in an environment, we need a spatial map so that
we can "get around." Tactual maps for the blind are very similar to conventional
printed maps, except the density of information must be reduced because symbols
must be large and well spaced to be identified by the sense of touch.

Environmental features can be classified as *point* features (e.g., a bus stop), *linear*
features (e.g., a road) and *areal* features (e.g., a park). Much research effort has
been expended to develop sets of discriminable symbols for maps. James and Armstrong (1976) have suggested a standard set of point and line symbols, as shown in
Figure 6–15. These have been used in the United Kingdom and other parts of the
world (Armstrong, 1978). James and Gill (1975) investigated areal symbols and identified five discriminable symbols as shown in Figure 6–16.[2]

Although a fair amount of research has been done on symbol discriminability,

[2]See Lederman and Kinch (1979) for a review of areal symbol research and Hampshire
(1979) for a review of methods for producing tactile graphic material.

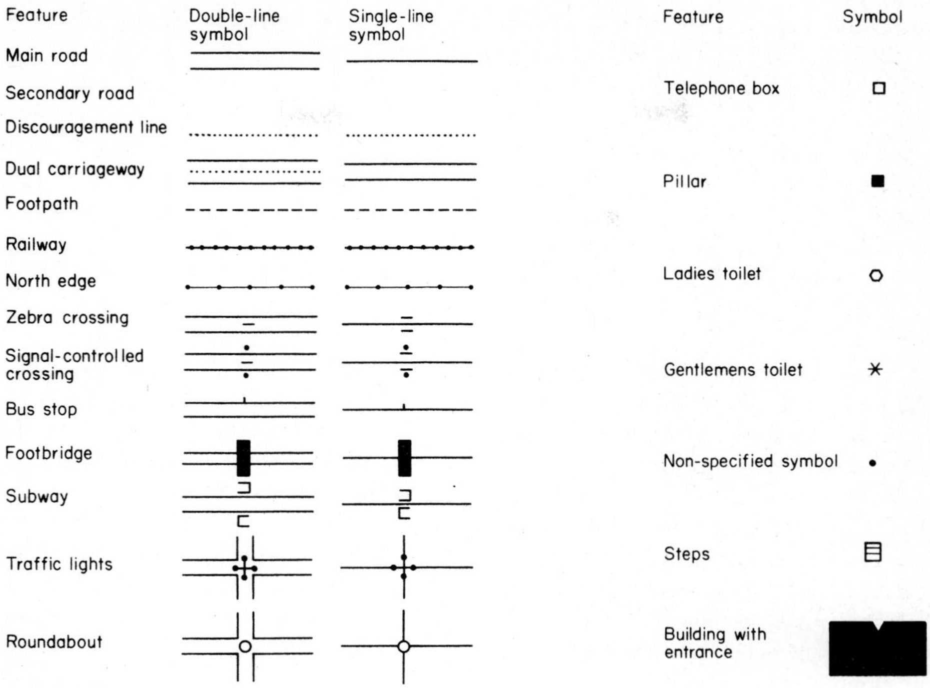

Feature	Double-line symbol	Single-line symbol		Feature	Symbol
Main road					
Secondary road				Telephone box	□
Discouragement line					
Dual carriageway				Pillar	■
Footpath					
Railway					
North edge				Ladies toilet	○
Zebra crossing					
Signal-controlled crossing				Gentlemens toilet	✳
Bus stop					
Footbridge				Non-specified symbol	•
Subway					
Traffic lights				Steps	
Roundabout				Building with entrance	

FIGURE 6-15
A standardized set of line and point symbols for tactual maps. (*Source: James and Armstrong, 1976, as shown in Armstrong, 1978, fig. 6.*)

there has been relatively little research studying the ability of blind persons to use spatial maps. What few studies there have been, however, seem to indicate that such maps can be used effectively to aid navigation in unfamiliar environments (James and Armstrong, 1975).

Tactual Graphs   One can hardly read a newspaper or magazine today without encountering a graphic display of data. Be they about the decline in college entrance examination scores or the increase in the federal deficit, graphs are being used with increasing frequency. Communicating such information to the blind requires careful

FIGURE 6-16
Discriminable areal symbols for use on tactual maps. (*Source: James and Gill, 1975, adapted from Armstrong, 1978, fig. 1.*)

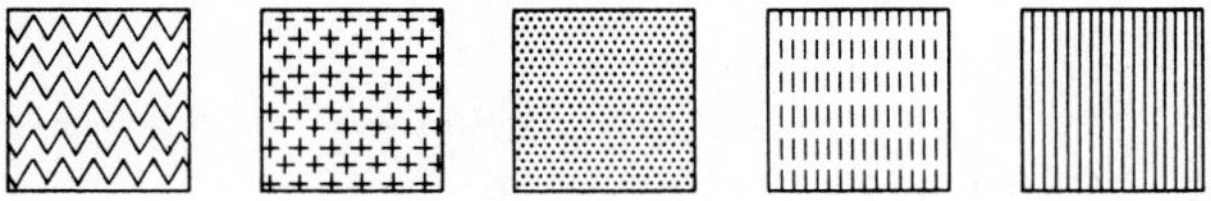

design to optimize the information transfer and reduce confusion. Apparently, it is fairly easy for blind people to learn to read two-dimensional tactual graphs such as the one shown in Figure 6–17 (Lederman and Campbell, 1982); however, the presence of a raised-background grid can cause confusion. Lederman and Campbell (1982) found that aligning and attaching a raised grid back to back with the raised data lines of a graph facilitated determining the *X* and *Y* values of a point of intersection. The subjects felt the data line with one hand and the grid, on the underside, with the other. Barth (1984), on the other hand, used a simpler procedure to facilitate performance. He used raised lines for the data and incised lines (groves) for the grid, intergrated together on one graph.

Tracking-Task Displays   Although the task of tracking is discussed in Chapter 9, we want to illustrate here the use of tactual displays for such purposes. Jagacinski, Gilson, and their colleagues have conducted a series of tracking studies, using the tactual display shown in Figure 6–18 (Jagacinski, Miller, and Gilson, 1979; Burke, Gilson, and Jagacinski, 1980; Jagacinski, Flach, and Gilson, 1983). Both single- and dual-task compensatory tracking situations were studied. In the single-task situation, the addition of quickening (see Chapter 9 for further discussion of quickening) to the tactual display produced tracking performance comparable to that obtained with traditional visual displays. In the dual-task situation subjects performed two tracking tasks simultaneously. The primary task was to control, by using one hand, the vertical movement of a target which was always shown on a visual display. The secondary task was to control, by using the other hand, the horizontal movement of the target which was displayed on either the visual display or the quickened tactual display. Performance was better with the visual-tactual primary-secondary task combination than with the visual-visual combination. Further, the superiority of the tactual

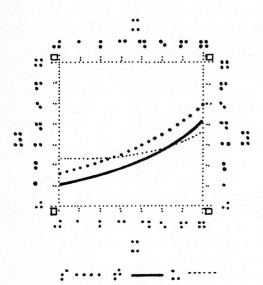

FIGURE 6-17
An example of a tactual graph for the blind. (*Source: Lederman and Campbell, 1982; fig. 1. VI.*)

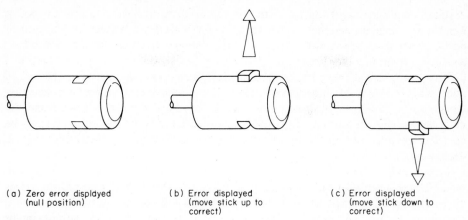

(a) Zero error displayed
(null position)

(b) Error displayed
(move stick up to
correct)

(c) Error displayed
(move stick down to
correct)

FIGURE 6-18
The tactual display used to present directional information to subjects in a tracking task. The correct response to condition *b* was to move the control stick up; to condition *c*, to move the stick down. (*Source: Jagacinski, Miller, and Gilson, 1979, fig. 1.*)

display was not due to the addition of quickening, but was attributed to its utilization of a different sensory modality which, it is believed, resulted in less competition for information processing resources.

Discussion

Relative to the other senses, the cutaneous senses seem generally suitable for transmitting only a limited number of discrete stimuli. Although we would then not expect tactual displays to become commonplace, there are at least two types of circumstances for which future developments of such displays would seem appropriate. First, such displays offer some promise of being useful in certain special circumstances in which the visual and auditory senses are overburdened; in such instances tactual displays might be used, such as for warning or tracking purposes. Second, as we have seen, they have definite potential as aids to blind persons.

OLFACTORY SENSE

In everyday life we depend on our sense of smell to give us information about things that otherwise would not be easily obtainable, for example, the fragrance of a flower, the smell of fresh-brewed coffee, and the odor of sour milk. The olfactory sense is simple in its construction, but its workings remain somewhat of a mystery. The sense organ for smell (called the *olfactory epithelium*) is a small, 4- to 6-cm^2, patch of cells located in the upper part of each nostril. These *olfactory cells* contain *olfactory hairs* which actually do the detecting of different odors. The cells are connected directly to the olfactory areas of the brain.

There is little doubt that the sensation of odor is caused by airborne molecules of a volatile substance entering the nose. Research workers, however, are still not in agreement on just what property of these molecules is responsible for the kind of odor perceived, or whether the sense of smell is composed of distinct primary odors, and if so, what they are (Engen, 1982).

Many people think that the sense of smell is very good—in one way it is, and in other ways it is not. The nose is apparently a very sensitive instrument for detecting the presence of odors; this sensitivity depends on the particular substance and the individual doing the sniffing. Ethyl mercaptan (a skunk odor), for example, can be detected when as little as 0.5 ml of the substance is dispersed in 10,000 liters of air (Engen, 1982).

Surprisingly, however, our sense of smell is not outstanding when it comes to making absolute identifications of specific odors (we do far better comparing odors on a relative basis). The number of different odors we can identify depends on a number of factors including the types of odors to be identified and the amount of training. Desor and Beauchamp (1974), for example, found that untrained subjects could identify approximately 15 to 32 common stimuli (coffee, paint, banana, tuna, etc.) by their odors alone. With training, however, some subjects could identify up to 60 stimuli without error. This may seem high compared with the visual or auditory sense, but it is not. These odor stimuli were complex ''whole odors'' as opposed to simple chemical odors. The analogy with the visual sense is the difference between simple color patches and complex pictures or faces. We can identify literally hundreds of different faces visually.

When it comes to identifying odors that differ only in intensity, we can identify about three or four different intensities (Engen, 1982). Thus, overall, the sense of smell is probably not to be counted on for identifying a specific stimulus from among many, but it can be effective in detecting the presence of an odor.

OLFACTORY DISPLAYS

Olfactory displays, needless to say, have not found widespread application. Part of the reason for this is that they cannot be depended on as a reliable source of information because people differ greatly with respect to their sensitivity to various odors; a stuffy nose can reduce sensitivity; people adapt quickly to odors so that the presence of an odor is not sensed after a little while of exposure; the dispersion of an odor is hard to control; and some odors make people feel sick.

Despite all these problems, olfactory displays do have some useful applications—primarily as warning devices. The gas company, for example, adds an odorant to natural gas so that we can detect gas leaks in our homes. This is no less an information display than if a blinking light were used to warn us of the leak.

The author has seen an unusual sign in a computer room which reads, ''If the red light is blinking or you smell wintergreen, evacuate the building.'' The fire prevention system in the building releases carbon dioxide when a fire is detected. When the system was intially activated, the people in the building did not detect the presence of the carbon dioxide, and several people lost consciousness from the lack of oxygen. A wintergreen odor was added to the gas as a warning.

In the prior examples odor was added to a gas to act as a warning that the gas was being released, intentionally in the case of carbon dioxide and unintentionally in the case of natural gas. Another example of an olfactory display uses odor to signal an emergency not associated with the gas itself. Several underground metal mines in the United States use a "stench" system to signal workers to evacuate the mine in an emergency. The odor is released into the mine's ventilation system and is quickly carried throughout the entire mine. This illustrates one general advantage of olfactory displays; they can penetrate vast areas which might not be economically reached by visual or auditory displays. Note, however, this is somewhat of a unique situation, consisting of vast areas (often hundreds of miles of tunnels) with no visual or auditory access and a closed ventilation system for controlling the dispersion of the odorant.

Olfactory displays will probably never become widespread in application, but they do represent a unique form of information display which could be creatively integrated into very special situations to supplement more traditional forms of displays.

REFERENCES

Adams, S., and Trucks, L. (1976). A procedure for evaluating auditory warning signals. In *Proceedings of the 6th Congress of the International Ergonomics Association and Technical Program for the 20th Annual Meeting of the Human Factors Society*. Santa Monica, CA: Human Factors Society.

American National Standards Institute (ANSI) (1966). *Standard specification for octave, half-octave, and third-octave band filter sets* (ANSI S1.11–1966), New York.

Armstrong, J. (1978). The development of tactual maps for the visually handicapped. In G. Gordon (ed.), *Active touch*. Elmsford, NY: Pergamon.

Barth, J. (1984). Incised grids: Enhancing the readability of tangible graphs for the blind. *Human Factors*, 26, 61–70.

Bliss, J. C., Katcher, M. H., Rogers, C. H., and Shepard, R. P. (1970). Optical-to-tactile image conversion for the blind. *IEEE Transactions on Man-Machine Systems*, 11(1), 58–65.

Burke, M., Gilson, R., and Jagacinski, R. (1980). Multi-modal information processing for visual workload relief. *Ergonomics*, 23, 961–975.

Caelli, T., and Porter, D. (1980). On difficulties in localizing ambulance sirens. *Human Factors*, 22, 719–724.

Davies, D. and Jones, D. (1982). Hearing and noise. In W. T. Singleton (ed.), *The Body at Work*, pp. 365–413. New York: Cambridge University Press.

Deatherage, B. H. (1972). Auditory and other sensory forms of information presentation. In H. P. Van Cott and R. G. Kinkade (eds.), *Human engineering guide to equipment design*. Washington: Government Printing Office.

Desor, J., and Beauchamp, G. (1974). The human capacity to transmit olfactory information. *Perception and Psychophysics*, 16, 551–556.

Dobelle, W. (1977). Current status of research on providing sight to the blind by electrical stimulation of the brain. *Visual Impairment and Blindness*, 71, 290–297.

Dodds, A., Clark-Carter, D., and Howarth, C. (1984). The Sonic Pathfinder: An evaluation. *Journal of Visual Impairment and Blindness*, 78, 203–206.

Engen, T. (1982). *The perception of odors*. New York: Academic.

Fidell S., Pearsons, K., and Bennett, R. (1974). Prediction of aural detectability of noise signals. *Human Factors*, 16, 373–383.

Frost, B., and Richardson, B. (1976). Tactile localization of sounds: Acuity, tracking moving sources, and selective attention. *Journal of the Acoustical Society of America*, 59, 907–914.

Gales, R. Hearing charactertistics. (1979). In C. M. Harris (ed.), *Handbook of noise control* (2d ed.). New York: McGraw-Hill.

Garfield, E. (1983, July 11). The tyranny of the horn—Automobile, that is. *Current Contents*, no. 28, pp. 5–11.

Geldard, F. A. (1957). Adventures in tactile literacy. *American Psychologist*, 12, 115–124.

Geldard, F. A. (1972). *The human senses* (2d ed.). New York: Wiley.

Guyton, A. (1969). *Function of the human body* (3d ed.). Philadelphia: Saunders.

Hampshire, B. (1979). The design and production of tactile graphic materials for the visually impaired. *Applied Ergonomics*, 10, 87–96.

Hawkes, G. (1962). *Tactile communication*, Rept. 61–11. Oklahoma City: Federal Aviation Agency, Aviation Medical Service Research Division, Civil Aeromedical Research Institute.

Heyes, A. (1984). The Sonic Pathfinder: A new electronic travel aid. *Journal of Visual Impairment and Blindness*, 78, 200–202.

Industrial noise manual (2d ed.). (1966). Detroit: American Industrial Hygiene Association.

Jagacinski, R., Flach, J., and Gilson, R. (1983). A comparison of visual and kinesthetic-tactual displays for compensatory tracking. *IEEE Transactions on Systems, Man, and Cybernetics*, 13, 1103–1112.

Jagacinski, R., Miller, D., and Gilson, R. (1979). A comparison of kinesthetic-tactual and visual displays via a critical tracking task. *Human Factors*, 21, 79–86.

James, G., and Armstrong, J. (1975). An evaluation of a shopping center map for the visually handicapped. *Journal of Occupational Psychology*, 48, 125–128.

James, G., and Armstrong, J. (1976). *Handbook on mobility maps*, Mobility Monograph no. 2. University of Nottingham.

James, G. and Gill, J. (1975). A pilot study on the discriminability of tactile areal and line symbols. *American Foundation for the Blind Research Bulletin*, 29, 23–33.

Jenkins, W. O. (1947). The tactual discrimination of shapes for coding aircraft-type controls. In P. M. Fitts (ed.), *Psychological research on equipment design*, Research Rept. 19. Army Air Force, Aviation Psychology Program.

Kahn, M. J. (1983) Human awakening and subsequent identification of fire-related cues. *Proceedings of the Human Factors Society, 27th Annual Meeting*. Santa Monica, CA: Human Factors Society, pp. 806–810.

Kay, L. (1973). Sonic glasses for the blind: A progress report. *AFB Research Bulletin*, 25, 25–28.

Kay, L. Bui, S., Brabyn, J., and Strelow, E. (1977). A single object sensor: A simplified binaural mobility aid. *Visual Impairment and Blindness*, 71, 210–213.

Kirman, J. (1973). Tactile communication of speech: A review and an analysis. *Psychological Bulletin*, 80, 54–74.

Kryter, K. (1985). *The effects of noise on man* (2d ed.). Orlando, FL: Academic.

Lederman, S. (1978). Heightening tactile impressions of surface texture. In G. Gordon (ed.), *Active touch*. Elmsford, NY: Pergamon.

Lederman, S., and Campbell, J. (1982). Tangible graphs for the blind. *Human Factors*, 24, 85–100.

Lederman, S., and Kinch, S. (1979). Texture in tactual maps and grapics for the visually handicapped. *Visual Impairment and Blindness*, 73, 217–227.

Licklider, J. C. R. (1961). *Audio warning signals for Air Force weapon systems* (TR 60–814). USAF, Wright Air Development Division, Wright Patterson Air Force Base.

Linvill, J. G. (1973). *Research and development of tactile facsimile reading aid for the blind (the Optacon)*. Stanford, CA: Stanford Electronics Laboratories, Stanford University.

Loomis, J. (1981). Tactile pattern perception. *Perception, 10*, 5–28.

Mansur, D. (1984). *Graphs in sound: A numerical data analysis method for the blind* (UCRL-53548). Berkeley, CA: Lawrence Livermore National Laboratory.

Miller, G. A. (1947). Sensitivity to changes in the intensity of white noise and its relation to masking and loudness. *Journal of the Acoustical Society of America, 19*, 609–619.

Minnix, R. (1978). The nature of sound. In D. M. Lipscomb and A. C. Taylor, Jr. (eds.), *Noise control: Handbook of principles and practices*. New York: Van Nostrand Reinhold.

Moore, B. (1982). *An introduction to the psychology of hearing* (2d ed.). New York: Academic.

Mudd, S. A. (1961). The scaling and experimental investigation of four dimensions of pure tone and their use in an audio-visual monitoring problem. Unpublished Ph.D. thesis, Lafayette, IN: Purdue Univeristy.

Mulligan, B., McBride, D., and Goodman, L. (1984). *A design guide for nonspeech auditory displays* (SR-84-1). Pensacola, FL: Naval Aerospace Medical Research Laboratory.

Munson, W. A. (1947). The growth of auditory sensitivity. *Journal of the Acoustical Society of America, 19*, 584.

Peterson, A. P. G., and Gross, E. E., Jr. (1972). *Handbook of noise measurement* (7th ed.). New Concord, MA: General Radio Co.

Pollack, I. (1952). Comfortable listening levels for pure tones in quiet and noise. *Journal of the Acoustical Society of America, 24*, 158.

Pollack, I., and Ficks, L. (1954). Information of elementary multidimensional auditory displays. *Journal of the Acoustical Society of America, 26*, 155–158.

Richardson, B., and Frost, B. (1979). Tactile localization of the direction and distance of sounds. *Perception and Psychophysics, 25*, 336–344.

Richardson, B., Wuillemin, D., and Saunders, F. (1978). Tactile discrimination of competing sounds. *Perception and Psychophysics, 24*, 546–550.

Riesz, R. R. (1928). Differential intensity sensitivity of the ear for pure tones. *Physiological Review, 31*, 867–875.

Russell, L. (1969). *Pathsounder instructor's handbook*. Cambridge: Massachusetts Institute of Technology, Sensory Aids Evaluation and Development Center.

Shower, E. G., and Biddulph, R. (1931). Differential pitch sensitivity of the ear. *Journal of the Acoustical Society of America, 3*, 275–287.

Stevens, J. (1982). Temperature can sharpen tactile acuity. *Perception and Psychophysics, 31*, 577–580.

Stevens, S. S., and Davis, H. (1938). *Hearing, its psychology and physiology*. New York: Wiley.

Talamo, J. D. (1982). The perception of machinery indicator sounds. *Ergonomics, 25*, 41–51.

Vallbo, A., and Johansson, R. (1978). The tactile sensory innervation of the glabrous skill of the human hand. In G. Gordon (ed.), *Active touch*. Elmsford, NY: Pergamon.

Van Cott, H. P., and Warrick, M. J. (1972). Man as a system component. In H. P. Van Cott and R. G. Kinkade (eds.), *Human engineering guide to equipment design* (rev. ed.). Washington: Superintendent of Documents.

Van der Molen, M., and Keuss, P. (1979). The relationship between reaction time and intensity in discrete auditory tasks. *Quarterly Journal of Experimental Psychology, 31*, 95–102.

Wegel, R. L., and Lane, C. E. (1924). The auditory masking of one pure tone by another and its probable relation to the dynamics of the inner ear. *Physiological Review, 23,* 266–285.

Weiss, C., and Kerschner, R. (1984). A case study: Nuclear power plant alarm system human factors review. In D. Attwood and C. McCann (eds.), *Proceedings of the 1985 International Conference on Occupational Ergonomics,* pp. 81–85. Rexdale, Ontario, Canada: Human Factors Association of Canada.

CHAPTER 7

SPEECH COMMUNICATIONS

From a human factors perspective, speech is an information "display," a form of auditory information. The source of this information is usually a human, but it can also be a machine. With our current level of technology, we can synthesize speech very inexpensively. Many of us have experienced synthesized speech in such consumer products as cameras and automobiles, and telephone companies are making greater use of synthesized speech to reduce the burden on operators.

Usually the receiver of speech (be it generated by a human or a microchip) is also a human, but it can be a machine. Although not as advanced as speech synthesis, speech recognition by computer systems is a reality. In this context, speech serves a control function, telling the computer system what to do, and so we defer our discussion of speech recognition to Chapter 10 which deals with control devices.

It is not feasible in this chapter to deal extensively with the many facets and intricacies of speech perception. For an introductory overview, consult Goldstein (1984) or Cole (1980). For our purposes, we first review briefly the nature of speech and some of the criteria used to evaluate it. Most of the chapter deals with how the components of a speech information system affect speech intelligibility. Finally, we close the chapter with a discussion of some human factors aspects of synthetic speech.

THE NATURE OF SPEECH

The ability of people to speak is, of course, integrally tied to and dependent on the respiratory process—if you do not breathe, you cannot speak. Speech sounds are generated by the sound waves that are created by the breath stream as it is exhaled from the lungs and modified by the speech organs. The various organs involved in speech include the lungs, the larynx (with its vocal cords), the pharynx (the channel between the larynx and the mouth), the mouth or oral cavity (with tongue, teeth, lips, and velum), and the nasal cavity. The vocal cords consist of folds that can open and close very rapidly, thus causing frequent stoppage of the breath stream. (The opening between the folds is called the *glottis*.) In speech, the rate of opening and closing of the vocal folds controls the basic frequency of the resulting speech sounds. As the sound wave is transmitted from the vocal cords, it is further modified in three resonators: the pharynx, the oral cavity, and the nasal cavity. By manipulating the mandible (in the back of the oral cavity) and the articulators (the tongue, lips, and velum) we are able to modify the sound wave to utter various speech sounds. In addition, by controlling the velum and the pharyngeal musculature we can influence the nasal quality of the articulation.

Types of Speech Sounds

The basic unit of speech is the phoneme. A *phoneme* is defined as the shortest segment of speech which, if changed, would change the meaning of a word. The English language has about 13 phonemes with vowel sounds (e.g., the *u* sound in *put* or the *u* sound in *but*) and about 24 phonemes with consonant sounds (e.g., the *t* sound in *tie,* the *g* sound in *gyp,* or the *g* sound in *gale),* and a couple of other phonemes called *diphthongs* (e.g., the *oy* sound in *boy* or the *ou* sound in *about*). Other languages have phonemes that do not exist in English and vice vera.

Phonemes are built up into syllables which form words that compose sentences. As such, phonemes are analogous to letters in printed text—although there are more phonemes than there are letters. One characteristic of phonemes is that their acoustical sound properties can change when they are paired with other phonemes, yet we usually do not perceive the change. For example, can you detect the subtle difference in the *d* sound of *di* and *du*? Acoustically the sounds are quite different, although we perceive the *d* sounds as virtually the same.

Depicting Speech

The variations in air pressure with any given sound, including speech, can be represented graphically in various ways, as shown in Figure 7–1. One such representation is a *waveform* (Figure 7–1a) which shows the variation in air pressure (intensity) over time. Another form of representation is a *spectrum* (Figure 7–1b) which shows, for a given period, phoneme, or word, the intensity of the various frequencies occurring during that sample. The third form of representation is called a *sound spectrogram* (Figure 7–1c). Here frequency is plotted on the vertical axis, time is on the horizontal axis, and intensity is depicted by the degree of darkness of the plot.

Intensity of Speech

The average intensity, or speech power, of individual speech sounds varies tremendously, with the vowels generally having much greater speech power than the consonants. For example, the *a* as pronounced in *talk* has roughly 680 times the speech power of *th* as pronounced in *then*. This is a difference of about 28 dB. Unfortu-

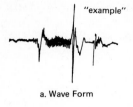

"example"

a. Wave Form

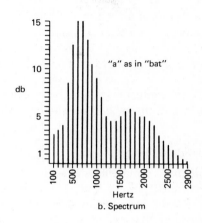

"a" as in "bat"

b. Spectrum

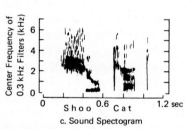

c. Sound Spectogram

FIGURE 7-1
Three methods of graphically representing speech. (*Source: c. Kiang, 1975.*)

nately, the consonants, with their lower speech power, are the important phonemes for speech intelligibility.

The overall intensity of speech, of course, varies from one person to another and also for the same individual. For example, the speech intensity of males tends to be higher than for females by about 3 to 5 dB. Speech intensity usually varies from about 45 dBA for weak conversational speech; to 55 dBA for conversational speech; to 65 dBA for telephone-lecture speech; to 75 dBA for loud, near-shouting speech; to 85 dBA for shouting (Kryter, 1985).

Frequency Composition of Speech

As indicated above, each speech sound has its own unique spectrum of sound frequencies, although there are, of course, differences in these among people and individuals can and do shift their spectra up and down the register depending in part on the circumstances, such as in talking in a quiet conversation (when frequency and intensity are relatively low) or screeching at the children to stay out of trouble (when frequency and intensity typically are near their peak). In general, the spectrum for men has more dominant components in the lower frequencies than for women. These differences are revealed in Figure 7–2 which shows long-time average speech spectra for men and women.

CRITERIA FOR EVALUATING SPEECH

In the process of designing a speech communication system, one needs to establish the criteria or standards that the system should meet to be acceptable for the intended uses. Criteria are also required to evaluate the effects of noise and ear protectors on speech communications and to measure the effectiveness of hearing aids. The major criterion for evaluating a speech communication system is intelligibility, but there are others as well, including quality or naturalness of the speech.

Speech Intelligibility

Intelligibility is simply the degree to which a speech message (e.g., a list of words) is correctly recognized. Intelligibility is assessed by transmitting speech material to individuals who are asked to repeat what they hear or to answer questions about what they heard. Various speech intelligibility tests exist which differ in terms of the types of materials presented [e.g., nonsense syllables, phonetically balanced (PB) word lists that contain all the various speech sounds, and sentences]. In general, in a given situation, intelligibility is highest for sentences, less for isolated words, and lowest for nonsense syllables. In fact, in everyday conversational speech about half the words are unintelligible when they are taken out of their fluent context and presented alone—even when one listens to one's own voice (Pollack and Pickett, 1964). In a normal rate of speaking, trying saying, "Did you go to the store?" "Dijoo"? or "Did you"? Intelligibility, therefore is very dependent on context and expectations; without the context of the sentence, the individual words may not be recognized.

Speech Quality

Speech quality, or naturalness, goes beyond intelligibility. It is important in situations where it is desirable to recognize the identity of a speaker, such as on the telephone. Speech quality is also important as one determinant of user satisfaction with a communication system. For example, mobile telephones for automobiles have

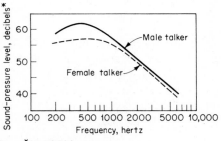

FIGURE 7-2
The octave-band long-term average speech spectrum for male and female talkers. (*Source: Adapted from Kryter, 1972, fig. 5-2, p. 164.*)

become relatively inexpensive and available, and telephones for the home that allow a person to talk and hear from anywhere in the room (called *speaker phones*) are also popular. Automatic telephone answering machines are also becoming common. In all these cases the quality of the speech heard over the system is often the critical factor that accounts for the purchase of one system over another.

Speech quality is defined in terms of preference. Usually samples of speech are presented over the system, and people are asked to either rate the quality (excellent, fair, poor, unacceptable) or to compare it to some standard and indicate which would be preferred. This type of methodology is often used to evaluate synthetic speech as well.

COMPONENTS OF SPEECH COMMUNICATION SYSTEMS

A speech communication system can be thought to consist of the speaker, the message, the transmission system, the noise environment, and the hearer. We discuss each in terms of its effects on intelligibility of speech communications, and we provide some guidance for improving intelligibility within the system.

Speaker

As we all know, the intelligibility of speech depends in part on the character of the speaker's voice. Although, in common parlance, we refer to such features of speech as *enunciation,* research has made it possible to trace certain specific features of speech that affect its intelligibility. Bilger, Hanley, and Steer (1955), for example, found that the speech of "superior" speakers (as contrasted with less intelligible speakers) (1) had a longer "syllable duration," (2) spoke with greater intensity, (3) utilized more of the total time with speech sounds (and less with pauses), and (4) varied their speech more in terms of fundamental vocal frequencies.

The differences in intelligibility of speakers generally are due to the structure of their articulators (the organs that are used in generating speech) and the speech habits people have learned. Although neither factor can be modified very much, it has been found that appropriate speech training usually can bring about moderate improvements in the intelligibility of speakers.

Message

Several characteristics of the message affect its intelligibility, including the phonemes used, the words, and the context.

Phoneme Confusions   Certain speech sounds are more easily confused than others. Hull (1976), for example, found letters within each of the following groups to be frequently confused with one another: DVPBGCET, FXSH, KJA, MN. Kryter (1972) specified two groups of consonants which in noise tend to be confused within sets but not between sets: MNDGBVZ and TKPFS. As you can see, the groupings are not consistent. In general, avoid using single letters as codes where noise is present.

Word Characteristics   Intelligibility is greater for familiar words than for unfamiliar words. Therefore, in designing a speech communication system, the receiver should be familiar with the words that the sender will use. In general, intelligibility is higher for long words than for short words. Part of the reason is that if only parts of a long word are intelligibile, then often a person can figure out the word. With short words, there is less opportunity to pick up pieces of the word. An application of this principle is the use of a word-spelling alphabet (alpha, bravo, charlie, delta, . . .) in place of single letters (A, B, C, D, . . .) in the transmission of alphanumeric information.

Context Features   As we already discussed, intelligibility is higher for sentences than for isolated words because the context supplies information if a particular word is not understood. Further, intelligibility is higher when words are arranged in meaningful sentences (*the boat is carrying passengers*) than when the same words are arranged in nonsense sentences (*boat the carrying passengers is*). In addition, the more words in the vocabulary being used, the lower will be the intelligibility under conditions of noise. For example, Miller, Heise, and Lichten (1951) presented subjects with words from vocabularies of various sizes (2, 4, 8, 16, 32, and 256 words plus unselected monosyllables). These were presented under noise conditions with signal-to-noise ratios ranging from -18 to $+9$ dB. (The signal-to-noise ratio is simply the difference in sound-pressure level between the signal and the noise; negative numbers indicate that the noise is more intense than the signal.) The results are shown in Figure 7–3; the percentage of words correctly recognized was very strongly correlated with the size of the vocabulary that was used.

To improve intelligibility, especially under noisy conditions, the following guidelines should be observed when messages are designed for a speech communication system: (1) keep the vocabulary as small as possible; (2) use standard sentence constructions with information always transmitted in the same order (e.g., bin location, part number, part name, quantity); (3) avoid short words, and rather than use letters, use a word-spelling alphabet; and (4) familiarize the receiver with the words and sentence structure to be used.

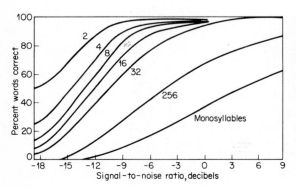

FIGURE 7-3
Intelligibility of words from vocabularies of different sizes under varying noise conditions. The numbers 2, 4, 8, etc., refer to the number of words in the vocabulary used. (*Source: Miller, Heise, and Lichten, 1951. Copyright © 1951 by the American Psychological Association and reproduced by permission.*)

Transmission System

Speech transmission systems (such as telephones and radios) can produce various forms of distortion, such as frequency distortion, filtering, amplitude distortion, and modifications of the time scale. If intelligibility (and not fidelity) is important in a system, then certain types of distortion (especially amplitude distortion) still can result in acceptable levels of intelligibility. Since high-fidelity systems are very expensive, it is useful to know what effects various forms of distortion have on intelligibility to be able (it is hoped) to make better decisions about the design or selection of communication equipment. For illustrative purposes, we discuss the effects of filtering and amplitude distortion.

Effects of Filtering on Speech   The filtering of speech consists basically in blocking out certain frequencies and permitting only the remaining frequencies to be transmitted. Filtering may be a fortuitous, unintentional consequence, or in some instances an intentional consequence, of the design of a component. Most filters eliminate frequencies *above* some level (a *low-pass* filter) or frequencies *below* some level (a *high-pass* filter). Typically, however, the cutoff, even if intentional, is not precisely at a specific frequency, but rather tapers off over a range of adjacent frequencies. Filters that eliminate all frequencies above, or all frequencies below, 2000 Hz *in the quiet* will still transmit speech quite intelligibly, although the speech does not sound natural. The distortion effects on speech of filtering out certain frequencies are summarized in Figure 7–4. This shows, for high-pass filters and for low-pass

FIGURE 7-4
Effect on intelligibility of elimination of frequencies by the use of filters. A low-pass filter permits frequencies below a given cutoff to pass through the communication system and eliminates frequencies above that cutoff. A high-pass filter, in turn, permits frequencies above the cutoff to pass and eliminates those below. (*Source: Adapted from French and Steinberg, 1947.*)

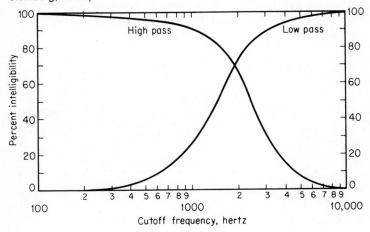

filters with various cutoffs, the intelligibility of the speech. The filtering out of frequencies above 4000 Hz or below about 600 Hz has relatively little effect on intelligibility. But look at the effect of filtering out frequencies above 1000 Hz or below 3000 Hz! Such data as those given in Figure 7–4 can provide the designer of communications equipment with some guidelines about how much filtering can be tolerated in the system.

Effects of Amplitude Distortion on Speech   *Amplitude distortion* is the deformation which results when a signal passes through a nonlinear circuit. One form of such distortion is *peak clipping,* in which the peaks of the sound waves are clipped and only the center part of the waves are left. Although peak clipping is produced by electronic circuits for experiments, some communication equipment has insufficient amplitude-handling capability to pass the peaks of the speech waves and at the same time provide adequate intensity for the lower-intensity speech components. Since peak clipping impairs the quality of speech and music, it is not used in regular broadcasting; but it is sometimes used in military and commerical communication equipment. *Center clipping,* on the other hand, eliminates the amplitudes below a given value and leaves the peaks of the waves. Center clipping is more of an experimental procedure than one used extensively in practice. The amount of clipping can be controlled through electronic circuits. Figure 7–5 illustrates the speech waves that would result from both forms of clipping.

The effects of these two forms of clipping on speech intelligibility are very different, as shown in Figure 7–6. Peak clipping does not cause major degradation of intelligibility even when the amount of clipping (in decibels) is reasonably high. On the other hand, even a small amount of center clipping results in rather thorough garbling of the message. The reason for this difference in the effects of peak and center clipping is that when the peaks are lopped off, we reduce the power of the vowels, which are less critical in intelligibility, and leave the consonants essentially unscathed. But when we cut out the center amplitudes, the consonants fall by the wayside, thus leaving essentially the high peaks of the vowels. Since intelligibility is relatively insensitive to peak clipping, the communications engineers can shear off the peaks and repackage the waveforms, using the available power to amplify the weaker but more important consonants.

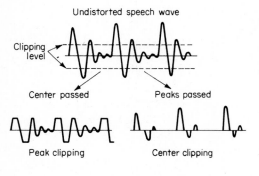

FIGURE 7-5
Undistorted speech wave and the speech waves that would result from peak clipping and center clipping. (*Source: Miller, 1963, p. 72.*)

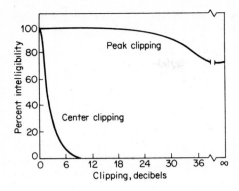

FIGURE 7-6
The effects on speech intelligibility of various amounts of peak clipping and of center clipping. (*Source: Licklider and Miller, 1951.*)

Noise Environment

Noise, be it external in the environment or internal to the transmission system, is the bane of speech intelligibility. Various approaches have been developed to evaluate the effects of noise on speech communications. The indices that have resulted from a few of these approaches are described briefly, namely, the Articulation Index (AI), the Preferred-Octave Speech Interference Level (PSIL), and the Preferred Noise Criteria (PNC) curves.

Articulation Index The Articulation Index was developed to predict speech intelligibility given a knowledge of the noise environment. It is much easier to measure the noise environment than it is to directly measure intelligibility in the environment. The AI, therefore, is a cost-effective method for indirectly assessing speech intelligibility. There are actually several articulation indices, all being a weighted sum of the differences between the sound-pressure level of speech and noise in various octave bands. One such technique which uses the one-third octave-band method (ANSI, 1969; Kryter, 1972) is described below and illustrated in Figure 7–7:

1 For each one-third octave band, plot on a worksheet, such as in Figure 7–7, the band level of the speech peak reaching the listener's ear. (The specific procedures for deriving such band levels are given by Kryter.) An idealized spectrum for males is presented in Figure 7–7 as the basis for illustrating the derivation of the AI.

2 Plot on the worksheet the band levels of steady state noise reaching the ear of the listener. An example of such a spectrum is presented in Figure 7–7.

3 Determine, at the center frequency of each of the bands on the worksheet, the difference in decibels between the level of speech and that of the noise. When the noise level exceeds that of speech, assign a zero value. When the speech level exceeds that of noise by more than 30 dB, assign a difference of 30 dB. Record this value in column 2 of the table given as part of Figure 7–7.

4 Multiply the value for each band derived by step 3 by the weight for that band as given in column 3 of the table given with Figure 7–7, and enter that value in column 4.

5 Add the values in column 4. This sum is the AI.

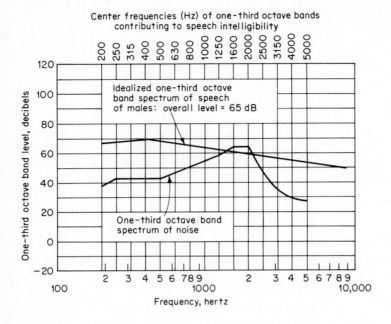

FIGURE 7-7
Example of the calculation of an articulation idex (AI) by the one-third
octave-band method. In any given situation the difference (in decibels)
between the level of speech and the level of noise is determined for each
band. These differences are multiplied by their weights. The sum of these
is the AI. (*Source: Adapted from Kryter, 1972.*)

The relationships between AI and intelligibility (percentage of material correctly
understood) for various types of materials are shown in Figure 7–8. Using the AI of
0.47 computed in Figure 7–7, we would predict that about 75 percent of a 1000-
word vocabulary would be correctly understood, while almost 90 percent of a 256-

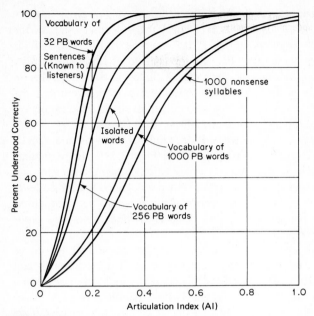

FIGURE 7-8
Relationship between the AI and the intelligibility of various types of speech test materials. (*Source: Adapted from Kryter, 1972, and French and Steinberg, 1947.*)

word vocabulary would be understood. If, on the other hand, complete sentences known to the receiver were used, we would expect near 100 percent intelligibility.

In general, expected satisfaction with a speech communication system is related to the AI of the system, as shown in Table 7–1.

Preferred-Octave Speech Interference Level This index, reported by Peterson and Gross (1978), can be used as a rough estimate of the effects of noise on speech

TABLE 7–1
RELATIONSHIP BETWEEN ARTICULATION INDEX AND EXPECTED USER SATISFACTION WITH A COMMUNICATION SYSTEM

AI	Rating of satisfaction
<0.3	Unsatisfactory to marginal
0.3–0.499	Acceptable
0.5–0.7	Good
>0.7	Very good to excellent

Source: Beranek (1947).

reception. PSIL is simply the numeric average of the decibel levels of noise in three octave bands centered at 500, 1000, and 2000 Hz. Thus, if the decibel levels of the noise were 70, 80, and 75 dB, respectively, the PSIL would be 75 dB. The PSIL is useful when the noise spectrum is relatively flat, but the PSIL loses some of its utility if the noise has intense components outside the range included in the PSIL, has an irregular spectrum, or consists primarily of pure tones. An older index called the *Speech Interference Level* (SIL) is an average of the decibel levels of different octave bands, namely, 600 to 1200, 1200 to 2400, and 2400 to 4800 Hz.

Some indication of the relationship between PSIL (and SIL) values and speech intelligibility under different circumstances is shown in Figure 7–9. In particular, this shows the effects of noise in face-to-face communications when the speaker and the hearer are at various distances from each other. For the distance and noise level combinations in the white area of the figure, a normal voice would be satisfactory; but for those combinations above the "maximum vocal effort line," it would probably be best to send up smoke signals.

The subjective reactions of people to noise levels in private and large offices (secretarial, drafting, business machine offices, etc.) were elicited by Beranek and Newman (1950) by questionnaires. The results of this survey were used to develop a rating chart for office noises, as shown in Figure 7–10. The line for each group represents PSILs (baselines) that were judged to exist at certain subjective ratings (vertical scale). The dot on each curve represents the judged upper limit for intelligibility. The judged PSIL limit for private offices [normal voice at 9 ft (2.7 m)] was slightly above 45 dB, and for larger offices [slightly raised voice at 3 ft. (0.9 m)] was 60 dB.

The judgments with regard to telephone use are given in Table 7–2. These judg-

FIGURE 7-9
Voice level and distance between talker and listener for satisfactory face-to-face communication as limited by ambient noise level (expressed in PSIL and SIL). For any given noise level (such as PSIL of 70) and distance [such as 8 ft (2.4 m)] it is possible to determine the speech level that would be required (in this case a "shout"). (*Source: Adapted from Webster, 1969, fig. 19, p. 69.*)

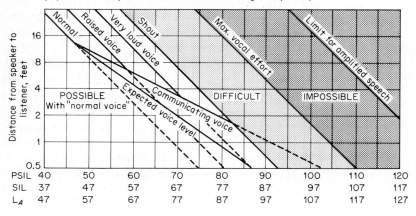

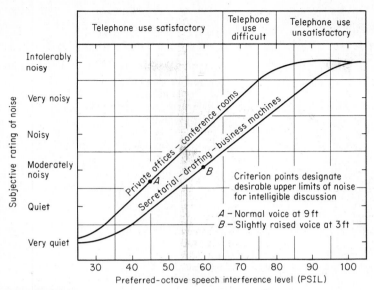

FIGURE 7-10
Rating chart for office noises. (*Source: Beranek and Newman, 1950, as modified by Peterson and Gross, 1978, to reflect the current practice of using octave bands with centers at 500, 1000, and 2000 Hz.*)

ments were made for long-distance or suburban calls. For calls within a single exchange, about 5 dB can be added to each of the listed levels, since there is usually better transmission within a local exchange.

Criteria for background noise in various communication situations have been set forth by Peterson and Gross, based on earlier standards by Beranek and Newman. These, expressed as PSILs, are given in Table 7–3.

Preferred Noise Criteria Curves   Beranek, Blazier, and Figwer (1971) developed a set of curves to evaluate the noise environment inside office buildings and in rooms and halls in which speech communications are important. The curves are

TABLE 7–2
EFFECTS OF PSIL ON TELEPHONE USE

PSIL, dB	Telephone use
<60	Satisfactory
60–75	Difficult
>80	Impossible

Source: Peterson and Gross, 1978, p. 38.

TABLE 7–3
MAXIMUM PERMISSIBLE PSIL FOR CERTAIN TYPES OF ROOMS AND
SPACES

Type of room	Maximum permissible PSIL (measured when room is not in use)
Secretarial offices, typing	60
Coliseum for sports only (amplification)	55
Small private office	45
Conference room for 20	35
Movie theater	35
Conference room for 50	30
Theaters for drama, 500 seats (no amplification)	30
Homes, sleeping areas	30
Assembly halls (no amplification)	30
Schoolrooms	30
Concert halls (no amplification)	25

Source: Peterson and Gross, 1978, table 3–5, p. 39.

based on the older Noise Criteria (NC) curves originally developed by Beranek in 1957. A set of PNC curves is shown in Figure 7–11. To use these curves, the noise spectrum of the area being evaluated is plotted on the chart of curves, and each

FIGURE 7-11
Preferred noise criterion (PNC) curves. (*Source: Beranek, Blazier, and Figwer, 1971.*)

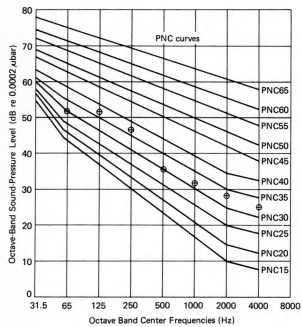

TABLE 7–4
RECOMMENDED PNC LEVELS FROM STEADY BACKGROUND NOISE
ACCORDING TO TYPE OF SPACE FOR INDOOR ACTIVITIES INVOLVING
SPEECH COMMUNICATIONS

Type of space	Recommended PNC level
Concert halls, opera houses	10–20
Large auditoriums, theaters, churches	Not to exceed 20
Small auditoriums, theaters, churches	Not to exceed 35
Bedrooms	25–40
Private or semiprivate offices, classrooms	30–40
Large offices, retail stores, restaurants	35–45
Shops, garages, power plant control rooms	50–60

Source: Beranek, Blazer, and Figwer, 1971.

octave-band level is compared with the PNC curves to find the one that penetrates the highest PNC level. The highest PNC curve penetrated is the PNC rating for the environment. As an example, the noise spectrum of an office is shown in Figure 7–11 by several encircled crosses. This noise would have a PNC rating of 37 since that is the value of the highest PNC curve penetrated by the noise spectrum. Table 7–4 lists a few recommended PNC levels for various purposes and spaces.

Reverberation   *Reverberation* is the effect of noise bouncing back and forth from the walls, ceiling, and floor of an enclosed room. As we know from experience, in some rooms or auditoriums this reverberation seems to obliterate speech or important segments of it. Figure 7–12 shows approximately the reduction in intelligibility that is caused by varying degrees of reverberation (specifically, the time in seconds that it takes the noise to die down). This relationship essentially is a straight-line one.

Hearer

The hearer, or listener, is the last link in the communication chain. For receiving speech messages under noise conditions, the hearer should have normal hearing, should be trained in the types of communications to be received, should be reasonably able to withstand the stresses of the situation, and should be able to concentrate on one of several conflicting stimuli.

A discussion of the effects of hearing protectors on speech intelligibility is presented in Chapter 16 on noise.

SYNTHESIZED SPEECH

Advances in technology made possible the synthesis of human speech at relatively low cost and with excellent reliability. The availability of this technology has generated a great deal of human factors research aimed at determining when synthesized speech is most appropriate to use, what aspects of the system influence human performance and preference, and what can be done to improve such systems. We briefly introduce the principal types of speech synthesis systems, and we discuss some of the human factors aspects of such systems. A good overview of the area is presented by Simpson et al. (1985) and by McCauley (1984).

Types of Speech Synthesis Systems

The traditional method of reproducing speech (as opposed to synthesizing it) is to record it as an analog signal on tape or record for later playback. The major problem with this technology is that moving parts, inherent in a record or tape playback system, wear out and break down. Further, only messages recorded in advance can be played back, and it takes time to access specific messages on the tape or record. A more advanced technology, but one similar in concept, involves digitizing speech and storing it in computer memory, which eliminates the moving-parts problem. Digitizing involves sampling the speech signal very rapidly and storing information about each sample (such as its intensity) for later decoding back to speech. The problem with this technique is that massive amounts of storage are required to store the digitized information. It is not uncommon to sample a speech signal 8000 or more times *per second*. Depending on how much information is encoded in each of the samples, to store a dozen or so English words could require 1 million bits of memory or more. Although the quality of analog and digital recordings is very good, the reliability and storage problems have made them somewhat impractical for most applications. To get around these problems, two methods have been developed that synthesize speech: analysis-synthesis and synthesis by rule. We discuss each briefly; but for a more indepth introduction, see Michaelis and Wiggins (1982).

Analysis-Synthesis In this technique the human vocal mechanism is simulated by using an electronic model to produce speech sounds. The model uses various mechanisms, such as filters and modulators, to "mold" the sounds. These filters, modulators, etc., have parameters that can be changed to produce various patterns of speech sounds. Rather than encode the actual speech wave, as would be the case

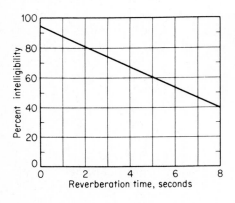

FIGURE 7-12
Intelligibility of speech in relation to reverberation time. The longer the reverberation of noise in a room, the lower the intelligibility of speech. (*Source: Adapted from Fletcher, 1953.*)

with digitized speech, the speech is analyzed and only the changing parameters needed to direct the speech production model are encoded. The principal advantage is that considerably less computer memory is required to store the speech information than with straight digitization. As with digitized speech, however, messages can be formed from only those words or phrases that were previously encoded and stored. Further, when individual words are linked, they can sound very awkward and unnatural because of a lack of *coarticulation,* or the natural blending and modification of speech sounds caused by words and phonemes that precede and follow a particular sound. Consider the word *bookcase.* A speech synthesizer linking the two words *book* and *case* would pronounce the *k* sound from *book* and the *k* sound from *case.* When a human says *bookcase,* only one *k* sound is made.

Synthesis by Rule   Rather than encoding model parameters from actual speech as in the case of analysis-synthesis, synthesis-by-rule systems determine the appropriate parameters for the speech production model at the time the speech is produced. The system has stored dictionaries of elementary speech segments and sets of rules (hence the name *synthesis by rule*) for combining them (including coarticulation rules) and for stressing particular sounds or words that produce the prosody of speech. *Prosody* is the rhythm or singsong quality of natural speech. The main advantage of this technique is that very large vocabularies are possible with relatively small amounts of computer memory. These sorts of systems make possible direct translation of text into speech. The problem is that the voice quality is usually not as good as that achieved by analysis-synthesis systems.

Uses of Synthesized Speech

Synthesized speech has taken on a "whiz-bang" high-tech connotation. Many consumer products can be purchased with built in speech synthesizers, including automobiles ("Your door is ajar."), cameras ("Out of film."), microwave ovens ("What power level?"), and wrist-watches ("It is 2 o'clock and 15 minutes."). Several children's toys use the technology for educational purposes. In the field of aviation, synthesized speech has been used for cockpit warnings (Werkowitz, 1980) and interactive training systems (McCauley and Semple, 1980). Telephone companies have made wide use of synthesized speech to present telephone numbers to callers. The possibility of using synthesized speech to aid the handicapped has also not gone unnoticed. Devices include machines for the visually handicapped that convert printed materials (typewritten or typeset) directly into speech at rates of up to 200 words per minute (Kurzweil Computer Products, Cambridge, MA), signs that talk when scanned by a special infrared receiver, and systems for providing a voice for the vocally handicapped. The widespread use of synthesized speech has introduced some interesting human factors considerations in terms of performance with, preference for, and appropriate use of synthesized speech.

Performance with Synthesized Speech

Human factors specialists are concerned about how well people can perform when they are presented messages in synthesized speech. This often boils down to questions of how intelligible synthesized speech is compared to natural speech and how well people can remember synthesized speech messages.

Intelligibility of Synthesized Speech   As we alluded earlier, the intelligibility of synthesized speech can be worse than that of natural speech depending on the type of system used to produce the sound and the sophistication of the speech production model built into the system. Waterworth (1984) tested four synthesizers and found that intelligibility for the best one was almost as good as that of natural speech, but that the worst system was unaccceptably poor (less than 15 percent of the words correctly recognized). Often for familiarized subjects and limited vocabularies used under high positive signal-to-noise ratios, a good system can be 99 to 100 percent intelligible (Simpson et al., 1985). Brabyn and Brabyn (1982), for example, report intelligibility scores of 91 to 99 percent for talking signs ("drinking fountain") used to guide the visually handicapped. Interestingly, bilinguals, but not native English-speaking subjects, had a harder time understanding the synthesized speech (91 percent intelligibility) than they did the natural speech (99 percent intelligibility). People apparently adapt rather quickly to synthesized speech; and with just a little exposure to a system's voice, intelligibility increases significantly (Thomas, Rosson, and Chodorow, 1984).

Memory for Synthesized Speech   A fairly common finding is that people have more trouble remembering messages spoken in synthesized speech than in natural speech (Luce, Feustel, and Pisoni, 1983; Waterworth and Thomas, 1985). It is generally agreed that listening to synthesized speech requires more processing capacity than does listening to natural speech. Apparently this is due to encoding difficulties that disrupt working memory and the transfer of information to long-term memory. (For a discussion of processing capacity and memory see Chapter 3.) It appears, however, that once encoded, synthesized speech is stored just as efficiently as is natural speech. Luce, Feustel, and Pisoni (1983) found that increasing the interword interval from 1 to 5 s during the presentation of a list of words almost doubled the number of words recalled. In a similar vein, Waterworth (1983) reported that increasing the pause between groups of digits improved their recall.

Preference for Synthesized Speech

A broad consideration is whether people really want inanimate objects such as cars and cameras to talk to them. Creitz (1982), for example, reports that sales of a microwave oven that told the user what to do were so bad that the manufacturer stopped producing them.

The quality of synthesized speech can be quite poor even though the messages are intelligible. Conversely, a system can be rated as pleasant and yet be totally incomprehensible. Synthesized speech is often described as machinelike, choppy, harsh and grainy, flat, and noisy (Edman and Metz, 1983). Usually what is lacking is coarticulation (where words are blended) and natural intonation. New techniques are constantly being developed to improve the quality of synthesized speech, but preference and rated quality are very subjective and differ from person to person for the same system. In some applications, such as aircraft cockpit warnings, many pilots do not want a natural-sounding synthesized voice; they want a voice that sounds like a machine, to distinguish it from human voices in the cockpit (Simpson et al., 1985).

As an illustration of the subtle effect of intonation on preference, Waterworth (1983) compared various ways of saying digit strings on people's preferences. Various combinations of neutral, terminator (dropping pitch), and continuant (rising pitch) intonations were tested. Figure 7–13 shows both the original intonation pattern used by British Telecom to deliver telephone numbers on their system and the new, improved intonation pattern developed as a result of Waterworth's research.

Guidelines for Use of Synthesized Speech

It is probably a little premature to offer specific guidelines for using synthesized speech because the technology is evolving so rapidly. In Chapter 3 we present some general guidelines for deciding when auditory or visual displays are appropriate. Synthesized speech is one form of an auditory display, and as such, its use should conform to the guidelines in Chapter 3. The design of the messages for synthesized speech systems should also follow those mentioned earlier in this chapter for natural speech communication systems. In addition, the following guidelines were gleaned from a number of sources (Simpson and Williams, 1980; Thomas, Rosson, and Chodorow, 1984; Wheale, 1980) and deal with various applications of synthesized speech:

1 Voice warnings should be presented in a voice that is qualitatively different from other voices that will be heard in the situation.

2 If synthesized speech is used exclusively for warnings, there should be no alerting tones before the voice warning.

3 If synthesized speech is used for other types of information in addition to warnings, some means of directing attention to the voice warning might be required.

4 Maximize intelligibility of the messages.

5 For general-purpose use, maximize user acceptance by making the voice as natural as possible.

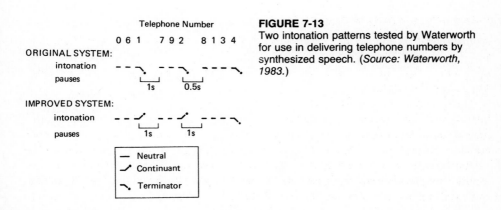

FIGURE 7-13
Two intonation patterns tested by Waterworth for use in delivering telephone numbers by synthesized speech. (*Source: Waterworth, 1983.*)

6 Consider providing a replay mode in the system so users can replay the message if they desire.

7 If a spelling mode is provided, its quality may need to be better than that used for the rest of the system.

8 Give the user the ability to interrupt the message; this is especially important for experienced users who do not need to listen to the entire message each time the system is used.

9 Provide an introductory or training message to familiarize the user with the system's voice.

10 Do not get caught up in "high-tech fever." Use synthesized speech sparingly and only where it is appropriate and acceptable to the users.

DISCUSSION

Speech communication by human or microchip will continue to be a dominant form of information display, so the human factors specialist must understand the basics of speech perception and the effects of various facets of the communication system on speech intelligibility. As synthesized speech technology evolves, whole new areas of application will open and the human factors problems and challenges will be there in force.

REFERENCES

American National Standards Institute (ANSI) (1969). *Methods for the calculation of the articulation index,* Rept. 535. New York.

Beranek, L. (1947). The design of speech communication systems. *Proceedings of the Institute of Radio Engineers,* 35, 880–890.

Beranek, L. (1957). Revised criteria for noise in buildings. *Noise Control,* 3(1), 19–27.

Beranek, L., Blazier, W., and Figwer, J. (1971). Preferred noise criteria (PNC) curves and their application to rooms. *Journal of the Acoustical Society of America,* 50, 1223–1228.

Beranek, L., and Newman, R. (1950). Speech interference levels as criteria for rating background noise in offices. *Journal of the Acoustical Society of America,* 22, 671.

Bilger, R., Hanley, T., and Steer, M. (1955). *A further investigation of the relationships between voice variables and speech intelligibility in high level noise,* TR for SDC, 104–2–26, Project 20–F–8, Contract N6ori–104, Lafayette, IN: Purdue University (mimeographed).

Brabyn, J., and Brabyn, L. (1982). Speech intelligibility of the talking signs. *Visual Impairment and Blindness,* 76, 77–78.

Cole, R. (ed.) (1980). *Perception and production of fluent speech.* Hillsdale, NJ: Erlbaum.

Creitz, W. (ed.) (1982). *Voice news.* Rockville, MD: Stoneridge Technical Services.

Edman, T., and Metz, S. (1983). A methodology for the evaluation of real-time speech digitization. *Proceedings of the Human Factors Society—27th Annual Meeting.* Santa Monica CA: Human Factors Society, pp. 104–107.

Fletcher, H. (1953). *Speech and hearing in communication.* Princeton, NJ: Van Nostrand Reinhold.

French, N., and Steinberg, J. (1947). Factors governing the intelligibility of speech sounds. *Journal of the Acoustical Society of America,* 19, 90–119.

Goldstein, E. (1984). *Sensation and perception* (2d ed.). Belmont, CA: Wadsworth.

Hull, A. (1976). Reducing sources of human error in transmission of alphanumeric codes. *Applied Ergonomics,* 7(2), 75–78.

Kiang, N. (1975). Stimulus representation in the discharge patterns of auditory neurons. In E. L. Eagles (Ed.) *The nervous system,* vol. 3. New York: Raven, pp. 81–96.

Kryter, K. (1972). Speech communication. In H. Van Cott and R. Kinkade (eds.), *Human engineering guide to equipment design.* Washington: Government Printing Office.

Kryter, K. (1985). *The effects of noise on man* (2d ed.). Orlando, FL: Academic.

Licklider, J., and Miller, G. (1951). The perception of speech. In S. S. Stevens (ed.), *Handbook of experimental psychology.* New York: Wiley.

Luce, P., Feustel, T., and Pisoni, D. (1983). Capacity demands in short-term memory for synthetic and natural speech. *Human Factors,* 25, 17–32.

McCauley, M. (1984). Human factors in voice technology. In F. Muckler (ed.), *Human factors review—1984.* Santa Monica, CA: Human Factors Society.

McCauley, M., and Semple, C. (1980). *Precision approach radar training system (PARTS)* (NAVTRAEQUIPCEN 79-C-0042-1). Orlando, FL: Naval Training Equipment Center.

Michaelis, P., and Wiggins, R. (1982). A human factors engineer's introduction to speech synthesizers. In A. Badre and B. Shneiderman (eds.), *Directions in human computer interaction.* Norwood, NJ: Ablex Publishing.

Miller, G. (1963). *Language and communication.* New York: McGraw-Hill.

Miller, G., Heise, G., and Lichten, W. (1951). The intelligibility of speech as a function of the context of the test materials. *Journal of Experimental Psychology,* 41, 329–335.

Peterson, A., and Gross, E., Jr. (1978). *Handbook of noise measurement* (8th ed.). New Concord, MA: General Radio Co.

Pollack, I., and Pickett, J. (1964). The intelligibility of excerpts from conversational speech. *Language and Speech,* 6, 165–171.

Simpson, C., McCauley, M., Roland, E., Ruth, J., and Williges, B. (1985). System design for speech recognition and generation. *Human Factors,* 27(2), 115–141.

Simpson, C., and Williams, D. (1980). Response time effects of alerting tone and semantic context for synthesized voice cockpit warnings. *Human factors,* 22, 319–320.

Thomas, J., Rosson, M., and Chodorow, M. (1984). Human factors and synthetic speech. *Proceedings of the Human Factors Society, 28th Annual Meeting.* Santa Monica, CA: Human Factors Society, pp. 763–767.

Waterworth, J. (1983). Effect of intonation form and pause duration of automatic telephone number announcements on subjective preference and memory performance. *Applied Ergonomics,* 14(1), 39–42.

Waterworth, J. (1984). The computer speaks—But what is it saying? *The Human Factor,* The newsletter of the Human Factors Division, British Telecom Research Laboratories, 6, 2.

Waterworth, J. and Thomas, C. (1985). Why is synthetic speech harder to remember than natural speech? In *CHI-85 Proceedings,* pp. 201–206. New York: Association for Computing Machinery.

Webster, J. (1969). *Effects of noise on speech intelligibility,* ASHA Rept. 4. Washington: American Speech and Hearing Association.

Werkowitz, E. (1980, May). Ergonomic considerations for the cockpit application of speech generation technology. In S. Harris (ed.), *Voice-interactive systems: Applications and payoffs.* Warminster, PA: Naval Air Development Center.

Wheale, J. (1980). Pilot opinion of flight deck voice warning systems. In D. Oborne and J. Levis (eds.), *Human factors in transport research,* volume 1. New York: Academic.

PART THREE

HUMAN OUTPUT AND CONTROL

HUMAN PHYSICAL ACTIVITIES

Most forms of human work require at least some limited type of physical activity, and some work requires substantial physical skill and/or effort. This chapter deals with such activities, including a discussion of certain relevant features of the human body, biomechanics, the measurement of certain physiological functions, and a sampling of some of the types of physical activities people can perform.

STRESS AND STRAIN

Before getting into such matters, however, we introduce the concepts of stress and strain, since they can be relevant to various aspects of human work, including physical work. In general terms, *stress* refers to some undesirable condition, circumstance, task, or other factor that impinges upon the individual, and *strain* refers to the effects of the stress. Some of the possible sources of stress are given in Figure 8–1, these being based on the work, the environment, and the circadian (diurnal) rhythm, the more specific sources being subdivided into physiological and psychological. (This particular figure, from Singleton, 1971, does not include social and interpersonal environmental sources of stress, which can have both physiological and psychological effects.) Stress can have either external or internal origins or some combination of these. The external origins can include the work activity itself, the physical or social aspects of the work situation, pressures from supervisors or peers, and other aspects of the work environment. Certain internal states of individuals, such as concern for consequences of failure, also can serve as internal sources of stress.

Granting that certain work-related factors would serve as external sources of stress to virtually anyone, it must be recognized that some such factors might be sources of stress for some individuals but not for others. These differences can arise from

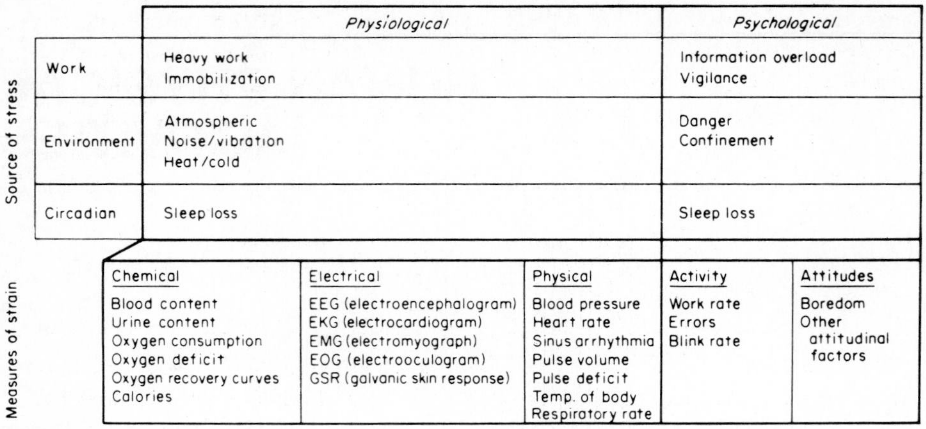

		Physiological			Psychological	
Source of stress	Work	Heavy work Immobilization			Information overload Vigilance	
	Environment	Atmospheric Noise/vibration Heat/cold			Danger Confinement	
	Circadian	Sleep loss			Sleep loss	
Measures of strain		Chemical	Electrical	Physical	Activity	Attitudes
		Blood content Urine content Oxygen consumption Oxygen deficit Oxygen recovery curves Calories	EEG (electroencephalogram) EKG (electrocardiogram) EMG (electromyograph) EOG (electrooculogram) GSR (galvanic skin response)	Blood pressure Heart rate Sinus arrhythmia Pulse volume Pulse deficit Temp. of body Respiratory rate	Work rate Errors Blink rate	Boredom Other attitudinal factors

FIGURE 8-1
Primary sources of stress and primary measures of strain as induced by stress. *(Source: Adapted from Singleton, 1971, fig. 4)*

various individual characteristics such as those of a physical, mental, attitudinal, or emotional nature. Individual differences in such characteristics might be reflected, for example, in differences in capacities to adjust or adapt to different external conditions. In addition, a person's perception of his or her ability to deal with the job and its external features could cause some people to experience stress but not others. As another example, the recognition of the importance of the consequences of one's work might trigger a stressful reaction on the part of some individuals.

In connection with physical activities, such as those dealt with in this chapter, stress is most directly related to the physical demands of the work and in some instances to the working conditions, such as high temperatures. The reduction of such sources of stress sometimes can be accomplished by modifying the work equipment, changing work schedules, modifying the environment, or other means.

FEATURES OF THE HUMAN BODY

The ability of people to perform physical activities is based on the structure of the body (the skeleton), the skeletal muscles, the nervous system, and the physiological features that control metabolism.

Skeletal Structure

The basic structure of the body is the skeleton, which consists of 206 bones. Certain bone structures serve primarily the purpose of housing and protecting essential organs of the body, such as the skull (which protects the brain) and the rib cage (which protects the heart, lungs, and other internal organs). The other (skeletal) bones—those of the upper and lower extremities and the articulated bones of the spine—are concerned primarily with the execution of physical activities, and it is these that are particularly relevant to our subject. The skeletal bones are connected at body joints, there being two general types of joints that are principally used in physical activities, namely *synovial* joints and *cartilaginous* joints. The synovial joints include hinge joints (such as the fingers and knees), pivot joints (such as the elbow, which is also a hinge joint), and ball-and-socket joints (such as the shoulder and hip). The primary examples of cartilaginous joints are the vertebrae of the spine, which make possible, collectively, considerable rotation and forward bending of the body.

Skeletal Muscle System

The bones of the body are held together at their joints by ligaments. The skeletal muscles (also called *straited,* or *voluntary,* muscles) consist of bundles of muscle fibers that have the property of contractility; the muscle fibers convert chemical energy to mechanical work.

The Nature of Muscles   Each muscle consists of many muscle fibers. These range in length from 0.2 in (0.5 cm) to about 5.5 in (14 cm). Some muscles have as many as 100,000 to 1,000,000 fibers. The ends of the fibers form sinews that combine to form the tendons which are connected to the skeletal bones in such a manner that, when the muscles are activated, the muscles apply mechanical leverage, like the cables of a crane.

Contractibility of Muscles   The individual fibers can be contracted to about half their length upon receipt of signals from the nerves (the neural system is discussed below). Each fiber contracts with a certain force, and the force applied by the complete muscle is the sum of the forces of the individual fibers. A muscle produces its greatest force at the beginning of its contraction (at its greatest relaxed length). As the muscle shortens, its force is reduced.

Neural Control of Muscles

Since the muscles are under the control of the nervous system, we briefly discuss the nervous system.

Nervous System   The complete nervous system includes the brain, the nerves of the spinal column, and a network of various types of nerves that fan out from the spinal column throughout the body. Two types of nerves are involved in the control of skeletal muscle activity: *motor* (or *efferent*) nerves and *sensory* (or *afferent*) nerves. The motor nerves carry impulses (what can be thought of as movement orders) to the muscles, causing the muscles to contract. Within a muscle the nerve is divided into constituent fibers, each of which controls the contraction of a group of muscle fibers.

The motor neuron consists of a cell body with a nucleus that is attached to a long nerve fiber that, in turn, ends in *motor end plates.* The impulse jumps from these end plates to the muscle fiber to trigger the muscle contraction. The impulses are electrochemical in nature, and they are not continuous but rather consist of successive, independent impulses. Since nerves require only a fraction of a second to recover, however, they can be "fired" in rapid succession.

The sensory nerves, in turn, transmit information back to the spinal column or farther to the brain. Various kinds of sensory nerves are relevant to muscular control. The proprioceptors, for example, transmit signals about the contraction of the muscles, thus providing feedback about such contraction. In addition, the kinesthetic nerves (which are located in the muscles, tendons, and joints) transmit signals about the position and posture of the body members. In general, these, and other, sensory nerves then provide the feedback that is essential to muscular control and that makes possible such activities as those of trapeze performers, heart surgeons, pool players, and pretzel twisters.

Control of Muscle Activity   To a considerable degree, individuals can exercise conscious control over the movements of their skeletal muscles and thus over the movements of their body members. However, certain physical responses are virtually automatic and take place without conscious control. Certain of these are built in to the repertoires of everyone; these are responses such as the reflex blinking of the eyelids (when some unexpected stimulus occurs near the eyes) or the knee jerk (when something strikes the kneecap). Other types of automatic responses are the consequence of experience or training and are thus "learned." Through experience and training we become "conditioned" to make certain kinds of movements without exerting conscious control. It is believed that such conditioning occurs because of the establishment of certain neural pathways as a consequence of repeated use of those pathways.

The stimuli that trigger the activation of certain muscular responses can be external or internal. For example, an assembler sees a part (an external stimulus) coming down along a conveyer and picks it up automatically without thinking about it. On the other hand, once a particular activity has been initiated (such as walking), the feedback from the various sensory nerves triggers a specific sequence of movements that makes it possible to keep walking without thinking about it.

In practice, many of our physical activities are carried out as the consequence of a combination of both conscious control and conditioned responses. A person driving an automobile, for example, receives external stimuli (both visual and auditory) and makes decisions about what actions should be taken, such as turning the steering wheel, applying the brake, or pressing the accelerator. The successful control of these devices is pretty much influenced by previous "learning." That is, the previously established neural pathways make it possible to execute the appropriate physical responses, while taking into account the feedback from relevant internal sensory signals. Many of the jobs people perform—especially skilled jobs—involve somewhat similar integration and coordination of sensory inputs in carrying out physical activities.

Metabolism

Metabolism is the chemical process by which foodstuffs are converted to two forms, namely, heat and mechanical energy. Some of the mechanical energy is used internally (as in the processes of respiration and digestion), and some is used externally (as in walking and working). In either case, excess heat that is generated must be dissipated.

The metabolic process is quite complicated and beyond the scope of this text, so we simply touch on a couple of aspects of the process. The nutrient substances involved include glucose, fat (fatty acids), and protein (amino acids). The conversion of these to energy consists of a chemical reaction that ends in the production of lactic acid. However, the lactic acid needs to be dissipated by being broken down into water and carbon dioxide. The first stage (the breakdown of glycogen into lactic acid) does not require oxygen and is said to be *anaerobic*. The second stage (the breakdown of lactic acid into water and carbon dioxide) is said to be *aerobic* because it does require oxygen.

At the initiation of physical activity, the muscles can utilize the glycogen which is already available. But the amount of glycogen and hence glucose available is small; so if the activity is continued, the body needs to replenish these nutrients from the blood, along with a supply of oxygen, which is required in the second stage. When an adequate supply of oxygen is provided, little or no lactic acid accumulates. If the level of activity requires more oxygen than is provided by the normal rate of blood flow through the cardiovascular system, the system adjusts itself to fulfill the increased demands, in particular by increasing the breathing rate to bring additional oxygen into the lungs and by increasing the heart rate to pump more blood through the "pipes" of the cardiovascular system. From the heart the blood is pumped through the lungs, where it picks up a supply of oxygen, which is then carted by the blood to the muscles where the oxygen is needed. With at least moderate rates of work, the heart rate and breathing rate are normally increased to the level that provides enough oxygen to perform the physical activities over a continuing period. However, when the amount of oxygen delivered to the muscles fails to meet the requirements (as when the level of physical activity is high), lactic acid tends to accumulate in the blood. If the rate or duration of physical activity results in continued accumulation of lactic acid, the muscles will ultimately cease to respond. If the rate of removal of lactic acid does not keep pace with its formation, additional oxygen must be supplied after cessation of the activity to remove the remaining lactic acid. This is referred to as the *oxygen debt*.

The oxygen debt is the amount of oxygen that is required by the muscles after the beginning of work, over and above that which is supplied to them by the circulatory system during their activity. This debt needs to be "repaid" after the cessation of work and is reflected in the elevated rate (i.e., above resting level) of oxygen consumption in the recovery process, as illustrated in Figure 8–2. This figure represents the "theoretical" pattern as it would be expected to occur under a steady state condition of work. However, as Bălănescu (1979) has demonstrated, the actual patterns may differ from this model, especially when the work activity varies during the work period. In any event, however, the oxygen that is originally borrowed at the beginning of the work period, or during it, is ultimately repaid.

Basal Metabolism   The *basal metabolic rate* is that which is required simply to maintain the body in a resting state, lying down. It frequently is measured in kilo-

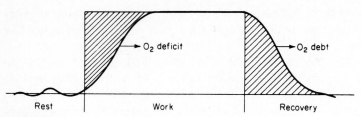

FIGURE 8-2
Illustration of the oxygen debt under conditions of steady state work.

calories.[1] The basal metabolism rate depends in part on the following:
size (large people tend to have higher metabolic rates), age (young people tend to
have higher rates), and sex (males tend to have higher rates). Grandjean (1981) reports the following approximate daily averages: for 155-lb (70-kg) males, 1700 kcal;
for 133-lb (60-kg) females, 1400 kcal. Most (approximately 70 percent) of the energy released from metabolism is in the form of heat.

Grandjean estimates that typical leisure activities have a caloric requirement of
about 600 kcal for males and 550 kcal for females. He estimates the minimum total
daily caloric requirements (including both basal and leisure requirements) to be about
2300 to 2400 kcal/d for males and 1900 to 2100 kcal/d for females.

The estimates made by other investigators are in this approximate range. These
values generally would also be applicable to people performing sedentary work that
does not involve any appreciable physical activity.

MEASURES OF PHYSIOLOGICAL FUNCTIONS

Of the many measures of physiological functions, certain are indicative of physiological strain. And more particularly certain measures are indices of the effects of work
and of the energy requirements of work. A few are discussed below.

Measures of Energy Expenditure

The measures of energy expenditure fall into two groups: gross body activity and
local muscle activity.

---

[1]The energy unit generally used in physiology is the kilocalorie (kcal) or large calorie
(abbreviated Cal, with a capital C, to distinguish it from the gram calorie which is abbreviated
cal). The *kilocalorie* is the amount of heat required to raise the temperature of 1 kg water from
15 to 16 °C. The relation of the kilocalorie to certain other units of energy measurement is:

1 kcal = 426.85 kg•m
1 kcal = 3087.4 ft•lb
1 kcal = 1000 cal
1 kcal = 4.2 kJ

Measures of Gross Body Activity   As indicated above, oxygen consumption is one indication of the energy requirements of the body. It follows that the direct measurement of oxygen consumption serves as one index of energy requirements. In this regard, one liter of oxygen is approximately equal to 5 Cal of energy; thus, oxygen consumption can be convered to kilocalories. Further, as oxygen requirements are increased, the cardiovascular system ''speeds up'' its functions, thus giving rise to several measures such as heart rate, blood pressure, respiratory rate, and stroke (pulse volume). (*Stroke volume* is the volume of blood pumped through the arteries for every heartbeat.)

The measurement of oxygen consumption of people at work is rather cumbersome, but it can be done, as shown by Figure 8–3. That figure shows the apparatus for monitoring heart rate as well as obtaining samples of oxygen. The measurement of heart rate can be accomplished with a few wires attached to the subject and a telemetry device about the size of a cigarette package. Although oxygen consumption can be estimated from the heart rate, the relationship between these two variables is not the same for everyone. Thus, to use heart rate to predict oxygen consumption, each individual needs to be ''calibrated.'' This is done by having the person perform a task in a simulated work environment in the laboratory at different levels of effort (e.g., walking on a treadmill at different speeds) while both heart rate and oxygen consumption are being measured. From such data, the relationship can be determined.

Another measure based on heart rate is the heart rate recovery curve as used by Brouha (1960), this being a curve of the heart rate measured at certain intervals after work (such as 1, 2, and 3 min). In general terms, the more strenuous the work activity, the longer it takes the heart rate to settle down to its prework level.

FIGURE 8-3
Apparatus for monitoring heart rate and obtaining sample oxygen consumption levels of a person while working on the job. *(Photograph courtesy of Human Factors Department of Eli Lilly and Company.)*

Since no single physiological measure is a completely adequate index of physiological strain, various attempts have been made to derive composite indices that would more adequately reflect the level of strain. Burger (1969), for example, proposes the use of an index of circulatory load. An approximation of such an index is the product of heart rate, mean blood pressure, and stroke volume. In turn, Dukes-Dobos et al. (1976) experimented with a combination of 18 cardiorespiratory measures and found a combination of 9 of them to be most sensitive to changes in the level of physical activity. The combination of these was used in the derivation of a cardiorespiratory variance score (CVS). However, such composite measures are still not completely adequate, and the problems of instrumentation in their measurement impose serious constraints on their use in many practical situations. Thus, it is still the most common practice to use single parameters such as heart rate or oxygen consumption as measures of physiological strain. As measures of the energy cost of work (the source of stress), oxygen consumption or calories frequently are used, usually expressed in units per minute or per hour.

Measures of Local Muscular Activity   One of the available measures of the physiological condition of individual muscles or muscle groups is electromyographic (EMG) recordings, which are also called *myograms*. These are inked tracings of the electric impulses that occur during work, and they provide estimates of the magnitude of voluntary muscular activity. Examples of such recordings are shown in Figure 8–4 for four muscles of one subject when a constant torque of 60 and 15 ft•lb is applied to a steel socket. Several problems have been associated with the recording, quantification, and interpretation of such recordings. In this regard Khalil (1973) has developed procedures for summing the action potentials of several muscles that are monitored simultaneously. The values given in Figure 8–4 show the individual and summed values for the four muscles. Although Khalil believes that his method of summing the recordings is suitable for measuring both static (i.e., isometric) and dynamic muscular exertion, Tichauer (1978) makes the point that the interpretation of myograms of dynamic tasks is much more complex than interpreting those for static work. He expresses the opinion that it is useless in most dynamic situations to

FIGURE 8-4
Electromyograms recorded for four muscles of a subject maintaining a constant torque of 60 and 15 ft·lb (8.3 and 2.1 kg·m). The sum of the four values is an index of the total amount of energy expended. *(Source: Adapted from Khalil, 1973, fig. 3.)*

Foot-pounds	Deltoid	Biceps	Triceps	Brachioradialis	Total
60	4.7	30.1	7.1	19.7	63.6
15	1.9	9.7	1.2	5.1	17.9

attempt the numeric quantification of such recordings, but indicates that their qualitative interpretatioan by experienced investigators can be very useful.

In this regard Örtengren et al. (1975) used the electromyographic technique to assess the amount of localized muscle fatigue associated with certain manual tasks in the Volvo automobile factory in Göteborg, Sweden, In particular, they were interested in the analysis of EMG data that would reveal "incidents" of muscular fatigue for various types of work activities. The incidents were identified by significant changes in the spectrum of the EMG recordings. Certain of the results are summarized in Table 8–1, in particular the mean number of such incidents per car.

The magnitude of these differences suggests that EMG recordings may have utility in identifying work activities that are conducive to high levels of muscular fatigue, with the thought that certain modifications in the work would reduce the level of such fatigue.

Other Measures of Physiological Functions

In addition to the measures of energy expenditure just discussed, a number of other measures of physiological functions are sometimes relevant to human factors. These can be thought of as measures of physiological strain. A number of these are listed in Figure 8–1. Other measures include the *flicker fusion frequency* (FFF), which is also called the *critical fusion frequency* (CFF—this is the lowest frequency at which a flickering on-off light appears to be constant); *evoked cortical potential* (ECP), which is somewhat like the EEG except that the electric potentials are measured on the scalp or upper neck; body fluid analysis (a complex chemical analysis of body fluids); eye and eyelid movements; and muscle tension.

WORKLOAD

The physical activities and environmental factors constitute the sources of physical stress of people at work (or even of people *not* at work!) and thus predetermine their workload. In discussing the workload involved in physical work, Grandjean (1981) makes the following points: (1) stress is greater with static work (such as holding an arm straight out) than with dynamic work (such as lifting a box), and (2) stress is

TABLE 8–1
INCIDENTS OF MUSCULAR FATIGUE
FOR THREE TYPES OF TASKS

Type of work	Mean number of incidents
Assembly	0.08
Wet rubbing: light	0.04
Wet rubbing: heavy	0.36

Source: Örtengren et al., 1975.

greater when the work is restricted to one or a limited number of muscles than when the work involves many muscles. Working in hot temperature also increases the workload.

Although workload frequently is measured in terms of energy consumed (as kilocalories per hour), Grandjean notes that in some circumstances this measure is not necessarily the best index of the effects of stress—such as when work involves only a restricted assortment of muscles. For such reasons he argues for the more general use of heart rate as a measure of workload.

Recognizing this measurement issue, we should get an overview of some of the factors associated with human work that influence workload.

Energy Consumption of Gross Body Activities

To get some feel for the numerical values of energy consumption for different kinds of physical activities, it is useful to present the physiological costs of certain everyday activities. The following examples (Edholm, 1967; Grandjean, 1981) are given: sleeping, 1.3 kcal/min; sitting, 1.6 kcal/min; standing, 2.25 kcal/min; walking (level, smooth surface), 2.1 kcal/min; light cleaning/ironing, 2.0 to 3.0 kcal/min; cycling [10 mi/h (16 km/h)], 5.2 kcal/min. Keep in mind the ubiquitous fact of individual differences.

Workload of Specific Work Activities

The energy consumption of several types of work is illustrated in Figure 8–5. The energy costs for these range from 1.6 to 16.2 kcal/min.

Workload and Methods of Work   The energy cost for certain types of work, however, can vary with the manner in which the work is carried out. The differential costs of methods of performing an activity are illustrated by the several methods of carrying a load as used in various cultures. Seven such methods were compared by Datta and Ramanathan (1971) on the basis of oxygen requirements; these methods and the results are shown in Figure 8–6. The requirement of the most efficient method (the double pack) is used as an arbitrary base of 100 percent. There are advantages and disadvantages to the various methods over and above their oxygen requirements, but the common denominator of the most efficient methods reported in this and other studies is maintenance of good postural balance, one that affects the body's center of gravity the least.

Workload and Work Posture   The posture of workers while performing some tasks is another factor that can influence the energy requirements. In this regard certain agricultural tasks in particular have to be carried out at or near ground level, as in picking strawberries. When such work is performed, however, any of several postures can be assumed. The energy costs of certain such postures were measured in a study by Vos (1973) in which he used a task of picking up metal tags placed in a standard pattern on the floor. A comparison of the energy expenditures of five different postures is given in Figure 8–7. This figure shows that a kneeling posture with hand support and a squatting posture required less energy than the other postures. (The kneeling posture, however, precludes the use of one hand and might cause knee discomfort over time.) On the basis of another phase of the study, it was shown that a sitting posture (with a low stool) is a bit better than squatting, but a sitting posture is not feasible if the task requires moving from place to place. Althought this particular analysis dealt with postures used on ground-level tasks, differences in postures used in certain other tasks also can have differential energy costs.

Workload and Work Rate Still another factor that affects the workload is the work rate, or pace. Given any particular task, it is generally true that—at some specific pace—the task can be carried out over an extended time without any appreciable physiological cost. For such reasonable paces the heart rate typically increases

FIGURE 8-5
Examples of energy costs of various types of human activity. Energy costs are given in kilocalories per minute. *(Source: Passmore and Durnin, 1955, as adapted and presented by Gordon, 1957.)*

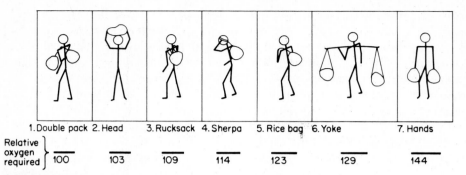

1. Double pack   2. Head   3. Rucksack   4. Sherpa   5. Rice bag   6. Yoke   7. Hands

Relative oxygen required

100   103   109   114   123   129   144

FIGURE 8-6
Relative oxygen consumption of seven methods of carrying a load, with the double-pack method used as a base of 100 percent. This illustrates that the manner in which an activity is carried out can influence the energy requirements. *(Source: Adapted from Datta and Ramanathan, 1971.)*

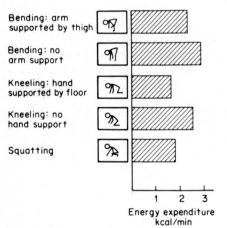

Bending: arm supported by thigh

Bending: no arm support

Kneeling: hand supported by floor

Kneeling: no hand support

Squatting

1   2   3
Energy expenditure kcal/min

FIGURE 8-7
Human energy expenditures (kilocalories per minute) for five postures used in task of picking up light objects from ground level. *(Source: Adapted from Vos, 1973, fig. 5.)*

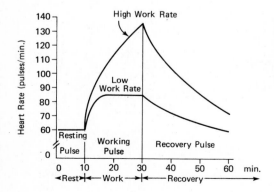

High Work Rate

Low Work Rate

Resting Pulse

Working Pulse

Recovery Pulse

Heart Rate (pulses/min.)

140 130 120 110 100 90 80 70 60 0

0   10   20   30   40   50   60   min.
◄Rest►◄── Work ──►◄── Recovery ──────►

FIGURE 8-8
Heart rate for two different work rates. At the higher work rate the heart rate continues to increase, whereas at the lower work rate the heart rate levels off at a plateau, or steady state. *(From Grandjean, 1981, fig. 61, p. 80.)*

somewhat and maintains that level for some time. However, if the rate of work is increased appreciably, the workload can cause a continued rise in heart rate and in other physiological costs. This is illustrated in Figure 8–8, which shows the heart rate for two work rates. Aside from the differences in the increases in the heart rates, this figure shows marked differences in the heart rate recovery curves.

Another example is shown in Figure 8–9, which illustrates the effect of different work paces for lifting cartons. An increase in the work pace was accompanied by an increase in energy expenditure. This figure also shows the effects of carton weight on energy expenditure.

Work Efficiency   The efficiency of a person performing work can be likened to that of a machine. The basic equation for measuring efficiency is

$$\% \text{ efficiency} = \frac{\text{work output}}{\text{energy consumption}} \times 100$$

Under optimum conditions human physical effort can be about 30 percent efficient. The remaining 70 percent of the energy consumption generally is converted to heat. However, with many activities some of the work output is wasted; that is, it is in a form that is not useful. Such wasted energy includes that attributed to static effort (as in holding or supporting things), to stooped or other awkward postures, or to inefficient equipment or methods. The estimated efficiencies of a few activities are as follows (Grandjean, 1981):

Activity	Efficiency (%)
Shoveling (stooped posture)	3
Shoveling (normal posture)	6
Using heavy hammer	15
Cycling	23
Walking without load	27

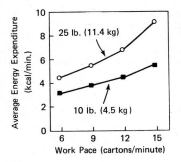

FIGURE 8-9
Effects of work pace on energy expenditure for a task of lifting cartons of two weights. *(Adapted from Hamilton and Chase, 1969. Reprinted from AIIE Transactions, Vol. 1, 1969. Copyright Institute of Industrial Engineers, 25 Technology Park/Atlanta, Norcross, GA, 30092.)*

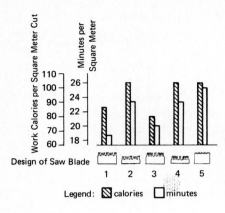

FIGURE 8-10
Calories expended and time spent in cutting a standard unit of work with five timber saws (unit of work: 1 m^2 of area cut). Saws 1 and 3 were clearly most efficient. *(From Grandjean, 1981, fig. 56, p. 74.)*

An example of the effects on efficiency of the equipment used is illustrated in Figure 8–10. This shows the work calories expended and time required when five types of timber saws are used in cutting slices or disks of $1 = m^2$ area. It is clear that saws 1 and 3 resulted in the lowest calories consumed per unit of useful work (per $1 = m^2$ cut). You may be able to relate this study to your own experiences. If you use a dull saw or the wrong type of saw, you can expend a lot of extra energy that is, in effect, wasted. Thus, in considering the notion of efficiency of work as based on the above equation, the work output component should be restricted to useful work.

To illustrate the concept of efficiency further, let us consider the matter of walking, as shown in Figure 8–11. This example brings us back to rate of work discussed before. Here we can see that the most efficient walking speeds for people with shoes and with bare feet are about 65 to 80 m/min (213 to 262 ft/min).

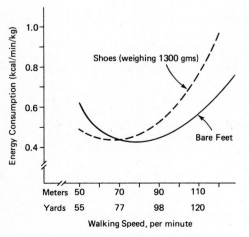

FIGURE 8-11
Energy consumption required for walking at various speeds. The lowest points on the curves represent the most efficient speeds. *(From Grandjean, 1981, fig. 58, p. 76.)*

Keeping Energy Expenditures within Bounds

If those who are concerned with the nature of human work activities (design engineers, industrial engineers, supervisors, administrators, etc.) want to keep energy costs within reasonable bounds, it is necessary for them to know both what those bounds should be and what the costs are (or would be) for specific activities (such as those shown in Figure 8–5).

Energy Costs of Grades of Work   To start, the definitions of different grades of work shown in Table 8–2 may be helpful.

In discussing energy expenditures, Lehmann (1958) estimates that the maximum energetic output a normal man can afford in the long run is about 4800 kcal/d; subtracting his estimate of basal and leisure requirements of 2300 kcal/d leaves a maximum of about 2500 kcal/d available for the working day. Although Lehmann proposes this as a maximum, he suggests about 2000 kcal/d as a more normal load. This is about 250 kcal/h, or 4.2 kcal/min. In turn, Grandjean (1981) states that most investigators agree that the norm for energy consumption during heavy work should be about 240 kcal/h, or 4.0 kcal/min, this being very close to the loads mentioned above.

Work and Rest   If we accept some ceiling (such as 4 or 5 kcal/min) as a desirable upper limit of the average energy cost of work (exclusive of basal requirements), it is manifest that if a particular activity exceeds that limit, there must be rest to compensate for the excess. In this connection Murrell (1965, p. 376) presents a formula for estimating the total amount of rest (scheduled or not scheduled) required for any given work activity, depending on its average energy cost. This formula (with different notations) is

$$R = \frac{T(K-S)}{K-1.5}$$

TABLE 8–2
GRADE OF PHYSICAL WORK BASED ON ENERGY EXPENDITURE LEVEL (ASSUMING
A REASONABLY FIT ADULT MALE)

Grade of work	Energy expenditure, kcal/min	Energy expenditure, 8 h (d·kcal)	Heart rate, beats per minute	Oxygen consumption, 1/min
Rest (sitting)	1.5	<720	60–70	0.3
Very light work	1.6–2.5	768–1200	65–75	0.32–0.5
Light work	2.5–5.0	1200–2400	75–100	0.5 –1.0
Moderate work	5.0–7.5	2400–3600	100–125	1.0 –1.5
Heavy work	7.5–10.0	3600–4800	125–150	1.5 –2.0
Very heavy work	10.0–12.5	480–6000	150–180	2.0 –2.5
Unduly heavy work	>12.5	>6000	>180	>2.5

Source: Adapted from American Industrial Hygiene Association, 1971. Reprinted with permission by American Industrial Hygiene Association.

in which R is rest required in minutes, T is total working time, K is average kilocalories per minute of work, and S is kilocalories per minute adopted as standard. The value of 1.5 in the denominator is an approximation of the resting level in kilocalories per minute. If we adopt as S a value of 4 kcal/min and want to figure R for a 1-h period (T = 60 min), our formula becomes

$$R = \frac{60(K-4)}{K-1.5}$$

Applying this to a series of values of K, we can obtain R values shown in the next to the top curve of Figure 8–12 (S = 4). The other curves (for values of S = 3, 5, and 6) are given for comparison when lower (S = 3 kcal/min) or higher (S = 5 or 6 kcal/min) standards of energy expenditure might seem appropriate. The lowest curve (for a value of S = 6 kcal/min), however, undoubtedly represents a level of activity that probably could not be maintained very long, except possibly by the hardiest among us. This general formulation needs to be accepted with a fair sprinkling of salt, in part because of individual differences in physical condition. Further, although the curves in Figure 8–12 swing down to the zero-rest-required line, we should keep in mind that this formulation deals only with the physiological costs of work. Because of *other* considerations, some rest must be provided for virtually any kind of continuous work, even though its physiological costs are nominal.

Work Limits of Local Muscle Groups

As indicated above, the restriction of work to a limited number of muscles is particularly stressful. However, such stress is related to the rate of muscle activity. If the rate of contraction of a muscle or muscle group is low enough, the contraction can be continued almost indefinitely. However, at a high rate of contraction, the muscle or mucle group can be worn to a frazzle in short order. Further, the time required for recovery of a muscle increases disproportionately with the degree of exhaustion.

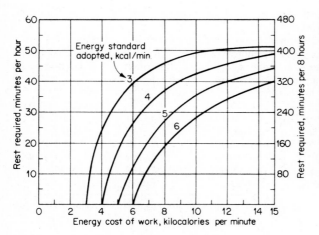

FIGURE 8-12
Total rest requirements for work activities of varying energy costs, for energy expenditure standards (ceilings) of 3, 4, 5, and 6 kcal/min; a generally accepted standard is 4 to 5 kcal/min. The rest requirements for maintenance of the adopted standard are given per hour (left) and per 8-h day (right). *(Source: Based on a formulation of Murrell, 1965, p. 376.)*

Therefore, care should be taken to minimize the continued use of small muscle groups, such as screwing caps on small bottles all day long by using the thumb and forefinger.

BIOMECHANICS[2]

Given the skeletal structure, the muscles, the nervous system, and the metabolic process, the human body is capable of quite a repertoire of physical activities. *Biomechanics* deals with the various aspects of physical movements of the body and body members. At a more specific level, *kinesiology* deals with the study of human motions as a function of the construction of the musculoskeletal system. This system can be viewed as consisting of a variety of levers. For example, the forearm has its fulcrum at the elbow, and one of the muscles of the upper arm provides the activating force just in front of the elbow to bend the forearm. There are three different types of levers, and various body movements represent all three types.

Types of Movements of Body Members

Certain of the basic types of movements that are performed by the body members are described below, along with their associated jargon in kinesiology:

- *Flexion:* Bending, or decreasing the angle between the parts of the body
- *Extension:* Straightening, or increasing the angle between the parts of the body
- *Adduction:* Moving toward the midline of the body
- *Medial rotation:* Turning toward the midline of the body
- *Lateral rotation:* Turning away from the midline of the body
- *Pronation:* Rotating the forearm so that the palm faces downward
- *Supination:* Rotating the forearm so that the palm faces upward

Some of these basic movements are illustrated in Figure 8–13, along with the following values for each (as based on a sample of 100 male college students): mean angle (in degrees) and 5th and 95th percentile angles (computed from the standard deviations for the sample). In this, as in other aspects of biomechanics, there are the ever-present individual differences, including the effects of physical condition and age.

Although specific movements of body members can be described in terms of these basic movements, in describing work activities it usually is preferable to do so in more operational terms. There are different ways in which movements can be so classified, one of them being given here:

- *Positioning* movements are those in which the hand or foot moves from one specific position to another, as in reaching for a control knob.

[2]For a more extensive treatment of biomechanics, especially as related to equipment design, see Damon, Stoudt, and McFarland (1966) and Tichauer (1978). A glossary of terms used in biomechanics is included in *American National Standard Industrial Engineering Terminology: Biomechanics,* ANSI Z94.1–1972, The American Society of Mechanical Engineers, New York.

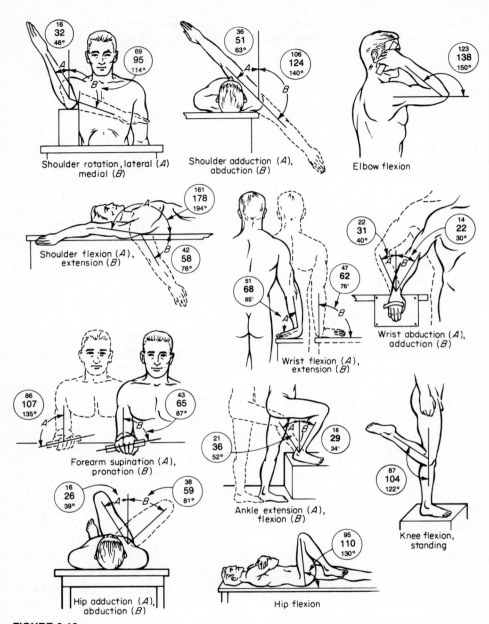

FIGURE 8-13
Range (in degrees) of rotation and movement of certain upper and lower extremities, based on a sample of 100 male college students. The three values (in degrees) given for each are the 5th percentile, the mean, and the 95th percentile. *(Source: Houy, 1983, table 1. Copyright by the Human Factors Society, Inc. and reproduced by permission.)*

- *Continuous* movements require muscular control adjustments of some type during the movement, as in operating the steering wheel of a car or guiding a piece of wood through a band saw.
- *Manipulative* movements involve the handling of parts, tools, control mechanisms, etc., typically with the fingers or hands.
- *Repetitive* movements are those in which the same movement is repeated, as in hammering, operating a screwdriver, and turning a handwheel.
- *Sequential* movements are several relatively separate, independent movements in a sequence.
- A *static* adjustment is the absence of a movement, consisting of maintaining a specific position of a body member for a time.

Various types of movements may be combined in sequence so that they blend one into another. For example, placing the foot on a brake pedal is a positioning movement, but this may be followed by a continuous movement of adjusting the amount of brake pressure to the conditions of the situation. Similarly, a continuous movement may include holding a position (a static adjustment) for a short time.

Performance Criteria for Physical Activities

In research relating to physical activities various types of criteria can be used. Our previous discussion dealt largely with physiological criteria, such as energy costs of work. However, for some purposes certain performance criteria are useful. Those are measures of how well the body member or body, when executing some physical activity, can perform some desired function. In the case of activities of the body members, the most common performance criteria are strength (i.e., the force that can be applied), endurance, speed, and accuracy. In the case of gross body activities (such as materials handling), performance criteria such as speed and endurance usually are used.

In connnection with the measurement of such criteria certain interesting devices have been developed. One of these is the force platform. This is a small platform on which a subject stands when carrying out some physical activity. By the use of some sensing elements below the platform (such as piezoelectric crystals) it is possible to sense and then automatically record the forces generated by the subject in each of three planes, namely, vertical, frontal, and transverse. The original force platform was developed by Lauru (1954). Such devices are sensitive to slight differences in physical movements and can thus lend themselves to use in comparing the three-dimensional forces in different activities. Recordings of the forces in the operation of manual and of electric typewriters are shown in Figure 8–14 for comparison. It has been proposed that such force-time recordings, as possible indices of energy expenditure, are nearly as accurate as metabolic measurements and thus can be used as a measure of physiological cost of a given motion (Brouha, 1960, p. 103). In fact, Brouha presents data for the oxygen cost of certain work activities that show high correlations (ranging from .83 to .96) with data from the force platform.

In some circumstances it is appropriate to use subjective criteria elicited from subjects—criteria such as ratings of subjective fatigue or tiredness, opinions about

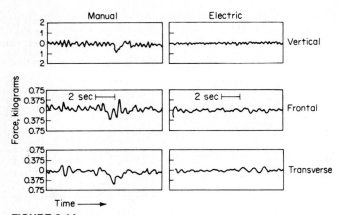

FIGURE 8-14
Forces in three dimensions (vertical, frontal, and transverse) in the
operation of a manual and an electric typewriter, recorded with a
force platform. *(Source: Brouha, 1960, p. 106.)*

optimum rates of work, or preferences for methods of performing specific functions.

The following sections summarize some research that deals with the effectiveness of the body members or of the body as a whole in performing various types of functions. These sections, then, are focused on the performance criteria mentioned above. Since it is not feasible to deal comprehensively with this topic, we cite examples of research that illustrate certain principles relating to the performance of physical activities.

STRENGTH AND ENDURANCE

Strength typically is associated with specific actions of the arm or legs. It can be measured under either of two conditions: *dynamic* conditions, when the body member is actually being moved (*isotonic* or *isokinetic* strength), and *static* conditions, when the force is applied against a fixed object, with no displacement of the body member (*isometric* strength). Since single tests of strength are not highly reliable (Stobbe and Plummer, 1984), it usually is desirable to test each person a few times (such as 3 or more) and average the results.

Strength tends to increase until about 30 years of age; then it levels off in middle age and decreases with old age (Hafez et al., 1982). However, the amount of decrease with age can be reduced by regular exercise. On the average the strength of women is about two-thirds that of men.

Arm Strength Although strength can be measured for various muscles or muscle groups, we use as an illustration a study of arm strength by Hunsicker (1955). He tested the arm strength of 55 subjects who made movements in each of several directions, with the upper part of the arm in each of five positions, as illustrated in Figure 8–15a. Some of the results are shown in Figure 8–15b, in particular the maximum strength of the 5th percentile of 55 males.

We can see that pull and push movements are clearly strongest, but that these are

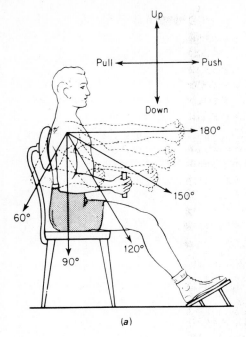

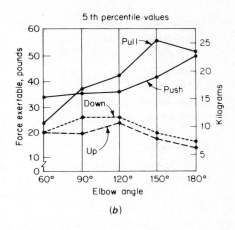

FIGURE 8-15
(*a*) Side view of subjects being tested for strength in executing push, pull, up, and down movements at each of five arm positions. (*b*) Maximum arm strength of 5th percentile of 55 male subjects. (*Source: Based on data from Hunsicker, 1955.*)

noticeably influenced by the position of the hand, with the strongest positions being at angles of 150 and 180°. The differences among the other movements are not great, but what patterns do emerge are the consequence of the mechanical advantages of such movements, considering the angles involved and the effectiveness of the muscle contractions in applying leverage to the body members. Although left-hand data are not shown, the strength of left-hand movements is roughly 10 percent less than that of movements of the right hand.

Endurance

If one considers the endurance of people to maintain a given muscular force, we can all attest from our own experience that such ability is related to the magnitude of the force. This is shown dramatically in Figure 8–16, which depicts the general pattern of endurance time as a function of force requirements of the task. It is obvious that people can maintain their maximum effort very briefly, whereas they can maintain a force of around 25 percent or less of their own maximum for a somewhat extended period (10 min or more). The implication of this relationship is fairly obvious—if it is necessary to require individuals to maintain force over a time, the force required should be well below each individual's own static force capacity.

Discussion of Strength and Endurance

The evidence seems to indicate that strength and endurance are substantially corre-lated. In a study by Jackson, Osburn, and Laughery (1984) dealing with this issue, the subjects were administered certain strength tests and, in addition, performed work

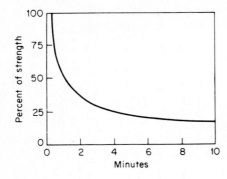

FIGURE 8-16
Endurance time as a function of force
requirements. *(Source: Kroemer, 1970, fig. 4, as
adapted from various sources.)*

sample tests that simulated certain mining and oil field job activities. They found that
the strength tests predicted quite well the performance on the work sample tests
which involved an ''absolute endurance component.'' Such results suggest that, in
the selection of people to perform demanding physical work requiring endurance, it
would be possible to use strength tests rather than endurance tests (that take more
testing time).

Note that a given level of strength and endurance is not fixed forever. Exercise
can increase these within limits, these increases sometimes being in the range of 30
to 50 percent above beginning levels.

SPEED OF MOVEMENTS

In many work situations people have to make some type of physical response on the
basis of stimuli they receive from their environments (from visual displays, auditory
signals, events, etc.). In some circumstances it is critical that the response be made
as quickly as possible (as in applying the brakes of an automobile when an emergency occurs). In other circumstances, however, some delay in the response action
will have no dire consequences (for example, the sky will not fall if a receptionist is
a bit slow in answering the telephone). It is in situations where time is critical that a
premium is placed on rapid response time. In some such instances the work situation
can be designed to capitalize on the response capabilities of people.

Components of Response Time

Total response time (sometimes called *reaction time*) depends on a number of variables (some are discussed below). To begin, however, let us dissect the various
elements of the response that comprise the total response time. At one level of description we can differentiate between the time to initiate a response following the
presentation of a stimulus (this being a more restrictive definition of reaction time)
and the time to make the response (sometimes called *movement time*). At the neurological level of description, Wargo (1967) estimates the time required for the various
sequences of the neurological processes to be as follows: receptor (sense organ) delays, 1 to 38 ms; neural transmission to the cortex, 2 to 100 ms; central process

delays, 70 to 300 ms; neural transmission to muscle, 10 to 20 ms; and muscle latency and activation time, 30 to 70 ms. These add up to a total ranging from 113 to 528 ms.

Reaction Time

Simple reaction time is the time to make a specific response when only one particular stimulus can occur, usually when an individual is anticipating the stimulus (as in conventional laboratory experiments). Reaction time is usually shortest in such circumstances, typically ranging from about 150 to 200 ms (0.15 to 0.20 s), with 200 ms being a fairly representative value. The value may be higher or lower depending on the stimulus modality and the nature of the stimulus (including its intensity and duration) as well as on the subject's age and other individual differences. There probably are relatively few situations outside the laboratory where *pure* simple reaction time exists, in the sense that there is a single, clear-cut stimulus and a single clear-cut response to be made.

In the case of *choice reaction time,* there are two or more possible responses. Because of the intervening decision process—the need to make a choice of responses depending on the stimulus—the reaction time typically increases in relationship to the number of stimuli and corresponding responses. Some indication of the influence of number of choices is given below (summarized from various sources by Damon, Stoudt, and McFarland, 1966, p. 239):

Number of choices	1	2	3	4	5	6	7	8	9	10
Approximate reaction time, s	0.20	0.35	0.40	0.45	0.50	0.55	0.60	0.60	0.65	0.65

Note that the increases in reaction time summarized above from empirical data do not entirely jibe with the theoretical predictions discussed in Chapter 3.

Expectancy Most data on simple and choice reaction times come from laboratories in which the subject is anticipating a stimulus. (And in some industrial circumstances people actually are waiting for a stimulus.) However, when stimuli occur infrequently or when they are not expected, the ante is raised. This was illustrated, for example, in a study by Warrick, Kibler, and Topmiller (1965) in which typists at their regular jobs were asked to press a button whenever a buzzer sounded, the buzzer going off only once or twice a week over 6 months. The reaction time (actually the total response time) to the "unexpected" signals averaged about 100 ms above that when the subjects had received an advance warning.

As another example to illustrate the point, Johansson and Rumar (1971) collected data in Sweden on the response time of 321 automobile drivers in applying the brake pedal following an auditory signal. This was done under two conditions, one in which the drivers were anticipating a signal within the next 10 km (6 mi) and the other under a "surprise" condition, with no advance warning. The mean response times for these two conditions were as follows:

Condition	Mean response time
"Anticipation"	0.54 s
'Surprise"	0.73 s

The response times for the surprise conditions for some subjects were over 2 s, and the investigators estimated that over half the drivers would take over 0.9 s (900 ms) to respond. In general, then, there is strong evidence to support the contention that reaction time is much longer when people are not expecting to have to make a response than when they are anticipating some signal or cue.

Other Factors That Influence Reaction Time   There are a number of other factors that can influence reaction time. Some of these, as referred to by Wickens (1984), are the following:

- *Stimulus modality:* For example, simple reaction time to auditory stimuli is about 30 to 50 ms faster than to visual stimuli (about 130 to 170 ms).
- *Stimulus intensity:* Reaction time tends to decrease in relation to the intensity up to a point, and then reaction time levels off.
- *Temporal uncertainty:* If there is some kind of warning in advance of the stimulus, any uncertainty of that warning tends to increase reaction time.
- *Expectancy:* This was discussed above.
- *Discriminability of the stimulus:* The more discriminable the stimulus, the faster the reaction time.
- *Compatibility:* The more compatible the stimulus and response the faster the reaction time.
- *Repetition:* The learning that accompanies repetition tends to reduce reaction time.
- *Accuracy required:* With choice reaction time, if a high degree of accuracy is required in making the correct response, reaction time tends to increase.

Movement Time

The time to complete the movement component of a response naturally depends on the nature of the movement. In the case of the types of movements characteristic of many simple control actions, Wargo (1967) estimates that a minimum of about 300 ms (.03 s) can be expected. Adding this to an estimated reaction time of 200 ms would result in a total response time of about 500 ms. However, various features of the movement response can affect the time required.

Direction of Movement   Because of the nature of our physical structures, motions can be made more rapidly in certain directions than in other directions. We use as an example data from Schmidtke and Stier (1960), who had subjects make positioning movements with the right hand in eight directions in a horizontal plane from a center starting point. Their results, shown in Figure 8–17, show the average times to make the movements in the eight directions. The pattern suggests that, in biomechanical terms, controlled arm movements that are primarily based on a pivoting of

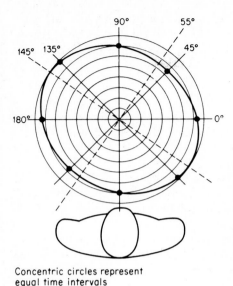

Concentric circles represent
equal time intervals

FIGURE 8-17
Average times of hand movements made in
various directions. Data were available for the
points indicated by black dots; the oval was
drawn from these points and represents
assumed, rather than actual, values between
the recorded points. The concentric circles
represent equal increments of time to provide a
reference for the average movement times
depicted by the oval. *(Source: Adapted from
Schmidtke and Stier, 1960.)*

the elbow (as toward the lower left or upper right) take less time than those that
require a greater degree of upper-arm and shoulder action (as toward the lower right
or upper left). Also evidence from other sources indicates that such movements are
also more accurate.

Distance of Movement   Another factor that influences movement time is the
distance of movement. Movement time is related to distance but is not proportional
to distance. This is illustrated in Figure 8–18. The subjects in the study moved a
sliding device to a marked position when a buzzer was sounded. Three different
distances were used: 2.5, 10, and 40 cm. Reaction time was virtually the same for
all distances. Movement time first increases rather sharply and then tends to flatten

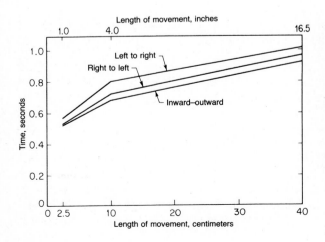

FIGURE 8-18
Times required for horizontal
positioning movements of
different lengths. Times given
are from the sounding of a
buzzer to completion of
movement. Reaction time,
which was essentially the
same for all movements
(about 0.25 s), is included in
time values. *(Source:
Adapted from Brown and
Slater-Hammel, 1949, and
Brown, Wieben, and Norris,
1948.)*

out. This can probably be attributed to the time required for acceleration to the maximum speed and (except where there is a mechanical stop) the time required for the secondary or corrective movement in bringing the body member to the precise stopping point.

Discussion of Response Time

We see that a variety of factors can influence response time. In one study, for example, Pattie (1973) investigated the time to activate four types of possible emergency cutoff switches as they might be used on agricultural tractors: clutch, toggle switch, horn rim, and rim blow. These devices reflect differences in the location, the type of response to be made, and even the body member (foot versus hand). The results, given in Table 8–3, show very distinct differences, ranging from 337 to 613 ms.

Even under the best circumstances there still is a lag in responding to situations in our environment. However, even if time is, as is said, of the essence, we should not throw up our hands in complete despair of the time lag in human responses. There are, indeed, ways of aiding and abetting people in responding rapidly to stimuli, for example, by using sensory modalities with shortest reaction time, presenting stimuli in a clear and unambiguous manner, minimizing the number of alternatives from which to choose, giving advance warning of stimuli if possible, using body members that are close to the cortex to reduce neural transmission time, using control mechanisms that minimize response time, and training the individuals. In more exotic circumstances, one can even bypass the human physical response by the direct use of electric muscle-action potentials for making control responses.

ACCURACY OF MOVEMENTS

The arms and hands are capable of performing a wide assortment of movements. Our discussion covers only a few much movements, for illustration purposes.

Blind Positioning Movements

When visual control of movements is not feasible, the individual needs to depend on the kinesthetic sense for feedback. Probably the most usual type of blind positioning

TABLE 8–3
MEAN TIMES TO ACTIVATE FOUR POWER CUTOFF
DEVICES ON A TRACTOR

Device	Mean time, ms
Clutch	613
Toggle switch (underneath steering wheel)	498
Horn rim (on steering wheel)	412
Rim blow (on under edge of steering wheel)	337

Source: Pattie, (1973).

movement is one in which the individual moves a hand (or foot) in free space from one location to another, as in reaching for a control device when the eyes are otherwise occupied. The very well-known study by Fitts (1947) probably provides the best available data relating to the accuracy of the *direction* of such movements in free space. He used an arrangement with targets positioned around the subject at 0, 45, 90, and 135° angles left and right, in three tiers, namely, a center (shoulder-level) tier and tiers 45° above and below the center tier. The blindfolded subjects were given a marker with a sharp point, when they pressed against each target when they tried to reach to it. A bull's-eye was scored 0, and marks in subsequent circles were scored from 1 to 5, marks outside the circles being scored 6.

Figure 8–19 shows the results. Each circle in this figure represents the subjects' accuracy in hitting the target in the corresponding position. The size of the circle is proportional to the average accuracy score for that target; the smaller the size, the better the accuracy. We can see that blind positioning movements are most accurate in the dead-ahead positions and least accurate in the side positions; also the center and lower tiers are slightly more accurate than the upper tier. Thus, control devices or other gadgets that are to be reached for blindly usually should be located in, or near, the straight-ahead position and at or below shoulder level.

Direction of Continuous Control Movements

Some tasks require continuous, freehand control during a movement but with high accuracy, as in sign painting, drawing, etching, etc. Such movements can be carried out with minimum tremor (and with greatest accuracy) when the movement is on the horizontal plane and in a lateral (left-right) direction with forearm pivoted at the

FIGURE 8-19
Relative accuracy scores for different areas in blind-positioning study by Fitts (1947). The position of the circles represents the location of targets, which ranged from 135° left (number 1) to 135° right (number 7), number 4 being straight ahead. The three tiers represent those up, center, and down. The size of each circle represents the relative number of errors, so small circles indicate greater accuracy. *(Source: USAF, AFHRL, Human Engineering Division.)*

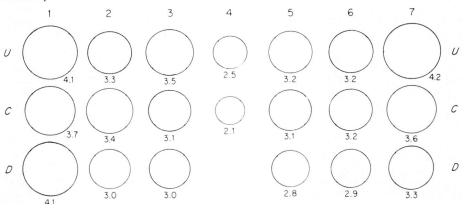

elbow. The advantage of this direction of movement is illustrated by the results of a study by Mead and Sampson (1972), in which they had subjects move a stylus (15 in long with a 4-in, 90° bend at the tip) along a narrow groove in any one of the four positions shown in Figure 8–20. As the subject moved the stylus in the directions indicated in that figure, any time the stylus touched the side of the groove, an "error" was electronically recorded. These errors, which can be viewed as measures of tremor, are also shown in Figure 8–20 and indicate that tremor was greatest during an in-out arm movement in the vertical plane (in which the tremor was up-down) and that it was least during a right-left arm movement in the horizontal plane (in which the tremor was in-out).

It is interesting that the results of this study are reasonably in line with the study of positioning movements as shown in Figure 8–17 since there was less tremor for movement *d* that was based on an essentially lateral component (left to right, with the arm moving at the elbow as a pivot) than movements *a* and *b* which consisted more of an in-out direction (involving more upper-arm and shoulder movement). Movement *c* (which resulted in relatively little tremor) probably benefited from the fact that the motion was based on forearm movement pivoted at the elbow.

Continuous control movements such as those involved in tracking tasks are discussed further in Chapter 9.

Static Muscular Control

Static muscular control is not a movement as such, but rather is the absence of a displacement of the body or body member. In such control, however, the muscles are very definitely involved, in that certain sets of muscles typically operate in opposition to each other to maintain equilibrium of the body or body member. Thus, if a body member, such as the hand, is being held in a fixed position, the various muscles controlling hand movement are in a balance that permits no net movement one way or the other. The tensions set up in the muscles to bring about this balance, however, require continued effort, as most of us who have attempted to maintain an

FIGURE 8-20
Directions and planes of arm movements with stylus as used in study of hand steadiness, with direction of hand tremor and number of "errors" (number of times the stylus touched the side of the groove) for each condition. *(Source: Adapted from Mead and Sampson, 1972, fig. 1 and table 1.)*

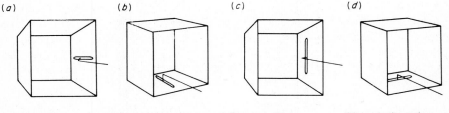

(*a*)
Plane: vertical
Arm movement: in-out
Tremor: up-down
Errors: 247

(*b*)
Plane: horizontal
Arm movement: in-out
Tremor: right-left
Errors: 203

(*c*)
Plane: vertical
Arm movement: up-down
Tremor: in-out
Errors: 45

(*d*)
Plane: horizontal
Arm movement: right-left
Tremor: in-out
Errors: 32

immobile state for any time can testify. In fact, it has been stated that maintaining a static position produces more wear and tear on people than some kind of adjustive posture.

Deviations from static postures are of two types: those called *tremor* (small vibrations of the body member) and those characterized by a gross drifting of the body or body member from its original position.

Tremor in Maintaining Static Position Tremor is of particular importance in work activities in which a body member must be maintained in a precise and immovable position (as in holding an electrode in place during welding). An interesting aspect of tremor is that the more a person tried to control it, the worse it usually is. The following four conditions help to reduce tremor:

1 Use of visual reference.
2 Support of body in general (as when seated) and of body member involved in static reaction (as hand or arm).
3 Hand position. There is less hand tremor if the hand is within 8 in above or below the heart level.
4 Friction. Contrary to most situations, mechanical friction in the devices used can reduce tremor by adding enough resistance to movement to counteract, in part, the energy of the vibrations of the body member.

MANUAL MATERIALS HANDLING

Many jobs and activities in life require manual materials handling (sometimes abbreviated MMH). This includes a wide variety of activities such as loading and unloading boxes or cartons, removing materials from a conveyer belt, stacking items in a warehouse, etc. The physical movements and associated demands involved in such activities are so varied that we can touch on only certain illustrative aspects. And remember that such factors as individual differences, physical condition, sex, etc., markedly influence the abilities of individuals to perform such activities.

Health Effects of Manual Materials Handling

Various short- and long-term health effects can be attributed to MMH. Some of these are (National Institute for Occupational Safety and Health, 1981) lacerations, bruises, and fractures; cardiovascular strain, such as increased heart rate and blood pressure; muscular fatigue; chronic bronchitis; musculoskeletal injury, especially to the spine; and back pain. With regard to injuries as such, the National Safety Council reports that injuries associated with MMH account for about 25 percent of all industrial injuries and result in about 12 million lost workdays per year and over $1 billion in compensation costs.

Back Injuries and Pain In recent years particular attention has been directed to back injuries and pain that might be caused by physical work activities. Although back injuries are easy to identify, the most insidious possible effect of materials handling concerns the long-term wear and tear on the spinal system. The disks between the vertebrae serve as cushions to bending of the spine. But these disks can become worn and thus can induce back pain. However, the specific sources of back pain frequently cannot be identified. And the fact that a person experiences back pain frequently cannot be attributed to the work itself, since many people would (and do) experience such troubles outside a working situation. In any event, since back troubles do plague people involved in materials handling, it is important that the work methods be such as to minimize the possibility of such troubles.[3]

Energy Costs of Manual Materials Handling   Still another consideration in the physical handling of materials is the energy cost of such work. Thus, an additional objective in developing work methods should be that of keeping energy costs within reasonable bounds.

Lifting Tasks

Lifting tasks, of course, make up a large proportion of MMH tasks. Several task-related variables can affect the physiological costs of work, the performance of workers, and the acceptability of tasks to workers. Of these several variables a few are discussed below.

Height and Range of Lift   Common practice provides for categorizing height of lift as follows: (1) floor to knuckle, (2) knuckle to shoulder, and (3) shoulder to reach (from shoulder to maximum reach above). The evidence indicates very strongly that reaches above the shoulder are most demanding in terms of physiological criteria and are least acceptable to workers (Mital,1984).

Further, Davies (1972) reports that the energy cost of lifting objects from the floor to about 20 in (51 cm) is about half again as much as lifting the same weight from about 20 to 40 in (51 to 102 cm). This is because of the additional effort of raising and lowering the body. On the basis of Davies' analysis, the most efficient lift range is between 40 and 60 in (102 and 152 cm). This suggests that, where feasible, work places should be designed to provide for the primary lifting to be within this range.

Methods of Lifting from the Floor   It has been rather common custom (and perhaps an article of faith) to recommend that lifting from the floor be carried out from a squatting position, with knees and hips bent and the back reasonably straight (sometimes called the *straight-back, bent-knee method* or *leg lift*). There had been inklings that this tended to minimize back problems. The "squat" method generally is in contrast with the "stoop" method (or back lift) in which the legs tend to be straight, with the back bending forward and doing most of the lifting. However, questions have been raised about the squat (or leg lift) method.

Garg and Saxena (1979), for example, compared the energy costs of these meth-

---

[3]The problem of back pain is so important in the industrial world that the journal *Ergonomics* published a special (and greatly expanded) issue on industrial back pain in Europe [*Ergonomics*, Jan. 1, 1985, 28(1)].

FIGURE 8-21
Energy costs of lifting a load from the floor as related to a measure of maximum acceptable workload based on the judgments of the subjects. (*From Garg and Saxena, 1979, fig. 2, p. 899.* Reprinted with permission by American Industrial Hygiene Association Journal.)

ods of lifting as related to a measure of maximum acceptable load (based on the subjects' estimates of workload in terms of weights lifted and frequency). Figure 8–21 shows that energy costs (in kilocalories per minute) were highest for the squat lift, with stoop lift and free-style lift being consistently lower (with the free-style lift the subject chose the method "perceived as most suitable"). Garg and Saxena suggest that the higher energy cost of the squat lift is due to the need to raise the body from a bent-knee posture. They also refer to other studies that indicate that the squat lift can contribute to greater compressive forces of the lower back. Further, Troup et al. (1983) report that pressure within the abdominal cavity is less when the trunk is flexed (stoop lift) than when the trunk is more vertical (as in the squat lift).

Although there are inklings that the stoop lift may have some advantages over the squat lift, one should be careful of overgeneralizations because of the ubiquitous presence of interacting variables. For example, if the load in question can fit between the legs, a squat lift usually places less stress on the back than a stoop lift. In line with this, it is reasonable to suggest that very heavy, small objects probably should be raised with a squat lift.

Frequency of Lifts Everyday experience tells us that we can tolerate occasional exertion (as in lifting) much better than frequent exertion. This pattern is confirmed by numerous investigations. In one study, for example, Mital (1984) had women and men with considerable industrial work experience perform various types of lifting tasks. He elicited their estimates of maximum weights they believed they could lift at various frequencies per minute if that rate were continued over a regular 8-h workday. By using their estimates for a frequency of once a minute as a base of 100 percent, the estimates for other frequencies were related to that base. The results are summarized in Figure 8–22. The results from two other comparative studies were very similar, thus lending confidence to the data. In addition, however, Mital found consistent patterns of increased heart rate and oxygen consumption associated with increased frequencies of lifting. The oxygen uptake just about doubled between 1 and 12 lifts per minute, for both males and females.

Discussion of Lifting Tasks Other factors also can be relevant to lifting tasks, such as the size of the object handled, the availability of handles, the amount of

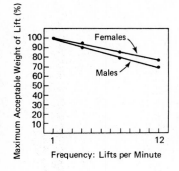

FIGURE 8-22
Relationship between frequency of lifts and maximum
weights judged by subjects to be acceptable over an 8-h
workday. The percentage values in once-a-minute
estimates are used as the base. *(Adapted from Mital, 1984,
figs. 3 and 4.)*

horizontal movement, etc. Some indication of the relationships among certain such
variables is reflected in the pages of research data from Snook (1978).

In connection with the possible effects of many variables on lifting tasks, we
would hope for some basis of synthesizing the effects of various combinations. One
such analysis is provided by Davies (1972, based on Frederich, 1959) and illustrated
in Figure 8–23, in which the energy consumption is presented for various combinations of weights for four lifting ranges. Clearly the most efficient lifting is with a
weight of about 40 lb (18 kg) for a lift range from about 40 to 60 in (102 to 152
cm). For other lift ranges, other weights result in most efficient lifting.

On the basis of a more comprehensive synthesis of data on lifting, the National
Institute of Occupational Safety and Health (NIOSH) (1981) has developed a guideline (in the form of a formula) for estimating the ''risk'' of various combinations of
task variables. This formula is given in Appendix C. Underlying this formula are the
concepts of *maximum permissible limit* (MPL) (the maximum limit that should be
allowed) and *action limit* (the limit above which some ''administration control'' of

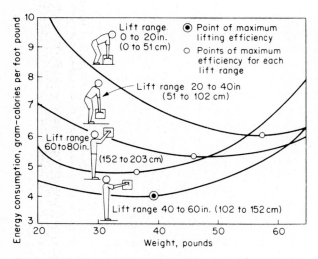

FIGURE 8-23
Energy consumption in
lifting per unit of work for
various weights and
specified lift ranges. The
most efficient lifting is with a
weight of about 40 lb (18kg)
for a lift range from about
40 to 60 in (102 to 152 cm).
*(Source: Frederick, 1959,
as presented by Davies,
1972, fig. 3.)*

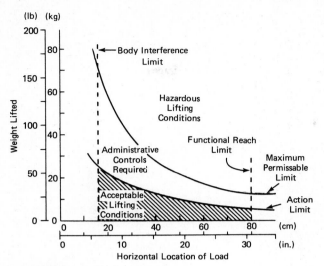

FIGURE 8-24
Levels of risk for lifting tasks as related to horizontal location of
load and weight lifted for infrequent lifts from floor to knuckle
height. *(This figure is derived from a guideline formula
developed by NIOSH, 1981.)*

the task is required). The formula provides for estimating the action limit (AL).
Figure 8–24 illustrates the application of this formula for tasks from floor to knuckle
height; it shows the AL and MPL and the three areas delineated (1) acceptable lifting
conditions, (2) administrative controls required, and (3) hazardous lifting conditions.

Still another scheme for assessing the workload of lifting tasks is that developed
by Ayoub, Selan, and Liles (1983). A formula that need not be given here results in
a job severity index (JSI) for any given lifting task. It takes into account certain task
variables (such as weight, frequency of lift, container size, and range of lift) along
with certain "job exposure" factors (such as length of the workweek and work shift).
The procedure has the further advantage of providing a basis for selecting job can-
didates.

In the event lifting tasks are found to be too demanding (by some procedure such
as that of NIOSH or the JSI), two basic strategies are possible: modify the task and
provide rest periods.

Carrying Tasks

As with lifting tasks, there are marked differences in the weights people find acceptable, depending on the frequency with which the activity is carried out. Aside from
individual differences, there are systematic differences between males and females.
As an example, Figure 8–25 presents data from Snook for males and females on the
maximum weights reported as acceptable by 90 percent of the subjects, for two distances of carry. The effects of frequency are particularly noticeable, with carrying

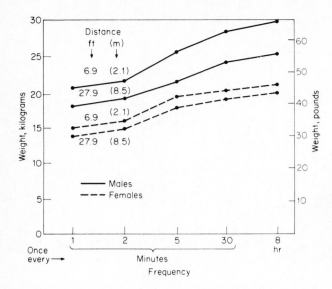

FIGURE 8-25
Maximum weights chosen as acceptable by 90 percent of a sample of male and female industrial workers for carrying loads at knuckle height, at different frequencies, for two distances. *(Source: Snook, 1978, table 10, p. 980.)*

distance having some additional effect on the level of acceptable loads. That figure presents data for carrying at knuckle height. When the load is carried at elbow height, the acceptable weight levels are somewhat lower.

Pushing Tasks

Additional data on manual handling tasks from Snook deal with acceptable levels of weights in pushing tasks. The data shown in Figure 8–26 also reveal systematic differences in the maximum acceptable loads depending on the frequency with which the tasks are performed, with appreciable differences in distances pushed, but only slight sex differences (in fact, for the shortest distance there was no difference by sex). This example represents data for pushing at average shoulder height. Pushing at elbow or knuckle height appears not to influence the levels reported as acceptable.

DISCUSSION

The physical work activities of people cover a wide gamut, ranging from those that involve very careful, skilled work to those that involve the expenditure of substantial physical energy. In the case of light, skilled work, the primary human factors focus is on the design of the work situation to capitalize on the types of skills that people possess. In the case of physically demanding jobs (in which the human being is used as a major energy source), the focus is more on the physiological costs of the work.

In this regard, the measurement of most physiological variables typically is time-consuming and costly, so sometimes it is not feasible. However, some encouraging research by Fleishman, Gebhardt, and Hogan (1984) indicates that people are really pretty good at estimating physical effort. A perceived-effort scale they have developed has been found to serve as a reasonably reliable and valid basis for predicting

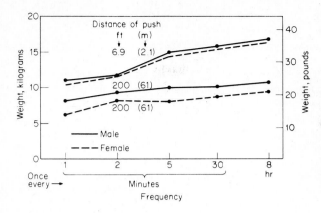

FIGURE 8-26
Maximum weights chosen as acceptable by 90 percent of a sample of female and male industrial workers for a pushing task at shoulder height, at different frequencies, for two distances. *(Source: Snook, 1978, tables 5, 6, 7, 8, pp. 975–978.)*

work costs. Thus, at least in many situations, it would be possible to use such an index as a yardstick for estimating physiological work costs.

REFERENCES

American Industrial Hygiene Association (AIHA). (1971). *Ergonomics guide to assessment of metabolic and cardiac costs of physical work.* Akron, OH: AIMA.

Ayoub, M. M., Selan, J. L., and Liles, D. H. (1983). An ergonomics approach for the design of manual materials handling tasks. *Human Factors, 25*(5), 507–515.

Bălănescu, B. F. (1979). Some aspects concernng the dynamic evaluation of oxygen consumption in exercise tests. *Ergonomics, 22*(12), 1337–1342.

Brouha, L. (1960). *Physiology in industry.* New York: Pergamon.

Brown, J. S., and Slater-Hammel, A. T. (1949). Discrete movements in the horizontal plane as a function of their length and direction. *Journal of Experimental Pyschology, 39,* 84–95.

Brown, J. S., Wieben, E. W., and Norris, E. B. (1948, September). *Discrete movements toward and away from the body in a horizontal plane* (Contract N5 ori-57, Rept. 6). USN, ONR, SDC.

Burger, G.C.E. (1969). Heart rate and the concept of circulatory load. *Ergonomics, 12*(6), 857–864.

Damon, A., Stoudt, H. W., and McFarland, R. A. (1966). *The human body in equipment design.* Cambridge, MA: Harvard.

Datta, S. R., and Ramanathan, N. L. (1971). Ergonomics comparison of seven modes of carrying loads on the horizontal plane. *Ergonomics, 14*(2), 269–278.

Davies, B. T. (1972). Moving loads manually. *Applied Ergonomics, 3*(4), 190–194.

Dukes-Dobos, F. N., Wright, G., Carlson, W. S., and Cohen, H. H. (1976, July 11–16). *Cardio-pulmonary correlates of subjective fatigue.* Technical Program for the 20th Annual Meeting of the Human Factors Society, pp. 24–27.

Edholm, O. G. (1967). *The biology of work.* World University Library, New York: McGraw-Hill.

Fitts, P. M. (1947). A study of location discrimination ability. In P. M. Fitts (ed.), *Psychological research on equipment design,* Research Rept. 19. Army Air Force, Aviation Psychology Program.

Fleishman, E. A., Gebhardt, D. L., and Hogan, J. C. (1984). The measurement of effort. *Ergonomics, 27*(9), 947–954.

Frederick, W. S. (1959). Human energy in manual lifting. *Modern Materials Handling, 14*(3), 74–76.

Garg, A., and Saxena, U. (1979). Effects of lifting frequency and technique on physical fatigue with special reference to psychophysical methodology and metabolic rate. *American Industrial Hygiene Association Journal, 40,* 894–903.

Gordon, E. E. (1957, November). The use of energy costs in regulating physical activity in chronic disease. *A.M.A. Archives of Industrial Health, 16,* 437–441.

Grandjean, E. (1981). *Fitting the task to the man.* New York: International Publications Service.

Hafez, H. A., Gidcumb, C. F., Reeder, M. J., Beshir, M. V., and Ayoub, M. M. (1982). Development of a human atlas of strengths. *Proceedings of the Human Factors Society, 1982.* Santa Monica, CA: Human Factors Society, pp. 575–579.

Hamilton, B. J., and Chase, R. B. (1969). A work physiology study of the relative effects of pace and weight in a carton handling task. *AIEE Transactions, 1,* 106–111.

Houy, D. R. (1983). Range of joint movement in college males. *Proceedings of the Human Factors Society, 1983,* vol. 1. Santa Monica, CA: Human Factors Society, pp. 374–378.

Hunsicker, P. A. (1955), August). *Arm strength at selected degrees of elbow flexion,* Tech. Rept. 54–548. U.S. Air Force, WADC.

Jackson, A. S., Osburn, H. G., and Laughery, K. R. (1984). Validity of isometric strength tests for predicting performance in physically demanding tasks. *Proceedings of the Human Factors Society, 1984,* vol. 1. Santa Monica, CA: Human Factors Society, pp. 452–454.

Johansson, G., and Rumar, K. (1971). Drivers' brake reaction times. *Human Factors, 13*(1), 23–27.

Khalil, T. M. (1973). An electromyographic methodology for the evaluation of industrial design. *Human Factors, 15*(3), 257–264.

Kroemer, K. H. E. (1970). Human strength: Terminology, measurement, and interpretation of data. *Human Factors, 12*(3), 297–313.

Lauru, L. (1954). The measurement of fatigue. *The Manager, 22,* 299–303 and 369–375.

Lehmann, G. (1958).Physiological measurements as a basis of work organization in industry. *Ergonomics, 1,* 328–344.

Mead, P. G., and Sampson, P. B. (1972). Hand steadiness during unrestricted linear arm movements. *Human Factors, 14*(1), 45–50.

Mital, A. (1984). Comprehensive maximum acceptable weight of lift database for regular 8-hour work shifts. *Ergonomics, 27*(11), 1127–1138.

Murrell, K. F. H. (1965). *Human performance in industry.* New York: Reinhold.

National Institute for Occupational Safety and Health (NIOSH). (1981, March). *Work practices guide for manual lifting* (NIOSH PB82–178998). Cincinnati.

Örtengren, R., Andersson, G., Broman, H. Magnusson, R., and Petersén, I. (1975). Electromyography: Studies of localized muscle fatigue at the assembly line. *Ergonomics, 18*(2), 157–174.

Passmore, R., and Durnin, J. V. G. A. (1955). Human energy expenditure. *Physiological Reviews, 35,* 801–875.

Pattie, C. (1973, May). Simulated tractor overturnings: A study of human responses in an emergency situation. Ph.D. thesis. W. Lafayette, IN.: Purdue University.

Schmidtke, H., and Stier, F. (1960). Der aufbau komplexer bewegungsabläufe aus elementarbewegungen. *Forschungsberichte des landes Nordrhein-Westfalen, 822,* 13–32.

Singleton, W. T. (1971). The measurement of man at work with particular reference to arousal. In W. T. Singleton, J. G. Fox, and D. Whitfield (eds.), *Measurement of man at work*. London: Taylor and Francis, pp. 17–25.

Snook, S. H. (1978). The design of manual handling tasks. *Ergonomics,* 21(12), 963–985.

Stobbe, T. J., and Plummer, R. W. (1984). A test-retest criterion for isometric strength testing. *Proceedings of the Human Factors Society, 1984,* vol. 1. Santa Monica, CA: Human Factors Society, pp. 455–459.

Tichauer, E. R. (1978). *The biomechanical basis of ergonomics*. New York: Wiley.

Troup, J. D. G., Leshinen, T. P. J., Stalhammar, H. R., and Kuorinka, I. A. A. (1983). A comparison of intraabdominal pressure increases, hip torque, and lumbar vertegral compression in different lifting techniques. *Human Factors,* 25(5), 517–525.

Vos, H. W. (1973). Physical workload in different body postures, while working near to, or below ground level. *Ergonomics,* 16(6), 817–828.

Wargo, M. J. (1967). Human operator response speed, frequency, and flexibility: A review and analysis. *Human Factors,* 9(3), 221–238.

Warrick, M. J., Kibler, A. W., and Topmiller, D. A. (1965). Response time to unexpected stimuli. *Human Factors,* 7(1), 81–86.

Wickens, C. D. (1984). *Engineering psychology and human performance*. Columbus, Ohio: Merrill.

CHAPTER 9

HUMAN CONTROL OF SYSTEMS

The types of systems and mechanisms people control in their jobs and everyday lives vary tremendously from simple light switches to complex power plants and aircraft. Whatever the nature of the system, the basic human functions involved in its control remain the same. The human receives information, processes it, selects an action, and executes the action. The action taken then serves as the control input to the system. In the case of most systems, there typically is some form of feedback to the person regarding the effects of the action taken.

In this chapter we discuss two topics that involve all the human functions of control. In a sense this chapter can be viewed as a transition from the display-oriented chapters (4, 5, 6, and 7) to the chapters related more to controls (10 and 11). Compatibility, our first topic, deals with the relationships between controls and displays and affects the ease and adequacy with which people can select and carry out appropriate actions from several alternatives. Tracking, our second topic, is a special form of human control that is involved when the control required is continuous and must conform to some external input signal. It is a task that often involves complex information processing and decision-making activities to direct proper control of a system, and this task is greatly influenced by the displays and dynamics of the system being controlled.

COMPATIBILITY

We introduced the concept of *compatibility* in Chapter 3 and distinguished four types: conceptual, spatial, movement, and modality compatibility. (It would be a good idea to review that material now.) In this chapter we deal with spatial and movement compatibility as they relate to the relationships between controls and displays.

234

Compatibility refers to the degree to which relationships are consistent with human expectations. As such, expectations have profound impact on human performance. Where compatibility relationships are designed into the system, (1) learning is faster, (2) reaction time is faster, (3) fewer errors are made, and (4) user satisfaction is higher. Although people can and do learn to use systems that are out of sync with their expectations, they do so at a price. Such systems place a greater information processing demand on an operator. Especially under conditions of stress, a person may make the most "natural" (or compatible) response; if this is not the correct response, an error or accident could result.

Two things should be kept in mind about the use of compatibility relationships: (1) some are stronger than others; that is, some expectations are shared by greater proportions of a population than are others. (2) In some circumstances it may be necessary to violate one compatibility relationship to take advantage of another one in the design of some systems; sometimes it is impossible to avoid this. This last fact is illustrated by the results of a study by Bergum and Bergum (1981) in which 93 percent of subjects expected upward movement of a pointer on a vertical scale to represent an increase, while 71 percent of the same group expected the numbers on the scale to increase from top to bottom!

Spatial Compatibility

There are many variations of spatial compatibility; most deal either with the *physical similarities* between displays and their corresponding controls or with the *arrangement* of displays and their controls.

Physical Similarity of Displays and Controls   Sometimes there exists the opportunity to design related displays and controls so there is reasonable correspondence of their physical features, and perhaps also of their modes of operation. Such a case was well illustrated years ago in a classic study by Fitts and Seeger (1953). In this study three different displays and three different controls were used in all possible combinations. The displays consisted of lights in various arrangements. As a light went on, the subject was to move a stylus along a corresponding channel to turn the light off. The three displays and controls are illustrated in Figure 9–1. Different groups of subjects used each of the nine combinations of stimulus-response panels; performance was measured in terms of reaction time, errors, and information lost. The results, also presented in Figure 9–1, showed that performance was best when the stimulus panel physically resembled the response panel (i.e., combinations $S_a = R_a$, $S_b = R_b$, and $S_c = R_c$).

Physical Arrangement of Displays and Controls

Both experiments and rational considerations lead one to conclude that, for optimum use, corresponding displays and controls should be arranged in corresponding patterns. This aspect of compatibility has been put to the test by a down-to-earth gadget used morning, noon, and night (and sometimes in between)—the arrangement of the

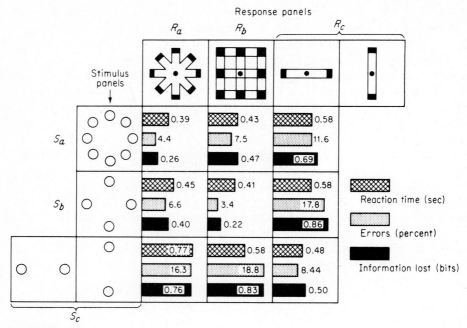

FIGURE 9-1
Illustrations of signal (stimulus) panels and response panels used in study by Fitts and
Seeger. The values in any one of the nine squares are the average performance measures
for the combination of stimulus panel and response panel in question. The compatible
combinations are S_a-R_a, S_b-R_b, and S_c-R_c, for which results are shown in the diagonal cells.
(Source: Fitts & Seeger, 1953.)

burner controls on a four-burner stove. Several studies have investigated this problem. Chapanis and Lindenbaum (1959) and Ray and Ray (1979) presented various
arrangements of controls and burners (such as those shown in Figure 9–2) to subjects
and asked them to turn on specific burners. The number of errors was recorded. The
results from both studies are shown in Table 9–1. As can be seen, the results of the
two studies are in complete agreement with respect to the relative rankings of those
arrangements tested by both. Taking both studies together, we see that arrangement
1, with the offset burners, was clearly the best. When burners are aligned, arrangement 2 appears to be superior.

In a further study dealing with burner controls, Shinar and Acton (1978) asked
subjects to indicate which of the unmarked controls they thought controlled each of
the burners (the burners were aligned rather than being offset). Referring to Figure
9–2, we see that the most frequently chosen arrangement was 3, being chosen by 31
percent of the subjects. Arrangement 2, which resulted in fewer errors than arrangement 3 in the other studies, was chosen by only 25 percent of the subjects. This
suggests that the arrangements people choose may not always result in the optimum
levels of performance. It is probably best, therefore, to use actual performance rather
than rely on subjective preference or choice to decide on arrangements between controls and displays.

TABLE 9–1
PERCENTAGE OF ERRORS IN EXPERIMENTAL
USE OF BURNER CONTROLS ON STOVES
SHOWN IN FIGURE 9–2

Design	Chapanis and Lindenbaum (1959)	Ray and Ray (1979)
1	0	Not tested
2	6	9
3	10	16
4	11	19
5	Not tested	12

Source: Chapanis and Lindenbaum (1959) and Ray and
Ray (1979).

Movement Compatibility

In several different types of circumstances movement compatibility becomes important. The following are some examples:

• Movement of a control device to *follow* the movement of a display (moving a
level to the right to follow a right movement of a blip on a radarscope)
• Movement of a control device to *control* the movement of a display (tuning a
radio to a particular wavelength)

FIGURE 9-2
Control-burner arrangements of a simulated stove used in experiments by Chapanis and
Lindenbaum, and by Ray and Ray. *(Source: Adapted from Chapanis & Linderbaum, 1959.)*

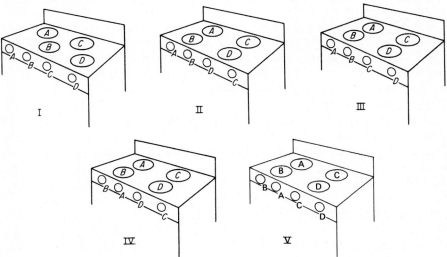

• Movement of a control device that produces a specific system response (turning a steering wheel to the right to turn right)

• Movement of a display indication without any related response (the clockwise turn of the hands of a clock)

People's expectations regarding movement relationships are often referred to as *population stereotypes,* with some stereotypes being stronger than others. Movement compatibility relationships depend in part on features of the controls and displays as well as their physical orientation to the user (such as whether they are in the same or different planes relative to the user). On the basis of research and experience, certain principles of movement compatibility have merged, some of which are discussed now.

Rotary Controls and Rotary Displays in Same Plane In the case of fixed rotary scales with moving pointers, the principle that has become firmly established is that a clockwise turn of the control should be associated with a clockwise turn of the pointer, and that such rotation generally should indicate an increase of the value in question. Conversely, a counterclockwise movement of the control and display pointer should be associated with a decrease of the value.

For moving scales with fixed pointers Bradley (1954) postulated that the following principles generally would be desirable:

1 The scale should rotate in the same direction as its control knob (i.e., there should be a direct drive between control and display).

2 The scale numbers should increase from left to right.

3 The control should turn clockwise to increase settings.

Unfortunately, it is not possible to incorporate all three of these principles into a single fixed-pointer moving-scale display and rotary control assembly. As shown in Figure 9–3, only two of the principles can be followed at the same time in these types of situations. Since one of the principles must therefore be sacrificed, which should it be? To answer that question, Bradley tested various control-display assemblies, including the four shown in Figure 9–3. The following criteria were used to evaluate the designs: starting errors (an initial movement in the wrong direction), setting errors (incorrect settings), and rank-order preferences of the subjects. Some of the results are shown at the bottom of Figure 9–3.

From the results it is clear that direct-drive linkage between the control and display (assemblies *A* and *B*) is the most important principle to preserve. The next most important principle appears to be that the scale numbers increase from left to right. Of lesser importance is that a clockwise control movement results in an increased setting.

Rotary Controls and Linear Displays in the Same Plane In the case of rotary controls and fixed-scale linear displays, the control can be placed above, below, to the left, or to the right of the display. Various compatibility principles have been

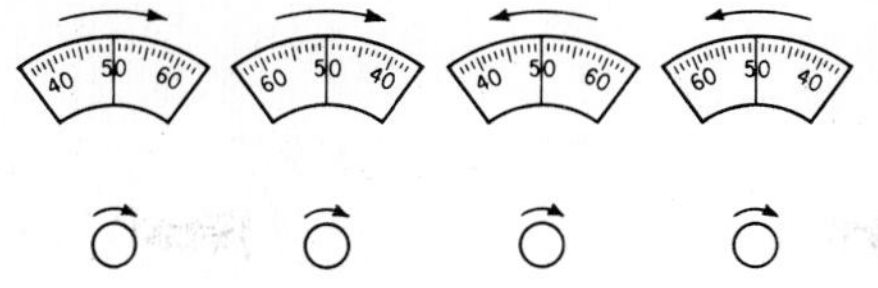

Assembly	A	B	C	D
Drive	Direct	Direct	Reversed	Reversed
Scale numbers increase	Left to right	Right to left	Left to right	Right to left
With clockwise knob movement setting will:	Decrease	Increase	Increase	Decrease

	A	B	C	D
Starting errors	13	11	87	106
Setting errors	0	9	1	8
Preference (number of times ranked "first")	31	22	17.5	1.5

FIGURE 9-3
Some of the moving-display and control-assembly types used in a study by Bradley. The various features of these related to three desirable characteristics are given below the diagrams; crosshatching indicates an undesirable feature. With the usual display orientation all three desirable features are not possible. Some data on three criteria are given at the bottom of the figure, indicating the general preferability of A. *(Source: Adapted from Bradley, 1954.)*

defined for such circumstances. The first is *Warrick's principle* (Warrick, 1947) which can be described as the expectation that the pointer on the display will move in the same direction as that side of the control which is nearest to it. This principle applies only when the control is located to the side of the display.

Brebner and Sandow (1976) formulated two other principles that apply to the use of rotary controls with linear displays (specifically to vertical displays). The first is called the *scale-side principle,* that is, the expectation that the pointer will move in the same direction as the side of the control knob which is on the same side as the scale markings on the display. This principle operates when the control is at the top, bottom, or side of a vertical display. The second principle is called *clockwise-for-increase principle* which states that people will turn a rotary control clockwise to increase the value on the display no matter where the control is located relative to the display.

Warrick's principle, the clockwise-for-increase principle, and the scale-side principle were compared and contrasted in experiments by Brebner and Sandow (1976) and by Petropoulos and Brebner (1981) with various arrangements of displays and controls. Subjects were shown slides and were asked to indicate in which direction they would turn the knob to "move the indicator to 15." Figure 9–4 presents some of the conditions tested, the predictions based on the three principles, and the results

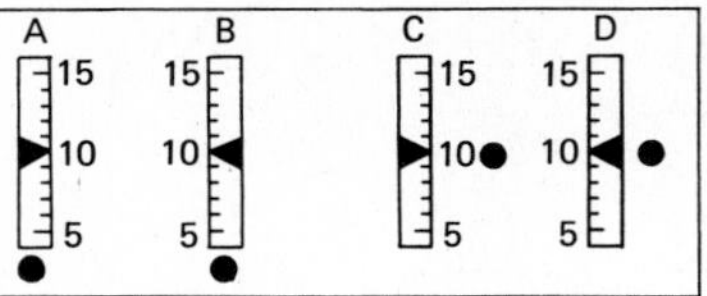

	A	B	C	D
Predictions:				
Warrick's principle	NA	NA	C	C
Scale-side principle	CC	C	CC	C
Clockwise-for-increase	C	C	C	C
Results: Percent choosing:				
Clockwise	43 (45)	80 (72)	73 (57)	86 (85)
Counterclockwise	57 (55)	20 (28)	27 (43)	14 (15)

NA = not applicable   C = clockwise   CC= counterclockwise

FIGURE 9-4
Four configurations of rotary controls and vertical linear scales.
Shown are the predicted stereotypes based on three principles. The
percentages choosing each direction of rotation to move the pointer to
15 are shown for two studies: Brebner & Sandow (1976) and in
parenthesis, Petropoulos & Brebner (1981).

obtained from the two studies. There was a major discrepancy between the two studies with regard to arrangement *C* which may have been due to differences in methodology. In general, however, it appears that when principles clash, as in arrangements *A* and *C,* the strength of the stereotype is weaker than when the principles are
congruent, as in arrangements *B* and *D*.

Rotary knobs can be placed above the displays to which they relate, but this is
generally considered to be a poor idea because the operator's hand will block the
display while using the control. Thus, if the control is placed below the display
(probably the best location since the hand will not block the display), arrangement *B*
in Figure 9–4 should be used. If the control is placed to the side, arrangement *D* is
recommended with the understanding that a left-hander can block the display when
operating the control.

Movement of Displays and Controls in Different Planes   Sometimes control
devices may be in a different plane from the displays with which they are associated.
Several studies have been conducted to investigate stereotypes associated with rotary
and stick-type controls and linear displays positioned in different planes. Holding
(1957), for example, found that with rotary controls, people's responses tended to be
based on two principles: (1) a general clockwise-for-increase principle and (2) a hel-
ical, or screwlike, tendency in which clockwise rotation is associated with movement
away from—and counterclockwise movement is associated with movement toward—
the individual.

Spragg, Finck, and Smith (1959) investigated the four combinations of control
and display movements shown in Figure 9–5 in connection with a tracking task. For
the horizontally mounted stick, on the vertical plane, the superiority of the *up-up*
relationship (control movement up associated with display movement up) over the
up-down relationship is evident in the results shown in Figure 9–5. For a vertically

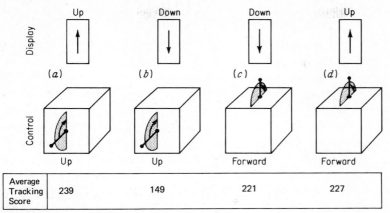

FIGURE 9-5
Tracking performance with horizontally mounted and vertically mounted stick controls and varying control-display relationships. *(Source: Adapted from Spragg, Finck, & Smith, 1959, data based on trials 9 to 16.)*

mounted stick, on the horizontal lateral-cutting plane, there was less difference between the *forward-up* and *forward-down* relationships.

Based on these studies and others, Grandjean (1981) recommended the movement compatibility relationships shown in Figure 9–6 for rotary and stick-type controls and linear displays located in various planes.

Movement Relationships of Rotary Vehicular Controls   In the operation of most vehicles there is no "display" to reflect the "output" of the system; rather, there is a "response" of the vehicle. In such instances, if the wheel control is in a horizontal plane, an operator tends to assume an orientation toward the forward point

FIGURE 9-6
Recommended movement relationships for rotary and stick-type controls and linear displays located in various planes. *(Source: Grandjean, 1981; fig. 113.)*

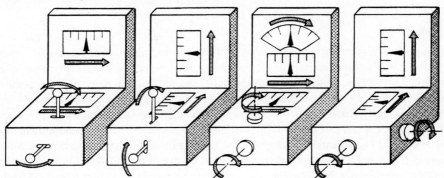

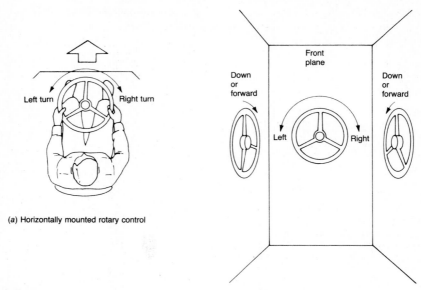

FIGURE 9-7
The most compatible relationships between the direction of movement of horizontally and vertically mounted rotary controls and the response of vehicles. *(Source: Adapted from Chapanis & Kinkade, 1972, figs. 8-6 and 8-7.)*

of the control, as shown in Figure 9–7*a* (Chapanis and Kinkade, 1972). If the wheel is in a vertical plane, the operator tends to orient to the top of the control, as shown in Figure 9–7*b*.

A rather horrendous control problem one of the authors has seen is with a shuttle car for an underground coal mine that has a control wheel that controls left and right turns. This wheel is on the right-hand side of the car relative to the driver when going in one direction, but when going in the other direction the driver moves to the opposite side to face the direction of travel, so that the wheel is then on the driver's left. The control relationships of this are so complicated that new drivers have major problems in learning how to control the cars.

Discussion

Although fairly clear-cut population stereotypes do exist for certain control-display relationships, these are by no means universal. (After years of soliciting population stereotype responses from students, engineers, and psychologists, one of the authors has yet to find a single person whose expectations correspond with those of the majority in every case.) When there is no strong population stereotype (or when relevant principles are in conflict), a designer still needs to make a design decision. One approach is to design control-display relationships to match existing relationships found in other systems likely to be used by the intended population. That is, standardization can substitute for a population stereotype. Another approach is to select a relationship that is logical and explainable. At least, then it will be easier to train people to use the system even if the logic was not apparent to them before training. When there is no clear-cut stereotype to adopt, there is no previous experience to follow, and there are no logical principles to use, then empirical tests of possible relationships should be carried out with the intended user population to serve as the basis for a design decision.

TRACKING

Tracking tasks require continuous control of something and are present in practically all aspects of vehicle control, including driving an automobile, piloting a plane, or steering and maintaining balance on a bicycle. Other examples of tracking tasks include following a racehorse with binoculars or tracing a line on a sheet of paper. The basic requirement of a tracking task is to execute correct movements at correct times. In some instances the task is paced by the person doing it, as in driving an automobile. In other instances the task is externally paced, in that the individual has no control over the rate at which the task has to be performed, as in following a racehorse with binoculars.

The analysis of tracking tasks can be extremely complex, and research on tracking has been primarily engineering-oriented, focusing on mathematical representations of the responses of well-trained operators. We do not deal extensively with such topics, but the interested reader can consult Kelley (1968) or Poulton (1974) for more extensive treatments of the field. In this chapter we discuss briefly some background regarding certain concepts involved in tracking and some inklings of the many variables associated with tracking performance.

Inputs and Outputs in Tracking

In a tracking task, an input, in effect, specifies the desired output of the system; for example, the curves in a road (the input) specify the desired path to be followed by an automobile (the output).

Inputs Inputs in a tracking task can be constant (e.g., steering a ship to a specified heading or flying a plane to an assigned altitude) or variable (e.g., following a winding road or chasing a maneuvering butterfly with a net). Such input typically is received directly from the environment and sensed by mechanical sensors or by people. If it is sensed mechanically, it may be presented to operators in the form of signals on some display. The input signal is sometimes referred to as a *target* (and in certain situations it actually is a target), and its movement is called a *course*. Whether the input signal represents a real target with a course or some other changing variable such as desired changes in temperature in a production process, it usually can be described mathematically and shown graphically as a function of time. While in most instances the mathematical or graphic representations do not depict the real geometry of the input in spatial terms, such representations still have utility for characterizing the input.

Inputs can take on complex patterns over time as anyone who has ever chased an elusive butterfly can tell you. Despite this, several elementary patterns can be distinguished which, when combined, form the most complex inputs. The elementary in-

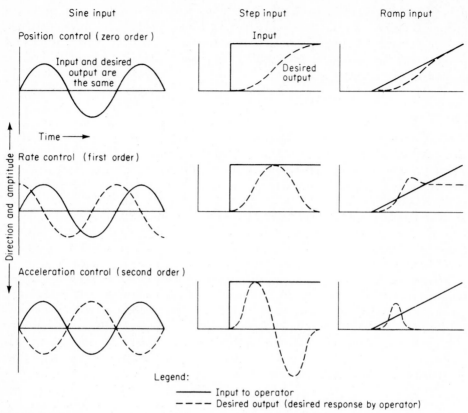

Sine input   Step input   Ramp input

Position control (zero order)

FIGURE 9-8
Tracking responses to sine, step, and ramp inputs which would be conducive to satisfactory
tracking with position, rate, and acceleration control. The desired response, however, is not often
achieved to perfection, and actual responses typically show variation from the ideal. In a
positioning response to a step input, for example, a person usually overshoots and then hunts
for the exact adjustment by overshooting in both directions, the magnitude diminishing until
arriving at the correct adjustment.

puts are the step input, ramp input, and sine input. Figure 9–8 shows these three
forms of input. A *step input* specifies a discrete change in value, such as an instruction from a control tower for an aircraft to change altitude or the need to increase
the temperature control of an oven at some time. A *ramp input* specifies a constant rate of change of some variable. Since velocity is the rate of change in position, maintaining a constant velocity (i.e., speed) of a car would be an example
of tracking a ramp input. A *sine input* is characterized by a sinusoidal (i.e., sine)
wave.

Outputs   The output is usually brought about by a physical response with a control mechanism (if by an individual) or by the transmission of some form of energy
(if by a mechanical element). In some systems the output is reflected by some indi-

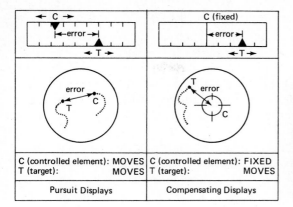

FIGURE 9-9
Illustrations of compensatory and pursuit tracking displays. A compensatory tracking display shows only the *difference* (error) between the target *T* and the controlled element *C*. A pursuit display shows the location (or other value represented) of both the target and the controlled element.

cation on a display, sometimes called a *follower,* or a *cursor;* in other systems it can be observed by the outward behavior of the system, such as the movement of an automobile. In either case it is frequently called the *controlled element.*

Pursuit and Compensatory Displays in Tracking

The input (the target) and the output (the controlled element) can be presented on either a *pursuit* display or a *compensatory* display, as illustrated in Figure 9–9. With a pursuit display both indications move, each showing its own location in space in relationship to the other. If the relative movement of the two elements is represented by a single value (a single dimension), they can be shown on a scale such as those in the upper part of Figure 9–9, whereas if this movement needs to be depicted in two dimensions (as in chasing a moving aircraft), the display might look like that in the lower left part of Figure 9–9. The task of the operator is to align the moving cursor with the moving target. In a compensatory display, using either one or two dimensions (shown to the right in Figure 9–9), only one of the two indicators moves (either the target or the controlled element) and the other is fixed. The task for the operator is to get the moving indicator to align with the fixed indicator.

When the two indicators are superimposed, by using either a compensatory or pursuit display, the controlled element is said to be *on target;* any difference represents error, and the function of the operator is to manipulate the controls to eliminate or minimize that error. With a pursuit display, the operator can determine whether the error is due to target movement or the movement of the controlled element, and further, she or he can see the target's course independent of any movements of the controlled element. With a compensatory display, however, only the absolute error, or difference between the target and controlled element, is shown. If the error begins to increase, the display does not indicate whether the target has moved, the controlled element has moved, or both have moved. A practical advantage of compensatory displays, however, is that they conserve space on an instrument panel because they do not need to represent the whole range of possible values or locations of the two elements.

Control Order of Systems.

Control order refers to what we might call the hierarchy of control relationships between the movement of a control and the output it is intended to control. The nature of control orders and their implications for tracking tasks are very intricate, and we do not go too deeply into this topic. To give some impression of control orders, however, let us begin with the concept of *position*. This refers to the output of the system, with the objective of the tracking task being to control the system so the output corresponds as closely as possible to the input. (The input, of course, specifies what the output should be.) Although we can readily envision the "position" of a vehicle in space, let us think of it in a broader context as a measure of the output. It might be shown as a moving pointer on a scale, a blip on a radar screen, an index of the revolutions per minute of a machine, or something else.

Position (Zero-Order) Control   In a *position-control* tracking task, the movement of the control device controls the output *directly,* such as moving a spotlight to keep it on an actor on a stage, following the movements of an athlete with a camera, or (in a laboratory task) following a moving curved line with a pen or other device. If the system involves a display, there is a direct relationship between the control movement and the display movement it produces.

Rate (First-Order) Control   With a *rate-control* system, the direct effect of the operator's movement is to control the rate at which the output is being changed. The accelerator of an automobile is a rate or first-order control device since it controls the speed (rate of change of position) of the automobile. In particular, the distance by which the pedal is depressed controls the speed, although there is a time lag between depressing the pedal and the speed controlled by the pedal. In turn, the speed of the automobile controls its position along the road. The operation of certain machine guns is controlled by hand cranks that control the rate at which the gun changes direction, and would involve rate control.

Acceleration (Second-Order) Control   *Acceleration* is the rate of change in the rate of movement of something. The operation of the steering wheel of an automobile is an example of acceleration control since the angle at which the wheel is turned controls the angle of the front wheels. In turn, the direction in which the automobile wheels point determines the rate at which the automobile turns. Thus, a given rotation of the steering wheel gives the automobile a corresponding acceleration toward its turning direction. Poulton (1974, p. 324) makes the point that the control of certain chemical plants or nuclear reactors may approximate an acceleration or second-order system.

Higher-Order Control   Certain systems have control systems that can be considered as *higher-order* systems, such as third- or fourth-order control. A third-order control, for example, would involve the direct control of the rate of change of the acceleration (jerk), that then controls the rate, that finally controls the position of whatever is being controlled. The control of a large ship could approximate a third-order or even a fourth-order control because of the linkages between the person steering and the actual movement and position of the ship and the mass (that is, the weight) of the ship. In continuous control processes that have a series of control linkages (such as a ship), the sequence of chain-reaction effects can be described in terms of mathematical functions, such as a change in the *position* of one variable changing the *velocity* (rate) of the next, the *acceleration* of the next, etc.

Control Responses with Various Control Orders

The operators of many systems are expected to make control responses to bring about the desired operation of the system as implied by the input (such as a road to be followed, the flight path of an aircraft, or the appropriate temperature control over time of an industrial process). In the absence of any scheme for helping the operator, such control can be complicated with higher-level control orders. Examples of the appropriate responses for sine, step, and ramp inputs with position, rate, and acceleration control systems are shown in Figure 9–8. In each case the dotted line represents the response over time (along the horizontal) that would be required for satisfactory tracking of the input in question. In general, the higher the order of control, the greater is the number of controlled movements that need to be made by an operator in response to any single change in the input, as illustrated in Figure 9–8.

As an example, consider a step input (the target jumps forward and stops) and the responses to it from using a joystick (the middle column of diagrams in Figure 9–8). In the case of a position control (zero order), the response is simple to move the stick forward the proper distance and hold it. With a rate control (first order), the stick is moved forward to impart a particular velocity to the controlled element. As the controlled element approaches the position of the stationary target, the stick must be returned to the null position (zero velocity) to stop the controlled element before the element overshoots the position of the target. In the case of an acceleration control (second order), things really get interesting, as shown at the bottom of Figure 9–8. With a step input, the control is moved forward, which imparts an acceleration (increasing speed) to the controlled element. The stick must then be brought back to the null position (zero acceleration) which leaves the controlled element moving at a constant velocity. To avoid overshooting the target, the stick must be moved in the opposite direction to decelerate the controlled element. The stick cannot be held there, however, or else the controlled element will stop and start accelerating in the opposite direction. The stick, therefore, must be returned to the null position so that no acceleration or deceleration is being imparted to the controlled element. And all this is needed just to move the controlled element forward and stop it!

Human Limitations in Tracking Tasks

Humans, unfortunately, are not very good at tracking tasks, especially when they involve higher-order control parameters. Wickens (1984) identifies specific information processing limitations that affect tracking performance, i.e., processing time, bandwidth, and anticipation.

Processing Time People do not process information instantaneously, hence there is a time delay between a change in a target and the initiation of the responses required to track the target. The magnitude of the time delay is dependent on the order of the system being controlled (McRuer and Jex, 1967). The time delay with zero- and first-order systems is about 150 to 300 ms. For a second-order system, the delay is on the order of 400 to 500 ms. These delays are harmful to tracking performance because the operator, in essence, is always chasing the target and is always somewhat behind it unless some means are provided to help the operator. We discuss a few of these techniques later in this chapter.

Bandwidth   *Bandwidth* refers to the upper limit of frequency with which corrective decisions can be made, and hence the term defines the maximum frequency of a random input that can be successfully tracked. This bandwidth is normally between 0.5 and 1.0 Hz (Elkind and Sprague, 1961). Since two corrections are required per cycle, a 1-Hz limit corresponds to a limit of roughly two corrections per second. This limit appears to be a central processing limit rather than a motor response limit because people have no difficulty in tracking predictable courses as high as 2 to 3 Hz (Pew, 1974).

Anticipation   Often operators must track targets by using systems that have time lags or that respond sluggishly to control inputs (ships and planes are examples). This requires that the operator anticipate future errors based on present conditions and then make control responses that are expected to reduce that anticipated future error. Unfortunately, humans are not very good at anticipating future outputs, especially for slow, sluggish systems. Part of this difficulty is due to the limitations inherent in working memory. Making the calculations, necessary to predict the future state of a higher-order system can stress all but the most experienced operators.

Factors That Influence Tracking Performance

The effectiveness of human control of tracking operations is influenced by a wide variety of factors such as the nature of the displays and of the controls and the features of the tracking system. A few such factors are discussed briefly below.

Preview of Track Ahead   In some tracking situations an individual has some preview of the track ahead, as in driving an automobile on a winding road that can be seen ahead. (In a blinding snowstorm or heavy fog, the individual does not have such a preview.) In general terms some preview of the input assists an operator in a tracking task. Although the nature of the task presumably can influence the possible benefit of a preview, Kvälseth (1979) indicates that such preview is most beneficial if the preview shows the portion of the track that immediately precedes the present position, rather than showing a lagged preview with a gap between what is previewed and the present position. The duration of the preview seems to be of less consequence than the opportunity to have at least some preview. Kvälseth (1978a) indicates that performance improves steadily as the preview span approaches approximately 0.5 s, beyond which the usefulness of the preview does not increase much. The primary advantage of preview is that it enables the operator to compensate for time lags. The 0.5-s preview corresponds well with the response time (lag) of operators using

TABLE 9–2
SUMMARY OF COMPARISON OF NUMEROUS STUDIES OF PURSUIT AND
COMPENSATORY TRACKING

| Control order | Number of studies | Difference in results reliably in favor of: | | Results inconclusive |
		Pursuit	Compensatory	
Zero order	45	29	0	16
Higher order	34	14	7*	13

*These studies used inappropriate experimental methods, and the results should probably be disregarded.
Source: Poulton, 1974.

higher-order systems. When there are other time lags in the system, larger previews
are expected to be beneficial.

Type of Display: Pursuit versus Compensatory In reviewing the research relative to the possible merits of visual analog pursuit versus compensatory displays,
Poulton (1974, p. 166) concluded that when there is a choice between the two, a
conventional pursuit (true motion) display is preferable to a compensatory (relative
motion) display. His argument was based on a summary of numerous studies, as
shown in Table 9–2. In the case of the seven studies in which compensatory tracking
was reported to be best and three of the studies involving zero-order control included
in the inconclusive column, Poulton argued that the results could be attributed to
inappropriate experimental designs and therefore the results should probably be disregarded. As an illustration of the advantage of pursuit displays, Figure 9–10 shows
the results of one study comparing the two display types (Briggs and Rockway,
1966).

One reason pursuit displays are generally better than compensatory displays is that
the operator can see the separate effects of target and controlled-element movements

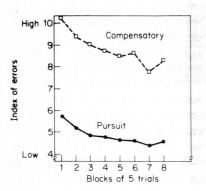

FIGURE 9-10
Comparison of errors of subjects when using a
compensatory versus a pursuit display in a tracking
task. *(Source: Adapted from Briggs & Rockway,
1966, fig. 2. Copyright © 1966 by the American
Psychological Association. Reprinted by permission.)*

on the error generated. This makes it easier to predict the target's course and to learn the consequences of various control actions on the movement of the controlled element. When the target is fixed (as, for example, when a ship is docking) and the displayed error is a consequence of only the movement of the controlled element, the advantage of a pursuit display is diminished (Wickens, 1984).

Another advantage of pursuit displays is that they involve greater movement compatibility. If a target suddenly moves to the left, the pursuit display shows a movement to the left and the correct response is a leftward movement to chase the target—a compatible arrangement. With a compensatory display, however, the left-moving target is displayed as a right-moving error, while the correct response remains a leftward control movement—obviously a less compatible arrangement than with a pursuit display.

Kvälseth (1978b) compared pursuit and compensatory displays by using digital displays (like counters) rather than the traditional analog displays. With digital displays, it is more difficult for users to visualize the target's course, since only a series of changing numbers is shown. As might be expected under such conditions, no difference in performance was found between the two types of displays.

Although pursuit displays generally result in better tracking performance than compensatory displays (at least with visual analog presentations), practical considerations may sometimes argue for the use of compensatory displays, especially because they may occupy less space on control panels.

Time Lags in Tracking   In a tracking task when an error is detected by the operator, three actions need to take place: (1) The operator must choose and execute a corrective response, (2) the system being controlled must then respond to the control input, and (3) the result must be displayed to the operator. Each of these activities takes time; i.e., each introduces a *time lag* into the task. In general, time lags tend to degrade operator performance. This is due in part to the greater demands placed on working memory and the increased need to anticipate future events when lags are present.

There are several types of time lags. *Response lag* is the time taken by the operator to make a response to an input. *Control system lag* is the time between a control action of an operator and the response of the system under control. For sluggish, higher-order systems, such as a ship or large aircraft, it may be several seconds before the system responds to an operator's control actions. *Display system lag* is the delay between the response of the system being controlled or a change in the target and the display of that response or change.

Control system lag has been a major focus of research on time lags. There are three basic types of control lag, these being illustrated in Figure 9–11 for a step input. Of these, *transmission time lag* simply delays the effect of a person's response; the output follows the control response by a constant time interval. An *exponential lag* refers to a situation in which the output is represented by an exponential function following a step input. And a *sigmoid time lag* is represented by the S-shaped curve of Figure 9–11.

The effects of various types of lags are intricately related to the various features of the tracking system and are not discussed here in detail. As an exmple of the

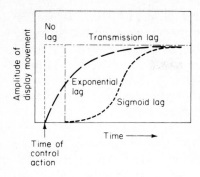

FIGURE 9-11
Illustration of three types of time lag following a step input. The dotted line represents the output when there is no lag, as (theoretically) might be the case when turning on a switch. The human response to a step input with most tracking systems tends to be somewhat similar to a sigmoid lag curve. *(Source: Poulton, 1974, fig. 11.9, p. 206.)*

effects of lags on performance, Kao and Smith (1978) present data that indicate that in a compensatory tracking task involving bimanual (two-handed) versus unimanual (one-handed) control, increasing time lags in the presentation of information on displays resulted in performance degradation for both forms of control with short and intermediate delays (up to 0.8 s); but with a longer display lag (1.5 s) bimanual performance was more adversely affected than unimanual performance. The results are shown in Figure 9–12.

Aside from some of the intricacies of lags with various types of tracking systems, Poulton (1974, p. 373) points out that all three types of control lag increase the error in tracking. Although such lags are generally undesirable, there sometimes are ways to minimize these effects. For example, Poulton (p. 378) indicates that a design engineer may be able to reduce the effective order of a system from acceleration control to rate control by introducing an approximately exponential lag. As another illustration of methods of minimizing the effects of lag, Rockway (1954) has shown that, with long (as opposed to short) delays, the control-response ratio (C/R ratio)

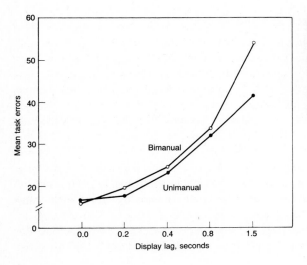

FIGURE 9-12
Illustration of the effects of various display lags in a compensatory tracking task performed with bimanual and unimanual control. The difference for the 1.5 s display lag is statistically significant. *(Source: Kao & Smith, 1978, fig. 2, p. 666.)*

can have some effect on tracking performance. The *C/R ratio* is the ratio of the amount of movement of the control device as related to the amount of corresponding movement of the display indication (see Chapter 10). In his study, Rockway found that high C/R ratios (1:3 and 1:6) resulted in performance degradation with long lags, whereas low C/R ratios (1:15 and 1:30) resulted in the maintenance or even improvement of performance with such lags. Thus, in some circumstances there are schemes that can be used to minimize the possible effects of time lags.

Specificity of Displayed Error in Tracking   In certain compensatory tracking systems the error (the difference between input and output) can be presented in varying degrees of *specificity*. Some such variations are shown in Figure 9–13 as they were used in a tracking experiment by Hunt (1961), including 3 categories of specificity (left, on target, and right); 7 categories; 13 categories; and continuous. The accuracy of tracking performance under these conditions (for two levels of task difficulty) indicates quite clearly that performance improved with the number of categories of information (greater specificity), this improvement taking a negatively accelerated form for tasks of both levels of difficulty. Although the results of other studies are not entirely consistent with these, the evidence suggests that in a tracking task, performance is facilitated by the presentation of more specific, rather than less specific, display information.

Paced versus Self-Paced Tracking   Most tracking tasks are self-paced, in that the person has control over the rate of the output, as in driving an automobile. In such an activity the driver usually can select the speed. With some tracking tasks, however, the pace is not under the individual's control. Poulton (1974, p.8), for example, refers to the fact that an airline pilot, when coming in to land, has to hold airspeed within close limits as well as keep the plane within a closely defined glide path. Tracking is easiest when the task is self-paced and increases in difficulty as the degree of external pacing is increased.

Procedures for Facilitating Tracking Performance

As we can see, humans have built-in time lags and limited bandwidths and are poor at anticipating future system states; in addition, control order, preview, and time lags influence tracking performance. These limitations and influences are especially important with higher-order systems (second-, third-, or fourth-order systems). Often the characteristics of such systems exceed the capabilities of the operators. Something, therefore, needs to be done to compensate for this disparity. A few such procedures are discussed below.

Aiding   One such procedure is the use of *aiding*. Aiding was initially developed for use in gunnery tracking systems and is most applicable to tracking situations of this general type, in which the operator is following a moving target with some device. Its effect is to modify the output of the control in order to help the tracker. In *rate aiding* a single adjustment of the control affects both the rate and the position components of the tracking system. Suppose we are trying to keep a high-powered telescope directed exactly on a high-flying aircraft by using a rate-aided system. When we fall behind the target, our control movement to catch up again automatically speeds up the *rate* of motion of our telescope (and thus, of course, its position). Similarly, if our telescope gets ahead of the target, a corrective motion automatically slows downs its rate (and influences its position accordingly). Such rate aiding simplifies the problem of quickly matching the rate of motion of the following device to that of the target and thus improves tracking performance. In *acceleration aiding* the control movement controls three variables of the controlled element, namely, acceleration, rate, and position.

The operational effect of aiding is to shift from the operator the mental operations that approximate those of differentiation, integration, and algebraic addition which are required in some tracking tasks, so the operator's choice is primarily that of amplification. (In effect, the operator simply has to figure out the ratio of the movement of the control to that of the controlled element.) The operational effects of aiding depend on a number of factors, such as the nature of the input signal, the control order, and whether the system is a pursuit or compensatory type. Thus, aiding should be used selectively, in those control situations in which it is uniquely appropriate.

FIGURE 9-13
Compensatory displays used in study of the effects of specificity of feedback of error information, and tracking performance using such displays. The feedback of error was presented by the use of lights (3, 7, 13, and continuous). *Source: Adapted from Hunt, 1961.)*

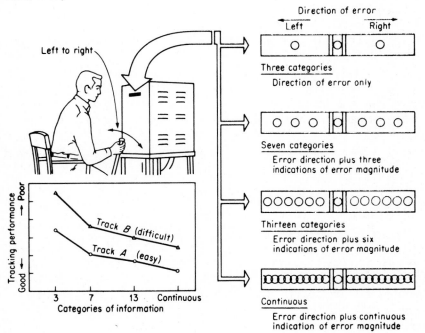

Predictor Displays   Still another procedure for simplifying the control of high-order systems (especially large vehicles such as submarines) is the use of predictor displays, as proposed primarily by Kelley (1962, 1968). In effect, predictor displays use a fast-time model of the system to predict the future excursion of the system (or controlled variable) and display this excursion to the operator on a scope or other device. The model repetitively computes predictions of the real system's future, based on one or more assumptions about what the operator will do with the control (e.g., return it to a neutral position, hold it where it is, or move it to one or another extreme). The predictions so generated are displayed to enable the operator to reduce the difference between the predicted and desired output of the system. A distinguishing feature of a predictor display is that it shows the present state of the system as well as the predicted future state.

An example of a predictor display is seen in Figure 9–14, which shows the present and predicted depth errors of a submarine over a 10-s period. A computer-generated stylized representation of an aircraft predictor display is shown in Figure 9–15. This display presents the present and predicted positions of the aircraft 8 s in the future during a landing approach.

Predictor displays offer particular advantages for complex control systems in which the operator needs to anticipate several seconds in advance, such as with submarines, aircraft, and spacecraft. The advantages in such situations have been demonstrated by the results of experiments such as the one by Dey (1972) simulating the control of a VTOL (vertical takeoff and landing) aircraft. The values of one index of the deviations from a desired "course" with and without the use of a predictor display were 2.48 and 7.92, respectively. The evidence from this and other experiments shows a rather consistent enhancement of control performance with a predictor display.

Quickening   Quickening is closely related to predictor displays in that the display presents information about where the system will likely be in the future given the action (or inaction) taken by the operator. Thus, the operator learns quickly (hence the name *quickening*) the future consequences of a control action. The distinction between a quickened display and a predictor display, however, is that a quickened display does not present information about the present position or state of

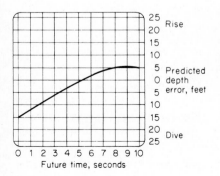

FIGURE 9-14
Example of a predictor display for a submarine. The display shows the predicted depth error, in feet, extrapolated to 10 s, assuming that the control device would be immediately returned to a neutral position. *(Source: Kelley, 1962.)*

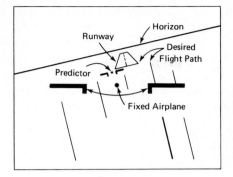

FIGURE 9-15
Stylized representation of an aircraft predictor display showing the present and predicted position of an aircraft 8 s in the future during a landing approach. *(Source: Jensen, 1981; fig. 1. Copyright by the Human Factors Society, Inc. and reproduced by permission.)*

the system, whereas a predictor display does. A quickened display looks like a regular display except it shows what the situation will be in the future rather than what the situation is at present. An operator using a quickened display can be lulled into a false sense of security. The fact that the quickened display indicates that the controlled element is on target does not necessarily mean that *at that moment* the controlled element is indeed on target. It is only an indication that in the future the element is likely to be on target—within the accuracy constraints of the equations used to predict the future state. Quickening can be of some benefit under specific conditions, but usually predictor displays will result in superior tracking performance.

Discussion The relative utility of the various methods of facilitating performance of tracking tasks becomes inextricably intertwined with the type of input (step, ramp, sine wave, etc.), with the control order (zero, first, second, etc.), and with the type of display (pursuit or compensatory). Aiding can be useful in certain circumstances but has limited general applicability. In connection with quickening, Poulton (1974, pp. 180–185) raised serious questions about the research strategies used in some studies and also referred to certain disadvantages of quickened displays. In general Poulton (pp. ix, 189) concluded that true motion predictor displays are likely to be far easier and safer to use for control systems of high order than quickened displays.

Note that quickened and predictor displays are not an either-or proposition. They can be combined in various proportions in a single display (Roscoe, Corl, and Jensen, 1981), and performance can be improved under some circumstances by combining them (Jensen, 1981).

DISCUSSION

The basic functions of those who are to control a system or mechanism are rooted in the nature of the process or operation in question. However, specific design features of the system or mechanism predetermine the specific operational demands imposed on the controller, in particular the inputs that must be received, the types of decisions to be made, and the control actions to be taken. Considerations of human factors during the design stage can contribute to the design of systems and mechanisms that can be controlled adequately by human beings and that can contribute to the well-being of those involved in subsequent control operations.

REFERENCES

Bergum, B. D., and Bergum, J. E. (1981). Population stereotypes: An attempt to measure and define. *Proceedings of the Human Factors Society 25th Annual Meeting.* Santa Monica, CA: Human Factors Society, pp. 662–665.

Bradley. J. V. (1954). *Desirable control-display relationships for moving-scale instrument,* Tech. Rept. 54–423. U.S. Air Force, WADC.

Brebner, J., and Sandow, B. (1976). The effect of scale side on population stereotype. *Ergonomics,* 19(5), 571–580.

Briggs, G. E., and Rockway, M. R. (1966). Learning and performance as a function of the percentage of pursuit tracking component in a tracking display. *Journal of Experimental Psychology,* 71, 165–169.

Chapanis, A., and Kinkade, R. G. (1972). Design of controls. In H. P. Van Cott and R. G. Kinkade (eds.), *Human engineering guide to equipment design.* Washington: Government Printing Office.

Chapanis, A., and Lindenbaum, L. (1959). A reaction time study of four control-display linkages. *Human Factors,* 1(4), 1–7.

Dey, D. (1972). The influence of a prediction display on a quasi-linear describing function and remnant measured with an adaptive analog-pilot in a closed loop. *Proceedings of Seventh Annual Conference on Manual Control* (NASA SP-281). Washington: National Aeronautics and Space Administration.

Elkind, J. I., and Sprague, L. T. (1961). Transmission of information in simple manual control systems. *IEEE Transactions on Human Factors in Electronics,* HFE-2, 58–60.

Fitts, P. M., and Seeger, C. M. (1953). S-R compatibility: Spatial characteristics of stimulus and response codes. *Journal of Experimental Psychology,* 46, 199–210.

Grandjean, E. (1981). *Fitting the task to the man.* London: Taylor & Francis.

Holding, D. H. (1957). Direction of motion relationships between controls and displays in different planes. *Journal of Applied Psychology,* 41, 93–97.

Hunt, D. P. (1961). The effect of the precision of informational feedback on human tracking performance. *Human Factors,* 3, 77–85.

Jensen, R. S. (1981). Prediction and quickening in perspective flight displays for curved landing approaches. *Human Factors,* 23, 355–363.

Kao, H. S. R., and Smith, K. U. (1978). Unimanual and bimanual control in a compensatory tracking task. *Ergonomics,* 21(9), 661–669.

Kelley, C. R. (1962, March). Predictor instruments look to the future. *Control Engineering,* 86.

Kelley, C. R. (1968). *Manual and automatic control.* New York: Wiley.

Kvälseth, T. O. (1978a). Effect of preview on digital pursuit control performance. *Human Factors,* 20(3), 371–377.

Kvälseth, T. O. (1978b). Human performance comparisons between digital pursuit and compensatory control. *Ergonomics,* 21(6), 419–425.

Kvälseth, T. O. (1979). Digital man-machine control systems: The effects of preview lag. *Ergonomics,* 22(1), 3–9.

McRuer, D. T., and Jex, H. R. (1967). A review of quasi-linear pilot models. *IEEE Transactions on Human Factors in Electronics,* 8, 231.

Petropoulos, H., and Brebner, J. (1981). Stereotypes for direction-of-movement of rotary controls associated with linear displays: The effects of scale presence and position, of pointer direction, and distances between the control and the display. *Ergonomics,* 24, 143–151.

Pew, R. W. (1974). Human perceptual-motor performance. In B. Kantowitz (ed.), *Human information processing.* Hillsdale, NJ: Erlbaum Associates.

Poulton, E. C. (1974). *Tracking skill and manual control.* New York: Academic.

Ray, R. D., and Ray, W. D. (1979). An analysis of domestic cooker control design. *Ergonomics,* 22, 1243–1248.

Rockway, M. R. (1954). *The effect of variations in control-display ratio and exponential time delay on tracking performance,* Tech. Rept. 54–618. U.S. Air Force, Wright Air Development Center.

Roscoe, S. N., Corl, L., and Jensen, R. S. (1981). Flight display dynamics revisited. *Human Factors,* 23, 341–353.

Shinar, D., and Acton, M. B. (1978). Control-display relationships on the four-burner range: Population stereotypes versus standards. *Human Factors,* 20(1), 13–17.

Spragg, S. D. S., Finck, A., and Smith, S. (1959). Performance on a two-dimensional following tracking task with miniature stick control, as a function of control-display movement relationship. *Journal of Psychology,* 48, 247–254.

Warrick, M. J. (1947). Direction of movement in the use of control knobs to position visual indicators. In P. M. Fitts (ed.), *Psychological research on equipment design,* Research Rept. 19. Army Air Force, Aviation Psychology Program.

Wickens, C. D. (1984). *Engineering psychology and human performance.* Columbus, OH: Merrill.

CHAPTER 10

CONTROLS

Human beings have demonstrated amazing ingenuity in designing machines for accomplishing things with less wear and tear upon themselves. Some of these machines perform more efficiently the functions that were once performed solely by hand, whereas others accomplish things that previously could only be dreamed of. Since most machines require human control, they must include control devices such as wheels, pushbuttons, or levers. It is through these control devices that people make their presence known to machines—that is, in addition to kicking the side of a candy machine. These devices must be designed to be suitable for the desired control actions in terms of sensory, psychomotor, and anthropometric characteristics of the intended users.

In this chapter we examine some of the more common types of control devices and some of the factors which influence their use by humans. We discuss hand controls, foot controls, data entry devices, and some other, more esoteric, control innovations. It is not our intent to recite a comprehensive set of recommendations regarding specific design features of such devices, although a partial set of such recommendations is given in Appendix B.

FUNCTIONS OF CONTROLS

The primary function of a control is to transmit control information to some device, mechanism, or system. The type of information so transmitted can be divided into two broad classes: discrete and continuous information. *Discrete* information is information that can represent only one of a limited number of conditions such as on-off; high-medium-low; boiler 1, boiler 2, boiler 3; or alphanumerics such as A, B, and C or 1, 2, and 3. *Continuous* information, on the other hand, can assume any value on a continuum, such as speed (as 0 to 60 km/h), pressure [as 1 to 100 lb/in^2 (0.07

to 7.03 kg/cm^2)], position of a valve (as fully closed to fully open), or amount of electric current (as 0 to 10 A). Visual display terminals (VDTs) often require the user to transmit to the system a special class of continuous information which we call *cursor positioning*. The information relates to the physical location of the cursor on the screen and is important for tasks such as selecting a word for editing, drawing a picture on the screen, or selecting a system function from a list of functions shown on the screen.

The information transmitted by a control may be presented in a display, or it may be manifested in the nature of the system response. The distinction between controls and displays is becoming more and more blurred with the advent of advanced technology. For example, pushbuttons are available that are, themselves, miniature dot-matrix displays (3.8 × 2.5 cm, or 1.5 × 1.0 in) which can be programmed to display messages or symbols (Micro Switch, 1984). VDTs can display pushbuttons and slider controls which can be activated by using touch screen technology, an example of computer-generated controls.

Generic Types of Controls

There are numerous types of control devices available today. Certain types of controls are best suited for certain applications. One simple way to classify controls is based on the type of information they can most effectively transmit (discrete versus continuous) and the force normally required to manipulate them (large versus small). The amount of force required to manipulate a control is a function of the device being controlled, the mechanism of control, and the design of the control itself. Electric and hydraulic systems typically require small forces to actuate controls, whereas direct mechanical linkage systems may require large forces. Table 10–1 lists some of the more common types of controls and the types of applications for which they tend to be most suited. Certain of these types of controls are illustrated in Figure 10–1.

FACTORS IN CONTROL DESIGN

Although certain types of controls might be considered more appropriate for one application than for another, the overall utility of the control can be greatly influenced by such factors as ease of identification, size, control-response ratio, resistance, lag, backlash, deadspace, and location. We discuss some of these factors here. Discussion of location of controls is deferred to Chapter 12.

Identification of Controls

Although the correct identification of controls is not critical in some circumstances (as in operating a pinball machine), there are some situations, however, in which their correct and rapid identification is of major consequence—even of life and death. For example, McFarland (1946) cites cases and statistics relating to aircraft accidents that had been attributed to errors in identifying control devices. For example, confusion between landing gear and flap controls was reported to be the cause of over

TABLE 10–1
COMMON TYPES OF CONTROLS CLASSIFIED BY TYPE OF INFORMATION
TRANSMITTED AND FORCE REQUIRED TO MANIPULATE

Force required to manipulate control	Type of information transmitted		
		Continuous	
	Discrete	Traditional	Cursor positioning
Small	Pushbuttons (including keyboards) Toggle switches Rotary selector switches Detent thumb wheels	Rotary knobs Multirotational knobs Thumb wheels Levers (or joysticks) Small cranks	Joysticks Trackballs Mice Digitizing tablets Touch tablets Light pens Touch screens
Large	Detent levers Large hand pushbuttons Foot pushbuttons	Handwheels Foot pedals Large levers Large cranks	

400 Air Force accidents in a 22-month period during World War II. It is with such circumstances in mind that control identification becomes important.

Gamst (1975) reports that, in some railroad locomotives, engineers intending to turn off the signal lights grasp the wrong control and actually turn off the fuel pumps instead, thus killing all the diesel engines in the train. Moussa-Hamouda and Mourant (1981) interviewed 405 automobile drivers with cars equipped with fingertip-reach controls attached to the steering column. Of the 405 drivers, 44 percent reported instances of inadvertent operation, for example, intending to operate the turn signal and instead turning on the windshield wipers.

The identification of controls is essentially a coding problem; the primary coding methods include shape, texture, size, location, operational method, color, and labels.

As discussed in Chapter 3, the choice of a coding method and the specific codes depends on the detectability, discriminability, compatibility, meaningfulness, and standardization of the codes selected. The utility of a coding method or specific set of codes typically is evaluated by such criteria as the number of discriminable differences that people can make (such as the number of shapes they can identify), bits of information transmitted, accuracy of use, and speed of use.

Shape Coding of Controls   The discrimination of shape-coded controls essentially involves tactual sensitivity. One procedure used to select controls that are not confused with one another is to present all possible pairs of controls to blindfolded subjects who, by touch alone, indicate whether the controls are the same or different. It is then possible to determine which shapes were confused with other shapes. Using a procedure similar to this, Jenkins (1947b) identified two sets of eight knob shapes, such that the knobs within each group were rarely confused with one another. These two sets of knobs were presented in Figure 6–13. In a similar vein, the U.S. Air

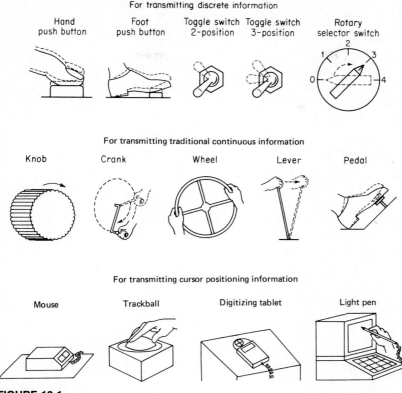

For transmitting discrete information

Hand push button Foot push button Toggle switch 2-position Toggle switch 3-position Rotary selector switch

For transmitting traditional continuous information

Knob Crank Wheel Lever Pedal

For transmitting cursor positioning information

Mouse Trackball Digitizing tablet Light pen

FIGURE 10-1
Examples of some types of control devices classified by the type of information they best transmit and the force required to activate them.

Force developed 15 knob designs which are not often confused with one another.[1] The knobs are shown in Figure 10–2. The knobs are divided into three classes based on the function they were designed to serve:

• *Class A: Multiple rotation.* These knobs are for use on continuous controls (1) which require twirling or spinning, (2) for which the adjustment range is one full turn or more, and (3) for which the knob position is not a critical item of information in the control operation.

• *Class B: Fractional rotation.* These knobs are for use on continuous controls (1) which do not require spinning or twirling, (2) for which the adjustment range usually is *less* than one full turn, and (3) for which the knob position is not a critical item of information in the control operation.

[1]A few of these knobs may be confused with each other, and such combinations should not be used together if identification is critical. These combinations were *ab, co, cd#, do#, eg#, kp, ln, lo, np,* and *op#.* Those with a number sign (#) were confused only with gloves on and were not confused without gloves. In terms of size, Hunt (1953) suggests for the maximum dimension not less than ½ in (1.3 cm) [except for class C where ¾ in (1.9 cm) is the minimum suggested] and not more than 4 in (10 cm). In height, not less than ½ in (1.3 cm) and not more than 1 in (2.5 cm) are recommended.

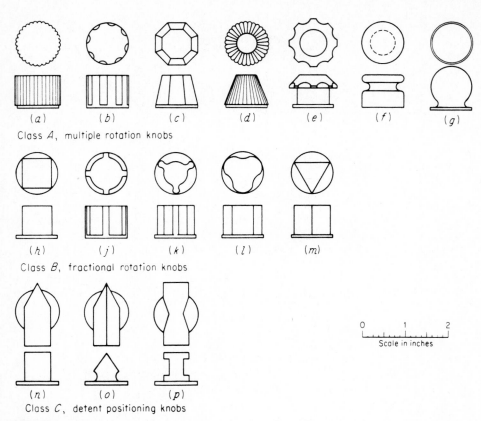

Class *A*, multiple rotation knobs

Class *B*, fractional rotation knobs

Class *C*, detent positioning knobs

FIGURE 10-2
Knob designs of three classes that are seldom confused by touch. *(Source: Adapted from Hunt, 1953.)*

- *Class C: Detent positioning.* These knobs are for use on discrete setting controls, for which knob position can be an important item of information in the control operation.

If, in addition to being individually discriminable by touch, the controls have shapes that are symbolically associated with their use, it is usually easier to learn what each knob is used for. In this connection, the U.S. Air Force developed a set of knobs that have been standardized for aircraft cockpits. These knobs, shown in Figure 10–3, besides being distinguishable from one another by touch, include some that also have symbolic meaning. It can be seen, for example, that the landing-gear knob resembles a landing wheel and that the flap control is shaped like a wing.

Texture Coding of Controls   In addition to shape, control devices can vary in surface texture. This characteristic was studied (along with certain other variables) in a series of experiments with flat cylindrical knobs such as those shown in Figure 10–4 (Bradley, 1967). In one phase of the study, knobs of this type [2-in (5.1-cm) diameter] were used, and subjects were presented with individual knobs through a

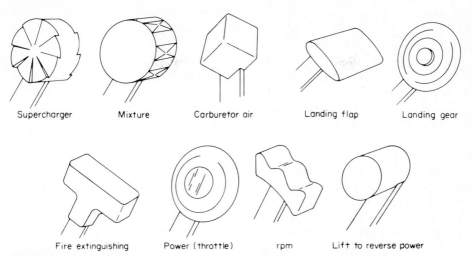

Supercharger   Mixture   Carburetor air   Landing flap   Landing gear

Fire extinguishing   Power (throttle)   rpm   Lift to reverse power

FIGURE 10-3
Standardized shape-coded knobs for United States Air Force aircraft. A number of these have
symbolic associations with their functions, such as a wheel representing the landing-gear control.
(Source: USAF, Human factors engineering, AFSC design handbook.)

curtained aperture and were asked to identify the particular design they felt. The
smooth knob was not confused with any other, and vice versa; the three fluted designs were confused with one another but not with other types; and the knurled
designs were confused with one another but not with other designs. Also with gloved
hands and with smaller-sized knobs (in a later phase of the study) there was some
cross-confusion among classes, but this was generally minimal. The investigator proposes that three surface characteristics can thus be used with reasonably accurate
discrimination, namely, smooth, fluted, and knurled.

Size Coding of Controls   Size coding of controls is not as useful for coding

FIGURE 10-4
Illustration of some of the knob designs used in study of tactual discrimination of surface
textures. Smooth: *A;* fluted: *B* (6 troughs), *C* (9), *D* (18); and knurled: *E* (full rectangular), *F* (half
rectangular), *G* (quarter rectangular), *H* (full diamond, *I* (half diamond), and *J* (quarter diamond).
(Source: Bradley, 1967.)

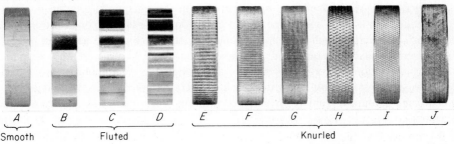

A   *B*   *C*   *D*   *E*   *F*   *G*   *H*   *I*   *J*
Smooth   Fluted   Knurled

purposes as shape, but there may be some instances where it is appropriate. When such coding is used, the different sizes should, of course, be discriminable one from the others. Part of the study by Bradley reported above dealt with the discriminability of cylindrical knobs of varying diameters and thickness. It was found that knobs that differed by ½ in (12.7 mm) in diameter and by ⅜ in (9.5 mm) in thickness could be identified by touch very accurately, but smaller differences sometimes resulted in confusion. Incidentally, Bradley proposes that a combination of three surface textures (smooth, fluted, and knurled), three diameters [¾, 1¼, and 1¾ in (19.1, 31.8, and 44.4 mm)], and two thicknesses [⅜ and ¾ in (9.5 and 19.1 mm)] could be used in all combinations to provide 18 tactually identifiable knobs.

Aside from the use of coding for individual control devices, size coding is part and parcel of ganged control knobs, where two or more knobs are mounted on concentric shafts with various sizes of knobs superimposed on each other like layers of a wedding cake. When this type of design is dictated by engineering considerations, the differences in the sizes of superimposed knobs need to be great enough to make them clearly distinguishable, as illustrated in Figure 10–13 on page 277 (Bradley, 1969).

Location Coding of Controls   Whenever we shift our foot from the accelerator to the brake, feel for the light switch at night, or grasp for a machine control that we cannot see, we are responding to *location coding*. But if there are several similar controls from which to choose, the selection of the correct one may be difficult unless they are far enough apart that our kinesthetic sense makes it possible for us to discriminate. Some indications about this come from a study by Fitts and Crannell as reported by Hunt (1953). In this study blindfolded subjects were asked to reach for designated toggle switches on vertical and horizontal panels, the switches being separated by 1 in (2.5 cm). The major results are summarized in Figure 10–5, which shows the percentage of reaches that were in error by specified amounts when the panels were in horizontal and vertical positions. The results have been simplified by combining several conditions. The results indicate, quite clearly, that accuracy was greatest when toggle switches were arranged vertically rather than horizontally. For vertically arranged switches, only a very small percentage of reaches were in error by more than 2.5 in (6.3 cm); therefore, controls should be separated by more than this distance so that errors of one switch will not result in the activation of an adjacent switch. For horizontally arranged switches, there should be at least 4 in (10.2 cm), and preferably more, between them if they are to be recognized by location without the aid of vision.

Operational Method of Coding Controls   In this method of coding controls, each control has its own unique method of operation. For example, one control might be a push-pull variety and another of a rotary variety. Each can be activated *only* by the movement that is unique to it. Moussa-Hamouda and Mourant (1981) compared various operational coding schemes for operating automobile windshield wiper/washer systems from fingertip-reach controls mounted on a stalk attached to the steering column. Figure 10–6 shows the various combinations tested, the average time required to operate the various configurations, and the percentage of responses that resulted in activation of the wrong function. The authors recommend the rotate-

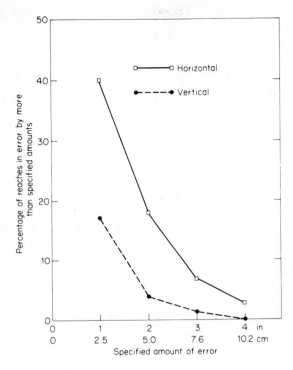

FIGURE 10-5
Accuracy of blind reaching to toggle switches (nine in a row on switch box) with switch box positioned horizontally and vertically. Each data point represents the mean of several conditions. *(Source: Adapted from Fitts & Crannell, as presented by Hunt, 1953.)*

away or rotate-toward configurations over the others. It is apparent that operational coding of controls would be inappropriate if there were any premium on control operation time or operating errors were of considerable importance. When operational coding is used, it is desirable that compatible relationships be utilized. By and large, this method of coding should be avoided except in those individual circumstances in which it seems to be uniquely appropriate.

Color Coding of Controls   Color coding was discussed in Chapter 4 in reference to visual displays. Color can be useful for identifying controls as well. A moderate number of coding categories can be used, and colors can often be picked that are meaningful, for example, red for an emergency stop control. One disadvantage, of course, is that the user must look at the control to identify it. Thus, color coding cannot be used in situations with poor illumination or in which the control is likely to become dirty. Color, however, can be combined with some other coding method, such as size, to enhance discriminability or increase the number of coding categories available.

Label Coding of Controls   Labels are probably the most common method of identifying controls and should be considered as the minimum coding requirement for any control. A large number of controls can be coded with labels and, if properly chosen, do not require much learning to comprehend. Extensive use of labels as the only means of coding controls is not desirable, however. Seminara, Gonzalez, and Parsons (1977) report that in most nuclear power plant control rooms, for example,

	Stalks	Wiper (2 speed) and Washer Functions	Mean Performance Time (s)	Percentage of Inadvertent Operation Errors
Rotate Away	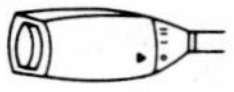	Wiper-rotates toward instrument panel Washer-button on end of stalk	0.90	4.6
Button		Wiper-button on end of stalk Wiper speed-rocker switch Washer-hand switch moves toward steering column	1.10	9.5
Hand Switch		Wiper-hand switch moves toward steering column Wiper speed-rocker switch Washer-button on end of stalk	1.00	4.9
Rotate Toward	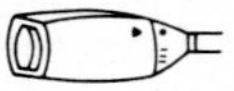	Wiper-rotates toward driver Washer-button on end of stalk	0.88	3.5
Side Switch		Wiper-side switch pushes toward column Washer-button on end of stalk	1.19	4.3

FIGURE 10-6
Comparison of various configurations of automobile windshield wiper/washer system controls. Shown are the operational coding methods used for the various functions, the average response time, and the percentage of responses that were incorrect. *(Source: Adapted from Moussa-Hamouda and Mourant, 1981.)*

there are literally walls of identical controls, distinguished only by labels. Workers often improvise their own coding methods including attaching uniquely shaped draft beer dispenser handles to the controls. Labels take time to read and thus should not be the sole coding method where speed of operation is important. If labels are used, they should be placed above the control (so that the hand will not cover them when the operator is reaching for the control) and should be visible to the operator before reaching for the control.

Discussion of Coding Methods In the use of codes for identification of controls, two or more code systems can be used in combination. Actually, combinations can be used in two ways. First, *unique combinations* of two or more codes can be used to identify separate control devices, such as the various combinations of texture, diameter, and thickness mentioned before (Bradley, 1967). Second, there can be completely *redundant codes,* such as identifying each control by a distinct shape *and* by a distinct color. Such a scheme probably would be particularly useful if accurate identification were especially critical. In discussing codes, we would be remiss if we failed to make a plug for standardization in the case of corresponding controls used in various models of the same type of equipment, such as automobiles and tractors. When individuals are likely to transfer from one model to another of the same general equipment type, the same system of coding should be used if at all possible. Otherwise, it is probable that marked "habit interference" will result and that people will revert to their previously learned modes of response.

Some general considerations that affect the choice of a control code should be kept in mind, such as whether the operators will have the luxury of being able to look at the control to identify it, whether they will have to continuously monitor a display, whether they will work in the dark, or whether they have a visual handicap. If vision is restricted, then only shape, size, texture, location, or operational method should be considered.

A second consideration is maintenance. When designing a system, including the controls, one must take into consideration the maintenance of that system. If shape, size, texture, or color coding is used, one must keep a stock of different spare controls and run the risk that the wrong control will be replaced if one is damaged. Thus, if there is a significant chance that controls will be damaged and have to be replaced, perhaps location or operational method coding should be considered.

We have been discussing coding methods mostly in terms of accuracy and speed of identification. Some coding methods, such as shape, size, and texture, may, however, influence the operation of the controls themselves. Carter (1978) compared the speed and accuracy of using knobs of different sizes and shapes. The study was aimed at designing underwater diving equipment; so the tests took place in cold water with subjects wearing gloves, and the knobs were located on different parts of the body. The relative speed and accuracy of operating the knobs was somewhat dependent on the location of the knobs, but there were clear differences in performance with knobs that differed in shape, size, or both. It is important to select control codes that are discriminable, compatible, and meaningful, but it is also important to consider the possible effect of the specific codes on the speed and accuracy with which the control is operated. Incidentally, controls that require the user to grip them tightly should not use shape codes with sharp edges or protruding points.

Control-Response Ratio

In continuous control tasks or when a quantitative setting is to be made with a control device, a specified movement of the control will result in a system response. The system response may be represented on a display or it may not, as in turning the steering wheel of a car. We call the ratio of the movement of the control device to the movement of the system response the *control-response ratio* (C/R ratio). In the past, the C/R ratio has been called the *control-display ratio* (C/D ratio). It was felt that control *display* was really not appropriate when no real display was present and that control *response* is a more general term. Movement of the control, display, or system may be measured in linear distance (in the case of levers, vertical dials, etc.) or in angles or number of revolutions (in the case of knobs, wheels, circular displays, etc.). A very *sensitive* control is one which brings about a marked change in the controlled element (display) with a slight control movement; its C/R ratio would be low (a small control movement is associated with a large display movement). The reciprocal of the C/R ratio is referred to as *gain* and is really the same thing as sensitivity. A control with a low C/R ratio has a high gain. Examples of low and high C/R ratios are shown in Figure 10–7.

C/R Ratios and Control Operation   The performance of human beings in the use of continuous control devices which have associated display movements can be affected by the C/R ratio. This effect is not simple, but rather is a function of the nature of the human motor activities involved in using such controls. In a sense, there are two types of human motions in such tasks. There is a *gross* adjustment movement (travel time or a sluing movement) in which the operator brings the controlled element (say, the display indicator) to the approximate desired position. This gross movement is followed by a *fine*-adjustment movement, in which the operator makes adjustments to bring the controlled element precisely to the desired location. (Actually, these two movements may not be individually identifiable, but there is typically some change in motor behavior as the desired position is approached.)

Optimum C/R Ratios   The determination of an optimum C/R ratio for any continuous control or quantitative-setting control task needs to take into account these two components of human motions. Where the control in question is a knob, the general nature of the relationships is reflected by the results of studies by Jenkins and Connor (1949) and is illustrated in Figure 10–8. That figure illustrates the essential features of the relationships, namely, that travel time drops off sharply with decreas-

FIGURE 10-7
Generalized illustrations of low and of high control response ratios (C/R ratios) for lever and rotary controls. The C/R ratio is a function of the linkage between the control and display.

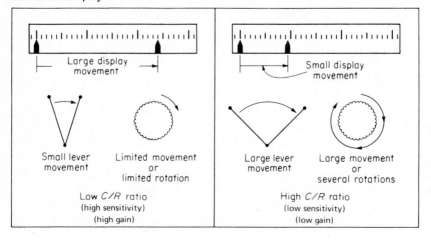

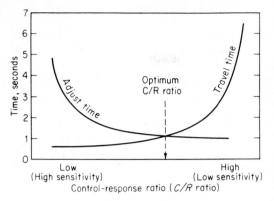

FIGURE 10-8
Relationship between C/R ratio and movement time (travel time and adjust time). While the data are from a study by Jenkins and Connor, the specific C/R ratios are not meaningful out of that context, so are omitted here. These data, however, depict very typically the nature of the relationships, especially for knob controls. *(Source: W. L. Jenkins & Connor, 1949.)*

ing C/R ratios and then tends to level off, and that adjustment time has the reverse pattern. Thus, if you were trying to find a radio station using a knob with a high C/R ratio (where it takes a couple of turns of the control to move the pointer on the dial a small distance), it would take you some time to get the pointer across the dial to the general location of the station (travel time). Once it was there, however, you could quickly "home in" on the station without overshooting it. With a low C/R ratio (where a small turn of the control sends the pointer across the dial), you could get the pointer near the station in no time, but it would take you a considerable time to "home in" because you would tend to overshoot the station. In determining the optimum C/R ratio, one must often trade off travel and adjustment time. The optimum ratio is somewhere around the point of intersection, as shown in Figure 10–8. Within this general range, the combination of travel and adjustment time usually would be minimized. Note that when a joystick or a lever is used instead of a control knob, the curve for travel time is almost a horizontal straight line. This is because it is almost as fast to move a stick through 180° as through a smaller angle. The optimum C/R ratio is then based on minimizing adjustment time only. However, not everyone agrees on the importance of C/R ratio in the design of joystick controls. Buck (1980), for one, found that with joysticks neither travel nor adjustment time was affected by the C/R ratio. Rather, the physical widths of the target, measured on the display and in the control (but not their ratio), were the primary factors affecting performance. Buck also found that the greatest proportion of performance time was spent in making fine-adjustment movements. This would reinforce the idea of selecting design parameters to minimize adjustment time.

The numeric value for the optimum C/R ratio is a function of the type of control (knob, lever, crank, etc.), size of the display, tolerance permitted in setting the control, and other system parameters such as lag. Unfortunately, there are no formulas for determining optimum C/R ratios. Rather this ratio should be determined experimentally for the control and display parameters being contemplated. To give some ballpark figures, Jenkins and Connor (1949) reported optimum C/R ratios for knobs in the range of 0.2 to 0.8; Chapanis and Kinkade (1972) reported optimum C/R ratios for levers in the range of 2.5 to 4.0.

Resistance in Controls

The major part of the feedback or "feel" of a control is due to the various types of resistance associated with the control. Control manipulation takes principally two forms: the amount of displacement of the control and the amount of force applied to the control. The operator of the control can sense both movement and force through the proprioceptive and kinesthetic senses. Force and movement, therefore, are the primary sources of control feedback.

To use most controls requires both force to overcome resistance and some displacement of the control (such as depressing a brake pedal or turning a steering wheel). Some controls, however, require only force or only displacement. A pure displacement control would have virtually no resistance to movement, and the only type of feedback would be the amount of movement of the body member. These are called *free-position,* or *isotonic,* controls. A *pure force,* or *isometric,* control, on the other hand, involves no displacement. The output is related to the amount of force or pressure that is applied to the control (these are sometimes called *stiff sticks*). Pure force controls are becoming more common with advances being made in electronics and servo control mechanisms.

It is difficult to specify which type of feedback, force or displacement, is most useful. However, some research and experience, plus a bit of conventional wisdom, suggest that in many circumstances a combination of the two is useful, such as in the operation of a steering wheel. In special circumstances, however, one type or the other may have an advantage. For example, from the scant amount of literature available, Poulton (1974) concludes that pure force controls are superior to a combination force-displacement (spring-centered) control when they are used with higher-order tracking control systems (see Chapter 9), to track a relatively fast-moving, gyrating target.

Types of Resistance   All control devices have some resistance, although in the case of free-positioning (isotonic) controls this is virtually negligible. Resistance can, of course, serve as a source of feedback, but the feedback from some types of resistance can be useful to the operator while the feedback from other types of resistance can have a negative effect. In some circumstances the designer can design or select those controls with resistance characteristics that will minimize possible negative effects and that possibly can enhance performance. This can be done in a number of ways, such as with servomechanisms, with hydraulic or other power, and with mechanical linkages. The primary types of resistance are elastic (or spring-loaded), static and coulomb friction, viscous damping, and inertia. We discuss each in turn.

• *Elastic resistance:* Such resistance (as in spring-loaded controls) varies with the *displacement* of a control device (the greater the displacement, the greater the resistance). The relationship may be linear or nonlinear. A major advantage of elastic resistance is that it can serve as useful feedback, combining both force and displacement in a redundant manner. Poulton indicates that for tracking tasks, spring-loaded controls are best for compensatory tracking in a position control system with a slow ramp target to track (see Chapter 9). With elastic resistance, the control automatically returns to the null position when the operator releases it; hence it is ideal for "deadman'' switches. By designing the control so that there is a distinct gradient in resistance at critical positions (such as near the terminal), additional cues can be provided. In addition, such resistance permits sudden changes in control direction but reduces the likelihood of undesirable activation caused by accidental contact with the control.

   • *Static and coulomb friction: Static friction,* the resistance to initial movement, is maximum at the initiation of a movement but drops off sharply. *Coulomb (sliding) friction* continues as a resistance to movement, but this friction force is not related to either velocity or displacement. This can be illustrated by pressing a pencil eraser firmly against a flat horizontal surface with one hand while trying to slide the eraser slowly by pushing it horizontally with the other hand. You have to exert some force before you can get the pencil (or control) to move; i.e., you have to overcome the static friction. Once the pencil is moving, however, a smaller amount of force, to overcome the coulomb friction, will keep it moving no matter how fast it is moving or how far you move it. If you try to change directions, you will notice a delay and some irregularity in the movements. With a couple of possible exceptions, static and coulomb friction tend to cause degradation in human performance. This is essentially a function of the fact that there is no systematic relation between such resistance and any aspect of the control movement (such as displacement, speed, or acceleration); thus, it cannot produce any meaningful feedback to the user of the control movement. On the other hand, such friction has the advantages of reducing the possibility of accidental activation and of helping to hold the control in place.

   • *Viscous damping:* Viscous damping feels like moving a spoon through thick syrup. The amount of force required to move the spoon (or control) to overcome viscous damping is related to the velocity of the control movement. The faster you move the control, the more viscous damping you have to overcome. Viscous damping generally has the effect of resisting quick movement and of helping to execute smooth control, especially in maintaining a prescribed *rate* of movement. The feedback, however, being related to *velocity* and not displacement, probably is not readily interpretable by operators. Such resistance, however, does minimize accidental activation.

   • *Inertia:* This is the resistance to movement (or change in direction of movement) caused by the mass (weight) of the mechanism involved. It varies in relation to *acceleration*. A force exerted on an inertia control will have little effect at first because it produces only an acceleration of the control (the greater the force, the greater the acceleration). The acceleration has to operate for a little while before the control will begin to move very much. Once the control begins to move, however, it is hard to stop. Inertia opposes quick control movements and thus probably would be a disadvantage in tracking. As Poulton points out, the human arm has enough inertia of its own. Inertia does, however, aid in smooth control and can aid in turning a crank at a fixed speed (Poulton). It also reduces the possibility of accidental activation of the control.

   Combining Resistances Almost all controls that move involve some forms of resistance, often more than one type. The effects of more than one type of resistance on control operation can be complex. Howland and Noble (1955), for example, show how time on target with a position control is affected by various combinations of

TABLE 10–2
EFFECTS OF VARIOUS COMBINATIONS OF CONTROL RESISTANCE ON TRACKING
PERFORMANCE

Condition	Type of control resistance present			Time on target, %
	Elastic*	Viscous damping†	Inertia‡	
1	Yes	No	No	60
2	Yes	Yes	No	52
3	No	Yes	No	44
4	No	No	No	41
5	No	No	Yes	35
6	No	Yes	Yes	35
7	Yes	Yes	Yes	34
8	Yes	No	Yes	27

*Elastic resistance: 0.02 in · lb/°
†Viscous damping: 0.02 (in · lb)/(° · s)
‡Inertia: 0.002 (in · lb)/(° · S^2)
Source: Howland and Noble, 1955.

control resistance. They used a knob control manipulated with the fingertips, and relatively small amounts of resistance were used. Table 10–2 summarizes the results. Elastic resistance alone resulted in the best performance, even better than no resistance at all. The addition of inertia by itself, or in combination with the other types of resistance, always resulted in a decrement in performance. The worst combination was elastic resistance plus inertia.

Ability of People to Judge Resistance   Regardless of what variety of resistance is intrinsic to a control mechanism, if it is to be used as a source of useful feedback, the meaningful differences in resistance have to be such that the corresponding pressure cues can be discriminated. Such discriminations were the focus of an investigation by Jenkins (1947a), with three kinds of pressure controls, namely, a stick, a wheel, and a pedal (like a rudder control in a plane). After some training and practice in reproducing specified forces, a series of trials was made by each subject. Measures of actual pressure exerted were then compared with the pressures that the subjects attempted to reproduce. The difference limens by pounds of pressure for the various types of controls are shown in Figure 10–9. The *difference limen* is the average difference that can just barely be detected; thus, two pressures have to differ by an amount greater than the limen to be detected as being different. The systematic drop-off of all the curves between 5 and 10 lb (2.3 to 4.5 kg) implies that if differences in pressure are to be used as feedback in operation of control devices of the types used, the pressures used should be around or above these values (and perhaps more for pedals because of the weight of the foot). The experimenter suggested that if varying levels of pressure discrimination are to be made, the equipment should provide a wide range of pressures up to 30 to 40 lb (13.6 to 18.1 kg). Beyond these pressures, the likelihood of fatigue increases as well as the likelihood of slower operation.

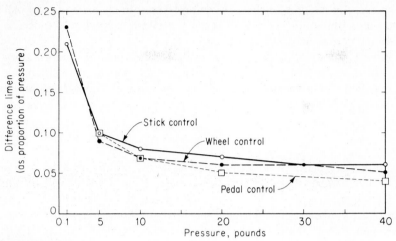

FIGURE 10-9
Difference limens for three control devices for various pressures that were to be
reproduced. (*Limen* is the standard deviation divided by the standard pressure.)
Source: W. O. Jenkins, 1947a.)

Deadspace

Deadspace in a control mechanism is the amount of control movement *around the
null position* that results in no movement of the device being controlled. It is almost
inevitable that some deadspace will exist in a control device. Deadspace of any consequence usually affects control performance, with the amount of effect being related
to the sensitivity of the control system. This is indicated in Figure 10–10 (Rockway,
1957). It can be observed that tracking performance deteriorated with increases in
deadspace (in degrees of control movement that produced no movement of the controlled device). But the deterioration was less with the less sensitive systems than
with more sensitive systems. This, of course, suggests that deadspace can, in part,
be compensated for by building in less sensitive C/R relationships.

Rogers (1970) found that increasing the deadspace in a control resulted in an

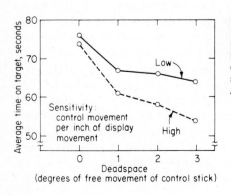

FIGURE 10-10
Relationship between deadspace in a control
mechanism and tracking performance for various
levels of control sensitivity. (High sensitivity
corresponds to 0.56 percent per inch. Low
sensitivity equals 3.33 percent per inch.) *(Source:
Adapted from Rockway, 1957.)*

almost linear increase in the time needed to acquire a target. In the experiment a ball had to be spun at a minimum velocity to move the response marker. Deadspace corresponded to the minimum velocity required.

Poulton (1974) suggests that deadspace may be more detrimental in compensatory tracking tasks than in pursuit, but with higher-order control systems a small amount of deadspace may be helpful when the control is spring-centered. This is because the control may not always spring back to exactly the same position when it is released.

Backlash

The best way to think of backlash is to imagine operating a joystick or lever with a loose hollow cylinder fitted over it. When the cylinder is moved to the right, for example, the stick touches it on the left. If you are moving the cylinder to the right and then reverse directions, the stick does not start to return to the left until it comes up against the right side of the cylinder. Until the stick contacts the cylinder, the operator's control movements have no effect on the system. In essence, then, backlash is deadspace *at any control position.*

Typically, operators cannot cope very well with backlash. This effect was illustrated by the results of an investigation by Rockway and Franks (1959) using a control task under varying conditions of backlash and of control gain (the reciprocal of C/R ratio). The results show that performance deteriorated with increasing backlash for all control gains but was most accentuated for high gains. The implications of such results are that if a high control gain is strongly indicated (as in, say, high-speed aircraft), the backlash needs to be minimized in order to reduce system errors; or conversely, if it is not practical to minimize backlash, the control gain should be as low as possible—also to minimize errors from the operation of the system.

DESIGN OF SPECIFIC HAND-OPERATED CONTROLS

The various factors discussed above relate to all types of controls, although some factors are more relevant to tracking controls than to other types. We now present a few examples of research investigating various design parameters of specific types of controls. These studies illustrate how specific design features affect the adequacy with which people can use the control for a particular purpose. Within the scope of this chapter, we can only present a smattering of the range of research carried out on control devices. We discuss briefly some aspects of cranks and handwheels, rotary knobs, and joysticks. We discuss data entry devices, cursor positioning devices, and foot-operated controls in later sections of this chapter.

Cranks and Handwheels

Cranks and handwheels frequently are used as a means of applying force to perform various types of functions, such as moving a carriage on a cutting tool or lifting objects. They can also be used as the control device for a tracking task. We present the results of some studies which illustrate the effect of size and friction on various aspects of operator performance. It should become clear that the optimum configuration depends on the aspect of operator performance one wishes to optimize.

Size of Cranks and Handwheels In a study by Davis (1949) several different sizes of cranks and handwheels were used to control the position of a pointer on a dial. In general, about one revolution of the crank was required in making the settings. The cranks and handwheels were mounted so that the plane of rotation was parallel to the frontal plane of the body. By using torques of 0, 20, 40, 60, and 90 in·lb (0, 0.36, 0.72, 1.08, and 1.62 kg·mm) it was possible to determine the time required to make settings under different friction-torque conditions that typically occur in the operation of such controls.

Figure 10–11 shows the average times for certain of the torque levels with various sizes of cranks or handwheels. A couple of points are illustrated by this figure. First, the average time for making settings was not affected much by the size of the crank or wheel when 0 in·lb of torque was used. For torques of 40 and 90 in·lb (0.72 and 1.62 kg·mm), however, the small cranks and wheels were definitely inferior to the larger ones. Second, the average setting times, in general, were longer for the higher torque cranks and wheels, this being especially true with the smaller diameter devices.

Work Output with Different Cranks The work output of subjects using three sizes of cranks and five levels of resistance was investigated by Katchmer (1957) with a view to identifying fairly optimum combinations. The cranks were adjusted to waist height for each subject, and the subjects were instructed to turn the crank at a

FIGURE 10-11
Average times for making settings with vertically mounted handwheels and cranks of various sizes under different torque conditions. *(Source: Adapted from Davis, 1949.)*

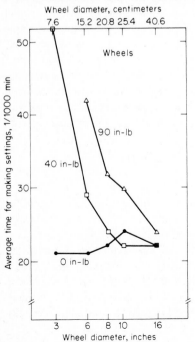

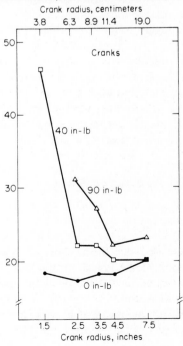

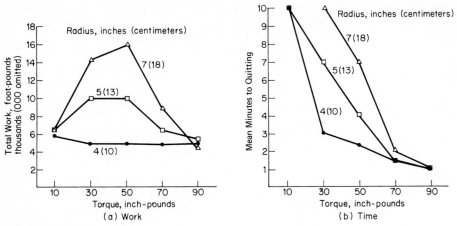

FIGURE 10-12
Average performance measures for subjects using 4-, 5-, and 7-in radius cranks under 10, 30, 50, 70, and 90 in lb torque resistance. *(Source: Adapted from Katchmer, 1957.)*

rapid rate until they felt they could no longer continue the chore (or until 10 min had passed). A summary of certain data from the study is presented in the graphs of Figure 10–12. In total foot-pounds of work (Figure 10–12*a*), the highest work levels occurred for the moderate torque loads [30 and 50 in·lb (0.54 and 0.9 kg·mm)] using the 5-in (13-cm) and 7-in (18-cm) radius cranks. Torque had no effect on the work output when the small 4-in (10-cm) crank was used. Looking at the average time that subjects worked before quitting (Figure 10–12*b*), we see that, again, the larger the radius of the crank, the longer the subjects would continue to turn it, except under the higher torque conditions. But now, in terms of time, the optimum torque is 10 in· lb (0.18 kg·mm), not 30 to 50 in·lb (0.54 and 0.9 kg·mm), as was the case with work output.

These graphs dramatically illustrate the point that a particular design may be optimum for one criterion but not for another (in this case, the different criteria being tolerance time and total work output). It is important, therefore, that the aspect or aspects of human performance one wants to maximize be specified before design parameters are chosen.

Concentrically Mounted Knobs

Space restrictions and other considerations sometimes argue for the use of concentrically mounted (or ganged) knobs, such as those illustrated in Figure 10–13. Granting situational advantages for such controls, there are also some possible disadvantages associated with them, especially the possibility of inadvertent operation of adjacent knobs. If the knobs are too thin, the fingers may scrape against and operate the knob behind; and if the diameter distance is too small, the fingers may overlap and operate the knob in front. The problem, then, is one of identifying the optimum dimensions of such knobs, to minimize such possibilities. In a study dealing with

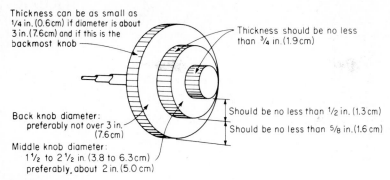

Thickness can be as small as
1/4 in. (0.6 cm) if diameter is about
3 in. (7.6 cm) and if this is the
backmost knob

Thickness should be no less
than 3/4 in. (1.9 cm)

Back knob diameter:
preferably not over 3 in.
(7.6 cm)

Should be no less than 1/2 in. (1.3 cm)

Should be no less than 5/8 in. (1.6 cm)

Middle knob diameter:
1 1/2 to 2 1/2 in. (3.8 to 6.3 cm)
preferably, about 2 in. (5.0 cm)

FIGURE 10-13
Dimensions of concentrically mounted knobs that are desirable in order to
allow human beings to differentiate knobs by touch. *(Source: Adapted from
Bradley, 1969.)*

this, Bradley (1969) used various combinations of such knobs and various performance criteria (errors, reach time, and turning time). He found that the dimensions
shown in Figure 10–13 were optimum.

Knobs for Producing Torque

Often, control knobs are used to apply fairly high levels of torque to equipment, for
example, in turning on or off a water faucet, tightening a clamp mechanism, or
turning a door knob. Kohl (1983) measured the maximum isometric force that females could exert on smooth aluminum knobs of various shapes and sizes [diameters
of 2.5 to 5.5 in (6.4 to 14.0 cm)]. Subjects performed the task under two conditions:
with greased hands and with hands covered with a nonslip compound. Figure 10–14
presents results after performance is averaged across the various knob sizes. It is no
surprise that greasy hands reduced the amount of force (torque) that could be applied
to the knobs. There was little effect of knob shape when the nonslip compound was
used; however, with greased hands, torque *decreased* as the number of sides on the
knob increased. As might be expected, the larger the knob diameter, the more torque
that could be developed, but this effect diminished as the number of sides on the
knob increased. Kohl cautioned that the triangular knob caused discomfort and therefore recommended the square shape.

Brullinger and Muntzinger (1984) performed a similar experiment, varying the
shape of knobs having a diameter of 80 mm (3.15 in). The subjects in this experiment were males, and they performed the task with clean knobs and with knobs
soiled with dirt and oil. Figure 10–15 shows the knob shapes used and the average
maximum torque produced with each. Larger differences in torque production were
found between shapes in this study than in Kohl's study. The control producing the
highest average torque (i.e., control 1, a rectangular bar) was not recommended
because of the discomfort caused by the edges digging into the palm of the hand.
Control 4, in soiled conditions, and control 5, in clean conditions, were recommended.

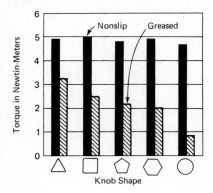

FIGURE 10-14
Effect of various shapes of knobs on maximum
isometric torque. Results are averaged over four
sizes of knobs (2.5-5.5 in; 6.4-14.0 cm). Subjects
were females with either greased hands or hands
covered with nonslip compound. *(Source: Adapted
from Kohl, 1983.* Reprinted with permission from the
Bell System Technical Journal, copyright 1983
AT&T.)

Both of these studies illustrate the need to consider multiple criteria in selecting
an optimal design. In both studies the control producing the highest level of torque
was not recommended because of the discomfort caused by its shape. A good design
must consider performance, user comfort, and user satisfaction.

Stick-Type Controls

At first, one might think that the length of a joystick or lever would dramatically
affect the speed or accuracy with which it could be used. This does not seem to be
true, however. Jenkins and Karr (1954), for example, found that varying the length
of a joystick from 12 to 30 in (30.5 to 76.2 cm) was relatively unimportant as long
as the C/R ratio was around 2.5 or 3.0. Hartman (1956), using a tracking task, found
only moderate effects on performance by varying joystick length from 6 to 27 in (15
to 69 cm). There was about a 10 percent advantage in using 18-in (46-cm) sticks
relative to other lengths.

FIGURE 10-15
Maximum isometric torque as a function of knob shape. Knobs were either
clean or soiled with dirt and oil. Subjects were males. Control 5 is coated
with resin. *(Source: Bullinger and Muntzinger, 1984.)*

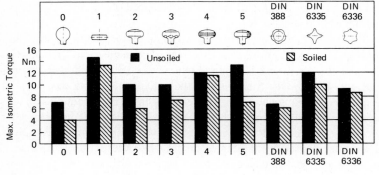

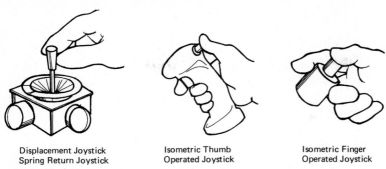

Displacement Joystick
Spring Return Joystick

Isometric Thumb
Operated Joystick

Isometric Finger
Operated Joystick

FIGURE 10-16
Joysticks used in study of repositioning time. Results are shown in Figure 10-17.
(Source: Adapted from Mehr, 1973.)

Mehr (1973) reports on a study that compared time to reposition a cursor given various types of joysticks. The joysticks used are shown in Figure 10–16. The first joystick was operated in two modes: as a *displacement joystick,* which remained in position when released, and as a *spring-return joystick.* The finger- and thumb-operated joysticks were force-operated isometric controls. The spring-return joystick and both isometric joysticks operated as first-order rate controls, while the displacement joystick operated as a zero-order position control. Figure 10–17 shows the mean repositioning time as a function of the length, measured in steps, of the repositioning movement (the display area was 1000×750 steps). The results showed that repositioning time was fastest with the finger-operated isometric control. The difference between the displacement and spring-return joysticks is difficult to explain because the presence or absence of the spring was confounded with whether the control was a zero- or first-order control. Notice that with the rate-controlled joysticks the repositioning time did not increase much, if at all, for movements longer than 500 steps.

Multifunction Hand Controls

Advances in technology have made some systems more complex and have endowed them with more functions. An excellent example is a high-performance fighter air-

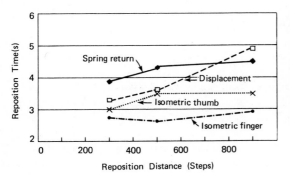

FIGURE 10-17
Repositioning time for the various types of joysticks shown in Figure 10-16. Spring return and both isometric joysticks were first-order rate controls. Displacement joystick remained in position when released and was a zero-order position control. All devices used optimized designs. *(Source: Adapted from Mehr, 1973.)*

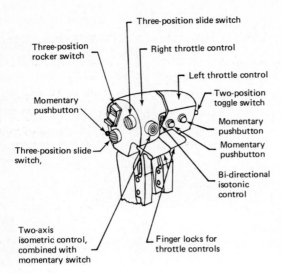

FIGURE 10-18
The left-hand control of the F-18 aircraft. (*Adapted from: Wierwille, 1984; fig. 2.*)

craft. A pilot must be able to control various aircraft, communication, and weapon functions in a demanding and hostile environment. Multifunction hand controls are becoming common in such aircraft. Figure 10–18, for example, shows the left-hand control of an F-18 aircraft. The control itself is actually two slide controls, one controlling the throttle of each engine, that can be operated together or separately or can be locked in any position. In addition, four control functions are operated by the thumb, two more are operated by the index finger, and five more are operated by the remaining three fingers. (By the way, the right-hand control, not shown here, contains seven control functions in addition to controlling the pitch and roll of the aircraft.)

Wierwille (1984) points out that the design of such multifunction hand controls is based on several principles: (1) The operator should not have to observe the control to operate it; (2) the hand should remain in contact with the primary controls throughout critical operations of the system; and (3) auxiliary controls should be able to be activated without loss of physical contact with the primary controls. Unfortunately, little research has been conducted on multifunciton controls, so we do not even know, for example, the maximum number of functions that can be efficiently controlled by an operator. As Wierwille points out, comprehensive testing, therefore, is mandatory whenever a multifunction control is used in a system.

FOOT CONTROLS

Far and away, hand controls are more widely employed than foot controls. One reason for this, as Kroemer (1971) points out, is that the general tenor of human factors handbooks implies that the feet are slower and less accurate than the hands. Kroemer hastens to add that this belief is neither supported nor discredited by experimental results. Besides this, foot controls often restrict the posture of the user. Having to maintain one foot (or both) on a control makes it more difficult to shift around in the seat or change position of the leg. Anyone who has taken a long drive on the open highway can verify this. Operating foot controls from a standing position requires people to balance all their weight on the other foot. Despite all this, foot controls will continue to have a place in the control tasks of seated operators.

Several important design parameters affect performance with foot controls, such as whether the controls require a thrust with or without ankle action, the location of the fulcrum (if the pedal is hinged), the angle of the foot to the tibia bone of the leg, the load (the force required), and the placement of the control relative to the user. As with hand controls, our purpose is not to provide detailed design guidance on all aspects of control design, but rather to illustrate how specific features have a direct bearing on how adequately the control is used. Various criteria have been employed to evaluate alternative designs, such as reaction time, travel time, speed of operation, precision, force produced, and subjective preference. Chapter 13 includes a discussion of the placement of foot controls relative to the operator; here we discuss design of foot controls and their placement relative to other controls.

Pedal Design Considerations

For illustrative purposes we present certain results of an experiment by Ayoub and Trombley (1967) in which one of the factors varied was the location of the fulcrum of the pedal. In one set of experimental conditions, the pedal, to be activated, had to be moved through an arc of 12°, no matter where the fulcrum was located (constant angle). The closer the fulcrum is to the heel, the farther the ball of the foot must travel to achieve a 12° arc. In a second set of conditions, the pedal, to be activated, had to be moved a constant distance [0.75 in (1.9 cm)]. In this case, the closer the fulcrum is to the heel, the smaller the arc will be to achieve the activation distance.

The mean travel times for various fulcrum locations are shown in Figure 10–19. It can be seen that for a constant angle of movement (of 12°), the optimum location

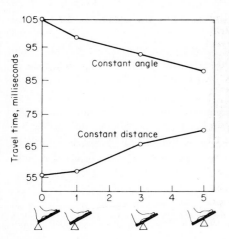

FIGURE 10-19
Mean travel time in pedal movement as related to location of the fulcrum for conditions of constant angle of movement (12°) and constant distance of movement. (0.75 in or 1.9 cm) *(Source: Adapted from Ayoub & Trombley, 1967, fig. 13. Copyright 1967 by American Institute of Industrial Engineers, Inc., 25 Technology Park, Atlanta, Norcross, Ga. 30071, and reproduced with permission.)*

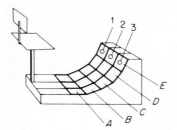

FIGURE 10-20
Arrangement of foot targets used in experiment by Kroemer in studying the speed and accuracy of discrete foot motions. *(Source: Kroemer, 1971, fig. 7.)*

of the fulcrum is forward of the ankle (about one-third of the distance between the ankle and the ball of the foot), whereas for a constant distance of movement the optimum location of the fulcrum is at the heel. However, as Kroemer points out, the somewhat inconsistent results of various experiments with pedals (in part based on differences in experimental conditions) preclude any general statements as to what pedal allows the fastest activation or highest frequency in repetitive operation under a variety of specified conditions.

Foot Controls for Discrete Control Action Foot control mechanisms usually are used to control one or a couple of functions. There are some indications, however, that the feet can be used for more varied control functions. In one investigation, for example, Kroemer developed an arrangement of 12 foot positions (targets) for subjects, as illustrated in Figure 10–20. Using procedures that need not be described here, he was able to measure the speed and accuracy with which subjects could hit the various targets with their feet. By and large, he found that after a short learning period the subject could perform the task with considerable accuracy and with very short travel time (averaging about 0.1 s). Although forward movements were slightly faster than backward or lateral movements, these differences were not of any practical consequence. His results strongly suggest that it might be possible to assign control tasks to the feet which previously have been considered in the domain of the hands. Anyone who can play, or has watched someone play, the 13 pedals of an organ can attest to the precision and timing of foot movements that can be attained with practice.

The time required to move a foot from one pedal to another is a function of the distance traveled and the size of the pedals. *Fitts's law* (Fitts, 1954), developed to predict hand movement time, states that movement time is a direct function of task difficulty and that task difficulty is directly related to the movement distance and inversely related to the size of the target. Thus the smaller the target and the longer the distance to be moved, the more difficult the task and the longer the movement time required. Drury (1975) used a modification of Fitts's law (Welford, 1968) to predict movement times for operating foot pedals. Drury's index of (task) difficulty (ID) was defined as

$$ID = \log_2 \frac{A}{W + S} + 0.5)$$

where A = movement amplitude (distance) to centerline of target
 W = target width
 S = shoe sole width

The equation developed to predict reciprocal (back-and-forth) (foot) movement time (RMT) was

$$RMT = 0.1874 + 0.0854 \text{ (ID)}$$

Single-movement time (SMT) can be estimated from reciprocal movement time (RMT) by

$$SMT = RMT/1.64$$

These equations appear to be accurate for coplanar foot movements (where pedals are in the same plane). For noncoplanar pedals, movement time can be double, as discussed below. Drury's equations are useful for predicting the effects of varying the size of a pedal and/or the distance between pedals on movement time. As it turns out, with typical spacing between pedals, the size of the pedal (over a reasonable range) has only a minimal effect on movement time. The closer the pedals are to each other, however, the greater the effect on movement time of varying the size of the pedals. Remember, reaction time must be added to movement time to obtain overall performance time estimates.

Automobile Brake and Accelerator Pedals   The most commonly used foot controls are the brake and accelerator controls of automobiles. Aside from their individual characteristics, an important factor in their use is their relative positions. In most automobiles the accelerator is lower than the brake pedal, thus requiring the lifting of the foot from the accelerator, its lateral movement, and then the depression of the brake. Several investigators, however, have found that movement time from accelerator to brake is shorter when the accelerator and brake are at the same level (Davies and Watts, 1970; Glass and Suggs, 1977; Snyder, 1976). Glass and Suggs, in fact, found that movement time was optimum when the brake was 1 to 2 in (2.5 to 5.1 cm) *below* the accelerator. In this configuration with the accelerator partially depressed, one could merely slide the foot to the brake without lifting it. The difference in time between the optimum placement and a typical truck configuration [with the brake 4 in (10.2 cm) above the accelerator] could reduce stopping distances by approximately 6 ft (1.8 m) with travel at 55 mi/h (88 km/h). Even more important, Glass and Suggs report that when the brake was even with or above the accelerator, several subjects caught their foot on the edge of the brake pedal during upward movement from the accelerator. This resulted in an additional 0.3 s, or 24 ft (7.3 m), when travel was at 55 mi/h (88km/h).

DATA ENTRY DEVICES

Our insatiable appetite for information has brought about a proliferation of machines and devices for storing, transmitting, and analyzing that information we hold so dear.

These machines, such as the typewriter, telephone, computer, and calculator, require alphabetic or numerical information, or both, as input. Several types of devices have been used for the input of alphanumeric information, such as knobs, levers, thumb wheel, and pushbutton keyboards. With very few exceptions, keyboard entry devices are generally superior to other data entry devices such as thumb wheels, knobs, and levers. Miner and Revesman (1962), for example, found that 10-key keyboards, such as those used on hand calculators, resulted in a 33 percent reduction in entry time and a 74 percent reduction in errors compared to using either knobs or levers to enter numeric data. Therefore, we focus our attention on keyboards because of their clear-cut advantages for data entry.

The speed and accuracy with which alphanumeric information can be entered by use of a keyboard, or any other device for that matter, is dependent on the quality of the data presented to the operator. As would be expected on the basis of common sense, data entry speed and accuracy are generally greater when the data presented to the operator are clear and legible. As discussed by Seibel (1972), speed and accuracy will also be greater if (1) the operator is familiar with the format of information to be entered; (2) upper- and lowercase characters are used for written text; and (3) long messages or strings of digits are divided into chunks, that is 123 456 rather than 123456.

Seibel also points out that there is no consistent difference in data entry rate between the usual alphanumeric and straight numeric data entry. In addition, messages composed of random sequences of just 10 of the possible alphabetic characters are not entered more rapidly or more accurately than random sequences of all 26 characters.

Chord versus Sequential Keyboards

Most data entry keyboards are *sequential* in the sense that individual characters are entered in a specific sequence, there being a specific key or other device for every character. On the other hand, in *chord* keyboards a single input unit (e.g., a number, letter, or word) requires the simultaneous activation of two or more keys. Chord keyboards, although far less common than sequential keyboards, are used with stenotype machines, pianos, and some mail-sorting machines.

Noyes (1983) presented a history of the developments and applications of chord keyboards. Most of the development and evaluation of chord keyboards took place during the 1960s and 1970s. The principal advantage of a chord keyboard is its suitability for one-handed alphanumeric data entry. Studies that have compared chord keyboards to sequential keyboards have generally found that the chord keyboard is equal to or better than the sequential keyboard (Bowen and Guinness, 1965; Conrad and Longman, 1965). Although chord keyboards seem to offer advantages over sequential keyboards, chord keyboards can be more difficult to learn; in particular, it is more difficult to learn the coding system (i.e., the combination of keys that represent individual items).

Noyes concludes that there probably is no real need for a general-purpose chord keyboard since sequential keyboards adequately fulfill the majority of our everyday requirements. There may, however, be special applications where chord keyboards could be useful, especially where one-handed alphanumeric data entry is required.

Keyboard Arrangement

Since most keyboards are sequential, we discuss briefly some aspects of alphabetic and numeric keyboard arrangements.

Alphabetic Keyboards   The standard typewriter keyboard (called QWERTY because of the sequence of letters in the topline) appears to have been intentionally designed to slow the operator's keystroke rate. The most commonly occurring letters (at least in English) are generally delegated to the weakest, slowest fingers, and frequently occurring letter combinations are on the same hand. You might ask, "How could they have been so stupid?" The answer is, they weren't. The QWERTY keyboard was, in fact, intentionally designed to slow the rate of keying. Back then, mechanical typewriters could not function as fast as people could stroke the keys, and keys often jammed. QWERTY slowed the operators and thus reduced the frequency of jammed keys. By the time technology caught up with human keying capabilities, it was too late to change. Various alternative arrangements of the conventional typewriter keyboard have been proposed. The most notable was by Dvorak (1943) over a quarter of a century ago. Figure 10–21 shows the QWERTY keyboard and the *simplified keyboard,* which is based on the Dvorak arrangement and is recognized by the American National Standards Institute as the official alternative keyboard arrangment to QWERTY (ANSI X4.22-1983). (In the original Dvorak keyboard, the numeric keys across the top row were in this order: 7-5-3-1-9-0-2-4-6-8.)

Overall, the research clearly shows that the Dvorak arrangement is superior to the QWERTY keyboard. Some claim 10 to 20 percent improvements in speed with less hand and finger fatigue. Norman (1983), however, contends that a 5 to 10 percent improvement is more realistic and questions how much of the improvement is actually realized on the job. Norman concludes that the superiority of the Dvorak arrangement is not great enough to justify switching from QWERTY and retraining millions of people to type with the new arrangement. The real question is not whether QWERTY should be scrapped, but rather whether both arrangements should be equally available on keyboards. In fact, there is a popular personal computer that is designed to switch between QWERTY and the simplified keyboard at the touch of a button. (The key caps, however, are arranged as in QWERTY.) In addition, several companies produce software packages that reprogram keyboards to conform to the simplified arrangement.

It seems safe to conclude that the QWERTY keyboard will remain the de facto standard for alphanumeric keyboards for a long time to come, because of the number of such keyboards in existence and the number of people experienced with them. However, it is probable that the simplified keyboard could become more popular as computer data entry proliferates and dual-arrangement keyboards become more common.

Qwerty Keyboard

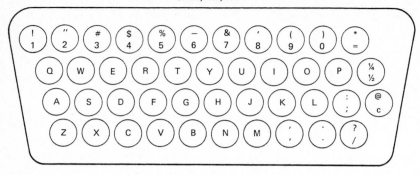

Dvorak Simplified Keyboard

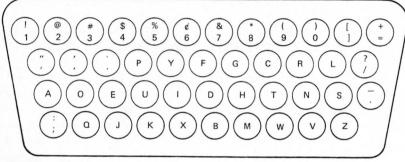

FIGURE 10-21
Current QWERTY and Simplified Keyboard arrangements. The Simplified Keyboard
is specified in American National Standards Institute standard ANSI X4.22-1983 and
is based on the Dvorak keyboard arrangment referred to in the text.

Numeric Keyboards   The two most common numeric data entry keyboards consist of three rows of three digits with the zero below, although there are two different
arrangements of the numerals. One arrangement is used on many calculators and the
other is used on pushbutton telephones. These two arrangements are shown in Table
10–3. There is little question that the highly practiced person will perform about
equally well with either arrangement. The difference in arrangement becomes more
important for people who only make occasional entries and for people who must
alternate between the different arrangements (and with the boom in pocket calculators
this "switch-hitter" group may be quite large). At least three studies (using postal
clerks, housewives, and air traffic controllers) have found the telephone arrangement,
while only slightly faster (approximately 0.05 s per digit), somewhat more accurate
than the calculator arrangement (Conrad, 1967; Conrad and Hull, 1968; Paul,
Sarlanis, and Buckley, 1965). Conrad and Hull, for example, report keying errors of
8.2 percent with the calculator compared to 6.4 percent with the telephone arrangement. Unfortunately for us switch-hitters, they found that these error rates were

TABLE 10–3
COMMON NUMERIC KEYBOARDS

Calculator			Telephone		
7	8	9	1	2	3
4	5	6	4	5	6
1	2	3	7	8	9
	0			0	

markedly better than those obtained from a group of subjects who alternated between the keyboards. This, of course, just reinforces the advantages of standardization in control display arrangements.

Although the telephone pushbutton arrangement results in more accurate performance than the calculator arrangement, how does it compare to the dial telephone? Pollard and Cooper (1978) found that even after almost 2 years of extensive experience with the pushbutton telephone, operators still made more errors than a control group of dial telephone operators (2.96 versus 1.60 percent). Data entry time, however, is significantly longer with a dial telephone.

Keyboard Feel

We cannot hope to discuss all the varied keyboard factors that can impact data entry performance. One variable, however, often mentioned in product review articles is the *feel* of the keyboard. Although it is not a scientific term, keyboard feel is a function of, among other things, key travel and resistance characteristics, auditory activation feedback, and hysteresis. *Hysteresis* refers to the tendency of a key switch to remain in the closed position even after a partial release of downward pressure; it is defined as the difference in travel distance between the closing and opening points of a switch. Too little hysteresis can produce key bounce and cause inadvertent insertions of extra characters; too much hysteresis can interfere with high-speed typing.

Brunner and Richardson (1984) compared three keyboards that differed in terms of a number of keying parameters, as shown in Table 10–4. The snap-spring keyboard was characterized by a rapid buildup of resistance on the downstroke with a sharp drop-off in midtravel at the point of switch closure. The elastomer keyboard exhibited a rapid buildup of resistance in the initial part of the downstroke which disappeared at the point of switch closure and then increased again during keystroke overtravel. Two versions of this keyboard were used. One had no auditory feedback when the switch closed, and the other had an electronic click timed to switch closure. The linear spring keyboard exhibited light initial resistance which doubled at the point of switch closure.

Among both occasional and expert typists, the linear-spring keyboard was least preferred and considered most fatiguing to use, presumably because it presented the least actuation resistance and the most ambiguous auditory feedback (it clicked when the switch closed and again when it opened). Figure 10–22 shows the mean interkeystroke interval (a measure of typing speed) and the mean number of insertion

TABLE 10–4
KEYBOARD CHARACTERISTICS USED BY BRUNNER AND RICHARDSON

Characteristics	Snap spring	Elastomer	Linear spring
Key travel	3.8 m	3.2 mm	3.8 mm
Switch closure	2.45 mm	1.78 mm	1.92 mm
Switch opening	1.45 mm	1.78 mm	1.895 mm
Hysteresis	1.0 mm	0.0 mm	0.025 mm
Auditory feedback	Snap at open and closure	a. None b. Click at closure	Up stop and down stop

Source: Brunner and Richardson, 1984.

errors made with the keyboards. As can be seen, the elastomer keyboards (with and without auditory feedback) resulted in the fastest keying times and low numbers of insertion errors. The linear spring keyboard with low resistance and negligible hysteresis resulted in the greatest number of insertion errors.

Membrane Keypads

Membrane keypads are becoming increasingly popular for use in many consumer products from microwave ovens to pocket calculators. Membrane switches usually consist of mechanical contacts separated by a very thin nonconductive layer. The nonconductive layer has spaces in it corresponding to the keys; when the top layer is depressed, the contacts positioned above and below the space close. Membrane keypads are inexpensive, afford considerable design flexibility, and are sealed to protect against contamination.

Such keypads, however, present some interesting human factors problems. First, key travel distance is virtually nonexistent. Whereas conventional pushbutton keyboards may have to be depressed 100 to 200 mils (1 mil = 1/1000 in) for activation, the typical membrane keypad can be activated by depressing it 10 to 20 mils. Thus, the familiar keystroke feedback is absent. Second, to help reduce accidental activation, membrane keypads often require more force to activate them than do conventional pushbuttons. For example, pushbuttons may require 50 to 90 g to activate, while membrane keypads may require 200 to 500 g. Third, the actual contact areas (where the spaces in the nonconductive inner layer are positioned) are often difficult to locate. Designers print graphics on the keypads or, as we will see, use other techniques to aid users in positioning their fingers on the contact areas.

Membrane switch technology has been applied most often to situations that require infrequent or intermittent data entry, such as entering a few numbers on a pocket calculator to balance a checkbook. Loeb (1983), however, investigated the use of a membrane keypad (with auditory feedback) for typing tasks, using subjects differing in typing ability. Figure 10–23 presents the average number of words typed per minute [(number of words typed − number of words in error) /total time in minutes]

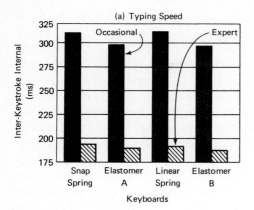

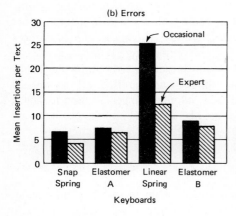

FIGURE 10-22
Effects of various types of keyboards on mean inter-keystroke interval and mean number of insertion errors for occasional and expert typists. Keyboards are described in Table 10-4. *(Source: Adapted from Brunner and Richardson, 1984; Tables 1 and 3.)*

for the various groups of subjects. There was very little difference between the conventional and membrane keyboards for the non-touch typists. The difference, however, was more pronounced for the touch typists (20 to 35 percent). Loeb did note that the advantage of the conventional keyboard over the membrane keyboard was reduced substantially, but not completely, as a function of limited practice on the

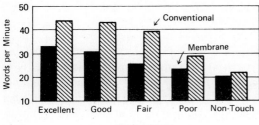

FIGURE 10-23
Comparison of typing performance using a conventional keyboard and a membrane keyboard. Words per minute (wpm) = (number of words typed − number of words in error)/total time in minutes. Categories of typists were defined as follows: Excellent = >60 wpm; Good = 50-60 wpm; Fair = 40-49 wpm; Poor = 26-39 wpm; Non-touch = 16-25 wpm. *(Source: Adapted from Loeb, 1983;* Table II. Reprinted with permission from the *Bell System Technical Journal,* copyright 1983 AT&T.)

membrane keyboard. It is possible, therefore, that with enough practice the membrane keyboard may be as good as a conventional keyboard for typing, at least as far as productivity is concerned.

The task of key pressing involves locating the key and activating it. Both actions require feedback to the user. Roe, Muto, and Blake (1984) conducted a study investigating various techniques for providing membrane keypad users with such feedback. Three feedback techniques were used in all possible combinations: (1) *auditory tone* for activation feedback, (2) *embossing* (each key outlined with a raised rim) for finger-position feedback, and (3) *snap domes* (a raised metal dome over each contact area which is depressed to activate the key) to supply good finger-position feedback and possibly some activation feedback. The results indicated that some form of activation feedback is especially important for reducing errors (virtually all the errors were errors of omission). Keypads incorporating the tone supplied the best activation feedback and resulted in the fewest number of errors. With the tone present, adding finger-position feedback with embossing, domes, or both resulted in a further, but equal, decrease in errors. With no tone present, the use of domes, with or without embossing, resulted in fewer errors than using only embossing. Without the tone it appeared, therefore, that embossing alone failed to provide activation feedback; the domes, however, did provide such feedback, although not as effectively as did the tone. The conclusion, therefore, is that both activation and finger-position feedback are important in membrane keypad use and should be designed into the device to aid the user.

Cursor Positioning Devices

Computer-based technology has introduced a new wrinkle in the data entry field. This technology creates a need for a method of pointing which enables users to indicate to the computer their selection of some element on the computer display or to move a cursor or pointer from one position on the screen to another. Tasks that involve this sort of activity include drawing pictures on the screen where the cursor serves as a kind of paint brush, selecting a choice from several presented on the screen by positioning the cursor over the item, or attaching the cursor to an object on the screen and moving it and the object to another location. A plethora of devices have been developed to facilitate these types of activities. Figure 10–1 presented a few of these. We briefly discuss some of them, indicating some of their human factors advantages and disadvantages.

Touch Screen   With a touch screen, the user simply points a finger at something on the screen and touches (or almost touches) the screen. This technique is easy to learn and requires the most natural response for the user—if you want something, point at it. It is, however, not well suited for drawing on the screen. Some other drawbacks are that the finger obscures what is being pointed at, that holding up the arm to the screen for long periods is fatiguing, and that the finger leaves a smudge on the screen which reduces character legibility. In addition, pointing is not very accurate. Beringer (1983) found, for example, that when people use a touch screen to select an item, they tend to hit below the target by $\frac{1}{8}$ to $\frac{1}{4}$ in (3.2 to 6.4 mm) and that accuracy was best for targets at the bottom of the screen and worst for those at the top. This appears to be due to the positioning of the arm and to parallax.

Touch Pad   The touch pad is a pad separate from the screen which is activated by pressing a finger or pointed instrument on it. The pad can be placed on a tabletop or held in the lap of the user. Points on the pad correspond to points on the screen. Compared to a touch screen, the touch pad separates the input from the system response. The user has to watch the screen while touching the pad, but the position of the pad reduces arm fatigue. Drawing pictures with the pad is compatible with drawing with a pencil on paper; but on most pads, the user cannot rest the hand on the pad because this will cause spurious inputs. Resolution, cursor control, and control-response ratio are important issues (Arnaut and Greenstein, 1985).

Light Pen   A light pen is a pen-shaped device with a cord coming out of the end, and it is attached to the computer. The pen is pressed against the screen and senses the CRT scanning beam as the beam illuminates a pixel on the screen. The light pen uses the same natural pointing response as is used with the touch screen and suffers from many of the same problems. In addition, to point, the user must find the pen and pick it up. The cord on the pen can also become tangled on the keyboard or monitor. Usually, pointing resolution is much better with a light pen than with a touch screen.

Digitizing Tablet   Digitizing tablets typically rest on a tabletop (although some are as large as a drafting table) and are similar to touch pads except that the tablets use a special electronic stylus to specify position. Digitizing tablets have higher resolution than touchpads, and the hand can rest upon the tablets during drawing without causing spurious inputs.

Mouse   A *mouse* is a hand-held device that is rolled on a tabletop and simultaneously controls the movement of a cursor on the screen. It is easy and fast to use. One disadvantage of the mouse is that a clear space near the computer screen is required to operate it; mice do not work well in a rat's nest of papers and open books. Freehand drawing is not as natural as drawing with a pencil-shaped stylus because the mouse is moved with arm and wrist movements rather than finger movements.

Other Cursor Pointing Devices   In addition to the more esoteric devices mentioned above, keyboards, joysticks, and trackballs are also used for cursor positioning. Keyboards (cursor keys for up, down, left and right) are notoriously slow and are not at all good for drawing. Their advantage is that the hands do not have to leave the keyboard during data entry. Joysticks and trackballs can be good pointing devices but are not well suited for drawing.

Comparison of Various Cursor Positioning Devices   Albert (1982) compared the devices listed in Table 10–5 on speed and accuracy of positioning a cursor. The rankings of the devices on these criteria are also shown in Table 10–5. The trackball,

TABLE 10–5
COMPARISON OF VARIOUS DEVICES ON SPEED AND ACCURACY OF CURSOR
POSITIONING

	Ranking	
Device	Speed	Accuracy
Touch screen	1 (fastest)	6.5 (worst)
Light pen	2	6.5
Digitizing tablet	3	2
Trackball	4	1 (best)
Force joystick	5	3
Position joystick	6	4
Keyboard	7 (slowest)	5

Source: Adapted from Albert, 1982, figs. 1 and 2. Copyright by the Human Factors Society, Inc. and reproduced by permission.

the most accurate device, was 100 percent more accurate than the touch screen, the least accurate device. The fastest device, the touch screen, was more than 8 times faster than the slowest device, the keyboard. These are impressive differences and illustrate the need to select the best device for the task at hand. Notice that there is a clear speed-accuracy tradeoff. The more accurate devices tended to be the slower devices; thus, the selection of the best device must take into account the relative importance on the task of speed and accuracy.

Whitfield, Ball, and Bird (1983) compared a touch screen with a touch pad on tasks requiring various levels of positioning accuracy. Using the touch screen resulted in faster performance but was less accurate than the touch pad. The authors concluded that the overall differences were small and that performance on both were comparable.

Card, English, and Burr (1978) compared performance using a mouse, isometric joystick, step keys (up, down, left, and right), and text keys (character, word, line, and paragraph) in a text-editing task. The results are shown in Table 10–6. The mouse was found to be the fastest and also to have the lowest error rate. This is in

TABLE 10–6
COMPARISON OF TEXT-EDITING
DEVICES

Device	Time, s	Error, %
Mouse	1.66	5
Joystick	1.83	11
Step keys	2.51	13
Text keys	2.26	9

Source: Card, English, and Burr, 1978.

contrast to the other studies mentioned which found speed-accuracy trade-offs among the devices tested.

Research on cursor positioning devices is still in its infancy. Specific design parameters of the devices and their relationship to the display have hardly been addressed. Such features as C/R ratio, physical size, methods for confirming a selection, and feedback on the selection made have not been systematically investigated. As technology advances, higher-resolution screens and devices will become possible with greater accuracy demands placed on the user. Researchers will have to scramble to keep up with the advancing technology.

SPECIAL CONTROL DEVICES

The hands and feet are traditionally the parts of the body delegated the task of operating controls. Microcomputer technology, however, has opened new vistas for control devices and applications. Our purpose here is to introduce some devices that are being developed or are in current use. The emphasis is on those devices which involve challenging human factors problems.

Teleoperators

The technological explosion of recent decades has made it possible to explore outer space and the depths of the ocean, thus exposing humans to new and hazardous environments. These and other developments require some types of extension of human control functions across distance or through some physical barrier, as in the control of lunar vehicles or of underwater activities. The control systems for performing such functions have been called *teleoperators* (Corliss and Johnson, 1968). Teleoperators are general-purpose human-machine systems that augment the physical skills of the operator. A person is always in the teleoperator control link on a real-time basis. Robots are not considered teleoperators because they operate autonomously and sometimes (as in the science fiction movie *2001: A Space Odyssey*) counter to human interests.

Corliss and Johnson put teleoperators in four categories: manipulators, prosthetic and orthotic devices, amplifiers, and walking machines. An example of a manipulator teleoperator is shown in Figure 10–24. Similar types of devices are commonly used today for handling dangerous materials (such as radioactive, explosive, or biologically toxic materials) or working in hostile environments such as involved in undersea work. One of the more spectacular examples is the 50-ft (15.2-m) remote manipulator arm of the U.S. space shuttle. The arm has been successfully used to retrieve wayward satellites and fix malfunctioning equipment on the shuttle itself.

Human Factors Considerations   The design of various types of teleoperators must, of course, take into account the biomechanics of human movement, anthropometrics, and the nature of the kinesthetic and proprioceptive information people use to control their movements. It is important for efficient teleoperator control that the device be designed to be compatible with the operator. In this regard, particular consideration needs to be given to the nature of the feedback given to the operator. In many teleoperator applications, the operators are denied direct physical contact with the objects being manipulated by the teleoperator device; thus the teleoperators lose their usual kinesthetic feedback. Feedback regarding the amount of force being exerted by the teleoperator can be given to the operator through the use of servo systems or direct mechanical linkage. Information on slip and shear force, however, is more difficult to convey in teleoperator systems (Feldman, 1982).

More and more teleoperator applications deny the operator direct visual access to the effector end of the teleoperator. In such cases, television cameras are commonly used to extend the operator's vision. In this regard, Kama (1965) reports that direct viewing of the object being manipulated is not really necessary; indirect viewing (i.e., TV) usually will suffice. Use of TV, however, does introduce major problems. TV distorts distance and depth cues; and when the cameras can be moved, zoomed, etc., operators can become visually disoriented. The use of remote sensing (visual and kinesthetic) can also introduce time delays between the operator's control action and the feedback received about the device's response. This is especially true if the operator is on earth controlling a manipulator in space.

An especially tricky aspect of teleoperators is the design of the controls used by the operator to direct the actions of the effector. A manipulator has multiple joints, each of which can be extended and rotated in a number of different planes. Designing a control configuration that will allow smooth, coordinated movements is a real challenge. Knowles (1962) suggests that the user of teleoperators needs to have some meaningful identification or integration with the equipment; or as Corliss and Johnson (1968) put it, the teleoperator controls should be "built along anthropomorphic

FIGURE 10-24
Experimental remote-handling equipment used in various studies.
(Source: USAF, AFHRL.)

lines." The problem with this is that teleoperators can perform movements impossible for humans to perform, such as rotating the wrist 360°. Often computers are enlisted to aid the operator, such as by using preprogrammed movement sequences that can be invoked by the operator as needed.

Speech-Activated Control

The use of speech to converse with computers is a stock item in science fiction books and movies. The advantages are obvious: The operator is not tied to a console, the operator's hands are free to do other chores (such as pour a cup of coffee), and, at least in science fiction, only minimal training of operators is required. Although such casual conversations with machines are not yet upon us, recent advances in speech recognition have spawned new and novel solutions to system problems.

Applications for Speech Recognition Speech recognition systems are usually substituted in place of keyboard entry or pencil-and-paper recording of data. The best applications appear to be where data entry must occur simultaneously with other activities. Entering destination data while sorting packages is an example; recording part numbers while assembling components is another. In addition, speech recognition is well suited to situations where the operator moves around to collect data. Inspectors on automobile assembly lines, for example, use speech to record defects as they walk around the car. One advantage of such a system is that the data are immediately fed into the computer, rather than having to be transcribed from handwritten notes. Speech recognition systems are also well suited for use by the handicapped for such activities as controlling motorized wheelchairs, or turning on and off lights, coffee pots, etc. (Damper, 1984). McCauley (1984) discusses other applications including aircraft cockpits and interactive training systems.

Types of Speech Recognition Systems Speech recognition systems can be classified along two dimensions. The first is whether the system is *speaker-dependent* or *speaker-independent*. Speaker-dependent systems, by far the most common, are trained to recognize a particular person's pronunciation of a limited vocabulary of words, phrases, or sentences. Speaker-independent systems can recognize words spoken by anyone. The second dimension relates to the constraints placed on how the speaker must utter the words. Speech recognition systems are classed as *isolated-word, connected-word,* or *continuous-speech* systems. In isolated-word systems, clearly the most common, the speaker must set off each utterance from the next with a pause. An utterance can be a word, phrase, or short sentence. This pause requirement is relaxed somewhat in connected-word systems; the speaker can string together several utterances without pausing between them, but each word in the string must be enunciated accurately. In continuous-speech systems, there are no pause constraints on the speaker; entire sentences can be uttered as in natural speech. Almost all commercial speech recognition systems today are of the speaker-dependent, isolated-word variety. Some speaker-independent systems are being used in telephone applications, but their vocabulary is very limited, often to the digits 0 to 9.

Connected-word and continuous-speech systems in use today also tend to be limited to small vocabularies and short, simple sentences (Reddy and Zue, 1983). Continuous-speech systems require about 10 times as much computational power as isolated-word systems (Pfauth and Fisher, 1983) and face some monumental recognition problems. As an illustration, consider the following sentence: "You gave the cat your dinner." Because of the phenomenon called *coarticulation,* the "t" of "cat" combines with the "y" of "your" and makes the sentence sound like "you gave the catcher dinner."

Current speaker-dependent, isolated-word systems recognize words in one of two fundamental ways. The most common involves comparison of the frequency spectrum of the voice signal (see Chapter 7) with the spectral characteristics of each vocabulary item stored in the computer until an acceptable match is found. The second approach focuses on isolating certain key features of an utterance, e.g., stops and fricatives (see Chapter 7), and comparing these to the stored vocabulary items. For a more detailed, but readable, explanation, see Levinson and Liberman (1981). A problem in practice is that background noise can render a valid command unrecognizable because of the distortion to the sound pattern transmitted to the system.

Human Factors Issues   There are several human factors related issues inherent in the speaker-dependent, isolated-word type of system (McCauley, 1984). First, the system can handle only a limited predefined vocabulary, with most currently available systems having vocabularies in the range of 20 to 200 items. The larger the vocabulary, generally, the lower the recognition accuracy and the slower the processing, all other things being equal. Several things can be done to increase vocabulary size and recognition accuracy. By placing constraints on what items can be said in what order, very large vocabularies can be generated. For example the first word or phrase in a voice command could be limited to one of, say, 50 preestablished items. When it is recognized, a second set of 50 items, composed of those items which are allowed to follow the first item, is searched. This process can be repeated, thus allowing for vocabularies of several hundred item strings (Coler et al., 1977). In addition, the specific vocabulary items can be selected to be maximally discriminable from one another (Green et al., 1983). A large package delivery service in the United States uses speech recognition to sort parcels. To minimize recognition errors, they use the names of professional football teams—Colts, Bears, Jets, Giants—to indicate the destinations of the parcels (Modern Materials Handling, 1983).

A second problem with isolated-word systems is that each word must be set off with pauses. If the speaker fails to insert a pause after each word or inserts a pause before the end of the word, the system very likely will fail to recognize it. The need to control speech pauses makes it mandatory that the operators of isolated-word systems be trained to discipline their voices and speech. Further, a sigh, cough, or mumble may be erroneously accepted as a valid response by the system. Speakers must be trained to control such extraneous verbal behavior.

In practice, not only must the operator be trained, but so must the system itself. As pointed out in Chapter 7, people have different speech patterns, inflections, frequency modulations, etc.; not everyone sounds the same. A speaker-dependent speech recognition system must develop a complete vocabulary for each operator.

This is usually done by having the operator repeat each item in the system's vocabulary a few times to develop a good representation of that item, for that speaker, in the computer's memory. Most systems today require around four repetitions (McCauley, 1984). This sort of training is very tedious and restricts the size of the vocabulary that can, in practical terms, be used. Imagine having to repeat distinctly each of 300 words 4 times!

People's voices change owing to fatigue and stress. The speech recognition system, trained to recognize a rested, calm (perhaps even bored) voice, may not recognize the speaker in a stress situation or after the voice becomes fatigued. Imagine controlling a piece of machinery with a speech recognition system. Your clothes become entangled, and you are being pulled into the machine. You yell, "Stop!" Unless you yelled "stop" in the same way as when you first trained the system, it is very likely that the machine will not respond. And yelling louder will not help.

Performance of Speech Recognition Systems In most reports, speech recognition systems turn out to be as good as or better than keyboard data entry, although error rates may be higher depending on the specific system, vocabulary, and task situation. Error rates of 1 percent are claimed by some systems, but this is usually under ideal conditions. Error rates can be, and usually are, higher in actual working situations. Speech recognition is not, however, the panacea for all data entry problems. Morrison et al. (1984), for example, found that doing text editing with a speech recognition system (in which the commands were spoken and the text was typed) was preferred less than was straight keyboard entry. Switching modalities during a command was inherently disruptive.

We can conclude that speech recognition has progressed to where it is a valuable method of data entry and is especially well suited to particular applications, but it will probably never totally replace the keyboard as the primary device for data entry.

Eye-Activated Control

Advances in equipment used to track eye and head movements have made it possible to convert eye movements to control input. To date, most eye-activated control applications have been in the military field. Other applications could be in control of teleoperators, selection of instruments on a panel, and visual search tasks. Calhoun, Arbak, and Boff (1984) describe one way such a system would operate. The user, wearing a special helmet to monitor head and eye position, would direct his or her line of sight to, say, a switch. The switch would then change (e.g., its color, brightness, etc.) to give feedback that the signal had been received by the system. To complete the switch activation, the user would make some type of input, perhaps a verbal response to a speech recognition system, to acknowledge or accept the switch selection.

Accuracy of eye positioning has been reported to be approximately ± 10 minutes of visual arc at the center of the visual field (Chapanis and Kinkade, 1972; Coluccio and Mason, 1970). This is very close to the 10 to 20 minutes of arc reported by Calhoun, Arbak, and Boff for successive eye fixations in an actual eye control system. Calhoun, Arbak, and Boff did find that accuracy was poorer for fixations in the lower part of the visual field. Eye control has some disadvantages, including the following: there is potential for overburdening of the visual system, especially in a visual dual-task situation; eye control is difficult in a vibrating or acceleratory environment because either one will produce eye movements which the operator cannot prevent; distractions which might cause operators to shift their gaze will degrade performance; the time required to sense and calculate the line of sight causes a corresponding delay in the feedback to the operator that an object has been selected; and the operator must hold the gaze for some short time to reduce the likelihood of inadvertent activation. Calhoun, Arbak, and Boff report median total activation times (from onset of some signal, through gaze fixation, to confirming selection with a consent switch) of 1.5 to 1.7 s.

DISCUSSION

We have discussed various aspects of controls and their influence on operator performance. One might think that the selection or design of controls has "all been done"; however, numerous examples of inappropriately designed or selected controls can still be found today, for example, in nuclear power plants, heavy-equipment cabs, and numerous consumer products. A wide variety of controls and control mechanisms are available off the shelf; but as we saw in this chapter, the best design often depends on the particular demands of the task and the criteria used to judge performance. For those who view controls as somehow less glamorous than other areas of human factors, we hope the discussions of cursor positioning devices, teleoperators, and speech recognition systems illustrate that exciting things are still happening in this area.

REFERENCES

Air Force System Command (1980, June). *Design handbook 1–3, human factors engineering* (3d ed.). U.S. Air Force.

Albert, A. (1982). The effect of graphic input devices on performance in a cursor positioning task. *Proceedings of the Human Factors Society 26th Annual Meeting*. Santa Monica, CA: Human Factors Society.

Arnaut, L., and Greenstein, J. (1985). Two experiments investigating the effects of control-display gain and method of cursor control on touch tablet performance. *Proceedings of the Human Factors Society 29th Annual Meeting*. Santa Monica, CA: Human Factors Society.

Ayoub, M. M., and Trombley, D. J. (1967). Experimental determination of an optimal foot pedal design. *Journal of Industrial Engineering, 17*, 550–559.

Beringer, D. (1983). The pilot-computer direct-access interface: Touch panels revisited. In R. Jensen (ed.), *Proceedings of the 2d Symposium on Aviation Psychology*. Columbus: Ohio State University, Aviation Psychology Laboratory.

Bowen, H. M., and Guinness, G. V. (1965). Preliminary experiments on keyboard design for semiautomatic mail sorting. *Journal of Applied Psychology, 49*(3), 194–198.

Bradley, J. V. (1967). Tactual coding of cylindrical knobs. *Human Factors, 9*(5), 483–496.

Bradley, J. V. (1969). Desirable dimensions for concentric controls. *Human Factors, 11*(3), 213–226.

Brunner, H., and Richardson, R. (1984). Effects of keyboard design and typing skill on user keyboard preferences and throughput performance. In D. Attwood and C. McCann (eds.),

Proceedings of the 1984 International Conference on Occupational Ergonomics, pp. 267–271. Rexdale, Ontario, Canada: Human Factors Association of Canada.

Buck, L. (1980). Motor performance in relation to control-display gain and target width. *Ergonomics,* 23, 579–589.

Bullinger, H., and Muntzinger, W. (1984). Development of ergonomically designed controls. In D. Attwood and C. McCann (eds.), *Proceedings of the 1984 International Conference on Occupational Ergonomics,* pp. 447–451. Rexdale, Ontario, Canada: Human Factors Association of Canada.

Calhoun, G., Arbak, C., and Boff, K. (1984). Eye-controlled switching for crew station design. *Proceedings of the Human Factors Society 28th Annual Meeting.* Santa Monica, CA: Human Factors Society.

Card, S., English, W., and Burr, B. (1978). Evaluation of mouse, rate-controlled isometric joystick, step keys, and text keys for text selection on a CRT. *Ergonomics,* 21, 601–613.

Carter, R. (1978). Knobology underwater. *Human Factors,* 20, 641–647.

Chapanis, A., and Kinkade, R. (1972). Design of controls. In H. P. Van Cott and R. Kinkade (eds.), *Human engineering guide to equipment design* (rev. ed.). Washington: Government Printing Office.

Coler, C., Plummer, R., Huff, E., and Hitchock, M. (1977). Automatic speech recognition research at NASA Ames Research Center. In M. Curran, R. Breaux, and E. Huff (eds.), *Voice technology for interactive real-time command/control systems applications, proceedings of a symposium.* Moffett Field, CA: NASA Ames Research Center.

Coluccio, T., and Mason, K. (1970). Measurement of target position in a realtime display by use of the Oculometer eye direction monitor. Paper presented at the Human Factors Society annual meeting.

Conrad, R. (1967). Performance with different push-button arrangements, *Het PTT-Bedrijfdeel,* 15, 110–113.

Conrad, R., and Hull, A. J. (1968). The preferred layout for numerical data-entry keysets. *Ergonomics,* 11(2), 165–173.

Conrad, R., and Longman, D. (1965). A standard typewriter versus chord keyboard—An experimental comparison. *Ergonomics,* 8, 77–88.

Corliss, W., and Johnson, E. (1968). *Teleoperator controls* (NASA SP-5070). Washington: NASA Office of Technology Utilization.

Damper, R. (1984). Voice-input aids for the physically disabled. *International Journal of Man-Machine Studies,* 21, 541–553.

Davies, B. T., and Watts, J. M., Jr. (1970). Further investigations of movement time between brake and accelerator pedals in automobiles. *Human Factors,* 12(6), 559–561.

Davis, L. E. (1949). Human factors in design of manual machine controls. *Mechanical Engineering,* 71, 811–816.

Drury, C. (1975). Application of Fitts' Law to foot-pedal design. *Human Factors,* 17, 368–373.

Dvorak, A. (1943). There is a better typewriter keyboard. *National Business Education Quarterly,* 12, 51–58.

Feldman, M. (1982, June). Evolving toward projected man. *Nuclear News,* 25(8), 127–132.

Fitts, P. (1954). The information capacity of the human motor system in controlling the amplitude of movement. *Journal of Experimental Psychology,* 47, 381–391.

Gamst, F. (1975). Human factors analysis of the diesel-electric locomotive cab. *Human Factors,* 17, 149–156.

Glass, S., and Suggs, C. (1977). Optimization of vehicle-brake pedal foot travel time. *Applied Ergonomics,* 8, 215–218.

Green, T., Payne, S., Morrison, D., and Shaw, A. (1983). Friendly interfacing to simple speech recognizers. *Behavior and Information Technology,* 2, 23–38.

Hartman, B. O. (1956). *The effect of joystick length on pursuit tracking,* Rept. 279. Fort Knox, KY: U.S. Army Medical Research Laboratory.

Howland, D., and Noble, M. (1955). The effect of physical constraints on a control on tracking performance. *Journal of Experimental Psychology,* 46, 353–360.

Hunt, D. P. (1953). *The coding of aircraft controls,* Tech. Rept. 53–221. U.S. Air Force, Wright Air Development Center.

Jenkins, W. (1947a). The discrimination and reproduction of motor adjustments with various types of aircraft controls. *American Journal of Psychology,* 60, 397–406.

Jenkins, W. (1947b). The tactual discrimination of shapes for coding aircraft-type controls. In P. Fitts (ed.), *Psychological research on equipment design,* Research Rept. 19. Army Air Force, Aviation Psychology Program.

Jenkins, W., and Connor, M. B. (1949). Some design factors in making settings on a linear scale. *Journal of Applied Psychology,* 33, 395–409.

Jenkins, W., and Karr, A. C. (1954). The use of a joy-stick in making settings on a simulated scope face. *Journal of Applied Psychology,* 38, 457–461.

Kama, W. N. (1965, April). *Effect of augmented television depth cues on the terminal phase of remote driving,* Tech. Rept. 65–6. U.S. Air Force, Aerospace Medical Research Laboratory.

Katchmer, L. T. (1957, March). *Physical force problems; 1. Hand crank performance for various crank radii and torque load combinations,* Tech. Memo 3–57. Aberdeen, MD: Aberdeen Proving Ground, Human Engineering Laboratory.

Knowles, W. B. (1962, August). *Human engineering in remote handling* (TDR 62–58) U.S. Air Force, AMRL.

Kohl, G. (1983). Effects of shape and size on knobs on maximal hand-turning forces applied by females. *The Bell System Technical Journal,* 62, 1705–1712.

Kroemer, K. H. E. (1971). Foot operation of controls. *Ergonomics,* 14(3), 333–361.

Levinson, S., and Liberman, M. (1981, April). Speech recognition by computer. *Scientific American,* 244, 64–76.

Loeb, K. (1983, July-August). Membrane keyboards and human performance. *The Bell System Technical Journal,* 6, 1733–1749.

McCauley, M. (1984). Human factors in voice technology. In F. Muckler (ed.), *Human factors review—1984.* Santa Monica, CA: Human Factors Society.

McFarland, R. A. (1946). *Human factors in air transport design.* New York: McGraw-Hill.

Mehr, M. (1973). Two-axis manual positioning and tracking controls. *Applied Ergonomics,* 4, 154–157.

Micro Switch. (1984). *Programmable display pushbutton PD series.* Freeport, IL.

Miner, F. J., and Revesman, S. L. (1962). Evaluation of input devices for a data setting task. *Journal of Applied Psychology,* 46, 332–336.

Morrison, D., Green, T., Shaw, A., and Payne, S. (1984). Speech-controlled text-editing: Effects of input modality and of command structure. Paper submitted to *International Journal of Man-Machine Studies.*

Moussa-Hamouda, E., and Mourant, R. (1981). Vehicle fingertip reach controls—Human factors recommendations. *Ergonomics,* 12, 66–70.

Norman, D. (1983, August 23). *The DVORAK revival: Is it really worth the cost?* La Jolla; Institute for Cognitive Science (C-015), University of California, San Diego.

Noyes, J. (1983). Chord keyboards. *Applied Ergonomics,* 14, 55–59.

Paul, L., Sarlanis, K., and Buckley, E. (1965). A human factors comparison of two data entry keyboards. Paper presented at *Sixth Annual Symposium of the Professional Group on Human Factors in Electronics.* IEEE.

Pfauth, M., and Fisher, W. (1983, September). Voice recognition enters the control room. *Control Engineering,* 30(9), 147–150.

Pollard, D., and Cooper, M. (1978). An extended comparison of telephone keying and dialing performance. *Ergonomics,* 21, 1027–1034.

Poulton, E. (1974). *Tracking skill and manual control.* New York: Academic.

Reddy, R., and Zue, V. (1983, November). Recognizing continuous speech remains an elusive goal. *IEEE Spectrum,* 20(11), 84–87.

Rockway, M. (1957, September). *Effects of variation in control deadspace and gain on tracking performance,* Tech. Rept. 57–326. U.S. Air Force, Wright Air Development Center.

Rockway, M., and Franks, P. (1959, January). *Effects of variations in control backlash and gain on tracking performance,* Tech. Rept. 58–553. U.S. Air Force, Wright Air Development Center.

Roe, C., Muto, W., and Blake, T. (1984). Feedback and key discrimination on membrane keypads. *Proceedings of the Human Factors Society 28th Annual Meeting.* Santa Monica, CA: Human Factors Society.

Rogers, J. (1970). Discrete tracking performance with limited velocity resolution. *Human Factors,* 12, 331–339.

Seibel, R. (1972). Data entry devices and procedures. In H. P. Van Cott and R. G. Kinkade (eds.), *Human engineering guide to equipment design.* Washington: Government Printing Office.

Seminara, J., Gonzalez, W., and Parsons, S. (1977). *Human factors review of nuclear power plant control room design* (EPRI NP-309). Palo Alto, CA: Electric Power Research Institute.

Snyder, H. (1976). Braking movement time and accelerator-brake separation. *Human Factors,* 18, 201–204.

Voice recognition back again, and better (1983, April 6). *Modern Materials Handling.* pp. 52–55.

Welford, A. (1968). *Fundamentals of skill.* London: Methuen.

Whitfield, D., Ball, R., and Bird, J. (1983). Some comparisons of on-display and off-display touch input devices for interaction with computer generated displays. *Ergonomics,* 26, 1033–1053.

Wierwille, W. (1984). The design and location of controls: A brief review and an introduction to new problems. In H. Schmidtke (ed.), *Ergonomic data for equipment design.* New York: Plenum.

CHAPTER 11

HAND TOOLS AND DEVICES

Some people still believe that the use of tools for pounding, digging, scraping and cutting is what distinguishes humans from apes. Actually, there is considerable evidence (Washburn, 1960) that such hand tools and other devices were used by pre-human primates almost a million years ago. Today hand tools are crafted for uncounted specific applications as well as for general-purpose activities. Until recently, however, human factors had largely ignored the design of hand tools and other hand-held devices, concentrating on more sophisticated equipment. Perhaps the attitude was that a million years of evolutionary experience would produce tools and devices uniquely adapted for human use. Actually, a million years is no guarantee of proper design. Indeed, many hand tools and devices are not designed for efficient, safe operation by humans—especially for repetitive tasks.

Improperly designed tools and devices have several undesirable consequences including accidents and injury. Ayoub, Purswell, and Hicks (1977), for example, reviewed injury data from several states in the United States and found that hand tool use was involved in 5 to 10 percent of all compensable injuries. Powered hand tools, however, accounted for only 21 to 29 percent of all hand tool related injuries. The most common tools implicated in injuries were knives, wrenches, and hammers. Such injury data, for the most part, represent single-incident traumatic events such as smashing a finger or cutting the palm of the hand. Other more insidious consequences of improper tool design are cumulative-effect traumas (some of which we discuss later), such as carpal tunnel syndrome, tenosynovitis, "trigger finger," ischemia, vibration-induced white finger, and even tennis elbow. Incidence rates of cumulative-effect traumas can be quite high in industries and jobs requiring repetitive use of hand tools. Armstrong et al. (1982), for example, found the overall incidence of cumulative-effect trauma in a poultry processing plant to be almost 13 cases per 200,000 work hours (200,000 work hours is equivalent to roughly 100 people working for 1 yr or 200 people working for 0.5 yr, and so on). In one department, thigh skinning, the incident rate was almost 130 cases per 200,000 work hours! These cumulative-effect traumas usually do not show up on accident injury reports but often lead to reduced work output, poorer-quality work, increased absenteeism, and single-incident traumatic injuries.

The proper design of hand tools and devices requires the application of technical, anatomic, kinesiological, anthropometric, physiological, and hygienic considerations. Tools cannot be designed in isolation. Often a redesign of the work space, for example, changing the height of the work surface or repositioning the workpiece relative to the worker, can compensate for a less than optimally designed tool.

An important aspect of hand tool use that is often overlooked is training. It is usually assumed that workers know how to use hammers, pliers, saws, wrenches, etc. Actually, nothing could be further from the truth. As a rule, workers do not always know the proper way to use tools. A good training program in tool usage can often reduce the incidence of injuries and cumulative-effect traumas.

In this chapter we concentrate on the human factors considerations in hand tool and device design. Our emphasis is on factors that contribute to cumulative-effect trauma and both the quality and quantity of productivity. We start with a discussion of the anatomy and functioning of the human hand. This allows us to formulate some general human factors principles of hand tool and device design. To close the chapter, a few examples of applying human factors principles to specific tools and devices are presented.

HUMAN HAND

The human hand is a complex structure composed of bones, arteries, nerves, ligaments, and tendons, as shown in Figure 11–1. The fingers are flexed by muscles in the forearm. The muscles are connected to the fingers by tendons which pass through a channel in the wrist. This channel is formed by the bones of the back of the hand on one side and the transverse carpal ligament (flexor retinaculum) on the other. The resulting channel is called the *carpal tunnel*. Through this tunnel passes a whole host of vulnerable anatomic structures including the radial artery and median nerve. Running over the outside of the transverse carpal ligament are the ulnar artery and ulnar nerve. This artery and this nerve pass beside a small bone in the wrist called the *pisiform bone*. All this may seem a bit clinical, but a clear understanding of the relationships of these structures is important to appreciate the consequences of improper hand tool design.

The bones of the wrist connect to the two long bones of the forearm—the *ulna* and the *radius*. The radius connects to the thumb side of the wrist, and the ulna connects to the little-finger side of the wrist. The configuration of the wrist joint permits movements in only two planes, each one at an approximately 90° angle to the other. The first plane allows palmar flexion or, when it is performed in the opposite direction, dorsiflexion, as shown in Figure 11-2a. The second movement plane, shown in Figure 11-2b, consists of either ulnar deviation or radial deviation of the hand.

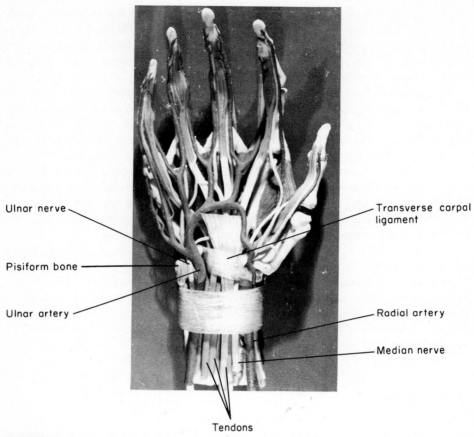

Ulnar nerve

Pisiform bone

Ulnar artery

Transverse carpal
ligament

Radial artery

Median nerve

Tendons

FIGURE 11-1
The anatomy of the hand as seen from the palm side. Shown are the blood vessels, tendons,
and nerves underneath the ligaments of the wrist: *(Source: Tichauer, 1978, fig. 50.)*

The ulna and radius of the forearm connect to the *humerus* of the upper arm, as
shown in Figure 11-3. The *bicep muscle* connects to the radius. When the arm is
extended, the bicep muscles will pull the radius strongly against the humerus. This
can cause friction and heat in the joint. The bicep is both a flexor of the forearm and
an outward rotator of the wrist. This can be seen by bending the arm 90° at the elbow
and rotating the wrist outward. Notice how the bicep muscle contracts and bulges.
Thus any movement that requires a strong pull and simultaneous inward rotation of
the hand should be avoided. As pointed out by Tichauer and Gage (1977), good
practical design can be observed in the operation of a corkscrew. A wine server pulls
hard at the cork while rotating the right forearm outward.

PRINCIPLES OF HAND TOOL AND DEVICE DESIGN

It is not possible to list or discuss all the many principles of hand tool design here.
The major principles are discussed as they relate to the biomechanics of the human

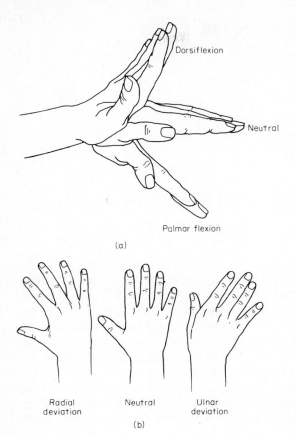

Dorsiflexion

Neutral

Palmar flexion

(a)

Radial deviation

Neutral

Ulnar deviation

(b)

FIGURE 11-2
Movements of the wrist joint about two axes.

hand. Also, it is not our intention to present detailed design specifications for specific hand tools. The interested reader can refer to Fraser (1980) and Eastman Kodak Co. (1983) for such data.

Maintain a Straight Wrist

The flexor tendons of the fingers pass through the carpal tunnel of the wrist. When the wrist is aligned with the forearm, all is well. However, if the wrist is bent, especially in palmar flexion or ulnar deviation (or both), problems occur. The tendons bend and bunch up in the carpal tunnel. Continued use will cause *tenosynovitis,* an inflammation of the tendons and their sheaths. A common type of motion which can lead to tenosynovitis is clothes wringing (Tichauer, 1978), in which, for example, the wringing is done by a clockwise movement of the right fist and counterclockwise action of the left. This same type of motion is involved in inserting screws in holes, manipulating rotating controls such as found on steering handles of motorcycles, and looping wire while pliers are used.

In addition to tenosynovitis, a condition called *carpal tunnel syndrome* can develop. Carpal tunnel syndrome is a disorder caused by injury of the median nerve

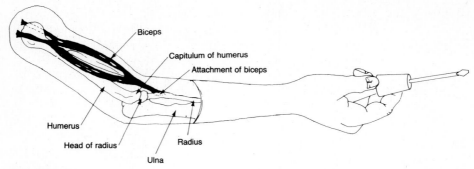

FIGURE 11-3
The elbow joint showing the connection of the bicep to the radius. *(Source: Tichauer, 1978, fig. 31.)*

where it passes through the carpal tunnel of the wrist. The symptoms include numbness, loss of feeling and grip, and finally muscle atrophy and loss of hand functions. This condition occurs 3 to 10 times more often in women than in men (Dionne, 1984), reflecting either a physiological-anatomical difference or only the fact that women are more likely to engage in work that involves repetitive bent-wrist motions. Incidence rates of carpal tunnel syndrome among workers depend, of course, on the particular job being performed. Armstrong (1983) cites studies showing rates of carpal tunnel syndrome of 3.4 and 25.6 cases per 200,000 work hours.

A key rule in hand tool use is to avoid ulnar deviation. The x-ray of a hand holding conventional pliers in Figure 11-4*a* shows a classic case of ulnar deviation. Redesigning the pliers, as shown in Figure 11-4*b,* so that the handles bend rather than the wrist, allows a more natural alignment of wrist and forearm. Tichauer (1976) reports the results of a comparison of two groups of 40 electronics assembly trainees, one using conventional straight pliers and the other using the redesigned bent pliers, during a 12-week training program. Figure 11-5 shows the incidence of carpal tunnel syndrome, tenosynovitis, and tennis elbow in the two groups over the 12 weeks of training. There was a sharp increase in symptoms in the 10th and 12th weeks for the group using the straight pliers while no such increase occurred in the group using the redesigned bent pliers.

The idea of bending the tool, not the wrist, has been applied to other things by John Bennett (Emanuel, Mills, and Bennett, 1980), who patented the idea of bent handles (19° ± 5°) for all tools and sports equipment. Examples of this design are shown in Figure 11-6 for a push broom and for a hammer. Knowlton and Gilbert (1983) compared the degrees of ulnar deviation induced by bent-handle and straight-handle hammers. Ulnar deviation was found to be 2.6 times greater when the straight-handle hammer was used. In addition, hand grip strength measured before and after driving twenty 3.5-in (8.9-cm) nails into a block of wood showed much greater decrements when the straight-handle hammer was used. Emanuel, Mills, and Bennett (1980), although not reporting the results of any systematic evaluation of bent-handle tools, did report the experiences of several individuals and organizations that have used tools with such handles, all with positive results. The U.S. Forest

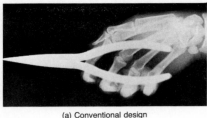

(a) Conventional design   (b) Redesigned pliers

FIGURE 11-4
X-rays of hand using conventional pliers in a wiring operation, *a,* and in using a redesigned model, *b*. The redesigned model is more anatomically correct. *(Source: Damon, 1965; Tichauer, 1966; photographs courtesy of Western Electric Company, Kansas City.)*

Service, for example, has tested the concept on a set of 19 types of tools such as knives, axes, hoes, shovels, and shears. Preliminary results indicate some decrease in fatigue and in general a preference for the bent handle over the straight handle. Krohn and Konz (1982) also found that people preferred a hammer handle with a slight bend (10°) over a traditional straight-handled hammer. The experiences of individuals with existing injuries or wrist problems suggest that the bent-handle design might be particularly useful for certain handicapped persons. As an aside, a softball bat with a bent handle has been manufactured and is legal for the game.

In addition to ulnar deviation, other types of wrist bending can also cause problems. Radial deviation, particularly if combined with pronation and dorsiflexion, in-

FIGURE 11-5
Comparison of two groups of trainees using different pliers. Shows percent of workers with tenosynovitis, epicondylitis (tennis elbow), and carpal tunnel irritation. *(Source: Adapted from Tichauer, 1976. Reprinted with permission from* Industrial Engineering *magazine, 1976, copyright American Institute of Industrial Engineers , Inc., 25 Technology Park/Atlanta, Norcross, GA 30092.)*

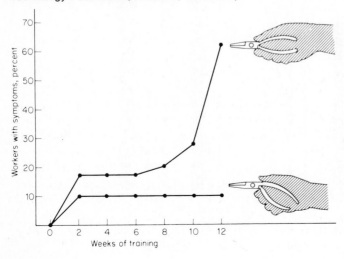

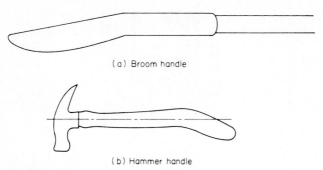

(a) Broom handle

(b) Hammer handle

FIGURE 11-6
Examples of the *Bennett* handle that helps the user keep the
wrist straight while using the tool. *(Source: Emanuel, Mills, &
Bennett, 1980, fig. 1.)*

FIGURE 11-7
Grip strength as a function of wrist and forearm position. Grip strength is the
average maximal grip sustained for 3 s, expressed as a percentage of neutral
supinated grip strength. *(Source: based on data from Terrell & Purswell, 1976,
table 1.)*

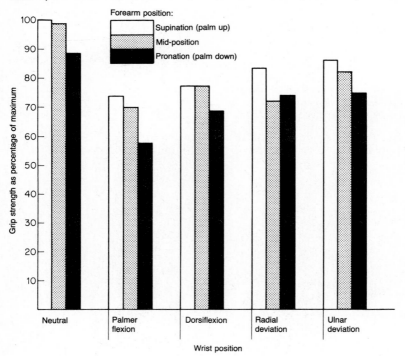

creases pressure between the head of the radius and the capitulum of the humerus in the elbow (see Figure 11–3). Many wire brushes have to be held in this way when they are used overhead. This can lead to *tennis elbow* *(epicondylitis)*, an inflammation of the tissue in the elbow region.

We have seen that ulnar deviation can cause tenosynovitis and that radial deviation can lead to tennis elbow. If that were not enough, Terrell and Purswell (1976) report that grip strength is reduced if the wrist is bent in any direction. Figure 11–7 shows the grip strength as a percentage of the maximum grip strength (achieved with the wrist in a neutral position with supinated forearm). Reductions in grip strength may increase the likelihood that the user will lose control of the tool or drop it, leading to an injury or poor-quality work. Attempts to maintain a strong grip will increase fatigue.

Figure 11–8 shows a man using a hacksaw in a manner that violates the straight-wrist principle. The man's right hand is in ulnar deviation while the left hand is in dorsiflexion with a touch of radial deviation. After a few hours of sawing like this, he will be in no shape to play the piano!

Avoid Tissue Compression Stress

Often in the operation of a hand tool or device, considerable force is applied with the hand as in squeezing pliers or scraping paint with a paint scraper. Such actions concentrate considerable compressive force on the palm of the hand. Particularly pressure-sensitive areas are those overlying critical blood vessels and nerves, specifically the ulnar and radial arteries. Figure 11-9*a* shows a hand holding a conventional paint scraper. The handle digs into the palm and obstructs blood flow through the ulnar artery. This obstruction of blood flow, or *ischemia,* leads to numbness and tingling of the fingers. Tichauer (1978) reports that affected workers will often take

FIGURE 11-8
Man using a hacksaw and violating the straight wrist principle. The right hand is in ulnar deviation while the left hand is in dorsiflexion. *(Source: Greenberg & Chaffin, 1977, fig. 35.)*

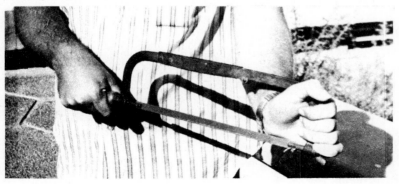

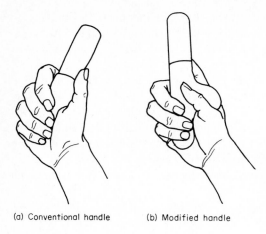

(a) Conventional handle   (b) Modified handle

FIGURE 11-9
A conventional paint scraper that presses
on the ulnar artery, and a modified handle
which rests on the tough tissues between
thumb and index finger, thus preventing
pressure on the critical areas of the hand.
(Source: Tichauer, 1967.)

temporary breaks from work to relieve the symptoms. Thrombosis of the ulnar artery
has also been reported (Tichauer, 1978).

If possible, handles should be designed to have large contract surfaces to distribute
the force over a larger area and to direct it to less-sensitive areas, such as the tough
tissue between thumb and index finger. Figure 11–9b, for example, shows an improved paint scraper handle that rests on the tissue between thumb and finger, thus
preventing pressure on the critical areas of the palm.

In a similar vein (no pun intended), the palm of the hand should never be used as
a hammer. Not only will such action damage the arteries, nerves, and tendons of the
hand, but also the shock waves generated may travel to other body regions such as
the elbow or shoulder.

Related to compression stress is the use of finger grooves on tool handles. As
anyone who has ever watched a professional basketball game knows, hands come in
many sizes. A person with thick fingers using a tool with finger grooves often finds
that the ridges of the grooves dig into the fingers. A small-handed person may put
two fingers into one groove, thereby squeezing the fingers together. For this reason
Tichauer (1978) recommends not using deep finger grooves or recesses in tool handles if repetitive large finger forces are required.

Avoid Repetitive Finger Action

Occasionally, if the index finger is used excessively for operating triggers, a condition known as *trigger finger,* a form of tenosynovitis, develops. The afflicted person
typically can flex but cannot extend the finger actively. The finger must be passively
straightened, and when it is, an audible click may be heard. The condition seems to
occur most frequently if the handle of the tool or device is so large that the distal
phalanx (segment) of the finger has to be flexed while the middle phalanx must be
kept straight (Tichauer, 1978).

As a rule, frequent use of the index finger should be avoided, and thumb-operated
controls should be used. The thumb is the only finger that is flexed, abducted, and
opposed by strong, short muscles located entirely within the palm of the hand. One

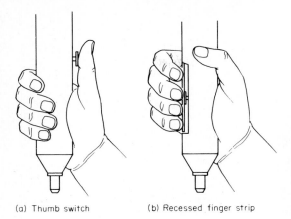

(a) Thumb switch   (b) Recessed finger strip

FIGURE 11-10
Thumb-operated and finger-strip-operated pneumatic tool. Thumb operation results in overextension of the thumb. Finger-strip control allows all the fingers to share the load and the thumb to grip and guide the tool.

must be careful, however, not to hyperextend the thumb such as shown in Figure 11–10a. This causes pain and inflammation. Preferable to thumb controls is the incorporation of a *finger-strip* control, as shown in Figure 11–10b, which allows several fingers to share the load and frees the thumb to grip and guide the tool.

The grip strength of the hand is related to the size of the object being gripped, as shown in Figure 11–11a. Maximum grip strength, for both males and females, occurs with a grip axis between 2.5 and 3.5 in (64 and 89 mm) (Greenberg and Chaffin, 1977). There is some evidence, however, that the optimum grip axis is also influenced by the shape of the object being gripped. Ayoub and Lo Presti (1971), for example, found that maximum grip strength on round objects occurred when the diameter of the object was about 1.6 in (41 mm). Therefore, if a tool handle is round, the optimum grip axis may be smaller than shown in Figure 11–11a. Figure 11–11b shows a pop-riveting gun which must be gripped in the fully open position—a distance of 6 in (152 mm) between the outside edges—and squeezed. This is nearly impossible for most women to perform.

Design for Safe Operation

Designing tools and devices for safe operation would include eliminating pinching hazards and sharp corners and edges. This can be done by putting guards over pinch points or stops to prevent handles from fully closing and pinching the palm of the hand. Sharp corners and edges can be rounded. Power tools such as saws and drills can be designed with brake devices so that the blade or bit stops quickly when the trigger is released. Proper placement of the power switch for quick operation can also reduce accidents with power tools. Each type of tool presents its own set of safety considerations. The designer must consider, in detail, how the tool will be used by the operator and how it is likely to be *misused* by the user.

Remember Women and Left-Handers

Women make up approximately 50 percent and left-handers make up approximately 8 to 10 percent of the world population (Barsley, 1970). Many hand tools and de-

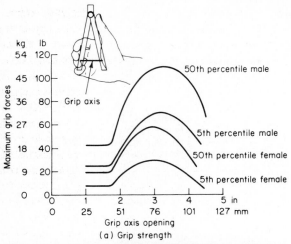

(a) Grip strength

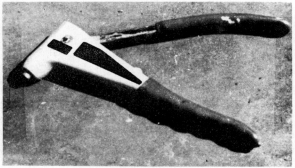

(b) Pop-riveting gun

FIGURE 11-11
(a) Maximum grip strength for various handle openings. Subjects were 50 male and 50 female electronics manufacturing employees. *(b)* A pop-riveting gun which to be operated must be grasped and squeezed in the fully open position. The outside edges of the handles are 6 in (150 mm) apart. *(Source: Greenberg & Chaffin, 1977, figs. 27 & 28.)*

vices are not designed to accommodate these populations. Ducharme (1977) reports that in the Air Force the average hand length of women is about 2 cm (0.8 in) shorter than that of the average male. Less than 1 percent of Air Force men have a hand that is as short as the average woman's hand. Further, grip strength of women is on the average only about two-thirds that of men (Konz, 1979). These differences obviously have implications for tool design.

Ducharme surveyed 1400 Air Force women working in the craft skills and asked them to rate the adequacy of the tools they used. Table 11–1 summarizes some of the major complaints and the tools that engendered them. Only those tools that were judged inadequate by at least 10 percent of the women in one or more craft fields are included. The percentages of women in each field who considered each tool inadequate are also shown in Table 11–1. From scanning the complaints it is obvious that almost all reported inadequacies can be traced to the fact that women have smaller hands and less grip strength than most men. With women assuming a greater role in traditionally male-dominated occupations, the design of hand tools and devices must reflect the anthropometric and ergonomic differences between men and women.

TABLE 11–1
TOOLS RATED INADEQUATE BY AIR FORCE FEMALES WORKING IN THE CRAFT SKILLS AND SOME OF THE REASONS GIVEN FOR THE INADEQUACY

Tool	Rated inadequate because	Percentage rating tool as inadequate
Wire strippers	Hard to hold in hand; handles too far apart to squeeze; too heavy; clumsy; too hard to squeeze; fingers get pinched	12[a]; 18[b]; 19[c]; 15[d]; 18[e]
Crimping tool	Handles too far apart; too hard to squeeze; not able to manipulate	14[a]; 14[c]; 25[d]
Soldering iron	Too heavy; handle too large; clumsy, too bulky	17[a]; 15[c]
Soldering gun	Too heavy; cannot reach trigger; hard to hold in hand	15[a]
Twist wire pliers	Too large to grip; handles too far apart	29[b]
Metal shears	Too large; need two hands to cut	22[f]
Rivet cutter	Too hard to squeeze; awkward	17[f]
Caulking gun	Hard trigger; awkward	11[g]

[a]Communication-electronic systems field.
[b]Missile electronics maintenance field.
[c]Avionics systems field.
[d]Aircraft accessory maintenance field.
[e]Mechanical/electrical field.
[f]Metalworking field.
[g]Structural/pavements field.
Source: Ducharme, 1977.

Another neglected group is left-handers. Actually, there are degrees of handedness from strong right-handed through ambidextrous to strong left-handed. Tools should be designed to be used in the operator's preferred hand. Often, designs preclude the use of tools by left-handers. For example, the drill shown in Figure 11–12 requires left-handers to operate the drill with the right hand. If a threaded fastener were provided in the right side of the drill housing, the handle could be moved for left-handed use.

Laveson and Meyer (1976) point out that some serrated knives have the beveled (cutting) edge on the left side. The left-hander presses down and to the left; the beveled edge then fails to cut, since the pressure is on the wrong side. A serrated bread knife, when used by a left-hander, will often tear rather than cut the bread. The difficulty could be corrected by double-edge beveling.

VIBRATION

Typically in the human factors literature, the topic of *vibration* refers to whole-body vibration such as heavy-equipment operators' experience as they drive over bumpy roads. We discuss this type of vibration in Chapter 17. Our focus here, however, is on hand vibration induced by such power tools as chain saws, pneumatic drills, grinding tools, and chipping hammers.

FIGURE 11-12
Drill with handle designed for right-handed operation. *(Source: Greenberg & Chaffin, 1977, fig. 25.)*

Vibration-Induced White Finger

In the early 1900s reports began to appear that linked hand-held vibratory tools with *vascular spasm* in the hands (Taylor, 1974). Since then numerous other reports and investigations have been published on what has become known as *vibration-induced white finger* (VWF), or *vibration syndrome*.[1]

VWF is a complex phenomenon whose exact pathology is not fully understood, but it probably involves damage to the nerves and smooth muscles of the blood vessels in the hand. Although the National Institute for Occupational Safety and Health (NIOSH) has linked regular, prolonged use of vibratory hand tools in certain occupations to vibration syndrome, little is known about what specific vibration parameters are the most necessary to control (NIOSH, 1983).

The primary symptom of VWF is a reduction in blood flow to the fingers and hand. This is caused by the smooth muscles of the blood vessels in the hand and fingers constricting and thereby reducing the flow of blood. Workers afflicted with VWF will have *vascular attacks* wherein the hand or fingers blanch (turn white). The feeling is the same as when the hand or foot "falls asleep." The person complains of "pins and needles" (i.e., tingling), numbness, or pain. These attacks seem to be brought on by cold. They are especially prevalent among workers who use vibratory tools in cold environments. The attacks can last minutes or hours.

---

[1]Several other names have also been used in the literature: *Raynaud's phenomenon of occupational origin, dead hand, dead finger, white finger, occupational vasomotor traumatic neurosis,* and *traumatic vasospastic disease.*

Workers afflicted with VWF appear to have reduced blood flow in the hands under normal conditions in addition to experiencing vascular attacks. Koradecka (1977), for example, examined pneumatic drill operators, manual grinder operators, and control groups working under similar microclimatic conditions but not exposed to vibration. He found that the exposed groups exhibited a 38 to 67 percent mean reduction in hand skin blood flow compared with control subjects.

One consequence of this reduced blood flow is a reduction in skin temperature. Figure 11-13 shows the mean finger skin temperature of vibration-exposed and control subjects. The vibration-exposed group starts out with a lower skin temperature, but when the hand is heated [10 min in 113°F (45°C) water], both groups attain the same temperature. Cooling the hands, however, restores the initial difference which remains throughout recovery. This lower skin temperature is often associated with a decrease in sensitivity and fine finger dexterity and a loss of grip.

Although most cases of VWF are not debilitating, advanced cases have been known to lead to gangrene of the fingertips (Walton, 1974). The prevalence of VWF, of course, varies with the type and amount of exposure and other individual differences. Wasserman et al. (1982), for example, studied 385 foundry and shipyard workers exposed to hand-arm vibration from pneumatic chipping hammers and grinders. Advanced symptoms of VWF (blanching of fingers) were found among 47 per-

FIGURE 11-13
Skin temperature of the fourth finger of the left hand during thermal testing of a group of workers occupationally exposed to hand vibration and a control group not exposed. *(Source: Koradeck, 1977, fig. 13.)*

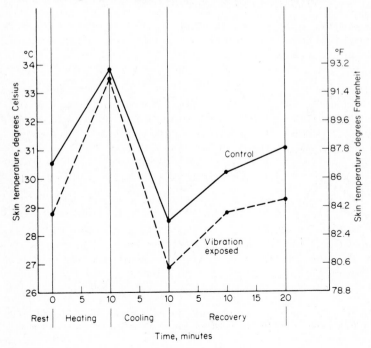

cent of the foundry workers and 19 percent of the shipyard workers. An additional 36 and 45 percent of the workers, respectively, showed less advanced symptoms (tingling and numbness). Not everyone exposed to hand-arm vibration develops VWF; some people, however, can show severe symptoms in less than a year. In the Wasserman et al. study, severe symptoms were found in 31 percent of the foundry workers exposed for 1.5 yr or less, in 41 percent of the workers exposed for 1.5 to 3 yr, and in 71 percent of the workers exposed more than 3 yr.

Vibration induced from hand-held tools has also been implicated in numerous other diseases, including neuritis, decalcification and cysts of the radial and ulnar bones, and pain and stiffness in the joints (see Wasserman et al., 1982, for a review).

A QUESTION OF STANDARDS

A number of proposals recommend safe exposure limits to hand-transmitted vibration. Figure 11–14 shows six such proposals for maximum vibration levels for continuous daily exposure. Clearly, the variation between them is so great that any one of them should be used with caution. The basis of these proposals is almost entirely the subjective assessment of what vibration levels are "noticeable," "unpleasant," etc.

FIGURE 11-14
Maximum vibration levels for continuous daily exposure: (1) Czechoslovakian Hygiene Regulation No. 33; (2) Teisinger and Louda, Czechoslovakia 1966; (3) USSR Hygiene Regulation 1955; (4) USSR Gataninas Proposal after 1955; (5) USSR Hygiene Regulation 1966; (6) Draft British Standard 1975 400 min/d. (*Source: Hempstock & O'Connor, 1977, fig. 1.*)

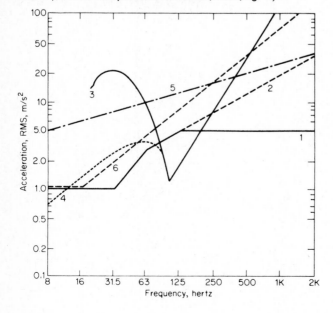

The International Standards Organization (ISO) has developed a draft standard setting forth methods for measuring hand-arm vibration and recommending exposure limits (ISO/DIS 5349–1982). Unfortunately, there are only limited medical and epidemiological data supporting the recommended limits and, in addition, accuracy of vibration measurement is still a problem. We should, therefore, probably accept the standard for what it is—a tentative recommendation offering provisional guidelines rather than definite damage risk criteria. The United States and several other countries have yet to accept the standard (NIOSH, 1983).

In 1971 the National Swedish Board of Occupational Safety and Health set maximum permissible vibration levels for chain saws sold in Sweden. The maximum permissible levels were decreased each year from a maximum 80 N in 1971 to 50 N in 1973.[2] Actually manufacturers were able to reduce the levels below 30 N by improved design and use of vibration-damping rubber elements between the body of the saw and the handles (Axelsson, 1977). The percentage of chain saw operators in northern Sweden with VWF decreased from 49 percent in 1967 to 28 percent in 1974. Of the 28 percent in 1974, some 19 percent showed significant improvement in the severity of VWF (Axelsson). Although other factors, such as better heating of workers' cabins and warmer working clothes, probably contributed to this reduction, Axelsson and others (Taylor, Pearson, and Keighley, 1977) however, believe that the reduction in the vibration level of the saws was the principal cause for the reduced incidence of VWF.

GLOVES

A chapter on hand tools and devices would not be complete without at least a brief discussion of gloves. Gloves are often used in conjunction with hand tools for protection against abrasions, cuts, punctures, and temperature extremes. Gloves come in an amazing number of varieties. In general, however, they can be distinguished in terms of the material used for construction (e.g., cotton, leather, vinyl, neoprene, asbestos, and even metal), the cut (i.e., gunn cut and clute cut, as shown in Figure 11–15, the design of the thumb (i.e., straight or wing, as shown in Figure 11–15, and the type of wristband (e.g., knit, band top, gauntlet, or extended length).

Effect on Manual Performance

All the research on the effects of wearing gloves on manual performance have concentrated on comparing gloves of different materials. No studies exist which evaluate glove design independent of the material. We do not know, therefore, whether there is any difference in performance while gunn or clute cut gloves are worn.

Research on the effects of various glove materials on manual performance indicates that the nature of the task influences the relative superiority of the various materials. On some tasks there will be no differences in performance when different gloves are worn, while on other tasks large differences will be found. In general, however, performance on tasks requiring fine motor control and tactile feedback will

---

[2]N, the newton, is a unit of force.

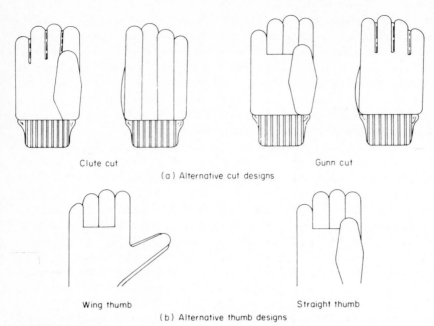

Clute cut   Gunn cut

(a) Alternative cut designs

Wing thumb   Straight thumb

(b) Alternative thumb designs

FIGURE 11-15
Common styles of work gloves distinguished by cut and thumb design.

be adversely affected by wearing gloves, as compared to bare-handed performance. An exception was noted in Chapter 6 wherein surface irregularities can be detected more accurately with thin cloth gloves than with bare fingers.

A study by Weidman (1970) is a good example of research in this area. Weidman compared four gloves [leather with canvas back, heavy-duty terry cloth, neoprene, and polyvinyl chloride (PVC)] and a bare-hand condition on performance of a maintenance-type task. The entire task was divided into five subtasks:

1 Using a key on a key ring to open a lock to get into a box
2 Replacing 16 television tubes by hand
3 Using a socket wrench and open-end wrench to remove 20 bolts
4 Using a screwdriver to remove four screws and replace two tubes
5 Using a trigger-type solder gun to remove two leads

The results are shown in Figure 11–16 as percentage increase in task completion times compared to the bare-hand condition. For subtasks 1 and 5 there was no statistically significant decrement, or differential decrement, caused by any type of glove. This was caused by the overall short time required to complete these subtasks (less than 50 s). For subtasks 2, 3, and 4 and total time, the results showed that performance did not differ between neoprene and bare-hand conditions. Leather, terry cloth, and PVC, however, were consistently inferior to both neoprene and bare hands, but did not differ from each other.

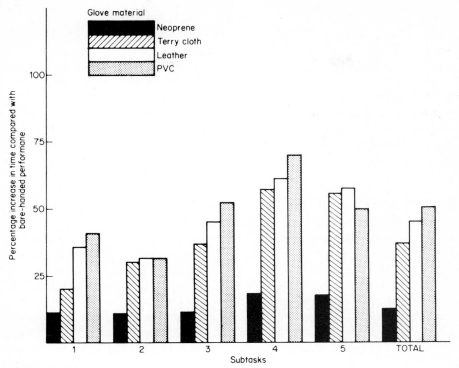

FIGURE 11-16
Performance (time to complete) on a maintenance-type task while wearing gloves
constructed of five different materials. Performance is shown as percentage increase
compared to barehanded performance. (See text for a description of the subtasks.)
(Source: Weidman, 1970.)

Other Considerations

Tichauer (1978) indicates that if a work glove is too thick between the fingers,
it may cause a worker to grasp a tool with insufficient force, resulting in the
tool slipping out of the hand. In addition, irritants entrapped unintentionally in the
glove may work into the skin, producing in some cases benign tumors. Chemicals might be soaked up in the glove material and cause dermatitis or other tissue
damage.

These potential hazards speak to a need for careful evaluation of the work environment and the selection of suitable gloves. Knitted wristbands can reduce the intrusion of potential irritants. Neoprene- or PVC-coated gloves can reduce absorption of
chemicals. The selection of proper glove material must also take into account the
type of mechanical hazards (cuts, punctures, abrasions) and thermal hazards likely to
be encountered on the job. Some materials are better suited to handle certain hazards
than are others (Coletta et al., 1976). Gloves must be given as careful thought as the
design or selection of the hand tools.

EVALUATION OF ALTERNATIVE DESIGNS

In recent years there has been an increase in the application of human factors to hand tool and device design. In this section we present a few examples. The emphasis is not on the innovation per se, but on the methodology used to develop and evaluate the new designs. In Chapter 2 we discussed laboratory and real-life research as well as types of criteria that can be used to measure possible effects of new designs. In several of the examples to follow, combinations of laboratory and real-life research methods were employed to provide a comprehensive evaluation of alternative designs. We illustrate the kinds of information sought and the insights provided by the various methodologies employed.

Toothbrush Design

Guilfoyle (1977) provides an informative description of the human factors development and evaluation of what ultimately became the Reach toothbrush shown in Figure 11–17. Although maybe not the most sophisticated of tools, the toothbrush is probably one of the most widely used hand-held tools in the world. Since the first bone shaft and hog-hair bristle toothbrush in 1780, the only major development had been the introduction of nylon filament bristles in the 1930s.

DuPont approached Applied Ergonomics, Inc., to design an improved toothbrush. The first step was to review relevant literature and conduct a series of consultations with dentists. From this preliminary investigation they discovered that no human factors research had been applied to toothbrush design and that the primary objective of a toothbrush was removal of plaque from the teeth. A secondary objective was gingival (gum) massage. Since there was little available literature on the public's dental care habits, a questionnaire was distributed to 300 adults. The information obtained helped the designers focus on basic dental care problems.

After analyzing the results of the questionnaire, the design team began collecting basic physical dimensions on existing toothbrushes and on the anthropometric characteristics of consumers (measurements of hands, teeth, and mouths). To obtain more information on how users handle a toothbrush, a series of time-motion film studies were made of people brushing their teeth. The design team studies how much time was devoted to brushing different mouth areas and the stroke directions used. The team also examined the way people manipulated the brush.

The time-motion studies made it clear that a contoured grip was undesirable because it hindered hand movements. A round handle was briefly considered but was dropped because it would not resist rotational forces. Another reason the round handle was dropped illustrates the need to consider the entire system into which a tool or device will fit. In this case, the diameter of the handle would have to be small enough to fit into the standard bathroom toothbrush holder. To do so would have made the handle too weak.

In addition to the time-motion studies, other laboratory studies related brushing time and bristle diameter to plaque removal. Based on the information obtained from all the varied sources, two prototype toothbrushes were designed and a sufficient number were manufactured for testing. Both prototypes had individual characteristics that required testing.

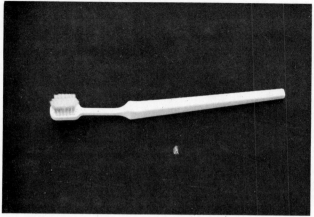

FIGURE 11-17
Reach toothbrush designed by Applied Ergonomics, Inc., and manufactured by Johnson and Johnson. Innovative features include a small bristle head and an angled (12°) and shaped handle with contoured thumb rest area.

These two prototypes were then tested against two other commercially available toothbrushes for removing plaque. Both prototypes proved superior. Interviews with the test subjects indicated a preference for the bristle head of one prototype and the handle design of the other. The features were blended into the final product. Plaque removal tests with the final design demonstrated superior removal compared to other conventionally designed brushes. Some of the features incorporated into the Reach toothbrush include a small bilevel bristle head to concentrate brushing on a small area, an angled (12°) and shaped handle for easier manipulation, and a contoured thumb area which makes brushing easier. (As an aside, the authors have both tried the Reach toothbrush and have found that it does take a little practice to learn to use it effectively.)

This design and evaluation process illustrates the use of past literature, expert (dentists') opinions, questionaire data, objective laboratory studies (time-motion and plaque-removal studies), and user opinion data to arrive at an improved tool design.

Multiple-Function Dental Syringe

Those of us who perhaps have not properly used our toothbrushes may have been on the receiving end of this tool. When a dentist fills a cavity, the dental assistant often works alongside, evacuating (with a suction vacuum) tooth debris being drilled by the dentist as well as providing air, water, and an air-water spray. Currently two instruments, a three-way air-water-spray syringe, and a separate suction device are used, as shown in Figure 11–18a. Evans et al. (1973) reasoned that reducing the number of instruments handled by the dental assistant by combining the suction and three-way syringe into a single four-way device would (1) free a second hand to

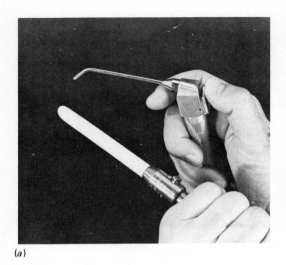

(a)

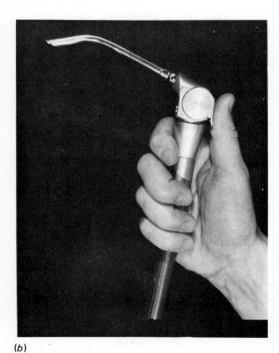

(b)

FIGURE 11-18
(a) The conventional three-way syringe (air, water, and spray) with separate suction used in oral dentistry. (b) The four-way (air, water, spray, and suction) device designed and evaluated by Evans et al. (*Source: Evans, Lucaccini, Hazell, & Lucas, 1973, figs. 1 & 3.*)

assist the dentist, (2) reduce the number of instruments in the mouth, (3) increase visibility, and (4) reduce patient discomfort.

The four-way device is shown in Figure 11–18*b*. The air and water streams are each operated by depressing small spring-loaded levers with the thumb. Spray is achieved by activating both levers simultaneously. Suction is controlled by rotating thumb wheels located to the sides of the levers. The thumb wheels are joined by a common shaft, providing duplicate control functions to the operator, thus accommodating left-handed as well as right-handed operators—a plus. The tip has a 45° angle bend and rotates to provide access to difficult-to-reach areas of the mouth.

The evaluation of the new instrument consisted of mechanical performance tests, mock clinical tests (simulated operations), and actual field tests (Evans et al., 1973). We briefly review each to illustrate the criteria used and the insights gained.

The mechanical performance tests were conducted to determine whether the device satisfied basic requirements for air, water, and suction pressure.

The mock (or simulated) clinical trials consisted of conducting restorative dental operations on a dental manikin instead of a patient. The purpose was to be sure that the instrument would function satisfactorily and safely before the live field tests were conducted. In addition, the reliability of the device under sustained use was evaluated.

Field tests were conducted to compare the three-way plus suction system with the new four-way device by actual dentists and dental assistants working on patients. At each of five locations, four days of training in the use of the new device were given. On the fifth day one operation was done with the conventional system, and one operation was performed by using the four-way device. Videotape and motion picture records were taken during each operation. The number of hand movements was counted, time to complete the cutting portion of the operation was recorded, energy costs for the dental assistant were calculated from samples of expired air, and finally operator preferences and suggestions were obtained. Here we see again the use of multiple criteria for evaluation and again the inclusion of subjective user-preference data. The results are shown in Table 11–2.

The interviews with dentists and dental assistants showed a highly favorable attitude toward the new device, but several suggestions for redesign were made including adding a plastic cover to the tip end to protect against chipping tooth enamel; reducing the size and length of the handle to increase ease of handling (especially for female users); replacing the on-off action control levers with variable controls; changing the direction of lever activation from down to up to conform to thumb motion; and separating and shape-coding the control levers to minimize inadvertent activation. These kinds of suggestions attest to the value of user inputs in system design and evaluation.

TABLE 11–2
COMPARISON OF TWO MULTIPLE FUNCTION DENTAL SYRINGES

Criteria	Four-way	Three-way plus suction
Hand movements	42	65
Cutting time, s	84	91
Energy expenditure, kcal/(min·kg)	0.023	0.023

Source: Evans et al., 1973.

Writing Instruments

Kao, in a series of studies, evaluated various types of writing instruments (1976, 1979) and alternative designs (1977). In two studies, ball-point pens, fountain pens, and pencils were compared on a number of criteria. An attitudinal survey of users (Kao,1976) showed that ball-point pens were most preferred, followed (in order) by pencils, and finally fountain pens. The survey addressed several subjective perceptions of the users including legibility, writing ease, and writing speed. In 1979, Kao followed up his subjective evaluation with laboratory tests. The dependent variables were objective measures of writing speed and writing pressure. He included felt-tip pens in the evaluation.

The results are important both for practical reasons (e.g., whether to invest $10 in a fountain pen or to pick up a $1.98 ball-point pen) and as an illustration of trade-offs that we often discover when evaluating alternative designs using multiple criteria. The results are shown in Figure 11–19. The time measure represents the time required to write 10 lowercase a's. The results indicate that for faster writing time the ball-point pen is the best instrument and the fountain pen the worst. On the other hand, for more comfortable and less fatiguing handwriting performance, the felt-tip pen is clearly the best and the ball-point is now the worst. So here we have a trade-off: speed versus fatigue. Given the relatively small differences in time between the instruments and large differences in pressure, Kao (1979) concludes that probably the felt-tip pen is the best all-round type of instrument for handwriting.

Some people might think that finding trade-offs among the designs being evaluated would make the evaluator's life more difficult—and it does. The consolation is knowing that a more complete understanding of the pros and cons of each design was achieved by using multiple criteria. If Kao had assessed only writing time, he might have erroneously concluded that the ballpoint pen was the most efficient instrument.

Kodak Disc Camera

The Eastman Kodak Company introduced the disc camera system in 1982, and the Human Factors Department of Kodak was a part of the product development program from the start. More than 50 experiments were conducted that utilized over 2000 subjects and assessed all aspects of the design and operation of the camera (Faulkner, 1982). The initial efforts centered on basic questions of customer needs, understanding of the circumstances under which people take pictures, and the typical problems people encounter. In one phase of the program, 30,000 prints taken by amateur photographers were examined to determine, among other things, the picture-taking frequency as a function of camera-to-subject distance and ambient light level (Faulkner and Rice, 1982). In another phase various camera configurations were tested for holdability (Caplan, 1982). The cameras differed in terms of the location of the shutter release, whether a cover and/or handle was provided, and other factors. Subjects simulated taking pictures with the various camera configurations, and the frequency with which they covered the lens or flash with their fingers was recorded. The worst configuration would have been expected to result in 14 times the number

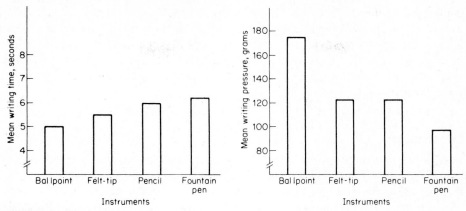

FIGURE 11-19
Comparison of writing speed and writing pressure with four writing instruments. Time represents the time required to write 10 lowercase a's. *(Source: Kao, 1979, figs. 1 & 2.)*

of degraded pictures as the best configuration owing to fingers blocking the lens and flash.

In a series of three experiments, subjects' preferences for various camera configurations were assessed (Faulkner, Rice, and Heron, 1983). The overall ratings from one of the studies are shown, along with the camera configurations tested, in Figure 11–20. Using various statistical procedures, the investigators identified three relatively distinct clusters among their subjects. The three subject clusters produced somewhat different ratings of the various configurations. One cluster of subjects appeared to be more concerned with superficial features such as pop-up flashes and lens covers. Another cluster seemed more concerned with ease of holding the camera in terms of comfort and steadiness. The third cluster rated the cameras more in terms of the location of the shutter release. No one configuration completely satisfied all three groups.

The development of the Kodak disc camera is a good example of a systematic

FIGURE 11-20
Preference ratings for various configurations of disk cameras. *(Source: Adapted from Faulkner, Rice, & Heron, 1983. Copyright by the Human Factors Society, Inc. and reproduced by permission.)*

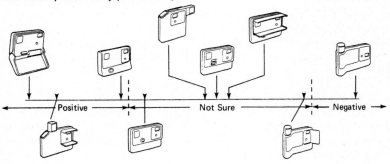

application of human factors to the design of a hand-held consumer product. Estimates, based on an analysis of more than 10,000 pictures taken with disc cameras, indicated that users can expect 25 percent more good to excellent pictures than users of typical cartridge-loading cameras (Faulkner, 1982). Much of this improvement is due to the incorporation of the results from the various human factors studies in the design of the disc camera.

DISCUSSION

In bygone centuries people used various tools and hand devices to accomplish certain objectives, such as making things or performing certain tasks, including clubbing their attackers. Various types of hand tools and other hand-held devices still serve—and will continue to serve—many purposes. In the future we can look forward to new hand tools designed for specialized applications. To be of real value, these new tools must be designed with full cognizance of the operational system (and its limitations and constraints) into which the device will fit. In addition, the designer must consider the human factors, ergonomic, and biomechanical aspects of the user-tool interface.

The evaluation of new tool designs will require creativity to devise methodologies for reliable and valid measurement of relevant criteria. Combinations of various research strategies, including laboratory, simulation, and real-life field testing, will be required. Although hand tools have been around for a million years, there is still considerable work to be done to adapt them optimally for human use.

REFERENCES

Armstrong, T. (1983). *Ergonomics guides: An ergonomics guide to carpal tunnel syndrome.* Akron, OH: American Industrial Hygiene Association.

Armstrong, T., Foulke, J., Joseph, B., and Goldstein, S. (1982). Investigation of cumulative trauma disorders in a poultry processing plant. *American Industrial Hygiene Association Journal,* 43(2), 103–116.

Axelsson, S. (1977). Progress in solving the problem of hand-arm vibration for chain saw operators in Sweden, 1967 to date. In D. Wasserman and W. Taylor (eds.), *Proceedings of the International Occupational Hand-Arm Vibration Conference* (no. 77–170). Cincinnati, OH: Department of Health, Education, and Welfare, National Institute for Occupational Safety and Health.

Ayoub, M., and Lo Presti, P. (1971). The determination of an optimum size cylindrical handle by use of electromyography. *Ergonomics,* 4(4), 503–518.

Ayoub, M., Purswell, J., and Hicks, J. (1977). Data collection for hand tool injury: An approach. In V. Pezoldt (ed.), *Rare event/accident research methodology.* Washington: National Bureau of Standards.

Barsley, M. (1970). *Left-handed man in a right-handed world.* London: Pitman.

Caplan, S. (1982). Designing new cameras for improved holdability. *Proceedings of the Human Factors Society 26th Annual Meeting.* Santa Monica, CA: Human Factors Society.

Coletta, G., Arons, I., Ashley, L., and Drennan, A. (1976). *The development of criteria for firefighter's gloves,* vol. 1: *Glove requirements* (no. 77–134–A). Cincinnati: Department of Health, Education, and Welfare, NIOSH.

Damon, F. (1965). The use of biomechanics in manufacturing operations. *The Western Electric Engineer,* 9(4).

Dionne, E. (1984, March). Carpal tunnel syndrome: Part I—The problem. *National Safety News,* pp. 42–45.

Ducharme, R. (1977). Women workers rate "male" tools inadequate. *Human Factors Society Bulletin,* 20(4), 1–2.

Eastman Kodak Co. (1983). *Ergonomic Design for People at Work,* vol. 1. Belmont, CA: Lifetime Learning Publications.

Emanuel, J., Mills, S., and Bennett, J. (1980). In search of a better handle. *Proceedings of the Symposium: Human Factors and Industrial Design in Consumer Products,* Medford, MA: Tufts University.

Evans, T., Lucaccini, L., Hazell, J., and Lucas, R. (1973). Evaluation of dental hand instruments. *Human Factors,* 15, 401–406.

Faulkner, T. (1982). Human factors in disc photography. *Third National Symposium on Human Factors and Industrial Design in Consumer Products.* Santa Monica, CA: Consumer Products Technical Group of the Human Factors Society, pp. 279–281.

Faulkner, T., and Rice, T. (1982). Human factors, photographic space, and disc photography. *Proceedings of the Human Factors Society, 26th Annual Meeting.* Santa Monica, CA: Human Factors Society, pp. 190–194.

Faulkner, T., Rice, T., and Heron, W. (1983). The influence of camera configuration on preference. *Human Factors,* 25, 127–141.

Fraser, T. (1980). *Ergonomic principles in the design of hand tools,* Occupational Safety and Health Series no. 44. Geneva, Switzerland: International Labour Office.

Greenberg, L., and Chaffin, D. (1977). *Workers and their tools.* Midland, MI: Pendell Publishing.

Guilfoyle, J. (1977). Look what design has done for the toothbrush. *Industrial Design,* 24, 34–38.

Hempstock, T., and O'Connor, D. (1977). Evaluation of human exposure to hand-transmitted vibration. In D. Wasserman and W. Taylor (eds.), *Proceedings of the International Occupational Hand-Arm Vibration Conference* (no. 77–170). Cincinnati: Department of Health, Education, and Welfare, NIOSH.

Kao, H. (1976). An analysis of user preference toward handwriting instruments. *Perceptual and Motor Skills,* 43, 522.

Kao, H. (1977). Ergonomics in penpoint design. *Acta Psychologica Taiwanica,* 18, 49–52.

Kao, H. (1979). Differential effects of writing instruments on handwriting performance. *Acta Psychologica Taiwanica,* 21, 9–13.

Knowlton, R., and Gilbert, J. (1983). Ulnar deviation and short-term strength reductions as affected by a curve-handled ripping hammer and a conventional claw hammer. *Ergonomics,* 26, 173–179.

Konz, S. (1979). *Work design.* Columbus, OH: Grid Inc.

Koradecka, D. (1977). Peripheral blood circulation under the influence of occupational exposure to hand-transmitted vibration. In D. Wasserman and W. Taylor (eds.), *Proceedings of the International Occupational Hand-Arm Vibration Conference* (no. 77–170). Cincinnati, OH: Department of Health, Education, and Welfare, NIOSH.

Krohn, R., and Konz, S. (1982). Bent hammer handles. *Proceedings of the Human Factors Society, 26th Annual Meeting.* Santa Monica, CA: Human Factors Society, pp. 413–417.

Laveson, J., and Meyer, R. (1976). Left out "lefties" in design. *Proceedings of the 20th Annual Meeting of the Human Factors Society.* Santa Monica, CA; Human Factors Society.

National Institute for Occupational Safety and Health (NIOSH). (1983). *Current intelligence bulletin 38: Vibration syndrome* (DHHS-NIOSH Publ. 83–110). Cincinnati.

Taylor, W. (1974). The vibration syndrome: Introduction. In W. Taylor (ed.), *The vibration syndrome*. New York: Academic.

Taylor, W., Pearson, J., and Keighley G. (1977). A longitudinal study of Raynaud's phenomenon in chain saw operators. In D. Wasserman and W. Taylor (eds.), *Proceedings of the Internatinal Occupational Hand-Arm Vibration Conference* (no. 77–170). Cincinnati: Department of Health, Education, and Welfare, NIOSH.

Terrell, R., and Purswell, J. (1976). The influence of forearm and wrist orientation on static grip strength as a design criterion for hand tools. *Proceedings of the 20th Annual Meeting of the Human Factors Society*. Santa Monica, CA: Human Factors Society.

Tichauer, E. (1966). Some aspects of stress on forearm and hand in industry. *Journal of Occupational Medicine,* 8(2), 63–71.

Tichauer, E. (1967). Ergonomics: The state of the art. *American Industrial Hygiene Association Journal,* 28, 105–116.

Tichauer, E. (1976, February). Biomechanics sustains occupational safety and health. *Industrial Engineering.* 8(2), 46–56.

Tichauer, E. (1978). *The biomechanical basis of ergonomics*. New York: Wiley.

Tichauer, E., and Gage, H. (1977). Ergonomic principles basic to hand tool design. *American Industrial Hygiene Association Journal,* 38, 622–634.

Walton, K. (1974). The pathology of Raynaud's phenomenon of occupational origin. In W. Taylor (ed.), *The vibration syndrome*. New York: Academic.

Washburn, S. (1960). Tools and human evolution. *Scientific American*, 203, 3–15.

Wasserman, D., Taylor, W., Behrens, V., Samueloff, S., and Reynolds, D. (1982). *Vibration white finger disease in U.S. workers using pneumatic chipping and grinding hand tools, I: Epidemiology* (DHHS-NIOSH Publ. 82–118). Cincinnati: NIOSH.

Weidman, B. (1970). *Effect of safety gloves on simulated work tasks* (AD 738981). Springfield, VA: National Technical Information Service.

PART FOUR

WORK SPACE AND ARRANGEMENT

CHAPTER 12

APPLIED ANTHROPOMETRY AND WORK SPACE

In our everyday experience we use all sorts of physical equipment and facilities, such as chairs, seats, tables, automated bank tellers, video games, and football helmets. In our use of such things we have found that their design features affect how suitable they are. Although the most obvious criterion by which we judge the design features of such things is comfort, there are other relevant criteria, such as those related to work performance, safety, and physical effects.

Some facilities people use and the methods of work that are followed have a major impact on posture. In this regard, the most important consequence of improper posture is with respect to spinal problems. As the spine is bent, there is increased pressure between the vertebrae. Although there are disks between the vertebrae that serve as cushions, these disks are subject to wear and tear (in medical terms, pathological degeneration), especially with continual improper posture. Grandjean et al. (1973) estimate that 50 percent of adults suffer backaches during at least one period of their lives, many such backaches being caused by damage to the disks. The back problems that can arise from improper posture argue for the proper design of seats and work stations and for the use of work methods that do not require improper posture.

ANTHROPOMETRY

Anthropometry deals with the measurement of the dimensions and certain other physical characteristics of the body; such measurements are, of course, relevant to the design of the things people use. There are two primary types: structural (static) and functional (dynamic) measurement. What is sometimes called *engineering anthropometry* is concerned with the application of both types of data to the design of the things people use. We briefly discuss structural and functional anthropometry before discussing how such data are used in the design of work spaces and equipment.

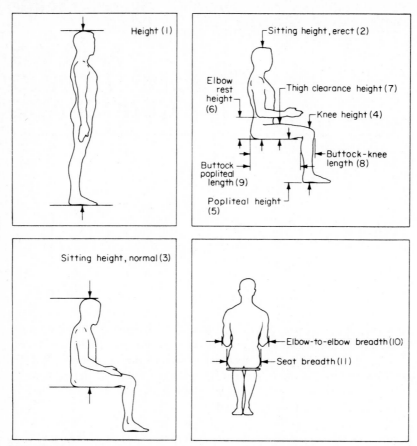

FIGURE 12-1
Diagrams of structural (static) body features measured in National Health Survey of anthropometric measurements of 6672 adults. See Table 12-1 for selected data based on the survey.

Structural (Static) Dimensions

Structural dimensions are measurements taken when the body is in a fixed (static) position. They consist of skeletal dimensions (between the centers of joints, such as between the elbow and the wrist) or of contour dimensions (skin-surface dimensions such as height or seat breadth). Many different body features can be, and have been, measured. The NASA *Anthropometric Source Book* (vol. 2, 1978), for example, illustrates 973 such measurements and presents data on certain of these measurements from 91 worldwide surveys. Many of these measurements, of course, have very specific applications, such as in designing helmets, earphones, or pince-nez glasses. However, measurements of certain body features have rather general utility, and summary data on some of these features are presented for illustrative purposes. These data come from a survey by the U.S. Public Health Service (1965) of a representative sample of 6672 adult males and females. The specific body features measured are shown in Figure 12–1; for each of these (plus weight), data on the 5th, 50th, and 95th percentiles are given in Table 12–1.

TABLE 12–1
SELECTED STRUCTURAL BODY DIMENSIONS AND WEIGHTS OF ADULTS

Body feature	Dimension, in						Dimension, cm*					
	Male, percentile			Female, percentile			Male, percentile			Female, percentile		
	5th	50th	95th	5th	50th	95th	5th	50th	95th	5th	50th	95th
1 Height	63.6	68.3	72.8	59.0	62.9	67.1	162	173	185	150	160	170
2 Sitting height, erect	33.2	35.7	38.0	30.9	33.4	35.7	84	91	97	79	85	91
3 Sitting height, normal	31.6	34.1	36.6	29.6	32.3	34.7	80	87	93	75	82	88
4 Knee height	19.3	21.4	23.4	17.9	19.6	21.5	49	54	59	46	50	55
5 Popliteal height	15.5	17.3	19.3	14.0	15.7	17.5	39	44	49	36	40	45
6 Elbow-rest height	7.4	9.5	11.6	7.1	9.2	11.0	19	24	30	18	23	28
7 Thigh-clearance height	4.3	5.7	6.9	4.1	5.4	6.9	11	15	18	10	14	18
8 Buttock-knee length	21.3	23.3	25.2	20.4	22.4	24.6	54	59	64	52	57	63
9 Buttock-popliteal length	17.3	19.5	21.6	17.0	18.9	21.0	44	50	55	43	48	53
10 Elbow-to-elbow breadth	13.7	16.5	19.9	12.3	15.1	19.3	35	42	51	31	38	49
11 Seat breadth	12.2	14.0	15.9	12.3	14.3	17.1	31	36	40	31	36	43
12 Weight†	120	166	217	104	137	199	58	75	98	47	62	90

†Weight give in pounds (first six columns) and kilograms (last six columns.).
*Centimeter values rounded to whole numbers.
Source: U.S. Public Health Service, 1965.

Body measurements vary as a function of age and sex and for different ethnic populations. In connection with age, for example, stature and related dimensions generally increase until the late teens or early twenties, remain relatively constant throughout early adulthood, and decline from early to middle adulthood into old age (Stoudt, 1981). (An exception is the length of the ears, which continue to grow throughout life.) Variations in population groups are illustrated in Figure 12–2, in particular the range of stature from the 5th to the 95th percentile for each group.

Functional (Dynamic) Dimensions

These dimensions are taken under conditions in which the body is engaged in some physical activity. In most physical activities (whether one is operating a steering

FIGURE 12-2
Range of variability (5th to 95th percentile) in stature of selected populations.
[*Source: Adapted from NASA Anthropometric source book (vol. I), fig. 11, p. 11-40.*]

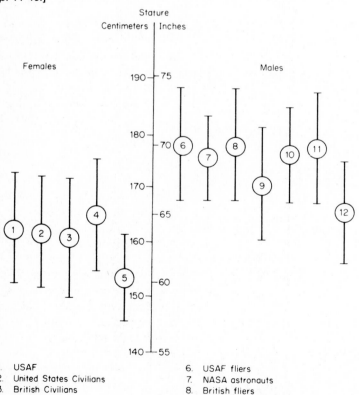

1.	USAF
2.	United States Civilians
3.	British Civilians
4.	Swedish civilians
5.	Japanese civilians

6.	USAF fliers
7.	NASA astronauts
8.	British fliers
9.	Italian military
10.	French fliers
11.	German Air Force
12.	Japanese civilians

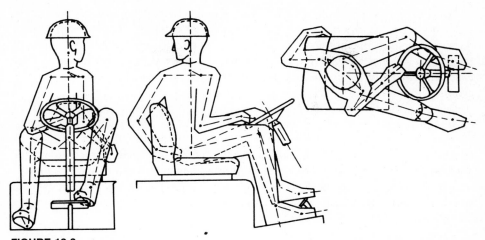

FIGURE 12-3
Three views of an operator of a forklift truck that illustrate the interactions of the body members.
Such views represent what is sometimes called somatography. (From North, 1980.)

wheel, assembling a mousetrap, or reaching across the table for the salt) the individual body members function in concert. The practical limit of arm reach, for example, is not the sole consequence of arm length; the limit is also affected by shoulder movement, partial trunk rotation, possible bending of the back, and the function to be performed by the hand. Figure 12–3 gives some impression of the possible interactions of various body members of a forklift truck operator. This figure is an illustration of what is sometimes called *somatography* in that it shows the three views of the operator: the frontal, side, and vertical.

USE OF ANTHROPOMETRIC DATA

Both structural (static) and functional (dynamic) anthropometric data have applications in the design of facilities and things people use. However, the design of many work situations should take into account the interaction of body members and thus should be based in part on functional data. In the use of anthropometric data for designing something, the data should be reasonably representative of the population that would use the item. In many instances the population of interest consists of "people at large," implying that the design features must accommodate a broad spectrum of people. When items are designed for specific groups (such as adult females, children, the elderly, football players, the handicapped, etc.), the data used should be specific for such groups in the country or culture in question. (However, there are many specific groups for which appropriate data are not yet available.)

Principles in the Application of Anthropometric Data

There are three general principles for applying anthropometric data to specific design problems; each applies to a different type of situation.

Design for Extreme Individuals   In designing certain features of our built physical world, one should try to accommodate all (or virtually all) the population in question. In some circumstances a specific design dimension or feature is a limiting factor that might restrict the use of the facility for some people; that limiting factor can dictate either a *maximum* or *minimum* value of the population variable or characteristic in question.

Designing for the *maximum* population value is the appropriate strategy if a given maximum (high) value of some design feature should accommodate all (or virtually all) people. Examples include heights of doorways, sizes of escape hatches on military aircraft, and strength of supporting devices (such as a trapeze, rope ladder, or workbench). In turn, designing for the *minimum* population value is the appropriate strategy if a given minimum (low) value of some design feature has to accommodate all (or virtually all) people. Examples include the distance of a control button from the operator and the force required to operate the control.

Sometimes there may be reasons for accommodating most, but not 100 percent, of the population. For example, it is not reasonable to have all doorways 9 ft (2.7 m) high to accommodate circus giants. Thus, it frequently is the practice to use the 95th and 5th percentiles of the distributions of relevant population characteristics as the maximum and minimum design parameters.

Designing for Adjustable Range   Certain features of equipment or facilities can be designed so they can be adjusted to the individuals who use them. Some examples are automobile seats, office chairs, desk heights, and footrests. In the design of such equipment, it frequently is the practice to provide for adjustments to cover the range from the 5th to the 95th percentile of the relevant population characteristic (sitting height, arm reach, etc.). The use of such a range is especially relevant if there could be technical problems in trying to accommodate the very extreme cases to cover 100 percent of the population); frequently the technical problems involved in accommodating the extreme cases are disproportionate to the advantages gained in doing so.

Designing for the Average   The human factors literature is strewn with pronouncements that there are very few (or no!!) individuals who are "average." Indeed, there are few, if any, individuals who are average on each of many anthropometric characteristics. Although the myth of the average person has been used as an argument against designing for such a will-o'-the-wisp concept, we would like to make a case here for the use of *average* values in the design of *certain* types of equipment or facilities, specifically those for which, for legitimate reasons, it is not appropriate to pitch the design at an extreme value (minimum or maximum) or feasible to provide for an adjustable range. As an example, the checkout counter of a supermarket built for the average customer probably would discommode customers less in general than one built either for the circus midget or for Goliath. This is not to say that it would be optimum for all people, but that, collectively, it would cause less inconvenience and difficulty than one which might be lower or higher.

Discussion of Anthropometric Design Principles   The discussion of the above principles refers to the application of anthropometric data for single dimensions (such

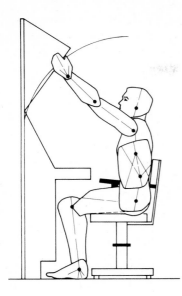

FIGURE 12-4
An articulated anthropometric scale model, such as used
in the design of work spaces. *(Source: Meyer, 1979,
fig. 5.)*

as height or arm reach). The design problem becomes more sticky when one needs
to take into account combinations of several dimensions. For example, the setting of
limits such as the 95th and 5th percentiles on each of several dimensions can eliminate a fairly high percentage of the population. For instance, Bittner (1974) found in
one situation the 95th and 5th percentile limits on each of 13 dimensions would
exclude 52 percent of the population instead of the 10 percent implied by the 95th
and 5th percentile limits of the individual dimensions. This occurs because body
dimensions are not perfectly correlated with each other. For example, people with
short arms do not necessarily have short legs. Not all the people excluded because
they are not within the 5th to 95th percentile range on one variable will be the same
people who are excluded on another measure. It is important, therefore, to consider
the relationships (correlations) between body dimensions in the design of things
based on combinations of dimensions.

In the application of anthropometric data, it is sometimes the practice to use physical models, such as the articulated model shown in Figure 12–4. Such models usually represent a specific percentile of the population. In addition, there are computer
programs available that assess the adequacy of tentative work-space designs in terms
of anthropometric considerations. We discuss some of these in Chapter 18.

It must be granted that tables of anthropometric data do not make for enthralling
bedtime reading (although they might help those who have insomnia). But when we
consider the application of such data to the design of the physical facilities and objects we use, we can see that such intrinsically uninteresting data play a live, active
role in the real, dynamic world in which we live and work.

In the application of anthropometric data to specific design problems, there can be
no nicely honed set of procedures to follow, because of the variations in the circumstances in question and in the types of individuals for whom the facilities would be
designed. As a general approach, however, the following suggestions are offered:

1 Determine the body dimensions important in the design (e.g., sitting height as a basic factor in seat-to-roof dimensions in automobiles).

2 Define the population to use the equipment or facilities. This establishes the dimensional range that needs to be considered (e.g., children, women, U.S. civilians, different age groups, world populations, different races).

3 Determine what principle should be applied (e.g., design for extreme individuals, for an adjustable range, or for the average).

4 When relevant, select the percentage of the population to be accommodated (for example, 90 percent, 95 percent) or whatever is relevant to the problem.

5 Locate anthropometric tables appropriate for the population, and extract relevant values.

6 If special clothing is to be worn, add appropriate allowances (some of which are available in the anthropometric literature).

WORK SPACES

There are many possible uses for anthropometric data in designing things for people. In this text we can illustrate only a few such uses. One of the most important is in the design of work spaces, including what are sometimes called *work-space envelopes*. (Remember that there are some existing work spaces that one cannot do anything about, such as the space under the kitchen sink where a plumber might have to work when fixing a stopped-up pipe.)

Work-Space Envelopes for Seated Personnel

The limits of the work-space envelope for seated personnel are determined by functional arm reach, which in turn is influenced especially by the direction of arm reach and the nature of the manual activity (i.e., the task or function) to be performed. Functional arm reach is also influenced by such factors as the presence of any restraints and by the apparel worn. Some examples of the effects of these variables are given for illustration.

Effects of Direction of Reach and Presence of Restraints on Work-Space Envelope The effect of these variables on the work-space envelope is illustrated by the results of a study by Roth, Ayoub, and Halcomb (1977). They measured the functional arm reach of subjects at various lateral angles from a dead-ahead seated position (from $-45°$ left to $+120°$ right) and at various levels (ranging from $-60°$ to $+90°$ from a seat reference point (SRP). Measurements were taken of the ''grip-center'' reach point at 114 such locations, under both restrained and unrestrained conditions. In the restrained conditions the shoulders of the subjects were held back against the seat back, whereas in the unrestrained condition the subjects could move their shoulders. The subjects were selected to be reasonably representative of adults in terms of height and weight.

Some of the results are shown in Figure 12–5, in particular the envelope for the 5th percentiles of males and females for each condition. Each line represents the 5th percentile value at specified levels above a seat reference point. (Data are given for only certain levels.) On the basis of the data for either sex, one could envision a three-dimensional space the outer surface of which would include the values repre-

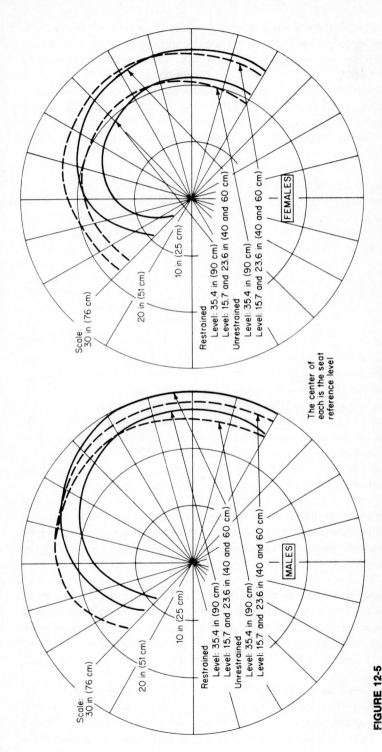

FIGURE 12-5
Functional arm reach for the 5th percentiles of males and females at specified levels above a seat reference point. The three-dimensional space represented by these data can be considered as forming a work-space envelope that would accommodate 95 percent of the population. (*Source: Roth, Ayoub, and Halcomb, 1977, for description of procedures used in obtaining data. Appreciation is expressed to Dr. M. M. Ayoub and his associates for special tabulations used in the preparation of this figure.*)

sented by the data in the figure. Such a three-dimensional space—a work-space envelope—would then represent the limits of convenient arm reach for the 5th percentile of the population, but such limits would, of course, accommodate 95 percent of the population.

As one might expect, the type of restraint can influence the functional arm reach. Garg, Bakken, and Saxena (1982), for example, in comparing three types of vehicle seat restraints, report a lap belt as permitting the farthest functional arm reach, with a crossed harness reducing average reach by 14 percent and a parallel harness reducing it by 24 percent.

Effects of Manual Activity on Work-Space Envelope As stated above, the nature of the manual activity to be performed influences the boundaries of the work-space envelope. For example, if an individual simply has to activate pushbuttons or toggle switches, a "fingertip" measurement is appropriate, as contrasted with the requirement to use knobs to grasp levers, for which a "thumb tip" measurement is used. (According to data summarized by Bullock, 1974, such measurements are about 5 or 6 cm or a couple of inches shorter than fingertip measurements.) In turn, a hand-grasp or griplike action (such as the grip-center measurement used by Roth, Ayoub, and Halcomb, 1977) limits the reach further (by 5 cm or more, according to Bullock).

Even different hand-grasp actions influence the space envelope, as demonstrated years ago by the classic study by Dempster (1955), who had male subjects use eight different hand grasps in an anthropometric study. (These involved grasping a handle-like device with the hand in one of eight fixed orientations, namely, supine, prone, inverted, and at five spatial angles.) Photographic traces of contours of the hand were made as the hand moved over a series of frontal planes spaced at 6-in (15-cm) intervals. From these a *kinetosphere* was developed for each grasp, showing the mean contours of the tracings as photographed from each of three angles—top (transverse), front (coronal), and side (sagittal). Although the kinetospheres for the several types of grip are not illustrated, they were substantially different. These were, however, combined to form *strophospheres* as shown in Figure 12–6. The shaded areas define the region that is common to the hand motions made with the various hand grips and therefore indicate those three-dimensional spaces within the work-space envelope within which the various types of hand grips in question could most adequately be executed by people.

Effects of Apparel on Work-Space Envelope The apparel worn by people can also restrict the movements of people and the distances they can reach and therefore can influence the size of the work-space envelope. In a survey of truck and bus drivers, for example, Sanders (1977) found that winter jackets restricted reach by approximately 2 in (5 cm).

Discussion of Work-Space Envelopes Work-space envelopes consist of the three-dimensional spaces that are reasonably optimum for seated persons who perform some type of manual activity. Thus (for example) control devices and other objects to be used usually should be located within such space. The reasonable limits

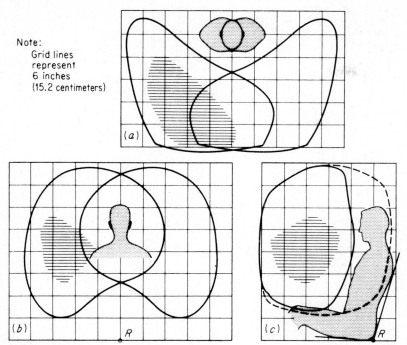

Note:
Grid lines
represent
6 inches
(15.2 centimeters)

FIGURE 12-6
Strophosphere resulting from superimposition of kinetospheres of range of hand
movements with a number of hand-grasp positions in three-dimensional space.
The shaded areas depict the region common to all hand motions (prone, supine,
inverted, and several different angles of grasps), probably the optimum region,
collectively, of the different types of hand manipulations. *(Source: Adapted from
Dempster, 1955.)*

of such space are determined by functional arm reach, which is influenced by such
variables as direction of arm reach, the nature of the manual activity, the use of
restraints, apparel worn, the angle of the backrest, and personal variables such as
age, sex, ethnic group, and handicaps. Whenever feasible, such spaces should be
designed with consideration for the personal characteristics of the population to use
the facility. It is fairly standard practice to design such space for the 5th percentile
of the using population, thus making it suitable for 95 percent of the population.

In connection with work-space envelopes for special populations, Rozler (1977),
for example, reports that with amputees who have a below-elbow prosthesis there is
an average decrease in usable workspace of 45 percent, and with an above-elbow
prosthesis the average decrease is 83 percent. Thus, with such special populations
the design of the work space requires particular (and sometimes individual) attention.

Minimum Requirements for Special Situations

People sometimes have to work in, move through, or even just fit into some restricted
or awkward spaces. For illustrative purposes examples of the clearances required for
certain types of work situations are given in Figure 12–7. Note that the dimensions

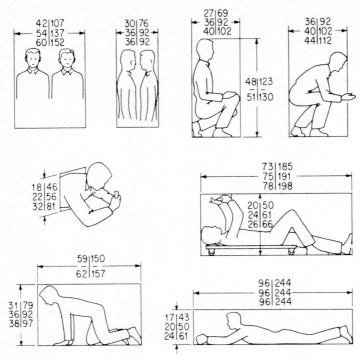

FIGURE 12-7
Clearances for certain work spaces that individuals may be required to work in or pass through. *Note:* The three dimensions given (inches at left, centimeters at right) are (from top to bottom in each case) minimum, best (with normal clothing), and with heavy clothing (such as arctic). *(Source: Adapted from Rigby, Cooper, and Spickard, 1961.)*

given include those for persons with heavy clothing. In most cases such clothing adds 4 to 6 in (10 to 15 cm) to the requirements and in the case of escape hatches 10 in (25 cm).

The work-space requirements of people can be stretched to include some unusual circumstances, such as the "nonwork" space needed by long-haul truck drivers, many of whom drive in pairs so that one can sleep while the other drives. The Federal Motor Carrier Safety Regulations presently specify the minimum space for sleeper berths in trucks. (Some such berths are sort of shelves behind the driver.) An investigation of the dimensions of the sleeping postures of 239 drivers was carried out by Sanders (1980). He reports the following dimensions of the preferred and

Sleeper berths	Length	Width
Preferred position	1.98 m (78 in)	0.84 m (33 in)
Prostrate position	2.04 m (80 in)	0.86 m (34 in)
Legal specifications	1.90 m (75 in)	0.61 m (24 in)

prostrate (face-down) postures of the 95th percentile drivers. The legal specifications are also given.

We can see that the postures of the subjects would not fit within the federal specifications, especially for the width of the berth. Such data can then serve as the basis for setting legal standards as well as for the more general objectives of designing for human use.

DESIGN OF WORK SURFACES

Within the three-dimensional envelope of a workplace, specific design decisions need to be made about various features of the workplace, including the location and design of whatever work surfaces are involved, such as benches, desks, tables, and control panels. This is not the place to provide guidelines for designing the myriad work surfaces that exist in this world, but some samples may be helpful to illustrate how anthropometric and biomechanical data can be used.

Horizontal Work Surfaces

The horizontal work surfaces to be used by seated and "sit-stand" workers generally should provide for manual activities to be within convenient arm reach. Certain *normal* and *maximum* areas were proposed by Barnes (1963) and earlier by Farley (1955) and have been used rather widely. These areas as defined by Barnes are shown in Figure 12–8 and are described as follows:

1 *Normal area*. This is the area that can be conveniently reached with a sweep of the forearm while the upper arm hangs in a natural position at the side.

2 *Maximum area*. This is the area that could be reached by extending the arm from the shoulder.

Related investigations by Squires (1956), however, have served as the basis for proposing a somewhat different work-surface contour that takes into account the dynamic interaction of the movement of the forearm as the elbow also is moving. It is for this reason that Das and Grady (1983b) recommend it over that proposed by Farley or Barnes. The area proposed by Squires is superimposed over the area proposed by Barnes in Figure 12–8. It is a bit more shallow that that proposed by Barnes but covers more area than that proposed by Farley. An additional argument in favor of Squires' model is that it requires less forward extension of the forearm, thus minimizing the stress on the elbow joint (Tichauer, 1978).

Although most office activities such as reading and writing are carried out on horizontal surfaces such as desks and tables, Eastman and Kamon (1976) propose that (where feasible) a slanted surface be used. In their study they found that subjects using slanted surfaces (12° and 24°) had better posture, showed less trunk movement, and reported less fatigue and less discomfort than when using horizontal surfaces. Drafting usually is performed on slanted surfaces. In this regard Grandjean (1981) points out that if a table is horizontal or is placed too low, the drafter compensates by excessive bending of the body, since the head cannot tilt more than about 30°. He goes on to recommend that such tables be adjustable as follows:

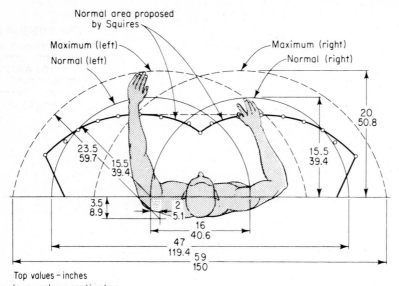

Top values – inches
Lower values – centimeters

FIGURE 12-8
Dimensions (in inches and centimeters) of normal and maximum working areas in horizontal plane proposed by Barnes, with normal work area proposed by Squires superimposed to show differences. *(Source: Barnes, 1963; Squires, 1956.)*

1 Height: 26 to 52 in (66 to 133 cm) (to accommodate both seated and standing personnel)
2 Angle: 0° to 75° from horizontal

Work Surface Height: Seated

Some of those who have experienced backaches, neck aches, and shoulder pains at work can testify that the height of a work surface can indeed bring on such aches and pains. If a work surface is too low, the back may be bent over too far; and if it is too high, the shoulders must be raised above their relaxed posture, thus triggering shoulder and neck discomfort. Although there are no universally accepted guidelines for setting work-surface heights in specific situations, a few general principles can still be stated.

1 When feasible, provision should be made for individuals to adjust the work-surface height to suit their own physical dimensions and preferences. This can be done in certain situations, for example, by having adjustable legs on tables and desks.

2 The work surface (or things or materials to be used) should be at a level that permits the arms to hang in a reasonably relaxed position from the shoulder, with the forearm usually being near horizontal or sloping down slightly; in any event, the forearm should not be required to angle upward very far.

3 The work-surface level should be such that it does not require excessive flexing (i.e., bending) of the spine.

TABLE 12–2
RECOMMENDATIONS BY AYOUB FOR SEATED WORK-SURFACE HEIGHTS FOR FOUR TYPES OF TASKS

Type of task (for seated person)	Male		Female	
	in	cm	in	cm
Fine work (e.g., fine assembly)	39.0–41.5	99–105	35.0–37.5	89–95
Precision work (e.g., mechanical assembly)	35.0–37.0	89–94	32.5–34.5	82–87
Writing, or light assembly	29.0–31.0	74–78	27.5–29.5	70–75
Coarse or medium work	27.0–28.5	69–72	26.0–27.5	66–70

Source: Ayoub, 1973.

4 The work-surface level should be adjustable for the type of manual activity to be performed.

In recent years some investigators have recommended reducing work-surface heights, generally to permit a relaxed posture of the upper arms with respect to desk heights. Bex (1971), on the basis of a European survey, reports that the most common heights have, in fact, been reduced from about 30 in (76 cm) in 1958 to about 28½ in (72 cm) in 1970. But on the basis of his own and other anthropometric data he argues for a further reduction of fixed desk heights to about 27 in (68.6 cm). However, he urges that, when feasible, adjustable desk heights be provided, the range to be from about 23 to 30 in (58 to 76 cm).

The use of low work surfaces, however, can conflict with the third general principle stated above that is aimed at minimizing lumbar flexion. The resolution of such a conflict generally would depend on the particular situation and task involved and might well involve certain trade-offs among the conflicting objectives.

The implications of the nature of the task with respect to work-surface height are emphasized by Ayoub (1973), who offers the guidelines shown in Table 12–2 for four types of tasks, these being based on average anthropometric dimensions.

Ayoub's recommendations for fine and precision work actually are in contradiction to the general recommendations that work surfaces be below elbow height. Specifically he recommends that the levels for fine and precision work be 6 and 2 in (15 and 5 cm) above elbow height, respectively. In such instances there should be provision for the arms to rest on the work surface and within close visual range.

Even though the work-surface height should be influenced in part by the nature of the task, individual preferences for work-surface height also are relevant. This was reflected by the results of a survey by Ward and Kirk (1970) of the preferences of British homemakers with respect to work-surface heights in the performance of three different types of tasks. The percentages preferring work surfaces at certain levels (relative to the elbow) in the performance of these tasks are given in Table 12–3. The mean preferred work-surface heights are also given, but there were considerable differences in individual preferences for the same task.

Aside from the implications of individual differences and task differences, certain other factors can influence work-surface height, such as the seat height (discussed

TABLE 12–3
PERCENTAGE OF SUBJECTS EXPRESSING PREFERRED LEVELS IN PERFORMING
THREE KITCHEN TASKS AND THE MEAN PREFERRED HEIGHT

Type of task	Level relative to elbow			Mean height
	Lower	Even	Above	
Working above surface (peeling vegetables, slicing bread, etc.)	54	14	32	23.7 in (60 cm)
Working on surface (spreading butter, chopping ingredients, etc.)	16	11	73	25.3 in (64 cm)
Exerting pressure (ironing, rolling pastry, etc.)	41	9	50	24.2 in (61 cm)

Source: Ward and Kirk, 1970.

later), the thickness of the work surface, and the thickness of the thigh, as shown in Figure 12–9. Dannhaus, Bittner, and Ayoub (1976) point out that the clearance between the seat and the underside of the work surface should accommodate the thighs of the largest users when the seat height is at the middle of its adjustability if it is adjustable. The combinations of variables involved make it almost impossible to design a fixed work-surface and seat arrangement that would be fully suitable for people of all sizes.

The combinations of variables that impinge on work-surface height lead back to the first principle given above: when feasible, there should be some provision for adjusting the height to the circumstance and the individual. Those can include the height from the floor, the use of footrests and platforms, and adjustable seat height.

FIGURE 12-9
Relationships of certain body dimensions and work surface and seat height. If feasible, the work-surface height should be adjustable for the individual's dimensions and preferences. *(Source: Bex, 1971, fig. 4.)*

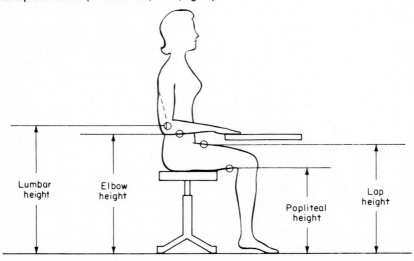

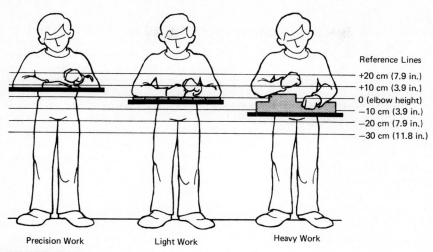

FIGURE 12-10
Relationship between elbow height (from floor) and recommended work-surface height for three types of work. The zero horizontal reference line is the elbow height of the individual, and the other lines represent levels above and below. Average elbow height reported by Grandjean for Europeans is 105 cm for males and 98 cm for females. *(Adapted from Grandjean, 1981, fig. 33.)*

Work-Surface Height: Standing

The critical features of work-surface heights for standing workers are in part the same as for seated workers, i.e., elbow height (in this case, from the floor) and the type of work being performed. Figure 12–10 shows recommended heights for precision work, light work, and heavy work as related to elbow height (Grandjean, 1981). For light and heavy work the recommended working heights are below elbow height, whereas that for precison work is slightly above (generally to provide elbow support for precise manual control). In turn, Figure 12–11 shows the relationship between

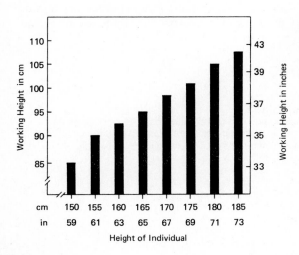

FIGURE 12-11
Working heights recommended for light work while standing as related to height of individuals. *(Adapted from Grandjean, 1981, fig. 32.)*

TABLE 12–4
RECOMMENDATIONS BY AYOUB FOR STANDING WORK-SURFACE HEIGHTS FOR
THREE TYPES OF TASKS

Type of task (for standing person)	Male		Female	
	in	cm	in	cm
Precision work, elbows supported	43.0–47.0	109–119	40.5–44.5	103–113
Light assembly work	39.0–43.0	99–109	34.5–38.5	87–98
Heavy work	33.5–39.5	85–101	31.0–37.0	78–94

Source: Ayoub, 1973.

the height of individuals and recommended working height for light work that would result in the appropriate relationship to the elbow. A specific set of recommendations expressed in terms of ranges of work-surface heights is given in Table 12–4. These data also reflect the desirability of relating the work-surface level to the type of work, with the level for heavier work being lower.

In all this, the theme of providing adjustments for individual differences comes through loud and clear. Although many work-surface heights do not lend themselves to on-the-spot adjustments in height, there sometimes are ways of matching facilities to people, as in selecting or building facilities for individuals (as countertops, work-benches, etc.), placing blocks under legs of benches or tables, having mechanically adjustable legs, or having low platforms (a few inches high) for people to stand on.

Work Surfaces for Standing or Sitting

It is sometimes desirable to provide the opportunity to perform a job in either a standing or a sitting posture or to permit workers to alternate their postures. In such instances the work-surface height should permit a relaxed position of the upper arm, and the chair and footrest should permit such a posture, as illustrated in Figure 12–12.

FIGURE 12-12
A work station that permits the worker to stand or sit and still maintain an appropriate relationship with the work surface. *(Adapted from Das and Grady, 1983a, fig. 4.)*

SCIENCE OF SEATING

Whether at work, at home, at horse races, on buses, or elsewhere, the members of the human race spend a major fraction of their lives sitting down. As we know from experience, the chairs and seats we use cover the gamut of comfort and discomfort. Improper design of chairs and seats also can affect the work performance of people and can contribute to backaches and back problems. The back problems triggered by improper posture arise principally from the pressures exerted by the vertebrae on the disks between them.

Principles of Seat Design

The various uses of chairs and seats (from TV lounge chairs to stadium bleachers) obviously require different designs, and the range of individual differences of those using them further complicates the design problem. Although there are problems due to the wide variety of seating facilities and individual user differences, certain principles have been generally accepted in seating design. Dangling questions about some of these principles, however, remain.

Back Support and Flexion A particularly important feature of seating deals with the effects on the lower (lumbar) section of the spine. First, let us look at the difference between two types of postures of the lumbar area as shown in Figure 12–13. In a kyphotic posture (kyphosis) the lower back is bent backward (i.e., convex); in a lordotic posture (lordosis) the back is bent forward (i.e., concave), as when the back is arched backward.

When people use a back support, it is generally recommended that it provide adequate lumbar support with a moderate concave curve for the lower back. The

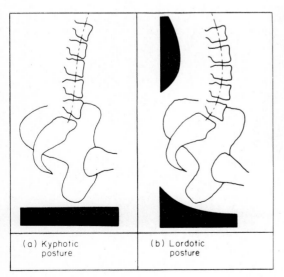

(a) Kyphotic
posture

(b) Lordotic
posture

FIGURE 12-13
Effect of seat design on the posture of the spine. *(a)* A kyphotic posture which would occur without a seat back or with an inadequate back; this leaning-forward posture accentuates the pressure between the vertebrae. *(b)* A lordotic posture which would result from adequate lumbar support and which would minimize pressure between the vertebrae. *(Source: Adapted from Hockenberry, 1979, fig. 1.)*

effect of such support is illustrated in Figure 12–14. This shows the preferred concave (i.e., lordotic) posture of the lumbar region with support from a curved back
support (*B*), as contrasted with the straight posture of the spine with a straight-backed
chair (*A*). To assist in providing lumbar support, it is common practice to have the
seat slope backward at a moderate angle (perhaps 3° to 5° or 6°) and to have the back
at an angle that permits the person to "lean against" the seat. The angle from the
seat to the back would range from about 95° or 100° to 120°. If the angle is adjustable, the range of adjustability should be within these values. If the seat back is
fixed, angles to the lower range usually would be preferable, although this would
depend on the purpose (lounge chairs, for example, usually would have a greater
angle than, say, office chairs).

Posture of the Back   In connection with posture of the spine, some years ago
the German orthopedic surgeon Schoberth (1962) reported that when a person is
standing, there typically is a very nearly vertical axis through the thigh and the pelvis
and a moderate lordosis of the lumbar area; in a seated posture, the thigh typically is
horizontal, the hip joint is flexed (bent), the pelvis has a sloping axis, and the lumbar
region is slightly kyphotic (convex). To add one more piece of background information, Keegan (1953) found by the use of x-rays that the most relaxed, or most
normal, sleeping posture (when a person sleeps on the side) is one in which there is
about a 45° flexion of the hip (that is, about a 135° angle from the spine). In wide-
awake circumstances, this approximates the posture of a person sitting on horseback
(with the legs sloping down at about a 135° angle from the spine).

Such features of "normal" posture led Mandal (1982) to propose some seating
practices that are at odds with certain commonly accepted notions. Although common

FIGURE 12-14
Effects on the posture of the lumbar region of the spine of (a) a
straight-backed chair and (b) a curved-backed chair. The curved
(convex) posture is preferable for persons who can use a back
support while performing their work.

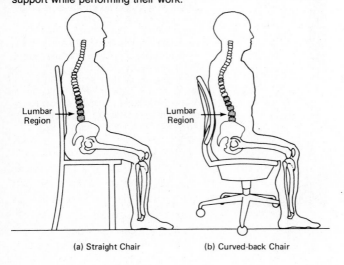

(a) Straight Chair   (b) Curved-back Chair

practice tends to emphasize the need for adequate lumbar support (to maintain some degree of lordosis), Mandal (1982) emphasizes the need to minimize lumbar flexion (that is, the angle or amount of flexion of the spine in the lumbar area). When the task makes it possible to use a back support, lumbar support can be provided by the seat design, as illustrated in Figure 12–14. However, some tasks require the person to maintain a vertical posture or to lean over a work surface (as when writing), thus precluding the use of the seat back. In such circumstances the physical arrangement can help to minimize lumbar flexion. Mandal points out that this can be done by proper adjustment of the seat height, slope of seat, and work-surface height. (Although Mandal was concerned primarily with seating for schools, his findings are relevant to other types of seating situations as well.)

Seat Height and Slope   A rather generally accepted principle of seat design is that the seat height should be low enough to avoid excessive pressure on the underside of the thigh. (Continued excessive pressure can reduce blood circulation to the lower leg.) In general this means that the seat should be lower than the distance from the floor to the thigh when the person is seated (i.e., popliteal height). It is fairly common practice to design seats for the 5th percentile of popliteal heights, which for males and females is 15.5 and 14 in (39 and 36 cm), respectively. Given that heels add an inch or more, many general-purpose chairs have seats heights around 17 or 18 in (43 or 46 cm). Adjustable seats should, of course, be provided when it is feasible.

However, Mandal (1982) questions the generality of this principle, arguing that, in some circumstances, low seats would result in more lumbar flexion than high seats. This effect is illustrated in Figure 12–15a, which shows that higher seats tend to result in postures that are more like the horseback posture than do lower seats. The horseback posture can also be facilitated by the use of forward-sloping seats, as illustrated in Figure 12–15b, and by the use of higher tables, as illustrated in Figure 12–15c.

The argument by Mandal for higher seats was supported by the results of a survey of seat height preferences of 80 young and middle-age people. The results of this survey are shown in Figure 12–16. All 80 subjects (represented by the dots) preferred seat heights higher than those recommended by the International Organization for Standardization (ISO) (1978). The subjects also expressed their preferences for table heights, with similar results: All preferred table heights higher than those recommended by the ISO for their respective individual heights. For both seat height and table height, the differences between expressed preferences and the ISO recommendations were about 6 to 10 in (15 to 25 cm).

Recognizing that Mandal's emphasis on lumbar flexion (as contrasted with lumbar support) is at odds with the common recommendation to use low seats to minimize under-thigh pressure, we suggest the following tentative guidelines: (1) Lower seats probably would be most suitable when people are relatively inactive and can maintain a seated posture with the back leaning against, and supported by, the seat back. (2) Higher seats probably would be most suitable for persons engaged in reasonably active manual work, those who have to lean over a table or desk (as when writing), and those who can alternate standing and sitting.

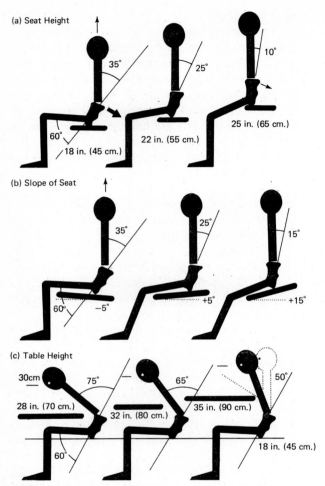

FIGURE 12-15
Effects of seat height, slope of seat, and table height on the
flexion of the lumbar (lower) section of the spine. *(From
Mandal, 1982, figs. 7, 8, and 9. Copyright by the Human
Factors Society, Inc. and reproduced by permission.)*

Seat Depth and Width   These features obviously depend in part on the type of
seat being considered. In the design of multipurpose seats for the public, however, a
couple of guidelines should be observed: (1) The depth should be set to be suitable
for small persons (to provide clearance for the calf of the leg and to minimize thigh
pressure), and (2) the width should be set to be suitable for large persons. On the
basis of comfort ratings for chairs of various designs, Grandjean et al. (1973) recommend that for multipurpose chairs depths not exceed 43 cm (16.8 in) and that the
width of the seat surface be not less than 40 cm (15.7 in), although such a seat width
[or perhaps one a bit wider, say 17 in (43 cm)] would do the trick for individual
seats. If people are to be lined up in a row or seats are to be adjacent to each other,
elbow-to-elbow breadth values need to be taken into account, with even 95th percen-

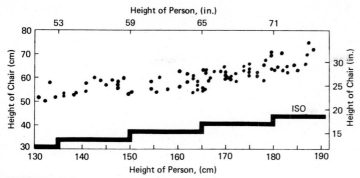

FIGURE 12-16
Preferred heights of chairs of 80 subjects (shown as dots) as related to
standards of the International Organization for Standards for persons of
various heights (shown by a solid line). *(From Mandal, 1982, fig. 11.
Copyright by the Human Factors Society, Inc. and reproduced by
permission.)*

tile values around 19 and 20 in producing a moderate sardine effect. (And for bun-
dled-up football observers the sardine effect is amplified further.) In any event these
are approximate minimum values for chairs with arms on them (for well-fed friends,
you should have even wider lounge chairs).

Contoured Seats   The human body has evolved in such a way that, when the
body is seated, the primary weight of the body can best be supported by the ischial
tuberosities (sometimes called the *sitting bones*) of the buttocks. Although the primary weight should be supported by these bones, Rebiffé (1969) expresses the opinion that seats should provide for a distribution of the weight over the entire buttocks,
with the pressure decreasing from these bones to the periphery of the buttocks. Such
a weight distribution can be aided by the use of contoured seats. Although the contouring of seats can help to produce such a distribution of weight over the buttocks,
it can have the disadvantage of restricting movement. It is commonly recognized that
static posture is not natural for human beings and that postural stress can be relieved
by moving around while sitting. Thus, to the extent that contouring of seats restricts
movement, it is a mixed blessing. A possible compromise is to limit the amount of
contour in the seat.

Discussion of Principles of Seating   There are at least a couple of constraints in
the application of seating principles to the design of seats for specific purposes: (1)
There is no single, inflexible set of principles that applies across the board to all
aspects of seating, and (2) in some circumstances one principle has to take prece-
dence over another (or one advantage has to be traded for another).

Recognizing these constraints, we offer the following summary of the principles
discussed above, along with a few other guidelines.

When back support can be used:

1 The seat back should provide support for the lumbar area.
2 The back should have moderate angle (10° to 30° from the vertical).

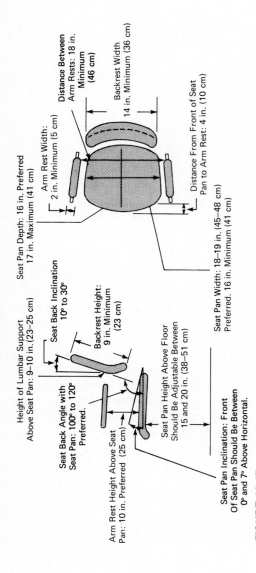

FIGURE 12-17
Recommended design features and dimensions of chairs for VDT operators. *(From AT&T Bell Telephone Laboratories, 1983, fig. 2-8, p. 19. Copyright 1985, Bell Telephone Laboratories, Inc., reprinted by permission.)*

Distance Between Arm Rests: 18 in. Minimum (46 cm)

Backrest Width 14 in. Minimum (36 cm)

Arm Rest Width: 2 in. Minimum (5 cm)

Seat Pan Depth: 16 in. Preferred 17 in. Maximum (41 cm)

Distance From Front of Seat Pan to Arm Rest: 4 in. (10 cm)

Height of Lumbar Support Above Seat Pan: 9–10 in. (23–25 cm)

Seat Back Inclination 10° to 30°

Backrest Height: 9 in. Minimum (23 cm)

Seat Back Angle with Seat Pan: 100° to 120° Preferred.

Seat Pan Height Above Floor Should Be Adjustable Between 15 and 20 in. (38–51 cm)

Seat Pan Width: 18–19 in. (45–48 cm) Preferred. 16 in. Minimum (41 cm)

Arm Rest Height Above Seat Pan: 10 in. Preferred (25 cm)

Seat Pan Inclination: Front Of Seat Pan Should Be Between 0° and 7° Above Horizontal.

3 The seat pan usually should slope back slightly.

4 The angle between the seat pan and back should be between 95° and 120°.

5 When feasible, the seat height and backrest should be adjustable.

6 For most multipurpose chairs, the seat height should be set for small people and the seat width for large people.

7 When feasible, provide for concentration of weight on the sitting bones with decreasing weight over the entire buttocks, as by the use of a moderately contoured seat pan.

When back support is not used (as when one is leaning over a table or desk):

8 Minimize lumbar flexion required by the task by use of a higher and possibly forward-sloping seat.

Seat Designs for Specific Purposes

A few examples are given to illustrate the influence on seat designs of the purposes in question.

Visual Display Terminal Chairs The tremendous upsurge in the use of VDTs has triggered considerable research and concern regarding the design of VDT work stations, including seats used in such work stations. On the basis of a review of seating research relevant to VDTs, the AT&T Bell Laboratories (1983) issued a set of recommended design features and dimensions of chairs for VDT operators. These recommendations are shown in Figure 12–17.

Multipurpose Chairs In a study by Grandjean et al. (1973), some 25 men and 25 women were asked to rate the comfort of 11 parts of the body when testing 12 different designs of multipurpose chairs. In addition, each subject compared every chair with every other one and rated the overall comfort by the paired-comparison method. The contours of the two most preferred chairs are shown in Figure 12–18, along with design recommendations made on the basis of the analysis of the results of all the data. The recommendations include foam rubber of 2 to 4 cm (about 0.75 to 1.5 in) on the entire seat.

Chairs for Resting and Reading The desirable features of chairs for relaxation and reading are, of course, different from those of chairs used for more active use. A study was carried out by Grandjean, Boni, and Krestzschmer (1969) in which they used a "seating machine" for eliciting judgments of subjects about the comfort of various seat designs. The seating machine consisted of features that could be adjusted to virtually any profile. Without summarizing all the results, they found that the angles and dimensions shown in Table 12–5 were preferred by more subjects than others for the two purposes of reading and resting. Profiles for two such chairs are

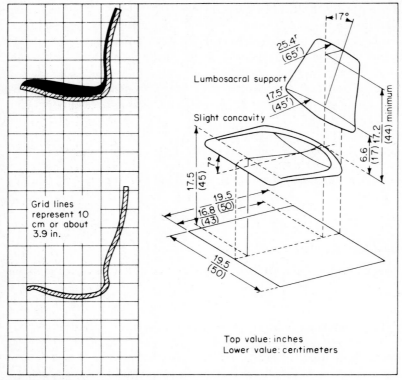

FIGURE 12-18
Contours of the two multipurpose chairs (of 12) judged to be most comfortable by 50 subjects and the design features recommended for multipurpose chairs based on the study. *(Source: Grandjean et al., 1973, figs. 2, 6, 13.)*

shown in Figure 12–19. Note in these the distinct angles of the backs and seats to provide full back support with particular support for the lower (lumbar) sections of the spine.

TABLE 12–5
RANGES OF DIMENSIONS OF SEATS
FOR READING AND RESTING
PREFERRED BY SUBJECTS

Dimension	Reading	Resting
Seat inclination, deg	23–24	25–26
Backrest inclination, deg	101–104	105–108
Seat height, cm	39–40	37–38
Seat height, in	15.3–15.7	14.6–15.0

Source: Grandjean, Boni, and Krestschmer, 1969.

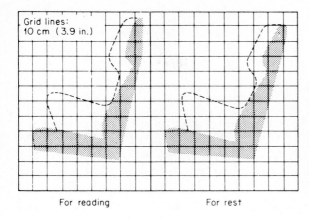

Grid lines:
10 cm (3.9 in.)

For reading   For rest

FIGURE 12-19
Profiles of seats proposed for reading and resting. The dotted lines correspond to the armrests and a possible outer contour. The shaded area shows the surface of the seat, including upholstering to be 6 cm (2.5 in) thick. *(Source: Grandjean, Boni, and Krestzschmer, 1969, fig. 2, p. 310.)*

WORKPLACE DESIGN

Although this is not the place to propose specific workplace designs for the butcher, the baker, and the candlestick maker, a couple of examples will illustrate the application of anthropometric data and principles to the design of complete workplaces.

Visual Display Terminal Work Stations

A VDT work station consists of a chair (discussed above), a table or desk, the VDT screen, a keyboard, and in some cases a document holder (for papers or other materials that include information to be entered in the keyboard) and footrest. The frequent complaints about discomfort from the use of VDTs typically arise from the way in which these components are integrated in the work station.

Two aspects of the use of VDTs are relevant to human factors considerations: the frequency or continuity of use by the individuals and the task to be performed. Many complaints about VDTs are made by individuals who use them rather continuously; the human factors features are of somewhat less importance for those who use VDTs only occasionally. There are three general types of VDT tasks:

1 Information input (use of a keyboard to enter verbal or numeric data into a computer or word processor, usually from documents)
2 Dialogue (such as interchanging information between different offices)
3 Data inquiry (as in obtaining airline flight schedules or stock quotations)

In the case of information input tasks, the physical location of source documents often is more important than the location of the screen. With dialogue tasks, the keyboard usually is not very important and may be placed off to the side. For data inquiry tasks, the paper documents (if any are used) can be placed off the the side, but the keyboard and screen should be directly in front of the user. Also, information input tasks usually involve more continual, repetitive work than the other tasks, thus adding to their human factors problems.

To assist in tracing features of VDT work stations that might contribute to complaints by users, Grandjean, Hünting, and Piderman (1983) conducted a survey of

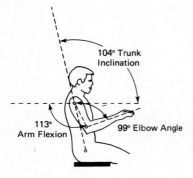

FIGURE 12-20
The mean body posture preferred by 68 VDT operators. This shows the means of the preferred angle values for the three angular features most relevant for the posture of the trunk and arms. *(From Grandjean, Hünting, and Piderman, 1983, p. 15. Copyright by the Human Factors Society, Inc. and reproduced by permission.)*

preferences of 68 VDT operators regarding certain specific aspects of such work stations. They found that three angular features were most relevant for providing postural comfort: angle of trunk inclination, arm flexion angle, and elbow angle. These are illustrated in Figure 12–20 along with the angles most preferred by the operators.

Various efforts have been made to develop a consolidated set of recommendations for the design of VDTs (AT&T Bell Laboratories, 1983; Grandjean, Hünting, and Piderman, 1983; and the National Research Council, 1983). These are all based on accumulated research data; and although there are substantial similarities among

FIGURE 12-21
One set of guidelines for the design of VDT work stations, showing various features and proposed ranges of adjustability and of acceptable angles of certain features. Dimensions are in inches, with centimeters in parentheses. *(Adopted largely from AT&T Bell Laboratories, 1983, fig. 1-2, p. 4. Features identified by an asterisk are from Miller and Suther, 1983. Copyright 1985, Bell Telephone Laboratories, Inc., reprinted by permission.)*

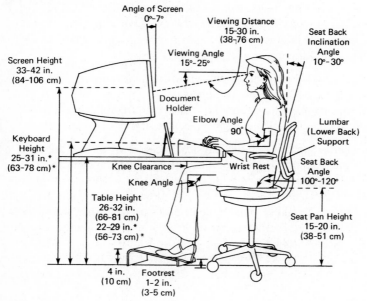

them, there also are certain differences (most of the differences are minor). For illustrative purposes the guidelines of the AT&T Bell Laboratories are shown in Figure 12–21 (along with supplementary data from another source). The figure identifies most of the features of VDT work stations and gives proposed angles of adjustability and acceptable angles of certain features. There is strong support by various investigators regarding the desirability of adjustable features of VDTs for persons who use them extensively (e.g., Shute and Starr, 1984). Some such features are illustrated in Figure 12–21. In addition, Miller and Suther (1983), based on a survey of people of various heights, urge the use of keyboards with an adjustable slope angle of 20° to 25°; higher slope angles were generally preferred by short people. (The mean preferred angle was 18°.)

A few suggestions of the National Research Council panel regarding VDT work stations are as follows:

- The screen should be in front of the operator and preferably near the keyboard.
- The screen and source documents preferably should be a few degrees below the straight-ahead line of sight.

FIGURE 12-22
A set of recommended design features for vehicle cabs. These features are designed to be suitable for persons between the 5th and 95th percentiles of operators. Dimensions are given in inches, with centimeters in parentheses. *(Adapted from Van Cott and Kinkade, 1972, fig. 9-11.)*

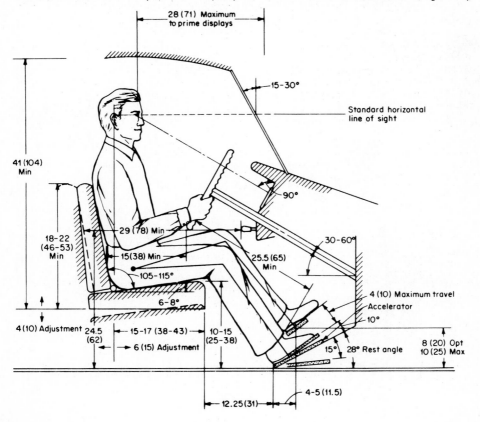

- The keyboard preferably should be detachable (for convenient placement by operator).
- A convenient writing surface should be provided.
- Leg room should be provided to enable variations in posture.
- Support should be provided for the hand, wrist, and arm for long-term use.

Although Figure 12–21 presents guidelines specifically for VDT work stations, several of its features also apply to other types of work stations, such as typing and certain control console tasks.

Vehicle Cabs

Vehicle cabs represent a type of "workplace" that requires special human factors design features. Some of the important features are seat height, depth, and back angle; forward and backward adjustability of the seat; leg and knee clearance; location of hand and foot controls; and visual field (the driver should have a good view of the road and traffic environment). One set of recommended design features of vehicle cabs is illustrated in Figure 12–22.

FIGURE 12-23
A couple of human factors problems in a nuclear power plant. In the photograph on the left, the control console is arranged so that certain controls could be accidentally activated by the knee or the hand used to brace the operator when reaching for the farthest controls. In the photograph on the right, certain of the displays are at a height that makes reading or lamp replacement a problem. Special stepladders are required. *(Copyright 1979, Electric Power Research Institute, EPRI Report NP–1118, "Human Factors Methods for Nuclear Control Room Design." Reprinted with permission.)*

DISCUSSION

The design of workplaces includes the work envelope, work surfaces (such as desks, tables, etc.), and seats (if used) as well as the design and location of equipment used. (The next chapter deals with some aspects of the specific arrangement of equipment components.) The design of workplaces clearly is rooted in the use of anthropometric data. Examples of work situations that were not well designed are easy to find. A couple of extreme examples are shown in Figure 12–23. (You may have run across some poorly designed workplaces, too.)

REFERENCES

AT&T Bell Laboratories (1983). *Video display terminals*. Short Hills, NJ.

Ayoub, M. M. (1973). Work place design and posture. *Human Factors*, 15(3), 265–268.

Barnes, R. M. (1963). *Motion and time study* (5th ed.). New York: Wiley.

Bex, F. H. A. (1971). Desk heights. *Applied Ergonomics*, 2(3), 138–140.

Bittner, A. C., Jr. (1974). Reduction in user population as the result of imposed anthropometric limits: Monte Carlo estimation (TP-74-6). Point Mugu, CA: Naval Missile Center.

Bullock, M. I. (1974). The determination of functional arm reach boundaries for operation of manual controls. *Ergonomics*, 17(3), 375–388.

Dannhaus, D., Bittner, A., Jr., and Ayoub, M. M. (1976). Seating, console, and workplace design. In M. M. Ayoub and C. A. Halcomb (eds.), *Improved seat console design: Final report*. Lubbock: Institute for Biotechnology, Texas Tech University.

Das, B., and Grady, R. M. (1983a). Industrial workplace layout design: An application of engineering anthropometry. *Ergonomics*, 26(5), 433–447.

Das, B., and Grady, R. M. (1983b). The normal working area in the horizontal plane: A comprehensive analysis between Farley's and Squires' concepts. *Ergonomics*, 26(5), 449–459.

Dempster, W. T. (1955, July). *Space requirements of the seated operator*, Tech. Rept. 55–159. U.S. Air Force, WADC.

Eastman, M. C., and Kamon, E. (1976). Posture and subjective evaluation at flat and slanted desks. *Human Factors*, 18(1), 15–26.

Electrical Power Research Institute (1979). *Human Factors Methods for Nuclear Control Room Design*, (NP-1118). Palo Alto, CA: Electrical Power Research Institute.

Farley, R. R. (1955). Some aspects of methods and motion study as used in development work. *General Motors Engineering Journal*, 2, 20–25.

Garg, A., Bakken, G. M., and Saxena, U. (1982). Effect of seat belts and harnesses on functional arm reach. *Human Factors*, 24(3), 367–372.

Grandjean, E. (1981). *Fitting the task to the man*. London: Taylor & Francis.

Grandjean, E., Boni, A., and Krestzschmer, H. (1969). The development of a rest chair profile for healthy and notalgic people. *Ergonomics*, 12(2), 307–315.

Grandjean, E., Hünting, W., Wotzka, G., and Shärer, R. (1973). An ergonomic investigation of multipurpose-chairs. *Human Factors*, 15(3), 247–255.

Grandjean, E., Hünting, W., and Piderman, M. (1983) VDT workstation design: Preferred settings and their effects. *Human factors*, 25(2), 161–175.

Hockenberry, J. (1979). Comfort = "the absence of discomfort." *C P News* (Newsletter of the Consumer Products Technical Group), Human Factors Society, 4(2).

International Organization for Standardization (ISO) (1978). *Chairs and tables for educational institutions: Functional sizes*, ISO/TC 136. Geneva, Switzerland: ISO General Secretariat.

Keegan, J. J. (1953). Alterations of the lumbar curve. *Journal of Bone and Joint Surgery,* 35, 589–603.

Mandal, A. C. (1982). The correct height of school furniture. *Human Factors,* 24(3), 257–269.

Meyer, R. P. (1979). Articulated anthropometric modes. *C P News* (Newsletter of the Consumer Products Technical Group), Human Factors Society, 4(2).

Miller, W., and Suther, T. W., III. (1983). Display station anthropometrics: Preferred height and angle settings of CTR and keyboard. *Human Factors,* 25(4), 401–408.

National Aeronautics and Space Administration (NASA). (1978). *Anthropometric source book* (vol. 1) *Anthropometry for designers;* (vol. 2) *A handbook of anthropometric data;* (Vol. 3) *Annotated bibliography* (NASA Ref. Pub. 1024).

National Research Council, Panel on Impact of Video Viewing on Vision of Workers (1983). *Video displays, work, and vision.* Washington: National Academy Press.

North, K. (1980). Ergonomics methodology—An obstacle or promoter for the implementation of ergonomics in industrial practice? *Ergonomics,* 23(8), 781–795.

Rebiffé, P. R.(1969). Le siège du conducteur: son adaptation aux exigences fonctionnelles et anthropométriques. *Ergonomics,* 12(2), 246–261.

Rigby, L. V., Cooper, J. I., and Spickard, W. A. (1961, October). *Guide to integrated system design for maintainability,* Tech. Rept. 61–424. U.S. Air Force, ASD.

Roth, J. T., Ayoub, M. M., and Halcomb, C. G. (1977, Oct. 17–20). Seating, console and workplace design: Seated operator reach profiles. *Proceedings of the Human Factors Society 21st Annual Meeting,* pp. 83–87.

Rozier, C. K. (1977). Three-dimensional work space for the amputee. *Human Factors,* 19(6), 525–533.

Sanders, M. S. (1977). *Anthropometric survey of truck and bus drivers: Anthropometry, control reach and control force.* Westlake Village, CA: Canyon Research Group.

Sanders, M. S. (1980). Sleep envelopes and sleeper berth requirements. *Human Factors,* 22(3), 313–317.

Schoberth, H. (1962). *Sitzhaltung, sitzschaden, sitzmöbel.* Berlin–Göttingen–Heidelberg: Springer.

Shute, S. J., and Starr, S. J. (1984). Effects of adjustable furniture on VDT users. *Human Factors,* 26(2), 157–170.

Squires, P. C. (1956). *The shape of the normal work area,* Rept. 275. New London, CT: Navy Department, Bureau of Medicine and Surgery, Medical Research Laboratory.

Stoudt, H. W. (1981). The anthropometry of the elderly. *Human Factors,* 23(1), 29–37.

Tichauer, E. R. (1978). *The biomechanical basis of ergonomics.* New York: Wiley.

United States Public Health Service. (1965, June). *Weight, height, and selected body dimensions of adults: United States, 1960–1962* (USPHS Pub. 1000, ser. 11, no. 8). Data from National Health Survey.

Van Cott, H. P., and Kinkade, R. G. (1972). *Human engineering guide to equipment design* (rev. ed.). Washington: Government Printing Office.

Ward, J. S., and Kirk, N. S. (1970). The relation between some anthropometric dimensions and preferred working surface heights in the kitchen. *Ergonomics,* 13(6), 783–797.

CHAPTER 13

PHYSICAL SPACE AND ARRANGEMENT

A large proportion of most people's lives is spent within built environments, ranging from ''local'' situations in which people find themselves (such as at workplaces, in a kitchen, or in an automobile), through intermediate types of situations (such as office buildings, homes, and theaters), to general environments (such as communities). Our common experience points up the effects that the designs of such space and facilities can have on people, including their performance, their comfort, and even their physical well-being.

An important aspect of such design is the arrangement of the ''components'' within the space and facilities people use. We use the term *component* to refer to virtually any physical features within a work situation, including those within a work space envelope (such as displays, controls, materials, etc.), or those elsewhere that are involved in a person's work activities (such as various machines or other work stations or work areas).

PRINCIPLES OF ARRANGEMENT OF COMPONENTS

It is reasonable to hypothesize that any given component has a generally optimum location for serving its purpose. This optimum is predicated on the human sensory, anthropometric, and biomechanical characteristics involved (reading a visual display, activating a foot pushbutton, etc.) or on the performance of some operational activity (such as reaching for parts, preparing food in a restaurant, or storing material in a warehouse). Preferably, of course, components should be in their optimum locations, but since this frequently is not possible, priorities and trade-offs sometimes must be established. These priorities and trade-offs, however, do not descend like manna from heaven but must be determined, usually on the basis of factors such as those discussed later.

Before we touch on a few methods that are used to figure out what should go where, let us set down a few general guidelines (in addition to the idea of the optimum location mentioned above) that may be helpful. Depending on the circumstance, these guidelines can be concerned with one or both of two separate but interrelated phases: that concerned with the *general location* of components (either individual components usually used by themselves or groups of related components) and that concerned with the *specific arrangement* of components within their "general" location (that is, the arrangement of components within a group of related components in that general location).

Importance Principle

This principle deals with *operational importance,* that is, the degree to which the performance of the activity with the component is vital to the achievement of the objectives of the system or some other consideration. The determination of importance usually is a matter of judgment.

Frequency-of-Use Principle

As implied by the name, this concept applies to the frequency with which some component is used, such as having the activation control of a punch press in a convenient location since it is used very frequently or having a copying machine near a typist.

Functional Principle

The *functional principle* of arrangement provides for the grouping of components according to their function, such as the grouping of displays, controls, or machines that are functionally related in the operation of the system. Thus, temperature indicators and temperature controls might well be grouped, and electric power distribution instruments and controls usually should be in the same general location.

Sequence-of-Use Principle

In the use of certain items, sequences or patterns of relationship frequently occur in the operation of equipment or in performing some service or task. In applying this principle, the items would be so arranged as to take advantage of such patterns.

Discussion

In putting together the various components of a system, no single guideline can, or should, be applied consistently across all situations. But, in a very general way, and in addition to the optimum premise, the notions of importance and frequency probably are particularly applicable to the more basic phase of locating components in a general area in the workspace; in turn, the sequence-of-use and functional principles tend to apply more to the arrangement of components within a general area.

The application of these various principles of arrangement of components gener-

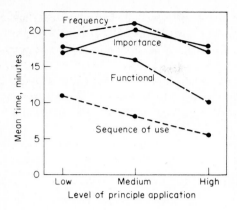

FIGURE 13-1
Time required to carry out a standard simulated task in the use of controls and displays arranged on the basis of four principles. In the case of each principle, three control panels were used, these varying in terms of the rated "level" or degree to which the principle has been applied in the panel design. *(Source: Adapted from Fowler, Williams, Fowler, and Young, 1968.)*

ally has had to be predicated on rational, judgmental considerations, since there has been little empirical evidence available regarding the evaluation of these principles. However, some data from at least one study cast a bit of light on this matter. The study in question, by Fowler et al. (1968), consisted of the evaluation of various control panel layouts in which the controls and displays had been arranged following each of the four principles described above. The panels included 126 standard military controls and displays. The arrangement of these following the four principles need not be described; but it should be added that, for each principle, three control panels were developed, varying in terms of three "levels" of application of the principle (based on a scoring scheme), these being high, medium, and low. The 200 male college student subjects used the various arrangements in a simulated task. Their performance was measured in terms of time and errors, the results for the time criterion being shown in Figure 13–1. This figure (and corresponding data regarding errors) showed a clear superiority for the sequence-of-use principle. Even the arrangements that were based on the low and medium application levels of this principle were better than, or equal to, the high levels for the other principles.

Although the sequence-of-use principle came out on top in this study, this principle obviously could be applied only in circumstances (as in this study) in which the operational requirements actually do involve the use of the components in question in rather consistent sequences. Where this is the case, this principle certainly should be followed.

METHODOLOGIES FOR ARRANGING COMPONENTS

Arranging components in a work space, be they tables and control panels in a room or controls and displays on a console, requires the availability of relevant data and the use of certain methods in applying the data.

Types of Data for Use in Arranging Components

The types of data that are relevant for use in arranging components generally fall into the following broad classes:

• **Basic data about human beings.** Anthropometric and biomechanical data are especially relevant, but other types of data may also be useful, such as data on sensory, cognitive, and psychomotor skills. Such data generally come from research undertakings and are published in various source books.

• **Task analysis data.** These are data about the work activities of people who are (or would be) involved in the specific system or work situation in question. (For our purposes we refer to the work activities as *tasks* regardless of the level of specificity of the activity.)

• **Environmental data.** This category covers any relevant environmental features of the situation, such as illumination, noise, vibration, motion, heat, traffic and congestion, etc.

The task data must be developed by the use of some systematic task analysis procedure. Certain methods of developing, organizing, and presenting task data are central to the processes of arranging components. Some of these methods are discussed and illustrated below. (Task analysis is discussed further in Chapter 18.)

How Not to Do It

Although it may appear intuitively obvious that a thorough understanding of the tasks to be performed is absolutely essential to arrange components in a work space, it is not always followed in practice. A consultant (one of your authors) had the occasion to assist in the layout of displays on a 32 × 8 ft (9.7 × 2.4 m) nuclear power plant console that consisted of four interrelated subconsoles. The consultant met with two engineers to discuss the console. A scale drawing of a preliminary design had been made, but no mock-up was available, nor was there any intention of having one made. The consultant asked the two engineers to "walk him through" the console so he could become familiar with it. The conversation proceeded, more or less as follows:

Engineers: These are the six *XYZ* concentrate indicators.

Consultant: What is the operator's task here? Does the operator have to read a specific value, compare values between indicators, or use these indicators while manipulating controls?

Engineers: We really don't know—that's Operations' responsibility. We will see that someone from Operations comes in later to answer your questions.

[The Operations office was 25 mi (40 km) from the office where the preliminary design was developed.]

Engineers: Let's continue. This is a digital readout that directs the operator's sequence of actions.

Consultant: How does the operator use that? What displays and controls are referred to by the readout? How much time does the operator have to respond to the display?

Engineers: We really don't know—that is Operations' responsibility.

Consultant (now becoming a little irritated): Exactly what did the two of you have to do with this console anyway?

Engineers: We designed it! Each of us started at a different end, and we laid out the displays going across. We then compared the layouts and selected the parts from each we liked the best.

Consultant: I think I need a cup of coffee now.

Gathering Basic Task Data

When a modification of an existing system is being developed, data relating to an existing model may be appropriate. Such data can be obtained by various methods, such as the use of film; observation; the use of eye-movement recordings (with eye cameras or other related devices); and interviews with experienced personnel (including questions to elicit their opinions, such as about frequency or importance of various activities or about the desirable arrangement of components).

However, in the case of new systems or facilities (without current counterparts), information about the activities to be performed (such as frequency, sequence, and other activity parameters) needs to be inferred from whatever tentative drawings, plans, procedures, or concepts are available.

Types of Task Data

Although a thorough task analysis yields all sorts of valuable information for designing a work space, we concentrate on two broad classes related to arranging components within the space. We assume that an analysis has been made to determine what components will be needed to do the tasks. That is, we assume that all the controls, displays, tables, chairs, telephones, bookshelves, and what-have-you have been enumerated. The objective now is to arrange these items in the space allocated.

The broad classes of task-related information useful in arranging components are (1) information on the use of the components individually and (2) information on the relationships between components as they are used.

Information Dealing with Components Individually

For each component, information is collected to determine the frequency with which each component is (or would be) used and how important or critical its use is considered to be. Ratings by experts are usually relied on to assess importance and sometimes are used to assess frequency as well. If data are available for an existing system, however, usually it is preferable to use such data for determining task frequency.

Often, importance and frequency are not perfectly correlated. That is, although frequently used components are often important, some important components are not used frequently—but when they are needed, they are very important. Because of this, it is often easier to deal with a composite frequency-importance index rather than with separate indices of frequency and importance. Usually the frequency data are converted to a scale comparable to the scale used to rate importance. For example, if importance is rated on a 5-point scale from "unimportant" to "extremely important," the frequency data can be converted to a 5-point scale from, perhaps, "seldom used" to "very frequently used." Several options are available for combining the

TABLE 13–1
ILLUSTRATION OF METHODS FOR COMBINING FREQUENCY AND IMPORTANCE DATA
INTO A COMPOSITE INDEX

	Dial				
	A	**B**	**C**	**D**	**E**
Raw frequency rating F	3	1	2	5	4
Raw importance rating I	3	5	1	3	4
Composite indices:					
$F + I$	6	6	3	8	8
$F \times I$	9	5	2	15	16
$2F + I$	9	7	5	13	12
$2F \times I$	18	10	4	30	32

two ratings: (1) add the values; (2) multiply the values; (3) differentially weight frequency and importance, and then add them; or (4) differentially weight frequency and importance, and then multiply them. The different methods may result in a somewhat different rank ordering of components. If there are marked differences in the rankings, a choice must be made on the basis of judgment, since there are no absolute guidelines to follow. Table 13–1 illustrates the results of this process for five dials by using hypothetical frequency and importance data.

An example of a different composite index is the *control accessibility index* developed by Banks and Boone (1981) to arrange control devices in a work space. Control devices, of course, need to be physically accessible to operators (as contrasted with displays, for example, which can be out of reach). Of the several variables that can affect the relative accessibility of controls, Banks and Boone developed their quantitative index of accessibility based on (1) frequency of use, (2) relative position of controls with respect to the operator, and (3) the operator's reach envelope. The index I_a was derived in part from ratings made by subjects who used mockups of three control panels for experimental purposes.

The details of the procedures and the resulting index need not be reported, but the results reflect the feasibility of deriving a quantitative index of accessibility for use in locating controls within a person's work space envelope or in its general vicinity.

Information Dealing with Relationships between Components

Relationships between components, be they people or things, are called *links*.

Types of Links Links fall generally into three classes; communication links, control links, and movement links. Communication and control links can be considered as functional. Movement links generally reflect sequential movements from one component to another. Some versions of the three types follow:

1 Communication links
    A Visual (person to person or equipment to person)

  B Auditory, voice (person to person, person to equipment, or equipment to person)

  C Auditory, nonvoice (equipment to person)

  D Touch (person to person or person to equipment)

2 Control links

  A Control (person to equipment)

3 Movement links (movements from one location to another)

  A Eye movements

  B Manual movements, foot movements, or both

  C Body movements

The kinds of information usually collected about links include how often the components are linked (e.g., how often display A is viewed immediately before or after display B or how often person 1 talks to person 2), in what sequence the links occur, assuming a fixed sequence exists (e.g., whether the operator views display A then E then F or whether the sequence is F then A then E), and the importance of the links. These sorts of data can be gathered over time, across specific tasks, or over several repetitions of a single task.

Summarizing Link Data   Link data are often summarized in a *link table*. An example of a link table is shown in Figure 13–2 for components in a hypothetical minicomputer laboratory. In this example the relationship of each component with any other one (i.e., each link) was rated by using the following scale:

FIGURE 13-2
Illustration of a chart of the link values of the relationships between pairs of components in a hypothetical minicomputer laboratory. *(Source: Cullinane, 1977, fig. 1.)*

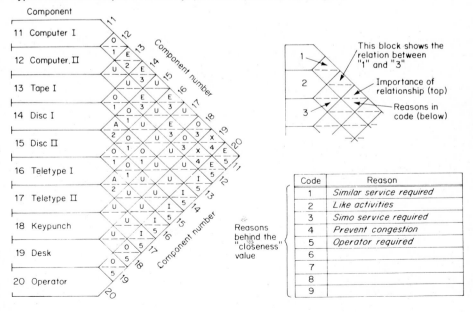

A: It is *absolutely essential* for the activities to be located close together.

E: It is *essential* for two activities to be located close together.

I: It is *important* that the two activities be located close together.

O: Ordinary closeness is *acceptable* for the two activities being considered.

U: A link does not exist, and it is *unimportant* whether two activities are placed together.

X: It is *undesirable* for two activities to be placed together.

These are shown in the upper triangular part of the square formed by the down-sloping line of one component and the up-sloping line of another (these being *A, E, I, O, U,* or *X*). The number in the lower triangular part of the square is a code explaining the reasons for the relationship. (In the example in Figure 13–2, the codes used for this purpose are in the inset in the lower right-hand corner.)

Graphic Representation of Link Data

The link table shown in Figure 13–2 is one method of presenting link data; however, it is somewhat mind-boggling and is not very effective for developing a picture of the problem at hand. Other graphical methods of presenting link data provide a clearer picture of the relationships between components. There are two distinct types of link diagrams (Geer, 1981): adjacency layout diagrams and spatial operational-sequence (SOS) diagrams.

Adjacency Layout Diagrams   Figure 13–3 shows an adjacency layout diagram of eye movements (links) between aircraft instruments during a specific maneuver. Generally, the thickness of the connecting lines is used to code the frequency and/or

FIGURE 13-3
Eye-movement link values between aircraft instruments during climbing maneuver with constant heading. *(Source: Jones, Milton, and Fitts, 1949.)*

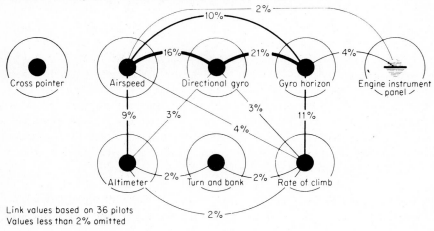

Link values based on 36 pilots
Values less than 2% omitted

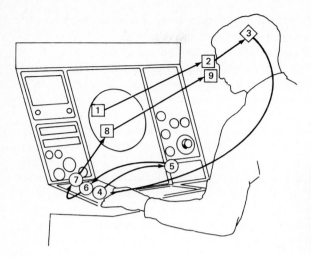

FIGURE 13-4
An example of a spatial
operational-sequence diagram
showing the sequence of sensory,
decision, and control activities
involved in tracking an aircraft.
(Adapted from Geer, 1981.)

importance of the link; the thicker the line, the more frequent or important the link. This method of presentation, however, does not indicate the sequence in which the components are used.

Spatial Operational-Sequence Diagrams   SOS diagrams graphically depict the actual sequence of operation overlaid on a pictorial representation of the work space. Figure 13–4 shows an SOS diagram for an operator tracking an aircraft on a cathode ray tube. The same sort of diagram could be developed to show the traffic flow paths for an operator responding to an emergency in a power plant control room or to show the eye movements of a pilot performing a specific maneuver. Obviously, SOS diagrams are most useful when there is a set sequence of operations performed with the components.

Arranging Components by Using Link Data

Probably the most common method of arranging components by using link data is through trial and error. The designer physically arranges scale drawings of the components, trying to maximize criteria that often conflict with one another. Trying to keep the most frequently used components in the most advantageous locations while arranging components to minimize the length of frequent or important links—and all the while keeping in mind sequences of operations and functional grouping—is probably as much art as it is science. There are, however, quantitative analysis techniques to aid the designer. No matter what technique is used, it is important that the resulting arrangement be validated by using a mock-up or simulator with actual operators carrying out actual or simulated tasks.

Quantitative Solutions to Arrangement Problems   Especially in simple problems of arranging components, it would be gilding the lily to apply sophisticated quantitative analytical methods. But with complex systems that have many components, some quantitative attack may well be justified. One such method is *linear programming*. This is a method that results in the optimizing of some criterion, or dependent, variable by manipulation of various independent variables. The optimum would be the minimum value of the criterion in some cases, and the maximum value of the criterion in other cases—whichever is the desired value in terms of the criterion.

An example of this technique draws on data from two sources: (1) data on frequencies with which a pilot made task responses, with eight controls, in flying simulated cargo missions in a C-131 aircraft, over 139 periods of 1 min (Deininger, 1958), and (2) data on the accuracy of manual blind-positioning responses in various areas, based on the study by Fitts (1947); and for this particular purpose accuracy data for 8 of the 20 target areas were used. The accuracy scores are average errors in inches in reaching to targets in the eight areas. Linear programming, when used in the analysis of the data by Huebner and Ryack (1961), involved the derivation of a *utility cost* rating for each of eight controls in each of the eight areas shown in Figure 13–5. Each one was computed by multiplying the *frequency* of responses involving each control (from Deininger) by the *accuracy* of responses in each area (from Fitts). For example, the values for the *C* control (cross-pointer set) for the eight areas range from 42.8 to 69.8, each such value being the product of the *frequency* value for the control (in this case 20) and the mean *accuracy* scores for the areas shown in Figure 13–5. For any possible arrangement of controls one can derive a *total cost,* this being the sum of the utility costs of the several controls in their locations specified by that arrangement. By linear programming it was possible to identify the particular arrangement for which the total cost was minimum. The resulting optimum arrangment is shown in Figure 13–5, with the individual controls simply being identified by a letter code. Note that each control is not in its *own* optimum location, but that *collectively* the derived arrangement is optimum in terms of the criterion of total cost. (Control *C*, mentioned above, was thus placed in space 6, even though its utility cost in that location was less.)

Another example of a quantitative approach for designing a facility is described

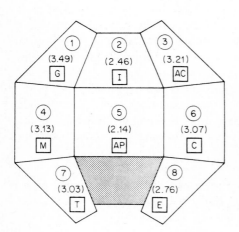

FIGURE 13-5
Perspective drawing of eight areas used in application of linear programming to the arrangement of eight aircraft controls. *(Source: Huebner and Ryack, 1961.)* Mean accuracy scores of blind-positioning movements are given in parentheses for the eight target areas. *(Source: Based on data from Fitts, 1947.)* The letters in boxes are symbols for the eight controls, their locations shown here representing the optimum linear programming solution.

by Bonney and Williams (1977), this dealing with the arrangement of control devices and other items at the "pulpit" of a bar and rod mill. Figure 13–6a shows the existing layout, the items being located on a wall panel, a left desk, a right desk, and the floor. Figure 13–6b shows the arrangement based on a computer program that was intended to optimize the collective locations of the items. Without going into details, the computerized arrangement was estimated to reduce limb movements from 35 to 31 m and provided for 98 percent of the operational time in postures and positions with a specified "comfort rating," as compared with 50 percent of the time for the existing arrangement.

Although quantitative solutions for design assignment problems may well be justified in connection with some complex design problems, Francis and White (1974, chap. 6) point out that in many circumstances such procedures are not warranted, and simpler approaches can produce substantially similar results. In connection with

FIGURE 13-6
(a) Existing arrangement of controls and other items of the "pulpit" of a bar and rod mill in Great Britain. *(b)* An improved arrangement developed with a computer program to optimize the location of the items. *(Source: Bonney and Williams, 1977, figs. 3 and 5.)*

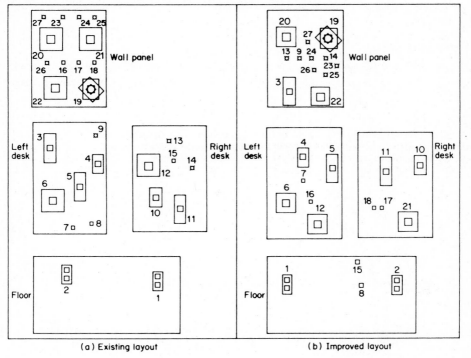

(a) Existing layout   (b) Improved layout

Key					
1. pedal 1	6. lever 4	11. lever 6	16. button 4	21. lever 9	26. lamp 7
2. pedal 2	7. button 1	12. lever 7	17. button 5	22. lever 10	27. lamp 8
3. lever 1	8. button 2	13. lamp 2	18. button 6	23. lamp 4	
4. lever 2	9. lamp 1	14. lamp 3	19. knob 1	24. lamp 5	
5. lever 3	10. lever 5	15. button 3	20. lever 8	25. lamp 6	

the design problem represented in Figure 13–5, for example, an approach such as the following might be used:

1 Put the control with highest frequency in the area with the lowest error (area 5).
2 Put the control with the second highest frequency in the area with the second lowest error (area 2).
3 Put the control with the third highest frequency in the area with the third lowest error (area 7).
4 Continue, to the eighth control.

GENERAL LOCATION OF CONTROLS AND DISPLAYS WITHIN WORK SPACE

As indicated above, it is reasonable to assume that any given component in a system or facility would have some reasonably optimum location, predicted on whatever sensory, anthropometric, biomechanical, or other considerations are relevant. Although the optimum locations of some specific components probably would depend on situational factors, some generalizations can be made about certain classes of components. Certain of these are discussed here.

Visual Displays

The normal line of sight is usually considered to be about 15° below the horizon. Visual sensitivity accompanied by moderate eye and head movements permits fairly

FIGURE 13-7
Isoresponse times within the visual field. Each line depicts an area within which the mean response time to lights is about the same. *(Source: Adapted from Haines and Gilliland, 1973, fig. 2. Copyright 1973 by the American Psychological Association and reproduced by permission.)*

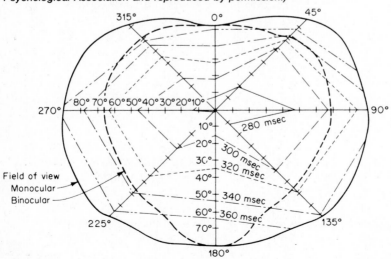

convenient visual scanning of an area around the normal line of sight. The area for most convenient visual regard (and therefore generally preferred for visual displays) has generally been considered to be defined by a circle roughly 10 to 15° in radius around the normal line of sight.

However, there are indications that the area of most effective visual regard is not a circle around the line of sight but rather is more oval. Such an indication is found in the research of Haines and Gilliland (1973), who measured subjects' times of response to a small light that flashed in various locations of the visual field. Figure 13–7 shows the mean response times to the lights in the various regions of the visual field. Each boundary on that figure indicates the region within which mean response time can be expected to be about the same (the *isoresponse* time region). The generally concentric lines tend to form ovals—but slightly lopsided ovals, being flatter above the line of sight than below.

The subjects in this study detected the lights even toward the outer fringes of this area, but the primary implications of this (and some other) research are that critical visual displays should be placed within a reasonably moderate oval around the normal line of sight.

Hand Controls

The optimum location of hand control devices is, of course, a function of the type of control, the mode of operation, and the appropriate criterion of performance (accuracy, speed, force, etc.). Certain preceding chapters have dealt with some tangents to this matter, such as the discussion of the work-space envelope in Chapter 12.

Controls That Require Force   Many controls are easily activated, and so a major consideration in their location is essentially the ease of reach. Controls that require at least a moderate force to apply (such as certain control levers and hand brakes in some tractors), however, bring in another factor—force that can be exerted in a given direction with, say, the hand in a given position. Investigations by Dupuis (1957) and by Dupuis, Preuschen, and Schulte (1955), dealt with this question, specifically the pulling force that can be exerted, when a person is seated, when the hand is at various distances from the body (actually, from a seat reference point). Figure 13–8, which illustrates the results, shows the serious reduction in effective force as the arm is flexed when it is pulled toward the body. The maximum force that can be exerted by pulling is about 57 to 66 cm forward from the seat reference point, and this span, of course, defines the optimum location of a lever control (such as a hand brake) if the pulling force is to be reasonably high.

The best location for cranks and levers (especially those to be operated continuously to and fro) is in front of the sitting or standing operator, so that the handle travels at about waist height in the sagittal plane (i.e., the vertical plane from front to back) passing through the shoulder.

Controls on Panels   Many controls are positioned on panels or in areas forward of the person who is to use them. Because of the anthropometric and biomechanical characteristics of people, controls in certain locations can be operated more effec-

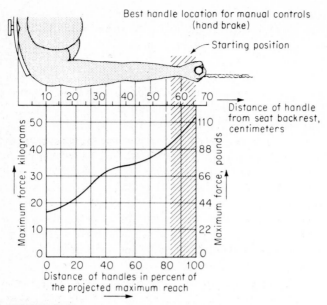

FIGURE 13-8
Relationship between maximum pulling force (such as on a hand brake) and location of control handle. *(Source: Dupuis, Preuschen, and Schulte, 1955.)*

FIGURE 13-9
Preferred vertical surface areas and limits for different classes of manual controls. *(Source: Adapted from Human Factors Engineering, 1980.)*

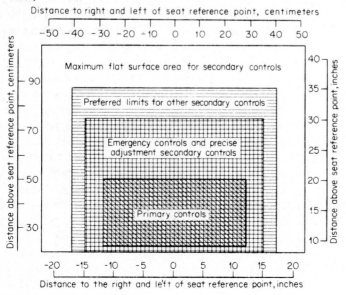

tively than those in other locations. Figure 13–9 shows one set of preferred areas for four classes of control devices as based on relative priorities (*Human factors engineering,* 1980). Although this proposed arrangement is based on data for military personnel, it would, of course, have more general applicability.

When there are many controls and displays to arrange in a console or panel, the use of angled side panels may place more of the controls within convenient access. The advantage of this was demonstrated empirically by Siegel and Brown (1958) in a study in which subjects, using a 48-in (122-cm) front panel with side panels at 35, 45, 55, and 65°, followed a sequence of verbal instructions to use the controls on the panels. A number of criteria were obtained, including objective criteria of average number of seat movements, average seat displacement, average body movements (number and extent), average number of arm extensions (partial and full), subjective criteria based on the subjects' responses of degree of ease or difficulty, judgments that the panels should be wider apart or closer together, and preference ranking for the four angles.

Only some of the data are presented, but they characterize the results generally. Figure 13–10 shows data for four of the criteria for the four angles. The criterion scales have been converted here to fairly arbitrary values, and only the "desirable" and "undesirable" directions are indicated. It can be seen, however, that all four criteria were best for the 65° side panels. The consistency across all criteria (these and the others) was quite evident.

Although Figures 13–9 and 13–10 provide general guidelines for locating controls, including consideration of relative priorities, sometimes the operational require-

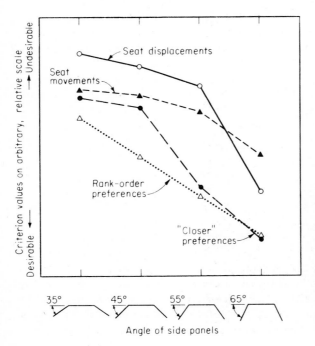

FIGURE 13-10
Representation of criteria from study relating to angles of side panels of console. The criterion values are all converted to an arbitrary scale for comparative purposes, but they all indicate the desirability of the 65° angle panels over the others. *(Source: Adapted from Siegel and Brown, 1958, courtesy of Applied Psychological Services, Wayne, PA.)*

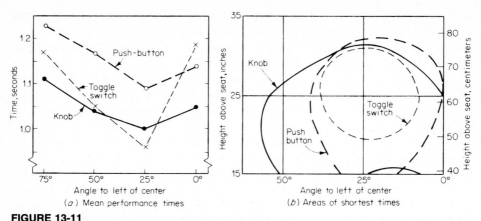

FIGURE 13-11
Data on time to activate pushbutton and toggle switch and time to reach knob, when the controls are in various positions. Data are for *left* hand and for close console. Part *a* gives mean times for controls positioned 25 in (63.5 cm) above seat reference level. Part *b* gives contours of the area for which times were *shortest* for the particular control (e.g., times within 5 percent of the minimum time for the control). *(Source: Adapted from Sharp and Hornseth, 1965.)*

ments of certain specific types of controls impose impossible constraints on their location. This was illustrated, for example, by the results of a study by Sharp and Hornseth (1965) in which seated subjects operated each of three types of controls (knobs, toggle switches, and pushbuttons) at each of 12 locations in each of three consoles (far, middle, and close). The controls of the close console were positioned for convenience of reach of individuals of small build (about the 5th percentile of males), and only data for this console are given here.

Figure 13–11*a* shows the time in seconds to activate the three types of controls when they are located at various angles from the center position, that part indicating that activation time was minimum for all three controls at about 25° from center. Figure 13–11*b* in turn shows the areas of the shortest performance times for each type of control (e.g., the area within which times were within 5 percent of minimum times for the control in question.) The smaller area for the toggle switch suggests that the selection of a location for such devices may be more critical than for the other devices if time is of the essence. However, as suggested by the investigators, a well-designed system preferably should not impose response-time requirements on operators in which differences of 0.1, 0.2, or 0.3 s are crucial.

Two-Hand Controls Some operations require the simultaneous use of controls by both hands. For example, in the operation of some metal-forming presses, the operators have to press two pushbuttons for safety reasons to keep the hands away from the press when it is activated. In some instances these pushbuttons (''palm'' buttons operated by the palm of the hand) are at eye level. This location was suspected of being responsible for a high rate of muscular strain and sprain injuries in an automobile plant (Nemeth and Blanche, 1982). In an investigation triggered by this suspicion, the experimenters had subjects perform a metal-processing task which required the use of dual palm buttons at eye level and at waist level. Electromyograph recordings of three muscle groups showed more than 4 times as much muscle activity with the eye-level buttons as with the waist-level buttons. Such a difference undoubtedly would later be translated to a lower level of muscular strain and sprain with waist-level buttons.

Foot Controls

Since only the most loose-jointed among us can put their feet behind their heads, foot controls generally need to be located in fairly conventional areas, such as those depicted in Figure 13–12. These areas, differentiated as optimal and maximal, for toe-operated and heel-operated controls, have been delineated on the basis of dynamic anthropometric data. The maximum areas indicated require a fair amount of thigh or leg movement or both, and preferably they should be avoided as locations for frequent or continual pedal use. Incidentally, Figure 13–12 is predicated on the use of a horizontal seat pan; with an angular seat pan (and an angled backrest) the pedal locations need to be manipulated accordingly (although such adjustments have been published, they are not given here).

FIGURE 13-12
Optimal and maximal vertical and forward pedal space for seated operators. *(Source: Adapted from Human Factors Engineering, 1980.)*

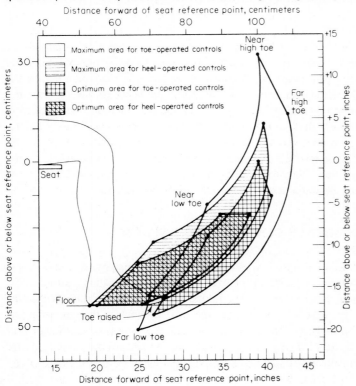

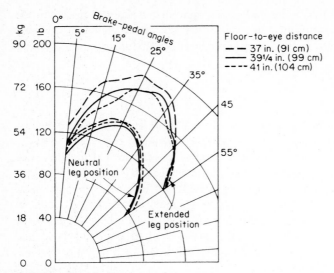

FIGURE 13-13
Mean maximum brake-pedal forces exerted by 100 Air Force
pilots in two leg positions and for various brake-pedal angles and
three floor-to-eye distances. (The seat level was adjusted for
three conditions so the eye level was at the three specified
distances above the floor.) *(Source: Adapted from Hertzberg
and Burke, 1971, fig. 6.)*

The areas given in Figure 13–12 generally apply to foot controls that do not
require substantial force. For applying considerable force, a pedal preferably should
be fairly well forward. This point is illustrated by Figure 13–13, which shows the
mean maximum brake-pedal forces for 100 Air Force pilots when the leg was in a
"normal" position and in an extended position (with the leg essentially forward from
the seat). The figure shows that the maximum brake-pedal forces were clearly greater
for the extended position for three seat levels used in the experiment. (The seat level
was adjusted so the "floor-to-eye" distances were standardized.) As an aside, Figure
13–13 also shows that the forces for both leg positions were greatest when the foot
angle on the pedal was between about 15 and 35° from the vertical.

SPECIFIC ARRANGEMENT OF CONTROLS AND DISPLAYS
WITHIN WORK SPACE

As indicated in the above discussion of the general arrangement of components,
certain general areas are most suitable (or even required) for locating various types
of components, such as visual displays, hand controls, and foot controls. Within the
constraints imposed by these considerations, the next process is that of arranging
components within the areas appropriate for them. In this process the arrangement of
groups of components can be based on the principles of sequence or function. Where
there are common sequences, or at least frequent relationships, in the use of com-

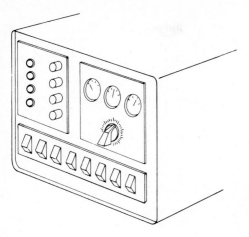

FIGURE 13-14
Example of controls and displays grouped
by function, in which the different groups
are clearly indicated.

ponents, the layout usually should be such as to facilitate the sequential process—as in hand movements, eye movements, etc. Where there are no fixed or common sequences, the components should be grouped on the basis of function. In such instances the various groups should be clearly indicated by borders, color, or other means. An example of such groups is shown in Figure 13–14. When relevant, compatibility principles should be followed in arranging components.

Even in existing systems it may be possible to add borders around groups of functionally related components to clarify their relationships. Such a procedure is illustrated in Figure 13–15 (page 382), which shows the original design of a control panel of a nuclear power plant, along with an enhancement of that same arrangement made by drawing borders around functionally related modules.

SPACING OF CONTROL DEVICES

Although we have talked about minimizing the distances between components, such as the sequential links between controls, there are obvious lower-bound constraints that need to be respected, such as the physical space required in the operation of individual controls to avoid touching other controls. Whatever lower-bound constraints there might be would be predicated on the combination of anthropometric factors (such as of the fingers and hands) and on the precision of normal psychomotor movements made in the use of control devices.

An illustration of the effects of such factors is given in Figure 13–16, which shows inadvertent "touching errors" in the use of knobs of various diameters as a function of the distances between their edges. In this instance the figure shows that errors dropped sharply with increasing distances between knobs up to about 1 in (2.5 cm), while beyond that distance performance improved at a much slower rate. When separate comparisons were made between knob centers (rather than edges), however, performance was more nearly error-free for knobs of ½-in (1.2-cm) diameter than for the larger knobs. This suggests that when panel space is at a premium, the

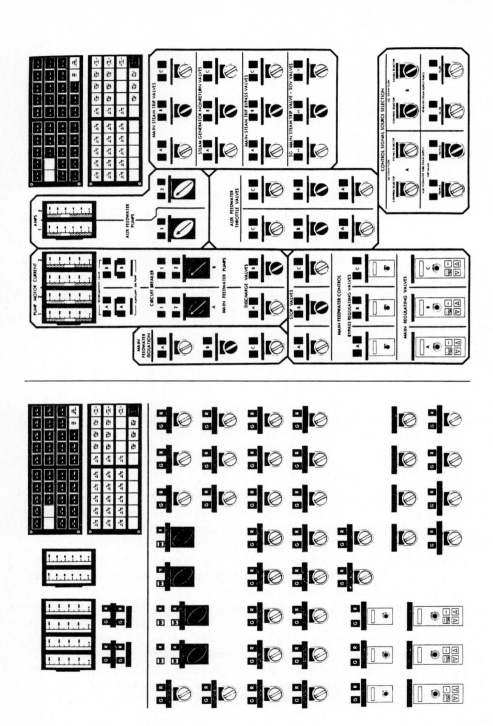

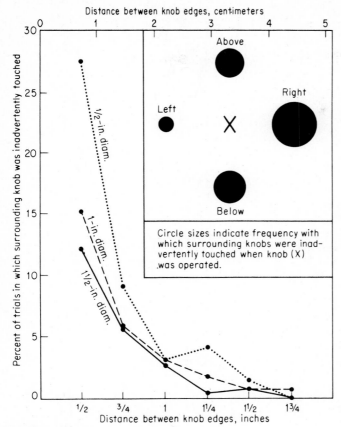

FIGURE 13-16
Frequency of inadvertent touching errors for knobs of various
diameters as a function of the distance between knob edges.
Areas of circles in inset indicate relatively the frequency with which
the four surrounding knobs were inadvertently touched when knob
x was being operated. *(Source: Bradley, 1969, fig. 2.)*

smaller-diameter knobs are to be preferred. By referring again to Figure 13–16, it
can be seen that touching errors were greatest for the knob to the right of the one to
be operated and were minimal for the one on the left.

On the basis of various studies of control devices Chapanis (1972, chap. 8) set

---

FIGURE 13-15 *(page 382)*
Simplified illustration of part of a control room panel of a nuclear power plant, showing the
original design (on the left) and an enhancement of that same arrangement made by drawing
lines around related groups of modules (on the right). Since the nature of the functional groups
is not relevant for illustrative purposes (and because of the reduced size), the labels of the
groups are not given. *(Copyright 1979, Electric Power Research Institute, EPRI Report NP–118,
"Human Factors Methods for Nuclear Control Room Design." Reprinted with permission.)*

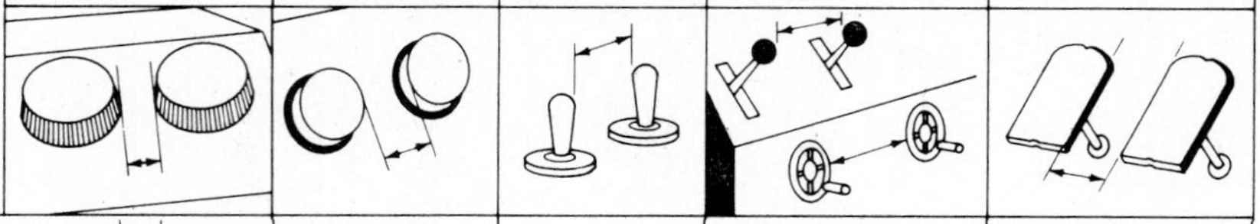

Number of body members and type of use		Knobs	Push buttons	Toggle switches	Cranks, levers	Pedals
1, randomly	in	2(1)	2(½)	2(¾)	4(2)	6(4)
	cm	5(2.5)	5(1.3)	5(1.8)	10(5)	15(10)
1, sequentially	in		1(¼)	1(½)		4(2)
	cm		2.5(.6)	2.5(1.3)		10(5)
2, simultaneously	in	5(3)			5(3)	
	cm	12.7(7.6)			12.7(7.6)	
2, randomly, sequentially	in		½(½)	¾(5/8)		
	cm		1.3(1.3)	1.8(1.6)		

FIGURE 13-17
Recommended separation (in inches and centimeters) between adjacent controls. Preferred separations are given for certain types of use with corresponding minimum separation in parentheses. *(Source: Adapted from Chapanis, 1972.)*

forth certain recommended distances (preferred and minimal) between pairs of similar devices, as given in Figure 13–17.

GENERAL GUIDELINES IN DESIGNING INDIVIDUAL WORKPLACES

In designing workplaces some compromises are almost inevitable because of competing priorities. In this regard, however, appropriate link values can aid in the trade-off process. Some general guidelines for designing workplaces that involve displays and controls are as follows (Van Cott and Kinkade, 1972, chap. 9):

- *First priority:* Primary visual tasks
- *Second priority:* Primary controls that interact with primary visual tasks
- *Third priority:* Control-display relationships (put controls near associated displays, compatible movement relationships, etc.)
- *Fourth priority:* Arrangement of elements to be used in sequence
- *Fifth priority:* Convenient location of elements that are used frequently
- *Sixth priority:* Consistency with other layouts within the system or in other systems

DISCUSSION

The built features of our environments include many "things" that we use: machines, computers, desks, other office equipment, storage spaces, kitchen cabinets, control panels, hospital laboratories, etc. The physical location and arrangement of such items (and of their specific features) clearly can affect the effectiveness and safety with which they are used as well as the comfort and physical well-being of

FIGURE 13-18
The control room of the Three-Mile Island nuclear power plant. *(Photograph courtesy of Dr. Thomas B. Malone, Essex Corporation.)*

the users. In many circumstances (especially where very few items or components are to be used) there is no particular problem in developing an arrangement that is reasonably satisfactory. In some situations, however (especially where there are many components), the development of a satisfactory arrangement can be very tricky. Then the use of systematic human factors analysis procedures and data usually is in order. An example of a very complex situation is given in Figure 13–18, this being the control room of the Three-Mile Island nuclear power plant. In connection with this installation, the chairperson of the presidential committee investigating the accident at Three-Mile Island is quoted as saying, "It looks to me like this is very bad human engineering." Subsequent investigations confirmed this opinion. No single picture can illustrate all the specific human factors deficiencies of the control system, but Figure 13–18 may at least give some impression of the problems of controlling such a facility and of the importance of locating and arranging the many components of systems so as to simplify the work demands of the people involved and to enhance the likelihood of satisfactory work performance.

REFERENCES

Banks, W. W., and Boone, M. P. (1981). A method for quantifying control accessibility. *Human Factors, 23*(3), 299–303.

Bonney, M. C., and Williams, R. W. (1977). CAPABLE: A computer program to layout controls and panels. *Ergonomics, 20*(3), 297–316.

Bradley, J. V. (1969). Optimum knob crowding. *Human Factors, 11*(3), 227–238.

Chapanis, A. (1972). Design of Controls. In H. P. Van Cott and R. G. Kinkade (eds.), *Human engineering guide to equipment design* (rev. ed.). Washington: Government Printing Office.

Cullinane, T. P. (1977). Minimizing cost and effort in performing a link analysis. *Human Factors, 19*(2), 151–156.

Deininger, R. L. (1958). *Process sampling, workplace arrangements, and operator activity levels*. Unpublished report, Engineering Psychology Branch, U.S. Air Force, WADD.

Dupuis, H. (1957). Farm tractor operation and human stresses. Paper presented at the meeting of the American Society of Agricultural Engineers, Chicago: December 15–18, 1957.

Dupuis, H., Preuschen, R., and Schulte, B. (1955). *Zweckmäbige gestaltung des schlepperführerstandes*. Dortmund, Germany: Max Planck Institutes für Arbeitphysiologie.

Electric Power Research Institute. (1979). *Human factors methods for nuclear control room design, vol. 1: Human factors enhancement of existing nuclear control rooms* (EPRI NP-1118). Palo Alto, CA: Prepared by Lockheed Missiles and Space Co., Inc.

Fitts, P. M. (1947). A study of location discrimination ability. In P. M. Fitts (ed.), *Psychological research on equipment design*, Res. Rept. 19. Army Air Force, Aviation Psychology Program.

Fowler, R. L., Williams, W. E., Fowler, M. G., and Young, D. D. (1968, December). An investigation of the relationship between operator performance and operator panel layout for continuous tasks, Tech. Rept. 68–170. U.S. Air Force AMRL (AD-692 126).

Francis, R. L., and White, J. A. (1974). *Facility layout and location: An analytical approach*. Englewood Cliffs, NJ: Prentice-Hall.

Geer, C. W. (1981). *Human engineering procedure guide* (AFAMRL-TR-81-35). Wright-Patterson AFB, Ohio.

Haines, R. F., and Gilliland, K. (1973). Response time in the full visual field. *Journal of Applied Psychology*, 58(3), 289–295.

Hertzberg, H. T. E., and Burke, F. E. (1971). Foot forces exerted at various aircraft brake pedal angles. *Human Factors*, 13(5), 445–456.

Huebner, W. J., Jr., and Ryack, B. L. (1961, March). *Linear programming and work place arrangment: Solution of assignment problems by the product technique*, Tech. Rept. 61–143. U.S. Air Force, Air Research and Development Command, WADD.

Human factors engineering (AFSC DH 1–3, U.S. Air Force Systems Command Design Handbook 1–3, 3d ed., rev. 1). (1980, June 25). Wright Patterson Air Force Base, OH: Aeronautical Systems Division.

Jones, R. E., Milton, J. L., and Fitts, P. M. (1949). *Eye fixations of aicraft pilots: IV. Frequency, duration, and sequence of fixations during routine instrument flight*, U.S. Air Force Tech. Rept. 5975.

Nemeth, S. E., and Blanche, K. M. (1982). Ergonomic evaluation of two-hand control location. *Human Factors*, 24(5), 567–571.

Sharp, E., and Hornseth, J. P. (1965, October). *The effects of control location upon performance time for knob, toggle switch, and push button*, Tech. Rept. 65–41. AMRL.

Siegel, A. I., and Brown, F. R. (1958). An experimental study of control console design. *Ergonomics*, 1, 251–257.

Van Cott, H. P., and Kinkade, R. G. (1972). Design of individual workplaces. In H. P. Van Cott and R. G. Kinkade (eds.), *Human engineering guide to equipment design* (rev. ed.). Washington: Government Printing Office.

PART FIVE

ENVIRONMENT

CHAPTER 14

ILLUMINATION

In many aspects of life we depend on the sun as our source of illumination, as in driving in the daylight, playing golf, and picking tomatoes. When human activities are carried on indoors, or at night, however, it is usually necessary to provide some form of artificial illumination. There does not, however, appear to be much of a difference between artificial illumination and natural daylight for performance of simple visual tasks, such as threading needles or proofreading (Santamaria and Bennett, 1981). The design of artificial illumination systems, as we shall see, does have an impact on the performance and comfort of those using the environment as well as on the affective responses of the people to the environment. Illuminating engineering is both an art and a science. The scientific aspects include the measurement of various lighting parameters and the design of energy efficient lighting systems. The artistic side comes into play in combining light sources to create, for example, a particular mood in a restaurant, to highlight a display in a store, or to complement a particular color scheme. It is not our intention here to make anyone an illuminating engineer. Rather, our aim is to familiarize you with the basic concepts in the field and to illustrate the importance of proper illumination from a human factors points of view. We leave the artistic aspects to the artists and concentrate on the scientific aspect of illumination. In recent years, much has been written about proper ambient lighting for areas where people work with visual display terminals (VDTs). In recognition of the special problems involved in designing a lighting environment for VDT use, we have devoted the last section of this chapter to those issues.

THE NATURE OF LIGHT

Light, according to the Illuminating Engineering Society (IES), is "radiant energy that is capable of exciting the retina (of the eye) and producing a visual sensation"

(IES Nomenclature Committee, 1979). The entire electromagnetic spectrum consists of waves of radiant energy that vary from about 1/1 billion of a millionth m to about 100 million m in length. This tremendous range includes cosmic rays; gamma rays; x-rays; ultraviolet rays; the visible spectrum; infrared rays; radar; FM, TV, and radio broadcast waves; and power transmission—as illustrated in Figure 14–1.

The visible spectrum ranges from about 380 to 780 nm. The *nanometer* (formerly referred to as a *millimicron*) is a unit of wavelength equal to 10^{-9} (one-billionth) m. Light can be thought of as the aspect of radiant energy that is visible; it is then basically psychophysical in nature rather than purely physical or purely psychological.

Color

Variations in wavelength within the visible spectrum give rise to the perception of color, the violets being around 400 nm, blending into the blues (around 450 nm), the greens (around 500 nm), the yellow-oranges (around 600 nm), and the reds (around 700 nm and above).

Just as the ear is not equally sensitive to all frequencies of sound, the eye is not equally sensitive to all wavelengths of light. Unlike the ear, however, the eye is composed of two basic receptors, rods and cones (see Chapter 4), each of which has its own sensitivity function. At high levels of illumination, the rods and cones both function (*photopic vision*) and the eye is most sensitive to light wavelengths around 550 nm (green). As illumination levels decrease, however, the cones cease to function, the rods take over the entire job of seeing (*scotopic vision*), and the eye becomes most sensitive to wavelengths around 500 nm (blue-green). This shift in sensitivity from photopic to scotopic vision is called the *Purkinje effect*. One practical application of this effect is that targets can be made blue-green to increase the probability of detection at night.

Light comes to us from two sources: *incandescent bodies* ("hot" sources, such as the sun, luminaires, or a flame) and *luminescent bodies* ("cold" sources, i.e., the objects we see in our environment, which reflect light to us). A hot light source that includes all wavelengths in about equal proportions is called *white light*. Most light

FIGURE 14–1
The radiant energy (electromagnetic) spectrum, showing the visible spectrum.
(Source: Adapted from Light and color, *1968, p. 5.)*

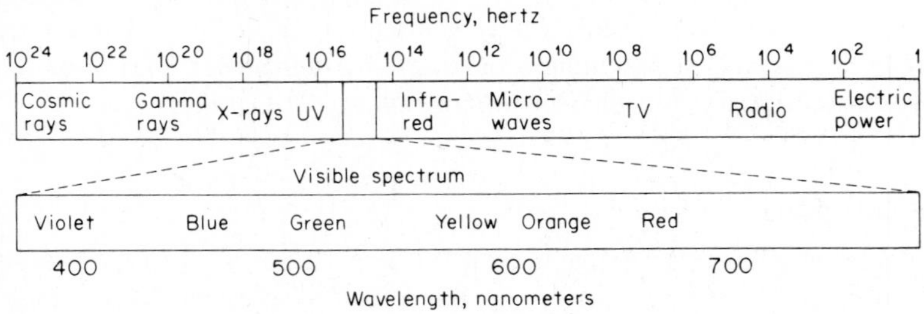

sources, as luminaires, have spectra that include most wavelengths but that tend to have more energy in certain areas of the spectrum than in others. These differences make the lights appear yellowish, reddish, bluish, etc. Illuminating engineers speak of the color temperature of a light source to describe its color appearance. The *color temperature* of a light source is the temperature at which the walls of a furnace must be maintained so that light from a small hole in it will have the same (or, in the case of fluorescent lamps, similar) color appearance as that of the light source. The noon sun, i.e., daylight, has a color temperature of 5500 K. Lower temperatures are more reddish, and higher temperatures are more bluish. Color temperature specifies only the color appearance of a light source and not the actual spectral composition of the light. Different spectral compositions can have the same color appearance but produce different colors from a surface illuminated by them. Therefore, color temperature is an incomplete and unreliable indicator of how objects will appear (in terms of color) when they are illuminated by the source. Color temperature does have value, however, for comparing the color appearance of similar types of light sources such as incandescent or fluorescent lights.

As light from a hot source falls upon an object, some specific combination of wavelengths is absorbed by the object. The light that is so reflected is the effect of the interaction of the spectral characteristics of the light source with the spectral absorption characteristics of the object. If a colored object is viewed under white light, it is seen in its *natural* color. If it is viewed under a light that has a concentration of energy in a limited segment of the spectrum, the reflected light may alter the apparent color of the object. We have probably all experienced, at one time or another, the difficulty of finding a familiar colored car in a parking lot illuminated by yellow sodium lights. The spectral composition of the yellow lights changes the perceived colors of the cars.

The light that is reflected from an object—and that produces our sensations of the object's color—can be described in terms of three characteristics: *dominant wavelength, luminance,* and *saturation.* The dominant wavelength gives rise to our sensation of *hue,* or color. Luminance, sometimes referred to as *value,* is associated with the relative amount of light reflected and gives rise to our sensation of *lightness,* or *brightness.* Although there is a general relationship between the luminance of light and the subjective response of lightness, not all colors that reflect equal total amounts of light energy are necessarily perceived as equal in lightness. This is due to the fact that the eye is differentially sensitive to various wavelengths, as we discussed earlier. Saturation is the predominance of a narrow range of wavelengths, or the degree of difference of a color from a gray with the same luminance. Saturation is also referred to as *purity,* or *chroma.*

Color Systems   It is very difficult to describe a particular color given the variations in hue, saturation, and lightness that are possible. Although difficult, it is often necessary to communicate such information without actually reproducing the color. For example, a design specification may have to describe the colors for, say, visual codes or labels. To assist in this communication process, various color systems have been developed to serve as standards for describing colors.

Dating at least to the seventeeth century, there have been dozens of attempts to order colors in some form of three-dimensional space (Hesselgren, 1984). Fortunately, the systems in use today have several features in common. Most systems are based on a color cone such as the one shown in Figure 14–2. Neutral-gray samples are located on the central vertical axis, going from black at the bottom to white at the top. The hues are arranged around the axis in the same sequence as the hues in the visual spectrum (rainbow), with purples being added where the reds and blues meet. The distance out from the central vertical axis represents the degree of saturation, with the intense hues on the periphery and the more "washed out" hues nearer the axis. The major color systems differ in the spacing of the gray samples between the black and white endpoints of the vertical axis, in the spacing of the hues around the hue circle, and in the location of the samples in each hue plane.

Several color systems consist of color plates or chips arranged in order based on the color cone. These systems include the Ostwald system (Container Corporation of America, 1942), the DIN color system (Richter, 1955), the natural color system (Hard and Sivik, 1981), and the Munsell system (Munsell Color Co., 1973). The Munsell system is probably the most widely known. It is intended to present the colors so that they are subjectively evaluated as differing from each other by equal amounts of the three basic color attributes. The system contains over 1000 color samples, each given a unique notation based on its hue (dominant wavelength) value (luminance), and chroma (saturation). For example, a strong red would be designated 7.5R/4/12. For a more detailed discussion of various color order systems, see Nimeroff (1968).

Another widely used system for describing color is the Commission Internationale de l'Eclairage (CIE) colorimetric system (CIE, 1971). The system is based on the fact that any color can be matched by a combination of three spectral wavelengths of light in the red, green, and blue regions. The CIE system determines the relative

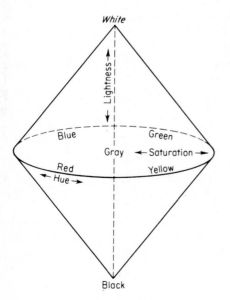

FIGURE 14–2
The color cone. Hue is shown on the circumference, lightness (from light to dark) on the vertical, and saturation on the radius from circumference to the center.

proportions of three imaginary light sources needed to produce a color. These imaginary light sources are mathematically derived and are aptly named *X, Y,* and *Z.* The *X* source corresponds, roughly, to the red region of the spectrum, the *Y* source to the green region, and the *Z* source to the blue region. By using techniques beyond the scope of this book, the relative proportion of each of these sources that produces a particular color is determined. [See Hardesty and Projector (1973) for computational procedures and examples.] These proportions, designated *x, y,* and *z,* are called the *color coordinates* of the color being matched. Since the three coordinates must add to 1.0, only two coordinates (*x* and *y*) are used to specify a color (*z* can be computed by subtraction). Figure 14-3 shows the CIE 1931 chromaticity diagram produced by combinations of the *x* and *y* color coordinates. All pure colors (that is, those which consist of a single wavelength) lie on the outer edge of the figure (the edge is called the *spectrum locus*). Figure 14-3 shows some of the wavelengths from 400 to 770 nm on the spectrum locus. Also indicated in the figure is the equal-energy point *E* which represents a perfectly white, colorless surface. The saturation of a color in-

FIGURE 14–3
The CIE 1931 chromaticity diagram with descriptions of the various color regions. Here *A, B,* and *C* represent the positions of standard sources roughly equivalent to gas-filled incandescent lamps, noon sunlight, and average daylight, respectively; *E* represents the equal-energy point corresponding to perfect white, colorless light. See text for discussion of the *x* and *y* coordinates. *(Source: Adapted from Hardesty and Projector, 1973.)*

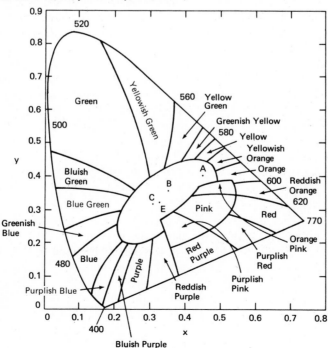

creases as one moves toward the spectrum locus and away from the equal energy point. The color names in Figure 14-3 describe the general color regions of the diagram; the colors actually blend from one to the other.

Note that several CIE chromaticity diagrams exist. The CIE 1931 diagram, the most frequently referenced, is for use with colors that occupy visual angles up to 4°. The CIE 1964 diagram is used for colors subtending larger visual angles. Both these diagrams cover only hue and saturation, which together make up the quality of *chromaticity*. Brightness is not addressed by either diagram.

Measurement of Light

There are many concepts and terms that relate to the measurement of light (*photometry*). We define a few of them here and show their interrelationships. One source of confusion to the photometric novice is the multitude of different measurement units used in the field. Part of this problem stems from the existence of two systems of measurement, the U.S. Customary System (USCS) and the International System of units (or SI units). We bow to the rest of the world and the scientific community and encourage the use of SI units. The USCS units are mentioned, however, and conversions are given to aid those of us who were weaned on footcandles and foot-lamberts.

The fundamental photometric quantity is *luminous flux*, which is the rate at which light energy is emitted from a source. The unit of luminous flux is the *lumen* (lm). Luminous flux is a somewhat esoteric concept which is similar to other flow rates such as gallons per minute. Time is implied in the unit of luminous flux. Thus one would say that light is emitted from a 100-W incandescent lamp at a rate of 1740 lm. The *luminous intensity* of a light source is measured in lumens emitted by the source per unit solid angle.[1] The unit of luminous intensity is the *candela* (cd). A 1-cd source emits 12.57 lm.

Consider a source of some luminous intensity emitting luminous flux in all directions. Imagine the source as being placed inside (at the center of) a sphere. The amount of light striking any point on the inside surface of the sphere is called *illumination*, or *illuminance*. It is measured in terms of luminous flux per unit area, as, for example, lumens per square foot (lm/ft^2) or lumens per square meter (lm/m^2). To obscure matters a bit, special names have been given to units of illuminance. One lumen per square foot is called a *footcandle* (fc), a USCS unit, whereas 1 lumen per square meter is called a *lux* (lx), an SI unit. One footcandle equals 10.76 lx; however, an accepted practice for some purposes is to consider 1 fc to equal 10 lx and to forget the fraction (Kaufman and Christensen, 1984).

Figure 14-4 illustrates the relationship among candelas, footcandles, and lux. Our 1-cd source emits 12.57 lm, and the total surface area of a sphere equals 12.57 times the radius squared (or $4\pi r^2$). At a radius of 1 m, therefore, our 1-cd source is distributing 12.57 lm evenly over the 12.57 m^2 of surface area. Thus, the amount of light on any point is 1 lm/m^2, or 1 lx. The same logic holds at a radius of 1 ft for the footcandle.

---

[1] A solid angle is measured in steradians (sr). There are 12.57 sr in a sphere.

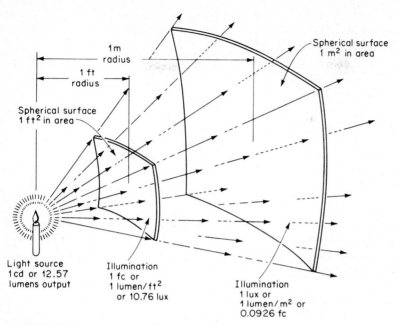

FIGURE 14–4
Illustration of the distribution of light from a light source following the inverse-
square law. *(Source:* Light measurement and control, *1965, p. 5.)*

The amount of illumination striking a surface from a point source follows the
inverse-square law:

$$\text{Illuminance (lx)} = \frac{\text{candlepower (cd)}}{D^2}$$

where D is the distance from the source in meters. At 2 m a 1-cd source would
produce ¼ lx, and at 3 m it would produce ⅑ lx.

We now have light striking a surface (*illuminance*), but what happens to light that
strikes a surface? Some of it is absorbed, and some of it is reflected. The light that
is reflected from the surface of objects is what allows us to "see" objects—their
configuration and color. The amount of light per unit area leaving a surface is called
luminance. The light leaving the surface may be reflected by the surface or emitted
by the surface, as would occur with a fluorescent light panel. The amount of light
can be measured in terms of luminous flux (lumens) or luminous intensity (candelas).
When the amount of light is measured in lumens and area is in square feet, the USCS
unit of luminance is the *foot-lambert* (fL). When the amount of light is measured in
candelas and the area is in square meters, the SI unit of luminance is *candela per
square meter* (cd/m²). Actually numerous measures of luminance are used in the
literature. Table 14-1 presents conversion factors for changing some to SI units.

TABLE 14-1
CONVERSION FACTORS FOR VARIOUS UNITS OF LUMINANCE

Unit	To convert to cd/m², multiply by:
Nit	1.0
Stilb	1×10^4
Apostilb	0.3183
Millilambert	3.183
Foot-lambert	3.426
Candles per square inch	1.55×10^3
Candles per square foot	10.76

The ratio of the amount of light (luminous flux) reflected by a surface (luminance) to the amount of light striking the surface (illuminance) is called the *reflectance*. The formula for reflectance depends on the type of units being used.[2] For a perfectly diffuse surface, the formula in SI units is

$$\text{Reflectance} = \frac{\pi \times \text{luminance (cd/m}^2)}{\text{illuminance (lx)}}$$

In USCS units it is

$$\text{Reflectance} = \frac{\text{luminance (fL)}}{\text{illuminance (fc)}}$$

Reflectance is expressed as a unitless proportion. From the SI formula we can see that if a perfect reflecting diffuse surface (i.e., reflectance = 1.0) were illuminated by 1 lx, the surface luminance would be $1/\pi$ cd/m². When the surface being considered is not perfectly diffuse, reflectance is replaced by the luminance factor. The *luminance factor* is the ratio of the luminance of a surface viewed from a particular position and lit in a specified way to the luminance of a diffusely reflecting white surface viewed from the same direction and lit in the same way.

LAMPS AND LUMINAIRES

The term *lamp* is a generic term for an artificial source of light. A *luminaire,* however, is a complete lighting unit consisting of a lamp or lamps together with the parts designed to distribute the light, to position and protect the lamps, and to connect the lamps to the power supply (IES Nomenclature Committee, 1979). A trip to a lighting store will quickly convince you that lamps come in hundreds of configurations and sizes and that luminaire designs are limited only by the taste of the designer, ranging from simple globes to elaborate crystal chandeliers.

---

[2]The reason for two reflectance formulas is that, in SI units luminance is measured in luminous flux per unit area per solid angle, while in USCS units luminance is measured in luminous flux per unit area.

In the next two sections we introduce some of the generic types of lamps and luminaires and discuss some of the implications of choosing one or another.

Lamps

There are two main classes of lamps: *incandescent filament lamps,* in which light is produced by electric heating of a filament or by combustion of gases within a thin mesh mantle, and *gas-discharge lamps,* in which light is produced by the passage of an electric current through a gas.

Gas-discharge lamps are of three types: high-intensity discharge (HID) lamps including mercury, metal halide, and high-pressure sodium lamps; low-pressure sodium lamps; and fluorescent lamps. In the case of fluorescent lamps, the radiation created by the electric current passing through the gas is invisible to the eye (it is ultraviolet radiation). This radiation, however, is used to excite a phosphor coating on the inner surface of the bulb, which in turn produces visible radiation. Different phosphors can create different colors of illumination.

Lamp Color   The various type of lamps differ markedly in terms of their characteristic spectral distribution, or what we plain folk call *color.* Figure 14-5 shows representative distributions for several types of lamps. These differences can affect task performance and the subjective impressions of the people in the illumination environment. The CIE (1974) has developed the general *color rendering index* (CRI) for quantifying the accuracy with which any specific light source renders colors. As pointed out by Boyce (1981), however, the CRI gives only gross indications of color rendering accuracy. Nonetheless, the CRI is a good indicator of expected performance in color judgment tasks that involve a wide range of colors. As the CRI increases, color judgment errors tend to decrease, as can be seen in Figure 14-6, which presents results of a color judgment task under various types of light sources with different CRIs.

On an even more detailed level, Delaney et al. (1978) tested subjects under various types of fluorescent illumination (cool white, deluxe cool white, warm white, deluxe warm white, etc.). They found that subjects made fewer color discrimination errors under deluxe lamps than under standard prime color lamps. In addition, subjects preferred the cool white color temperatures (5000 K) in terms of clarity and brightness.

Lion (1964) found that subjects performed three manipulative tasks under fluorescent lighting faster, and with less detrimental effect on accuracy, than under incandescent tungsten lighting. This, however, may be due to differences in the distribution of light rather than color. Fluorescent lamps tend to produce more diffuse lighting with less glare than do incandescent lamps.

Energy Considerations   In today's world, energy conservation has become a prime goal in system design, and this is especially true in design of lighting systems. The efficiency of a light source, called *lamp efficacy,* is measured in terms of the

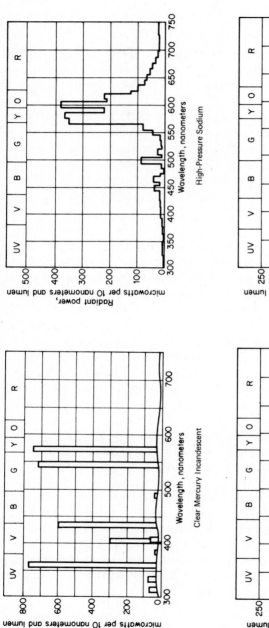
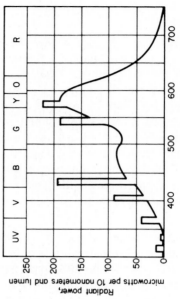
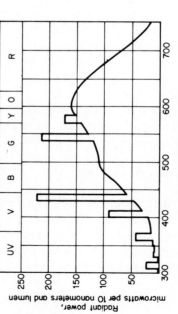

FIGURE 14-5
Representative spectral distributions (chromaticity diagrams) for clear mercury, high-pressure sodium, and two fluorescent lamps. (*Source:* IES Lighting Fundamentals Course, *1976, figs 4–5, 4–13, 4–23.*)

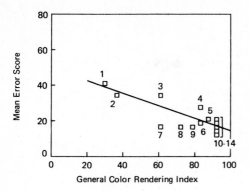

FIGURE 14–6
Mean error scores on the Farnsworth-Munsell 100-hue test performed under different light sources with different color rendering indices. The light sources are: 1, high-pressure sodium discharge; 2, high-pressure mercury discharge; 3, Homelite fluorescent; 4, Tri-band fluorescent; 5, Kolor-rite fluorescent; 6, natural fluorescent; 7, daylight fluorescent; 8, Plus-White fluorescent; 9, high-pressure mercury discharge with metal halide; 10 to 14, artificial daylight fluorescent. *(Source: Adapted from Boyce, 1981, fig. 4.14.)*

amount of light produced [lumens (lm)] per unit of power consumed [watts (W)]. Figure 14-7 shows some typical lamp efficacy values for various types of light sources.

Switching from incandescent to fluorescent lights can result in real dollar savings. For example, consider the comparison of two light sources (incandescent and fluorescent) shown in Table 14-2. The fluorescent tube represents a 41 percent energy savings, lasts 20 times longer, and gives 30 percent more light to boot! And who said there was no such thing as a free lunch?

Incidentally, don't be fooled into thinking all incandescent bulbs of the same wattage give off the same amount of light—they don't! Bulb manufacturers are now required to put the tested lumen output on their packages. For example, a long-life 100-W bulb may give off as little as 1470 lm compared to perhaps 1740 lm for a

FIGURE 14–7
Lamp efficacy ranges for common light sources. *(Source: IES Lighting Fundamentals Course, 1976, fig. 3–4.)*

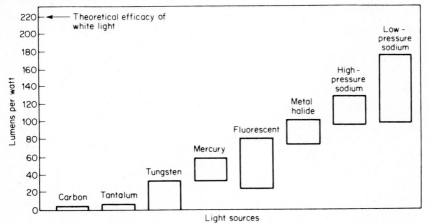

TABLE 14–2
COMPARISON OF TWO LIGHT SOURCES

	Incandescent	Deluxe fluorescent
Wattage	75 W	30 W (44-W total input)
Rated life	750 h	15,000 h
Amount of light	1,180 lm	1,530 lm

Source: IES Lighting Fundamentals Course, 1976.

standard 100-W bulb. Proper choice of light source often can result in substantial energy and cost savings—a worthy goal for any human factors design effort.

Luminaires

Luminaires are classified into five categories based on the proportion of light (lumens) emitted above and below the horizontal, as shown in Figure 14-8. In selecting a particular type of luminaire for use, consideration must be given to the pattern of light distribution, glare, task illumination, shadowing, and energy efficiency. Various devices can be incorporated into a luminaire to control the distribution of light, including lenses, diffusers, shielding, and reflectors. Choice of an efficient luminaire

FIGURE 14–8
Types of luminaires based on the proportion of lumens emitted above and below the horizontal.
(Source: IES Lighting Fundamentals Course, 1976, fig. 5–6.)

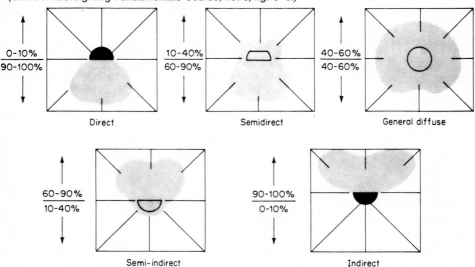

is a complex decision and should be made by a qualified, experienced person after an analysis of the lighting needs and physical environment has been made.

THE CONCEPT OF VISIBILITY

Visibility refers to how well something can be seen by the human eye. Visibility, therefore, involves human judgment. There is no device that can measure visibility directly; a human must always be involved in its determination. One key factor influencing visibility of a target is how well it stands out from its background, that is, its *contrast,* which was defined in Chapter 4. Although contrast is related to visibility, it is not the same as visibility. DiLaura (1978a) provides a simple illustration of this. Imagine a target with a contrast equal to 0.5 placed on a stage and lit up with a flashlight held in the balcony. It would be hardly visible. Now illuminate this same 0.5 contrast target with a stage light 10,000 times the intensity of the flashlight— same contrast, but hardly the same level of visibility. From this we can see that both contrast and luminance are important for visibility. Another factor is the size of the target; an elephant on the stage would be more visible than a mouse.

Visibility, we said, was how well something can be seen, but seeing can be defined in different ways. For example, *seeing the elephant* might mean ''detecting its presence on stage,'' or it could mean ''recognizing the elephant as an Indian elephant.'' These different definitions, or *information criteria,* obviously have a bearing on the visibility level of the target. One last variable we should mention is exposure time to the target. The more time you have to look at the elephant, the easier it becomes to recognize it as an Indian elephant (they have smaller ears.)

Incidentally, the visibility of a target should not be a function of the observer; visibility should be a measure of the task itself. We would like a measure of task visibility that does not change each time a different person uses it.

Measuring Task Visibility

To determine the level of visibility for a particular target, we must compare it against some standard level of visibility, much as we do when we determine length by comparing it to a standard meter stick. In the case of visibility, this standard must be specified in terms of the target, the illumination environment, and the information requirements of the task. The choice of these conditions is somewhat arbitrary. But because of the extensive research carried out by Blackwell (Blackwell, 1959, 1961, 1964, 1967; Blackwell and Blackwell 1968, 1971) the standard target has become a uniformly luminous disk which subtends a visual angle of 4 minutes (approximately 1.1 mm at a distance of 1 m) presented in pulses of $\frac{1}{5}$ s on a uniformly luminous screen. The information requirement is to detect the presence of the disk. Incidentally, the pulses are used to simulate the manner in which we usually acquire visual information (DiLaura, 1978b).

Blackwell sought to determine the *threshold of visibility* for the standard target, that is, the point at which subjects could detect the target 50 percent of the time when it was actually present. To do this, he had observers view a uniform screen having a given level of luminance. The standard 4-minute luminous disk was then presented at the center of the screen in ⅕-s pulses. The observer adjusted the physical contrast of the disk (i.e., increased or decreased the luminance of the disk) until it was judged to be just barely visible. An interesting phenomenon occurred when the screen luminance was changed. Blackwell found that the observers had to readjust the disk's contrast so that it was again at threshold. If screen luminance was decreased, the contrast of the disk had to be increased to bring it back to threshold. The higher the screen, or background, luminance, the lower was the contrast required to bring the disk to threshold.

Blackwell performed this experiment on approximately 100 young observers and determined the average relationship between background luminance and threshold disk contrast. This he called the *Visibility Reference Function*, or *Visibility Level 1* (VL 1); it is shown in Figure 14-9.

Blackwell performed this basic experiment under a variety of conditions, for example, using a dynamic (moving) target, telling and not telling subjects where and when the target might appear in the visual field, and requiring 99 percent probability of detection rather than just 50 percent. In each case the shape of the curve was the same, but the curve was displaced upward by a constant multiple of the Visibility Reference Function. The multipliers are as follows:

Dynamic presentation	2.78
Not knowing where or when	1.50
99 percent detection	1.90

Blackwell reasoned that with all three conditions present the curve would be approximately 8 times visibility level 1 (2.78 × 1.50 × 1.90 = 7.92). This he termed VL

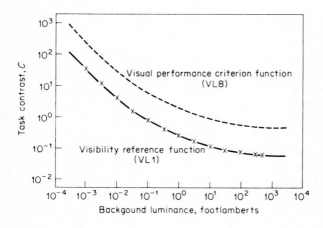

FIGURE 14–9
Two visibility-level (VL) curves, each representing combinations of contrast and luminance required for equal visibility. VL 1 (visibility reference function) is a basic curve relating to the luminance needed to achieve threshold visibility of a static task of discriminating a 4-minutes luminous disk exposed for ⅕ s. In turn, VL 8 (visual performance criterion function) shows the level adjusted for three factors: dynamic versus static target; not knowing versus knowing where and when the target will appear; and requiring a 99 versus 50 percent probabilty of detection. *(Source:* IES Lighting Handbook, *1972, fig. 3–24, as based on work of Blackwell.)*

8, and it is also shown in Figure 14–9. Incidentally, no one has ever empirically determined that under these three conditions the resulting curve would indeed be 8 times VL 1 (DiLaura, 1978c).

We now have a scale of visibility—visibility level (VL). Given the contrast of a 4-minute luminous disk and the luminance of its background, one can determine its VL (i.e., the degree to which the target is above threshold), expressed as a multiple of threshold visibility—VL 2, VL 3.4, etc. This is just fine if all we ever did in life was detect the presence of little luminous dots. The problem, of course, is to somehow relate practical real-life visual tasks to our reference target.

This bit of magic is done by using a *contrast-reducing visibility meter,* such as the Blackwell Visual Task Evaluator (VTE). The VTE superimposes a field of uniform luminance, called the *veiling field,* over the view of the target as seen through the device. By increasing the luminance of the veiling field and simultaneously decreasing the luminance of the target field, the target appears to disappear, as if a layer of mist had covered it. By performing some additional adjustments using the 4-minute disk reference, an index is obtained on the target task called *equivalent contrast $\bar{C}$.* Equivalent contrast can be interpreted as the contrast of the 4-minute luminous disk which makes it as difficult to detect as it is to ''see'' the practical (target) task. Whenever possible, equivalent contrast measurements should be carried out under reference lighting conditions established by a photometric sphere. In this situation, a practical target is placed at the center of a uniformly luminous sphere so that it is illuminated equally from all directions.

Follow this logic. Given a practical task, we can determine its equivalent contrast and measure the actual background luminance. Using Figure 14-9, we can locate the point corresponding to $\bar{C}$ and background luminance, and then we can determine how many times greater than VL 1 is $\bar{C}$ at the same background luminance. This gives us the VL for our practical task—under reference lighting conditions. The real lighting environment, however, is likely to illuminate the practical target so that it is more or less visually difficult than it was under the reference lighting conditions. With a little more magic, which we need not explain here, the VL of the task under its actual lighting environment can be determined. This is called the *effective VL* (VL_{eff}) and is a general metric for expressing a practical target's suprathreshold visibility in a real lighting environment (DiLaura, 1978b).

The procedure outlined above, however, is not without its critics (Brass, 1982; Rea, 1984). The VL curve may be a good representation of the response of the visual system to just perceptible points of light, but it may not be such a good representation of how we ''see'' more complex, practical tasks. It does not deal with suprathreshold perception or perception outside of the central fovea of the eye, nor does it take into account individual differences in visual sensitivity. Brass questions the slight slope in the VL curve between 10 and 3000 fL (31.8 to 9550 cd/m^2) and contends that this can result in recommended illumination levels of 60,000 fc (600,000 lx) for sewing black thread on black cloth—a value that is absurdly high.

This system of $\bar{C}$ and VL forms the basis for establishing recommended levels of illumination as proposed by the CIE. The Illuminating Engineering Society of North America also used this system for recommending illumination levels (*IES Lighting*

Handbook, 1972) but has since adopted a simpler approach, which we describe later (*IES Lighting Handbook,* 1981).

EFFECTS OF LIGHTING ON PERFORMANCE

We have already related bits and pieces of information about the performance effects of lighting, including accuracy of color judgment under various types of lamps. As pointed out by Boyce (1981), lighting itself cannot produce work output. What lighting can do is to make details easier to see and colors easier to discriminate without producing discomfort or distraction. Workers can then use this increased ease of seeing to increase output, if they have the motivation and ability to do so.

Field Studies

There have been numerous studies relating the amount of illumination to task performance. These studies can be categorized by the research setting in which they were performed (field versus laboratory) and the types of tasks investigated (real-world task versus standardized visual task). A study carried out by Stenzel (1962) is an example of using a real-world task in a field setting. He measured the output from a leather factory over a 4-year period during which the illumination was changed from 350 to 1000 lx. Output increased when the illumination level was increased. Knave (1984) presents an example of a study using a standardized visual task (visual acuity eye chart) in a field setting (a foundry). He reports that increasing illumination improved acuity as measured by a low-contrast eye chart.

   Field studies often lack sufficient experimental controls to rule out alternative explanations for the changes in performance under the lighting conditions tested. Often, more than just the lighting changes. Such things as supervision, the task itself, and the physical design of the workplace are often altered along with the change in the illumination. Further, the change in illumination often involves multiple lighting changes. Not only might the illumination be increased, but also different types of lamps and luminaires may be used with different spectral characteristics, along with changes in the number and position of the lamps. Under such conditions it is not possible to isolate the specific causal factor that changed performance. All we can say is that performance was not the same under the conditions tested.

Laboratory Studies

In an attempt to overcome the problems inherent in field studies, investigators have moved into the more controlled laboratory setting. Here greater control can be exerted over outside influences, and the specific lighting characteristics of interest can be manipulated independently of other lighting parameters. Here, too, investigators have used real-world and standardized visual tasks. An example of using real-world tasks in the laboratory is the work done by Merritt et al. (1983) in which they investigated the effect of amount of illumination on detection of important features in underground mining, including holes and rubble on the floor and a crack appearing in a rock wall. The work of Blackwell with his 4-minute luminous disk task is a good example of laboratory studies using a standardized visual task.

Laboratory studies are not without problems, however. First, subjects know they are in an experiment and are usually motivated to perform well. This high motivation, however, could mask the effects of subtle lighting changes. Standardized visual tasks performed in the laboratory are usually simple tasks, often of a reading variety, which do not require depth perception, texture perception, perception of moving objects, or peripheral perception. This limits the generalizability of the findings. Since laboratory studies purposely exclude extraneous variables which might influence performance, there is always the possibility that the effect found in the pristine laboratory will ''wash out'' when it is brought into the real world, where other factors can influence performance.

General Conclusions

Although field and laboratory studies have their unique problems, when similar findings emerge from studies carried out in both settings, we can have more confidence in the conclusions. Fortunately, such is the case with the effects of amount of illumination on performance. Although we cannot review all the studies in this area, Figure 14-10 presents representative results relating illumination (illuminance) level to time required to complete several industrial tasks (Bennett, Chitlangia, and Pangrekar, 1977). The general conclusion from such research is that increasing the level of illumination results in smaller and smaller improvements in performance until performance levels off. The point where this leveling off occurs is different for different tasks. Generally, the more difficult the task, i.e., the smaller the detail or the lower the contrast, the higher the illumination level at which this occurs. Boyce (1981) also notes that larger improvements in visual performance can be achieved by changing features of the task (i.e., increasing its size or contrast) than by increasing the illumination level, at least over any range of practical interest. Further, it is rarely possible to make performance on a visually difficult task (small size, low contrast) reach the same level as that on a visually easy task simply by increasing the illumination level.

This fact has lead to some controversy over just how important amount of illumination is for task performance. Results from several sources (including Bennett, Chitlangia, and Pangrekar, 1977; Hughes and McNelis, 1978) indicate that the age of the observer has a major impact on the results, with age taking its toll; the amount of illumination may be more important for older persons. Ross (1978), after reviewing numerous studies, concludes that above 34 to 68 cd/m^2 (10 to 20 fL) of background luminance, other variables, notably age and print quality, are more important determiners of performance than amount of illumination. Ross further concludes that increasing illumination above 500 lx (50 fc) results in little additional improvement in task performance. This can be seen in Figure 14-10.

Although performance levels off with increased illumination, it may be that people must expend additional effort at lower illumination levels to maintain their performance. In essence, although performance levels off, the difficulty of the task may continue to decrease with additional illumination. Boyce (1971) presents some evi-

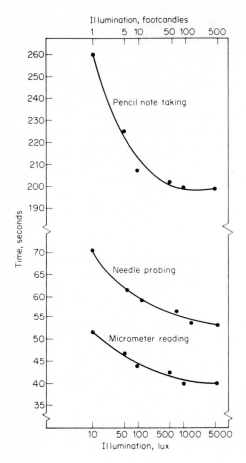

FIGURE 14–10
Relationship between amount of illumination and task completion time for three representative industrial tasks. *Source: Bennett, Chitlangia, and Pangrekar, 1977, figs. 1, 3, and 4.)*

dence to support this idea. Using a dual-task situation (see Chapter 3), he found that increasing task luminance from 40 to 80 cd/m^2 had no effect on the performance of the primary visual task, but did result in reduced errors on the secondary auditory task. This would indicate that higher levels of illumination may reduce the demands placed on workers' information processing systems and hence increase spare mental capacity. Simple laboratory tasks used to assess illumination effects may not be sensitive to changes in mental processing demands.

It may not always be wise, however, to provide high levels of illumination. Besides being wasteful of energy, too much illumination may create unwanted effects. Glare is one such effect which we discuss in the next section. Zahn and Haines (1971) offer another example. They varied the luminance level of a centrally located diffuse screen upon which a search task was presented. The luminance varied from 29.1 cd/m^2 (8.5 fL) to 23,325 cd/m^2 (6800 fL). They measured reaction time to peripheral lights and found that reaction time increased with increased central task luminance, the increase being greatest for the more peripheral locations. Thus, if peripheral detection is important, too much central task luminance may actually degrade performance.

Another example of the unpleasant effects of high levels of illumination comes from a study by Sanders, Gustanski, and Lawton (1974). "High" and "low" levels of illumination were created in the hallways of a university, and the noise level generated by the students waiting outside the classrooms was measured. The mean noise level under the high-illumination condition (20 large overhead fluorescent light panels lit) was 61.1 dBC, while under low illumination (two-thirds of the panels were turned off) the noise level generated was only 50.3 dBC. Here then, it appears, may be a silver lining in the energy crisis cloud—peace and quiet.

HOW MUCH IS ENOUGH?

The problem of determining the level of illumination that should be provided for various visual tasks has occupied the attention of illuminating engineers, psychologists, and others for many years. Over the years there has been a succession of recommendations, each claiming to provide adequate illumination for visual comfort and task performance. The recommended levels, however, continually increase. Current recommended levels are about 5 times greater than the levels recommended 30 years ago for the same tasks. The recommended levels increase as the efficiency of lighting improves.

One method for determining required illumination levels is based on the work of Blackwell which we discussed previously. This method is endorsed by the CIE (1972) and was used by the Illuminating Engineering Society (IES) until 1981. The approach involves determining the equivalent contrast of the task to be performed and then, by using the VL 8 curve in Figure 14-9, determining the appropriate background luminance required. Once the reflectance of the background is determined, it is an easy matter to determine the level of illumination required to produce the desired background luminance.

Because of uncertainties in the Blackwell approach, the IES in 1981 adopted a much simpler approach for determining minimum levels of illumination. The approach is based on consensus among lighting experts as to the level of illumination they considered adequate for the situation under study. The IES recognizes that the approach is an oversimplified solution to a difficult and complex problem; but if it is used with intelligence and judgment, this approach should aid in the design of effective lighting systems (Kaufman and Christensen, 1984).

The first step in the procedure is to identify the type of activity to be performed in the area for which illumination recommendations are sought. The *IES Lighting Handbook* (Kaufman and Christensen, 1984) contains extensive tables listing various types of tasks, each referencing one of the illumination categories shown in Table 14-3. Although we cannot reproduce the lists of tasks here, the column "Type of activity" in Table 14-3 contains enough information to give a sense of what is involved in each category. Note that categories *A* through *C* do not involve visual tasks.

Table 14-3 lists, for each category, a range (low–middle–high) of illuminances. To decide which of the three values to choose, one must compute a correction factor by using Table 14-4. For each characteristic in Table 14-4 (workers' ages, speed or accuracy, and reflectance of task background) a weight is assigned $(-1, 0, +1)$. These weights are then added algebraically to obtain the total weighting factor (TWF), (for example, $+1 - 1 = 0$). However, since categories A through C do not involve visual tasks, the *speed or accuracy* correction is not used for these cate-

TABLE 14-3
RECOMMENDED ILLUMINATION LEVELS FOR USE IN INTERIOR LIGHTING DESIGN

Category	Range of illuminances, lx (fc)	Type of activity
A	20–30–50* (2–3–5)*	Public areas with dark surroundings
B	50–75–100* (5–7.5–10)*	Simple orientation for short temporary visits
C	100–150–200* (10–15–20)	Working spaces where visual tasks are performed only occasionally
D	200–300–500† (20–30–50)†	Performance of visual tasks of high contrast or large size: e.g., reading printed material, typed originals, handwriting in ink and good xerography; rough bench and machine work; ordinary inspection; rough assembly
E	500–750–1000† (50–75–100)†	Performance of visual tasks of medium contrast or small size: e.g., reading medium-pencil handwriting, poorly printed or reproduced material; medium bench and machine work; difficult inspection; medium assembly
F	1000–1500–2000† (100–150–200)†	Performance of visual tasks of low contrast or very small size: e.g., reading handwriting in hard pencil on poor-quality paper and very poorly reproduced material; highly difficult inspection
G	2000–3000–5000‡ (200–300–500)‡	Performance of visual tasks of low contrast and very small size over a prolonged period: e.g., fine assembly; very difficult inspection; fine bench and machine work
H	5000–7500–10,000‡ (500–750–1000)‡	Performance of very prolonged and exacting visual tasks: e.g., the most difficult inspection; extra fine bench and machine work; extra fine assembly
I	10,000–15,000–20,000‡ (1000–1500–2000)‡	Performance of very special visual tasks of extremely low contrast and small size: e.g., surgical procedures

*General lighting throughout room.
†Illuminance on task.
‡Illuminance on task, obtained by a combination of general and local (supplementary) lighting.
Source: RQQ, 1980, table 1.

TABLE 14-4
WEIGHTING FACTORS TO BE CONSIDERED IN SELECTING THE SPECIFIC
ILLUMINATION LEVELS WITHIN EACH CATEGORY OF TABLE 14–3

Task and worker characteristics	Weight		
	−1	0	+1
Age	Under 40	40–55	Over 55
Speed or accuracy	Not important	Important	Critical
Reflectance of task background	Greater than 70%	30 to 70%	Less than 30%

Source: RQQ, 1980, table 2.

gories and the average reflectance of the room, walls, and floor are used rather than task background reflectance. The following rules are then applied to select the low, middle, or high value in the category.

For categories A through C;

$$\text{TWF} = -1 \text{ or } -2 \qquad \text{Use low value.}$$
$$\text{TWF} = 0 \qquad \text{Use middle value.}$$
$$\text{TWF} = +1 \text{ or } +2 \qquad \text{Use high value.}$$

For categories D through I;

$$\text{TWF} = -2 \text{ or } -3 \qquad \text{Use low value.}$$
$$\text{TWF} = -1, 0, \text{ or } +1 \qquad \text{Use middle value.}$$
$$\text{TWF} = +2 \text{ or } +3 \qquad \text{Use high value.}$$

This system can also be used if the equivalent contrast value of the task is known. A table is provided, but not reproduced here, that indicates the appropriate illumination category for any value of equivalent contrast. The calculation and use of the correction factors remain the same.

The IES (1982) also recommends minimum illumination levels for safety (as opposed to efficient visual performance of tasks). These values are shown in Table 14-5. Related to safety is the question of how much lighting is necessary to permit people to find their way out of a building during an emergency. Jaschinski (1982) simulated such an escape task under various levels of emergency lighting. He found that escape time and "spare mental capacity" worsened as the light level decreased below 2 lx. Where many elderly people are apt to be present, a minimum of 4-lx emergency lighting is recommended.

Although high levels of illumination might be warranted in some circumstances, there are definite indications that for certain tasks high levels of illumination can weaken information cues (Logan and Berger, 1961) by suppressing the "visual gradients" or the pattern density of the objects being viewed. An illustration of this

TABLE 14-5
RECOMMENDED LEVELS OF ILLUMINATION FOR SAFETY PURPOSES

| | Hazard requiring visual detection | | | |
| | Slight | | High | |
Activity level:	Low	High	Low	High
Recommended illuminance level (lx)	5.4	11	22	54

Source: IES, 1982, table 7.

effect is shown in Figure 14-11, which reveals the effects of general versus "surface-grazing" illumination. Thus, in some circumstances lower levels of illumination (albeit of the appropriate type) would be better than high levels. The illumination recommendations, for at least some tasks, may have to be tempered by experience.

DISTRIBUTION OF LIGHT

In addition to the overall level of illumination in the environment, one must also consider the distribution of light and its effects on visual comfort and task performance. The eye adapts to the ambient luminance it encounters (see Chapter 4). If there are great differences in luminance between areas in the work environment, the eye must adapt to these different levels as one shifts one's gaze from place to place. This *transient adaptation* reduces visibility momentarily until the eye fully adapts to the new luminance level. In addition, large differences in luminance can be a source of glare and cause visual discomfort.

FIGURE 14–11
Illustration of the effects of general versus surface-grazing illumination on the visibility of a loose thread on cloth *(Source: Faulkner and Murphy, 1973, figs. 3 and 4.)*

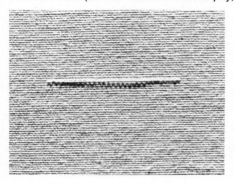

Luminance Ratio

The *luminance ratio* is the ratio of the luminances of any two areas in the visual field. The IES (1982) recommendations for maximum luminance ratios in office environments (these also hold for other types of work environments) are as follows:

Task : adjacent surroundings   3 : 1
Task : remote darker areas   10 : 1
Task : remote lighter areas   1 : 10

Brass (1982) contends, however, (1) that there are practically no real-life luminous environments providing overall luminance ratios of less than 20 : 1 and (2) that such ratios do not cause undue visual discomfort or transient adaptation problems. This would appear to be one area where intelligence and judgment should play a role in applying the IES recommendations.

Reflectance

The distribution of light within a room not only is a function of the amount of light and location of the luminaires, but also is influenced by the reflectance of the walls, ceilings, and other room surfaces. The amount of illuminance falling on a work surface, such as a desk, is made up of a direct component (i.e., amount of illuminance striking the surface directly from light sources) and a reflected component (i.e., the illuminance reflected from walls, ceilings, etc. that reach the work surface).

To maximize the illuminance on working surfaces, it is generally desirable to use rather light walls, ceilings, and other surfaces. However, areas of high reflectance in the visual field can become sources of reflected glare. For this and other reasons (including practical considerations), the reflectances of surfaces in a room (such as an office) generally increase from the floor to the ceiling. Figure 14-12 illustrates the IES recommendations, indicating for each type of surface the range of acceptable reflectance levels. Although Figure 14-12 applies specifically to offices, essentially the same reflectance values can be applied to other work situations, as in industry.

GLARE

Glare is produced by brightness within the field of vision that is sufficiently greater than the luminance to which the eyes are adapted so as to cause annoyance, discomfort, or loss in visual performance and visibility. *Direct glare* is caused by light sources in the field of view, and *reflected glare* is caused by light being reflected by a surface in the field of view. Reflected glare can be *specular* (as from smooth, polished, mirrorlike surfaces), *spread* (as from brushed, etched, or pebbled surfaces), *diffuse* (as from flat painted or matte surfaces), or *compound* (a combination of the first three). Figure 14-13 shows the three main types of reflections and one example of a compound reflection.

Glare is also classified according to its effects on the observer. Three types are

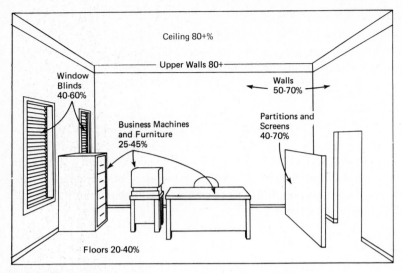

FIGURE 14–12
Reflectances recommended for room and furniture surfaces in an office.
(Source: IES, 1982, table 1.)

recognized: *discomfort glare* produces discomfort, but does not necessarily interfere with visual performance or visibility; *disability glare* reduces visual performance and visibility and often is accompanied by discomfort; and finally *blinding glare* is so intense that for an appreciable length of time after it has been removed no object can be seen.

Discomfort Glare

Discomfort glare is a sensation of annoyance or pain caused by high or nonuniform distributions of brightness in the field of view. Discomfort glare may be accompanied by disability glare, but it is a distinctly different phenomenon. The process by which glare causes discomfort is not well understood, but is believed to be related to the constriction of the pupil exposed to bright light sources. Visual discomfort from glare is, unfortunately, a common experience. Olson and Sivak (1984), for example, report

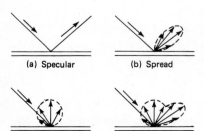

FIGURE 14–13
Types of reflection. *(a)* Specular—from a polished surface; *(b)* spread—from a rough surface; *(c)* diffuse—from a matte surface; *(d)* compound—diffuse and spread. *(Source: Kaufman and Christensen, 1984, figs. 2–26 and 2–27.)*

on the discomfort caused by even moderate-intensity glare sources reflected in auto-mobile rear-view mirrors.

There has been considerable research related to glare and its effects on the subjec-tive sensations of visual comfort and discomfort. Since discomfort glare is a subjec-tive experience, it must be assessed by asking people to rate the degree of discomfort felt while in the presence of a particular glare source. A common metric used for assessing either the degree of discomfort glare produced by a source or the sensitivity of a person to glare is the *borderline between comfort and discomfort* (BCD). The BCD is the luminance of a glare source that is judged by a person to be just necessary to cause that person feelings of discomfort. The higher the BCD, the less glaring was the source or the less sensitive the person was to glare. In one study, Bennett (1977b) varied background luminance, size of the glare source, and angle in the visual field and found the following correlations between the variables and BCD:

Background luminance	0.26
Size of glare source	−0.41
Angle in visual field	0.12

Thus, the higher the background luminance, the smaller the size of the glare source, and the higher the angle in the visual field, the less discomfort is produced. Bennett makes the point, however, that these three factors together account for only 28 per-cent of the variance in BCD judgments. Differences between observers accounted for as much as 55 percent of the variance.

In another study, Bennett (1977a) related various demographic factors to BCD judgments. Not surprisingly, he found that glare sensitivity increased with age. Fig-

FIGURE 14–14
Sensitivity to discomfort glare as a function of age. As BCD de-creases, sensitivity is increasing. Data are based on a 1° glare source size and a background luminance of 1.6 fL. *(Source: Ben-nett, 1977a, fig. 1.)*

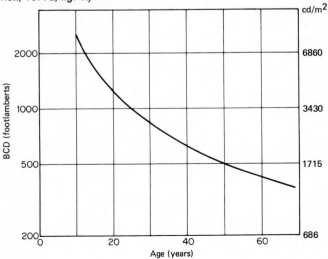

ure 14-14 shows the relationship between age and BCD found by Bennett (remember, as BCD goes down, glare sensitivity is increasing). Incidentally, he found that blue-eyed people are more sensitive to glare than brown-eyed people. People with brown eyes had a median BCD of 1300 fl (4459 cd/m^2) while blue-eyed people had a median BCD of 1100 fL (3773 cd/m^2). Freivalds, Harpster, and Heckman (1983) investigated the effect of corrective lenses on glare sensitivity. The order from greatest sensitivity to least sensitivity was: wearers of hard contact lenses, wearers of soft contact lenses, wearers of frame glasses, and those who did not wear corrective lenses.

The IES has adopted a procedure for computing a *discomfort glare rating* (DGR) for luminaires in an interior space (Kaufman and Christensen, 1984). The DGR is a numeric assessment of the capacity of a number of luminaires in a given arrangement for producing discomfort. The major variables considered in computing the DGR are the average luminance of the visual field and the luminance, size, and position of each light source. DGRs are more or less arbitrary numbers, but they can be converted to a *visual comfort probability* (VCP). The VCP is the percentage of people who would rate the lighting as comfortable when looking in a specific direction from a specific position in the environment. Usually the VCP is computed by assuming that people are looking horizontally from a seated position at the center of the rear of the room (the least advantageous location). The IES considers a VCP of 70 to be satisfactory from a discomfort glare standpoint. Since 70 percent of the people would be expected to rate the environment comfortable in the worst position, it is assumed that most occupants would be in comfortable visual locations.

Disability Glare

Glare that directly interferes with visibility and ultimately with visual performance is considered to be disability glare. Trotter (1982) distinguishes two causes of disability glare: intraocular veiling luminances and transient adaptation. Light that enters the eye is scattered within the eyeball by "inhomogeneities" in the lens and the fluid that fills the eyeball. This scattered light creates a "veiling luminance" on the retina and reduces the contrast of the target being viewed; in effect, it "washes out" the image. Any source of light, direct or reflected, in the field of view causes some veiling luminances on the retina. The effect on foveal vision is a function of the intensity of the glare source and the angle of the source to the line of sight. The smaller the angle, the greater the effect on visibility.

Such effects are illustrated by the results of a study in which the subjects viewed test targets with a glare source of a 100-W inside-frosted tungsten-filament lamp in various positions in the field of vision (Luckiesh and Moss, 1927–1932). The test targets consisted of parallel bars of different sizes and contrasts with their background. The glare source was varied in position in relation to the direct line of vision, these positions being at 5, 10, 20, and 40° with the direct line of vision, as indicated in Figure 14-15. The effect of the glare on visual performance is shown as a percentage of the visual effectiveness that would be possible without the glare source. With the glare source at a 40° angle, the visual effectiveness is 58 percent, this being reduced to 16 percent at an angle of 5°.

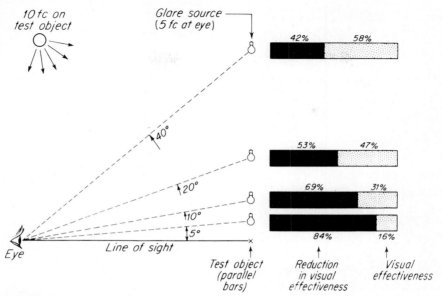

FIGURE 14-15
Effects of direct glare on visual effectiveness. The effects of glare become greater as the glare source gets closer to the line of sight. *(Source: Luckiesh and Moss, 1927–1932.)*

There is a tendency for the eyes to turn toward a bright light (such as an oncoming automobile headlight); this phenomenon is called *phototropism*. Store owners and department store window designers take advantage of this human tendency when they direct bright lights toward a particular part of the store (perhaps where the high-profit items are) or at a particular item in the window. This is but one example of the artistic aspects of illuminating engineering we talked about at the beginning of this chapter.

Phototropism, however, can have negative effects on task performance and safety if it draws the eyes away from the area of greatest importance. Figure 14-16, for example, illustrates this phenomenon. In Figure 14-16a, the lights are arranged to produce a reflected glare source of high luminance, and the eyes are drawn to that spot rather than to the middle of the machine where the cutting blade is located. By moving the luminaires, the glare source is removed (Figure 14-16b) and it is easier to concentrate on the cutting blade.

As a result of phototropism, the eyes are drawn to a bright glare source and become adapted to the higher luminance level. This is called *transient adaptation*. This in turn reduces the eyes' sensitivity to light. When the person looks back at the task, the eyes are less sensitive and visibility is reduced. It is like going into a dark movie theater on a bright sunny day, only the effect is not so pronounced. Windows are an especially good source of disability glare and transient adaption because the illumination levels can be 1000 times more intense than normal indoor light levels.

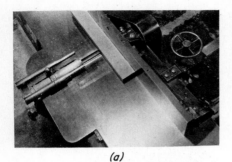

(a)

(b)

FIGURE 14–16
Effects of change of location of luminaires on wood planer. *(a)* Original position caused reflection from polished surface, which distracted attention from cutters *(b)* Relighting gives lower brightness reflection, which makes cutter the center of attraction. *(Source: Hopkinson and Longmore, 1959, by permission of the Controller of H. M. Stationery Office.)*

Reduction of Glare

• *To reduce direct glare from luminaires:* (1) Select luminaires with low DGR; (2) reduce the luminance of the light sources (for example, by using several low-intensity luminaires instead of a few very bright ones); (3) position luminaires as far from the line of sight as feasible; (4) increase the luminance of the area around any glare source so the luminance ratio is reduced; and (5) use light shields, hoods, visors, diffusing lenses, filters, or cross-polarizers.

• *To reduce direct glare from windows:* (1) Use windows that are set some distance above the floor; (2) construct an outdoor overhang above the window; (3) construct vertical fins by the window extending into the room (to restrict direct line of sight to the windows); (4) use light surrounds (to minimize contrast with light from the window); and (5) use shades, blinds, or louvers.

• *To reduce reflected glare:* (1) Keep the luminance level of luminaires as low as feasible; (2) provide a good level of general illumination (such as with many small light sources and use of indirect lights); (3) use diffuse light, indirect light, baffles, window shades, etc.; (4) position light source or work area so reflected light will not be directed toward the eyes; and (5) use surfaces that diffuse light, such as flat paint, nonglossy paper, and crinkled finish on office machines; avoid bright metal, glass, glossy paper, etc.

LIGHTING AND THE ELDERLY

It comes as no surprise to anyone over 40 that people's visual faculties decline with age. Most apparent is our tendency to become farsighted. (This tendency means not that one's far vision has improved, but rather that one's near vision has deteriorated.) However, certain other physiological changes in our visual system that occur with age have implications for lighting design (Boyce, 1981; Hughes and Neer, 1981; Wright and Rea, 1984).

For one, the lens of the eye continues its cellular growth throughout life. New cellular layers are continually being added to the outer parts of the lens. Cells at the center of the lens atrophy and harden; opacities develop, and the pliability of the lens decreases. In addition, the muscles controlling the diameter of the pupil begin to atrophy with age. This reduces the size of the pupil and decreases the range and speed with which the pupil can adjust to differing levels of illumination.

The thickening of the lens is the primary cause of farsightedness among the elderly. The lens can no longer conform to the shape necessary to focus a close object onto the retina; this defect can, however, be corrected with glasses. The increased opacity of the lens, coupled with the smaller pupil diameter, also reduces the amount of illumination reaching the retina. Weale (1961) determined that there is a 50 percent reduction in retinal illumination at age 50, compared to age 20, and this reduction increases to 66 percent at age 60. This reduced level of retinal illumination also plays a role in slowing the rate and reducing the level of dark adaptation among the elderly. Further, as the lens thickens, it yellows and thereby reduces the transmission of blue light through it. For this reason, elderly people have more trouble sorting or matching colors, making more errors in the blue-green and red regions than in the other regions of the hue circle (Verriest, Vandevyvere, and Vanderdonck, as cited in Boyce, 1981).

As the lens thickens, there is an increase in the scattering of light passing through the lens and in the intraocular fluids. There are widely differing estimates as to the degree of scatter that occurs with age. Wright and Rea (1984) cite studies showing a 2- to 10-fold increase between ages 20–30 and 60–70. This scattering, as with disability glare, produces a veiling luminance over the retinal image, effectively reducing its contrast. Further, scattering blurs the image on the retina (since the primary target is also scattered), thereby reducing visual acuity. This reduction in visual acuity cannot be corrected with glasses. The National Center for Health Statistics (1977) reported that the percentage of people with defective visual acuity (20/50 or poorer in the better eye with usual corrective lenses) increases from 0.7 percent at ages 35 to 44, to 14 percent at ages 65 to 74.

The reduced pupillary diameter response and the increased intraocular scattering of light combine to increase human sensitivity to glare with advancing age. Figure 14-14, for example, showed that at age 60 discomfort glare sensitivity is about 3 times greater than at age 20. The slower response of the pupil to changing levels of illumination also increases the magnitude and duration of transient adaptation effects.

What this all means is that lighting environments for the elderly must be designed with special care to ensure visual comfort and effective task performance. Specifically, the elderly need more light than the young to see, and they gain more from increased task illumination than do the young. The contrast of the task is also more critical to the elderly, who need higher contrasts in order to maximize their visual performance. Figure 14-17, for example, shows the dramatic increase in contrast necessary to allow people over 50 years old to see as well as people 20 to 30 years old.

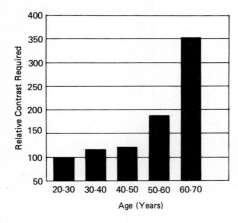

FIGURE 14–17
Effect of age on contrast needed to maintain visual performance. Age group from 20 to 30 years is considered as baseline, 100. *(Source: Based on data from Blackwell and Blackwell, 1971.)*

Glare control is also more important in environments used by the elderly. Lower luminance ratios are needed to reduce transient adaptation problems and visual discomfort. Full-spectrum light sources with good color rendering indices are recommended for color perception as well as for psychological and biological reasons (Hughes and Neer, 1981).

SPECIAL APPLICATION: LIGHTING FOR VIDEO DISPLAY TERMINALS

Video display terminals, i.e., computer terminals, are everywhere. Traditionally we think of VDTs in offices and maybe homes, but increasingly we find them on the manufacturing floor, in warehouse storage areas, and even in automobiles. Police cars in Los Angeles, for example, are equipped with a computer terminal in the front seat. We cannot hope to cover the lighting implications of all these various situations, but the general principles discussed here should apply in most circumstances.

VDTs differ from other more traditional sources of visual information in a number of ways. They emit light rather than depending on reflected light to be seen. Most VDTs have highly specular curved glass faces which present some interesting optical effects. And finally, the task surface is more vertical than with most other "desktop" tasks. All these factors must be considered in the setup of a lighting environment where VDTs will be used.

We focus our attention in this section on three main lighting considerations as they apply to VDTs: illumination level, luminance ratios, and screen reflections. In Chapter 4 we discussed other aspects of VDTs such as character font, color, contrast with the background, and flicker. Many of these factors, of course, interact with lighting to influence visual performance. In general, the less legible the characters on the screen, the more important is the quality of the lighting environment.

Illuminance Level

Since VDT screens emit light, they do not require ambient illumination in order to be read. But alas, people do not work by VDT alone. General ambient illumination is required to carry on other tasks performed in conjunction with VDT work, including reading handwritten notes, looking at the keyboard, watching the clock, and moving around without tripping over the wastepaper basket.

With most office visual tasks, the higher the level of ambient illumination (up to a point), the easier it is to see the task. The opposite is true for VDT screens. Ambient illumination reflects off the screen and makes it more difficult to see the information on the screen. Recommended illumination levels, therefore, are a compromise between the higher levels demanded for such tasks as reading handwritten, pencil copy and the low levels required to read the VDT screen. Numerous standards specify illumination requirements for VDTs (Helander and Rupp, 1984). The majority specify levels between 300 and 500 lx, although some specify lower levels. The IES (1982), for example, specifies 50 to 100 lx for general lighting where VDTs are used. AT&T (1983) recommends 200 lx measured in the horizontal plane near the center of the VDT screen. Where paper copy must be used in conjunction with VDT work, additional light directed on the paper copy (task lighting) is usually recommended. Depending on the difficulty of the visual task, 500 to 1500 lx may be required on the task itself (IES, 1982). One must be careful, however, not to create glare sources with the task lighting. Surveys of actual illumination levels found in VDT-equipped offices reveal that the majority are within the range of 300 to 500 lx (Stammerjohn, Smith, and Cohen, 1981; Laubli, Hunting, and Grandjean, 1982) with an occasional report of lower levels (Shahnavaz, 1982).

Shahnavaz and Hedman (1984) reported that for a group of night-shift VDT operators (where the ambient illumination levels were more constant) the amount of light (illuminance) striking the screen was related to the amount of change in visual accommodation after working for 6 h. Figure 14-18 shows the average screen illuminance for those workers who showed a 0.25 or less diopter change in accommodation and for those who showed more than a 0.25 diopter change. (A diopter is a measure of the focal length of a lens.) Two accommodation distances are shown. Higher screen illuminances were associated with less accommodation shift. Higher screen illuminance is indicative of higher ambient illumination. It may well be that with the higher levels of ambient illumination, operators looked around the room more while working on the VDTs and thus relaxed the muscles of the eye that control the shape of the lens. These results would give some support for keeping ambient illuminance levels in the recommended range of 300 to 500 lx, rather than reducing them to lower levels.

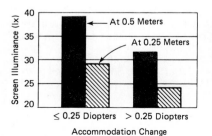

FIGURE 14–18
Average illumination of VDT screens for workers showing accommodation changes of 0.25 diopter or less or more than 0.25 diopter after 6 h of VDT work. Results are shown for two accommodation distances: 0.5 and 0.25 m. *(Source: Shahnavaz and Hedman, 1984, fig. 3.)*

Luminance Ratios

As we discussed previously, high luminance ratios between the task and adjacent and remote areas can cause transient adaptation problems and disability or discomfort glare. With VDTs, the problem is especially acute because of the low luminance levels of the VDT screen (assuming light characters on a dark background). The principal sources of high luminance in the office environment are task lighting, windows, and bare fluorescent bulbs used for general area lighting.

The general recommendations for luminance ratios are 1:3 between the screen and immediate surrounds and 1:10 for remote areas. Helander, Billingsley, and Schurick (1984) indicate, however, that the 1:3 ratio may be too stringent. Actual field measurements confirm that luminance ratios in offices often exceed 1:10 (Stammerjohn, Smith, and Cohen, 1981; Fellmann et al., 1982). Laubli, Hunting, and Grandjean (1982), for example, reported screen-to-source-document ratios of 1:10 to 1:87 with the median being 1:26. Screen-to-window ratios ranged from 1:87 to 1:1450 with a median of 1:300!

High luminance contrasts (ratios) have been associated with increased frequency of eye complaints among VDT users (Laubli, Hunting, and Grandjean, 1982) and temporary changes in accommodation (Shahnavaz and Hedman, 1984). Figure 14-19 presents the relationship between luminance contrast (screen to surrounds) and the percentage of VDT operators showing overaccommodation changes after 6 h of work. Where high contrasts (ratios) existed, over three-quarters of the operators showed accommodation changes.

Reducing high luminance ratios can be accomplished by increasing screen luminance, using negative-polarity displays (dark characters on light background), decreasing the luminance of the offending area, or increasing the overall level of illumination. Increasing the overall level of illumination reduces the discomfort caused by "hot spots" in the environment, but adds reflected luminance to the screen which creates additional problems.

Screen Reflections

Screen reflections are one of the most prevalent problems associated with VDT use. Laubli, Hunting, and Grandjean (1982), for example, reported that 45 percent of the

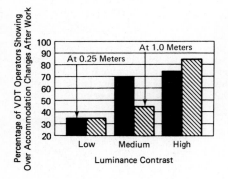

FIGURE 14–19
Percentage of VDT operators showing overaccommodation changes after 6 h of work in relation to luminance contrasts between the screen and surrounds. Contrast was computed as $(L_2 - L_1)/L_1$, where L_1 = luminance of screen and L_2 = luminance of surrounds. Low $<$ 100; medium = 100 to 150; high $>$ 150. *(Source: Shahnavaz and Hedman, 1984, fig. 7.)*

operators they interviewed complained of annoying reflections on their VDT screens. Typically reflections of windows, overhead lights, and even the operator are seen on the face of the screen. Figure 14-20 shows an example of typical screen reflections.

Reflections are generally specular or diffuse. Specular reflections produce mirror-like images. Such images are reflected off the smooth front surface of the screen. Since VDT screens are convex (with a radius of curvature of approximately 64 cm), distant objects reflected from the face of the screen are optically projected and appear to be about 32 cm behind the screen. Diffuse reflections produce a veiling luminance over parts of the screen and can wash out the information underneath. Such reflections are off the irregular phosphor surface behind the glass screen or, in some cases, off the front of the glass screen itself. To determine the source of screen reflections, either specular or diffuse, place a small mirror parallel to the screen, assume the operator's position(s), and slowly move the mirror over the surface of the screen. The source of the reflections will be clearly visible in the mirror.

Screen reflections affect VDT users in several ways. Reflections can be distracting (remember phototropism?) and annoying. Laubli, Hunting, and Grandjean (1982), for example, found a positive correlation between the luminance level of reflections and reported annoyance, but no relation between luminance and eye impairments. Reflections reduce the contrast between the characters and their background, making

FIGURE 14–20
Example of typical reflections on a VDT screen.

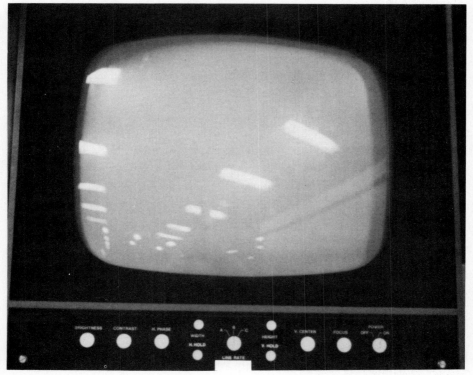

it more difficult to read information from the screen. Stammerjohn, Smith, and Cohen (1981) reported reflected luminance levels of up to 50 cd/m². Of the screens they evaluated, 17 percent had reflections that made it difficult to read characters on parts of the screen. Reflection can cause shifts in convergence between the characters on the screen and specular reflections optically located behind the screen. If the reflections are bright, they can also reduce visual sensitivity owing to accommodation shifts.

Methods of reducing screen reflections are directed at the source of the reflections, between the source and the screen, and at the screen itself. Techniques that attempt to correct the problem at the source include (1) covering windows, (2) repositioning sources of reflections, (3) reducing intensity or luminance of sources, and (4) using diffuse, indirect lighting. Reflections can be reduced by erecting shields or partitions between the source and the screen to block the light directed at the screen. Hoods can also be attached to the VDT. However, these tend to restrict the posture of the operator, and they can cast annoying shadows on the screen and keyboard. Correcting the problem at the screen involves using such techniques as (1) moving or tilting the screen so reflections are not in the field of view, (2) positioning the work station between rows of overhead lights rather than directly under them, (3) positioning the screen so it is perpendicular to windows rather than facing toward or away from them, (4) using reverse video (dark characters on a light background), and (5) applying antireflection treatment to the screen.

There are a number of antireflection techniques available for VDT screens (but they have their limitations):

1 *Etching or frosting* the front surface of the screen reduces specular reflections by making them diffuse, but does not reduce the diffuse reflections. In addition, the light from the characters on the screen is scattered, and the edges of the characters appear blurred, thus reducing legibility.

2 *Quarter-wave thin-film coatings* "ease" the transition of light from air to the glass, thereby reducing reflections (both specular and diffuse) from the front surface of the screen. These coatings do not, however, reduce diffuse reflections from the phosphor surface. An advantage is that the light emitted by the characters on the screen is not scattered, and character sharpness is not degraded. The disadvantage is that fingerprints and scratches defeat the system.

3 *Neutral-density filters* reduce the amount of light passing through them. This acts to reduce the luminance of the characters on the screen, but at the same time increases their contrast. The explanation is that the ambient light is reduced twice—once going through the filter to the phosphor surface and again when it is reflected from the phosphor surface back through the filter—whereas the light from the characters on the screen is reduced only once. Unless the surface of the filter is treated (etched, frosted, quarter-wave-coated), the filter itself can be a source of specular reflections.

4 *Circular polarizers* act as neutral-density filters but usually have a highly polished reflective front surface which makes them highly susceptible to specular reflections.

5 *Micromesh filters* look like a black nylon stocking stretched over the screen. Only light perpendicular to the mesh can pass through to the screen. This reduces

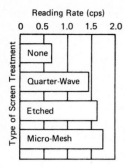

FIGURE 14–21
Effect of type of VDT antireflection treatment on reading rate. Reading rate, in characters per second (cps), was computed as follows: (number of characters read − number of errors and omissions)/time. *(Source: Adapted from Habinek et al., 1982.)*

both specular and diffuse reflections and enhances contrast. The problems with them are that the luminance of the characters on the screen drops off drastically when the screen is not viewed head on and that they tend to collect dirt.

One effect of neutral-density filters, circular polarizers, and micromesh filters is that they darken the screen. This can have the effect of increasing luminance ratios between the screen and surrounds above recommended levels. Cakir, Hart, and Stewart (1980), for example, found that using such filters increased luminance ratios between the screen and source document from 1:6 to 1:100!

Habinek et al. (1982) compared reading performance with, and preference for, three types of antireflection treatments: micromesh filters, quarter-wave coating, and etching. Under the high-reflection condition, reading rates were higher (as much as 2.5 times higher) with the filters than without, as shown in Figure 14-21. Subjects preferred the treated screens to the untreated screens with little differences in preference among the three treatments.

VDTs present some interesting challenges in lighting design. Successfully balancing illumination levels, luminance ratios, and screen reflections is a difficult task that often requires delicate trade-offs to be made. Actually the problems are not unique to VDTs; they are only exaggerated by them. Undoubtedly, many of the visual problems and complaints associated with VDTs are due to improper lighting conditions. Judicious application of human factors and lighting principles can go a long way toward alleviating these problems.

DISCUSSION

We have come a long way in this chapter and have introduced many concepts which were probably unfamiliar to most readers. Yet we have just barely scratched the surface of the field of illuminating engineering. To design a lighting system to supply the proper amount of illumination, with the proper spectral composition, without creating glare—and to do it all in an energy-efficient manner—is quite an achievement. Obviously, our goal in this chapter was not to develop this talent, but rather to acquaint you with the major concepts, to make you a more intelligent consumer of the expertise offered by illuminating engineers.

One last word of caution. It is easy to be swept away by the elegant formulas and sophisticated measuring techniques used by illuminating engineers. Remember, however, that often these techniques are based on simplifying assumptions and often the techniques ignore individual differences in visual capabilities. Their value lies more in making relative comparisons between lighting conditions than in specifying absolute requirements. Recommendations that are made must be tempered with good judgment and should be looked upon as first approximations that will undoubtedly require adjustment to fit the particular situation.

REFERENCES

AT&T (1983). *Video display terminals: Preliminary guidelines for selection, installation and use*. Short Hills, NJ: Bell Laboratories, Circulation Department.

Bennett, C. (1977a, January). The demographic variables of discomfort glare. *Lighting Design and Application*, pp. 22–24.

Bennett, C. (1977b). Discomfort glare: Concentrated sources parametric study of angular small sources. *Journal of the Illuminating Engineering Society*, 7(1), 2–14.

Bennett, C., Chitlangia, A., and Pangrekar, A. (1977). Illumination levels and performance of practical visual tasks. *Proceedings of the Human Factors Society 21st Annual Meeting*. Santa Monica, CA: Human Factors Society.

Blackwell, H. R. (1959). Development and use of a quantitative method for specification of interior illumination levels on the basis of performance data. *Illuminating Engineering*, 54, 317–353.

Blackwell, H. R. (1961). Development of visual task evaluators for use in specifying recommended illumination levels. *Illuminating Engineering*, 56, 543–544.

Blackwell, H. R. (1964). Further validation studies of visual task evaluation. *Illuminating Engineering*, 59(9), 627–641.

Blackwell, H. (1967). The evaluation of interior lighting on the basis of visual criteria. *Applied Optics*, 6(9), 1443–1467.

Blackwell, H. R., and Blackwell, O. M. (1968). The effect of illumination quantity upon the performance of different visual tasks. *Illuminating Engineering*, 63(3), 143–152.

Blackwell, O., and Blackwell, H. (1971). IERI report: Visual performance data for 156 normal observers of various ages. *Journal of the Illuminating Engineering Society*, 1, 2–13.

Boyce, P. (1971). *Illumination and the sensitivity of performance measures*, Rept. R412. Capenhurst, UK: Electricity Council Research Centre.

Boyce, P. (1981). *Human factors in lighting*. New York: Macmillan.

Brass, J. (1982, November). Discarding ESI in favor of brightness contrast engineering—A "wide-angle" view. *Lighting Design and Applications*, 12(11), 30–34.

Cakir, A., Hart, D., and Stewart, T. (1980). *Visual display terminals: A manual covering ergonomics, workplace design, health and safety, task organization*. New York: Wiley.

Commission Internationale de l'Eclairage (1971). *Colorimetry*, CIE Publ. 15. Paris.

Commission Internationale de l'Eclairage (1972). *A unified framework of methods for evaluating visual performance aspects of lighting*, CIE Publ. 19. Paris.

Commission Internationale de l'Eclairage (1974). *Methods of measuring and specifying color rendering properties of light sources*, CIE Publ. 13.2. Paris.

Container Corporation of America. (1942). *Color harmony manual*. Chicago: Container Corporation of America.

Delaney, W., Hughes, P., McNelis, J., Sarver, J., and Soules, T. (1978). An examination of visual clarity with high color rendering of fluorescent light sources. *Journal of the Illuminating Engineering Society*, 2(1), 74–84.

DiLaura, D. (1978a, February). Visibility, human performance, and lighting design—Equivalent contrast as a prelude to visibility level. *Lighting Design and Application,* p. 8.

DiLaura, D. (1978b, April). Visibility, human performance, and lighting design—Visibility level as a supra-threshold metric. *Lighting Design and Application,* pp. 50–51.

DiLaura, D. (1978c, June). Visibility, human performance, and lighting design—The choice and use of a visibility level. *Lighting Design and Application,* pp. 58–59.

Faulkner, T. W., and Murphy, T. J. (1973). Lighting for difficult visual tasks. *Human Factors,* 15(2), 149–162.

Fellmann, Th., Brauninger, U., Gierer, R., and Grandjean, E. (1982). An ergonomic evaluation of VDTs. *Behaviour and Information Technology,* 1, 69–80.

Freivalds, A., Harpster, J., and Heckman, L. (1983). Glare and night vision impairment in corrective lens wearers. *Proceedings of the Human Factors Society 27th Annual Meeting.* Santa Monica, CA: Human Factors Society. pp. 324–328.

Habinek, J., Jacobson, P., Miller, W., and Suther, T. (1982). A comparison of VDT antireflection treatments. *Proceedings of the Human Factors Society 26th Annual Meeting.* Santa Monica, CA: Human Factors Society. pp. 285–289.

Hard, A., and Sivik, L. (1981). NCS—Natural color system: A Swedish standard for color notation. *Color Research and Application,* 6, 129–138.

Hardesty, G., and Projector, T. (1973). *NAVSHIPS display illumination design guide.* Washington: Government Printing Office.

Helander, M., Billingsley, P., and Schurick, J. (1984). An evaluation of human factors research on visual display terminals in the workplace. In F. Muckler (ed.), *Human factors review: 1984,* pp. 55–129. Santa Monica, CA: Human Factors Society.

Helander, M., and Rupp, B. (1984). An overview of standards and guidelines for visual display terminals. *Applied Ergonomics,* 15, 185–195.

Hesselgren, S. (1984). Why colour order systems? *Color Research and Application,* 9(4), 220–228.

Hopkinson, R. G., and Longmore, J. (1959). Attention and distraction in the lighting of workplaces. *Ergonomics,* 2, 321–334.

Hughes, P., and McNelis, J. (1978, May). Lighting, productivity, and work environment. *Lighting Design and Application,* pp. 37–42.

Hughes, P., and Neer, R. (1981). Lighting for the elderly: A psychobiological approach to lighting. *Human Factors,* 23, 65–85.

IES lighting fundamentals course. (1976). New York: Illuminating Engineering Society.

IES lighting handbook (5th ed.). (1972). New York: Illuminating Engineering Society.

IES lighting handbook (application vol.). (1981). New York: Illuminating Engineering Society.

IES Nomenclature Committee. (1979). Proposed American national standard nomenclature and definitions for illuminating engineering (proposed revision of Z7.1R1973). *Journal Illuminating Engineering Society,* 9(1), 2–46.

Illuminating Engineering Society of North America (1982). *Office lighting* (ANSI/IES RP–1–1982). New York.

Jaschinski, W. (1982). Conditions of emergency lighting. *Ergonomics,* 25, 363–372.

Kaufman, J., and Christensen, J. (Eds.) (1984). *IES lighting handbook* (*reference vol.*). New York: Illuminating Engineering Society of North America.

Knave, B. (1984). Ergonomics and lighting. *Applied Ergonomics,* 15(1), 15–20.

Laubli, Th., Hunting, W., and Grandjean, E. (1982). Visual impairments in VDU operators related to environmental conditions. In E. Grandjean and E. Vigliani (eds.), *Ergonomic aspects of visual display terminals.* London: Taylor & Francis, pp. 85–94.

Light and color (TP-119). (1968, August). Nela Park, Cleveland, OH: Large Lamp Department, General Electric Company.

Light measurement and control (TP-118). (1965, March). Nela Park, Cleveland, OH: Large Lamp Department, General Electric Company.

Lion, J. S. (1964). The performance of manipulative and inspection tasks under tungsten and fluorescent lighting. *Ergonomics, 7*(1), 51–61.

Logan, H. L., and Berger, E. (1961). Measurement of visual information cues. *Illuminating Engineering, 56,* 393–403.

Luckiesh, M., and Moss, F. K. (1927–1932). The new science of seeing. In *Interpreting the science of seeing into lighting practice* (vol. 1). Cleveland: General Electric Co.

Merritt, J., Perry, T., Crooks, W., and Uhlaner, J. (1983). *Recommendations for minimal luminance requirements for metal and nonmetal mines* (contract J0318022). Pittsburgh: U.S. Bureau of Mines.

Munsell Color Co. (1973). *Munsell book of colors.* Baltimore, MD: Munsell Color Co.

National Center for Health Statistics (1977). *Monocular visual acuity of persons 4–74 years,* DHEW Publ. HRA 77–1646. Washington, D.C.: U.S. Department of Health, Education, and Welfare.

Nimeroff, I. (1968). *Colorimetry.* National Bureau of Standards Monograph 104. Washington: Government Printing Office.

Olson, P., and Sivak, M. (1984). Glare from automobile rear-vision mirrors, *Human Factors, 26,* 269–282.

Rea, M. (1984, June). What's happening to ESI? *Lighting Design and Application, 14*(6), 46–50.

Richter, M. (1955). The official German standard color chart. *Journal of Optical Society of America, 45,* 223–226.

Ross, D. (1978, May). Task lighting—yet another view. *Lighting Design and Application,* pp. 37–42.

RQQ. (1980). Selection of illuminance values for interior lighting design (RQQ Rept. 6). *Journal Illuminating Engineering Society, 9*(3), 188–190.

Sanders, M., Gustanski, J., and Lawton, M. (1974). Effect of ambient illumination on noise level of groups. *Journal of Applied Psychology, 59*(4), 527–528.

Santamaria, J., and Bennett, C. (1981, March). Performance effects of daylight. *Lighting Design and Application, 11*(3), 31–34.

Shahnavaz, H. (1982). Lighting conditions and workplace dimensions of VDU-operators. *Ergonomics, 25,* 1165–1173.

Shahnavaz, H., and Hedman, L. (1984). Visual accommodation changes in VDU-operators related to environmental lighting and screen quality. *Ergonomics, 27,* 1071–1082.

Stammerjohn, L., Smith, M., and Cohen, B. (1981). Evaluation of work station design factors in VDT operations. *Human Factors, 23,* 401–412.

Stenzel, A. (1962). Experience with 1000 lx in a leather factory. *Lichttechnik, 14,* 16.

Trotter, D. (1982). *The lighting of underground mines.* Clausthal-Zellerfeld, Germany: Trans Tech Publications.

Weale, R. (1961). Retinal illumination and age. *Transactions of the Illuminating Engineering Society, 26,* 95.

Wright, G., and Rea, M. (1984). Age, a human factor in lighting. In D. Attwood and C. McCann (eds.), *Proceedings of the 1984 International Conference on Occupational Ergonomics.* Rexdale, Ontario, Canada: Human Factors Association of Canada. pp. 508–512.

Zahn, J., and Haines, R. (1971). The influence of central search task luminance upon peripheral visual detection time. *Psychonomic Science, 24*(6), 271–273.

CHAPTER 15

ATMOSPHERIC CONDITIONS

Since the current model of the human organism is the result of evolutionary processes spanning millions of years, it has developed substantial adaptability to the environmental variables within the world in which we live, including its atmosphere. There are, however, limits to the human range of adaptability. In addition, science and technology are busy developing new kinds of environments for the human being, involving spacecraft, cold-storage warehouses, and underwater caissons, and causing changes in our natural environment as the unintentional by-product of civilization— such as smog and air pollution. Since we cannot cover all aspects of the ambient environment, we cover those that are of most general concern, in particular thermal conditions.

HEAT EXCHANGE PROCESS

The body has a thermal regulation system that tends to maintain a core temperature around 98.6°F (37.0°C). (The *core temperature* is the temperature of most of the internal organs.) Heat-sensitive nerves throughout the body (especially in the skin) transmit information to the brain which, in turn, sends out control signals to direct the regulatory mechanisms, to try to maintain the core temperature. Temperature regulation is brought about by the control of blood flow, the secretion of sweat, and sometimes shivering.

The primary source of body heat is the metabolic process that converts chemical energy (from food) to heat and mechanical energy (or work). Since, under most circumstances, we generate more heat than we need, the body seeks to lose that surplus heat. Under cold conditions, of course, the body tries to limit the heat loss. When all goes well with the heat regulation process, the body remains in a state of thermal equilibrium.

Methods of Heat Exhange

There are four methods of heat exchange between the body and its environment:

1 *Convection:* The transmission via the air (which technically is a fluid).
2 *Evaporation:* Of perspiration and, to the some degree, of the vapor exhaled from the lungs.
3 *Radiation:* The transmission of thermal energy between objects, as between the sun and the earth, or between ourselves and the walls, ceilings, and other surfaces (such as cold windows or hot metal furnaces) in our environment.
4 *Conduction:* The transmission of thermal energy by direct contact, as in walking barefoot on a cold floor or putting one's hand on a hot stove. Our clothing usually insulates us so well that this method is of negligible importance and is not considered in most methods of measuring heat exchange.

Heat Exchange Equation   The heat exchange process, as it is carried out in most circumstances, is set forth in the following equation:

$$S = M - E \pm R \pm C - W$$

where S = storage
M = metabolism
E = evaporation
R = radiation
C = convection
W = work

If the body is in a state of equilibrium, S (storage) is zero. In conditions of imbalance the body temperature can increase (if it is high enough, it can cause heat stroke or death) or decrease (which can also cause death in extreme conditions).

Environmental Factors Influencing Heat Exchange

The heat exchange process is, of course, very much affected by four environmental conditions: air temperature; humidity; air flow; and the temperature of surrounding surfaces (walls, ceilings, windows, furnaces, etc; sometimes such temperature is called *wall temperature,* for short). The interaction of these conditions is quite complex and cannot be covered in detail here. But we can point out that under high air temperature and high wall temperature conditions, heat loss by convection and radiation is minimized, so that heat *gain* to the body may result. Under such circumstances the only remaining means of heat loss is by evaporation. But if the humidity is also high, evaporative heat loss will be minimized, with the result that body temperature rises.

Such effects are illustrated in Figure 15-1. This shows, for each of five combinations of air and wall temperature, the percentage of heat loss by evaporation, radia-

[1]Under unusual circumstances the E (evaporation) and W (work) components can be positive instead of negative.

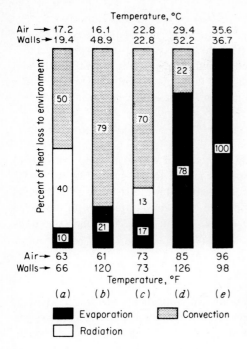

FIGURE 15-1
Percentage of heat loss to environment by evaporation, radiation, and convection under different conditions of air and wall temperatures. *(Source: Adapted from Winslow and Herrington, 1949.)*

tion, and convection. In particular, conditions *d* and *e* illustrate the fact that, with high air and wall temperatures, convection and radiation cannot dissipate much body heat, and the burden of heat dissipation is thrown on the evaporative process. But, as we know, evaporative heat loss is limited by the humidity; there is indeed truth to the old statement that "it isn't the heat—it's the humidity."

Zones of Heat Exchange

The physiological reactions of the body to various thermal conditions tend to fall into three categories, or zones, as illustrated in Figure 15-2 and characterized here:

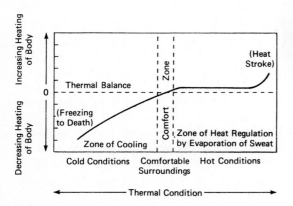

FIGURE 15-2
Zones of heat exchange: (1) zone of cooling; (2) comfort zone (zone of thermal balance); and (3) zone of heat regulation. *(From Grandjean, 1981, fig. 189, p. 333.)*

- *Zone of cooling:* The following effects occur when changing from a warm environment to a cool one: (1) The skin becomes cool; (2) the blood is routed away from the skin and more to the central part of the body, where it is warmed before flowing back to the skin area; (3) rectal temperature rises slightly but then falls; and (4) shivering and ''goose flesh'' may occur. The body may stabilize with large areas of skin receiving little blood. Extreme conditions can cause *hypothermia* and even death.
- *Comfort zone:* A condition in which the body is in a state of approximate thermal balance.
- *Zone of heat regulation:* The following effects occur when changing from a cool environment to a warm one: (1) More blood is routed to the surface of the body, resulting in higher skin temperature; (2) rectal temperature falls but with continued exposure rises; and (3) sweating may begin. The body may stabilize with increased sweating and increased blood flow to the surface of the body. Extreme conditions can cause *hyperthermia,* heatstroke, and even death.

Acclimatization to Heat and Cold

Acclimatization consists of a series of physiological adjustments that occur when an individual is habitually exposed to extreme thermal conditions, hot or cold, as the case may be. Although there is some question about the time required to become fully acclimatized, it is evident that much acclimatization to heat occurs within 4 to 7 days, and reasonably complete acclimatization occurs usually in 12 to 14 successive days of heat exposure. Quite a bit of acclimatization to cold occurs within a week of exposure, but full acclimatization may take months or even years. Even complete acclimatization, however, does not fully protect an individual from extreme heat or cold, although acclimatized individuals can tolerate extremes better than those who are not acclimatized.

MEASUREMENT OF THERMAL VARIABLES

There are, of course, ways of measuring the various atmospheric variables such as air temperature, wall temperature, humidity, and air movement. In addition, however, there are a number of indices of the combinations of such variables, including those that reflect the effects of such combinations on people. A few of these composite indices are discussed below.

Effective Temperature

There are two indices of effective temperature, both of which were developed under the sponsorship of the American Society of Heating, Refrigerating, and Air-Conditioning Engineers (ASHRAE, 1981). The original effective-temperature (ET) scale was intended to equate varying combinations of temperature, humidity, and air movements in terms of equal sensations of warmth or cold. For example, an ET of 70°F (21°C) characterized the thermal sensation of a 70°F (21°C) temperature in combination with 100 percent humidity. Equal thermal sensations [ET = 70° F (21°C)] can also be achieved with other (higher) temperatures in combination with other (lower) humidities [such as 81°F (27°C) with 10 percent humidity]. Although ET has been widely used, and has been a useful index, it overemphasizes the effects of humidity in cool and neutral conditions and underemphasizes the effects in warm conditions; also, it does not fully account for air velocity in hot-humid conditions.

Because of these limitations of the original ET, a newer index, identified as ET*, was developed. It is derived by a complicated formulation based on the effects of environmental variables on the physiological regulation of the body. Figure 15-3 includes the new ET* values. For the moment consider the dry-bulb temperatures (at the bottom of the figure) and the relative-humidity (RH) lines that angle upward from left to right. Follow any given dry-bulb temperature line, such as 70°F (21°C), up to the 50 percent RH line. That intersection represents the ET* that corresponds with the dry-bulb temperature, such as 70°F (21°C). At that intersection the ET* line (a broken line) angles upward to the left and downward to the right. All the combinations of dry-bulb and humidity values on the ET* line represent combinations that are considered to be equal in terms of the bases used in the development of the scale. (For our present purposes the other features of the figure can be disregarded.) The ET* scale as illustrated applies for lightly clothed, sedentary individuals in spaces with low air movement and areas in which the effects of radiation would be limited (specifically, areas in which the mean radiant temperature equals air temperature).

Note that the original ET scale had the loci of the ET lines at the intersection of corresponding dry-bulb temperatures in combination with 100 percent RH. Since the loci of the new ET* lines are at the 50 percent RH lines, the new ET* values would tend to be numerically higher than the older ET values for corresponding conditions. (Most of the later references to effective temperature are to the old ET scale rather than to the new ET* scale.)

Operative Temperature

This index takes into account the combined effects of radiation and convection, but not humidity and air flow. (An equation for deriving this index is included in the *ASHRAE Handbook,* 1981.)

Oxford Index

The *Oxford index,* or *WD* (wet-dry) *index,* is a simple weighting of wet-bulb and dry-bulb temperatures *w* and *d,* respectively, as follows:

$$WD = 0.85w + 0.15d$$

The *wet-bulb temperature* is a measure of air temperature under conditions of 100 percent RH, obtained with an aspirated wet-wick or sling thermometer.

The Oxford index has been found to be a reasonably satisfactory index to equate climates with similar tolerance limits.

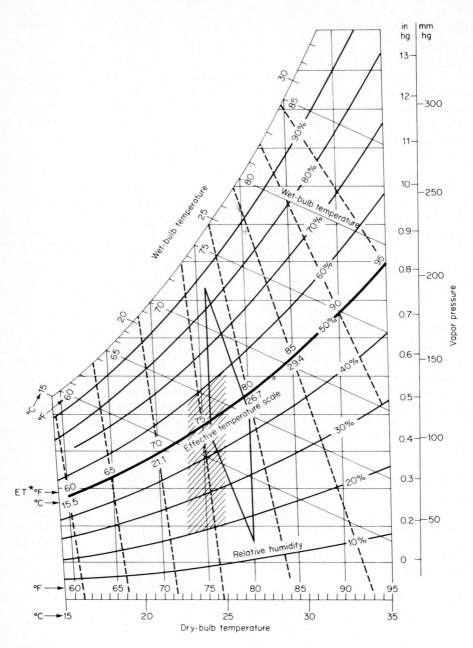

FIGURE 15-3
New effective temperature (ET*) scale. The broken lines represent the ET* values. See text for discussion. Vapor pressure is an index of humidity. Wet-bulb temperature is a measure of temperature under conditions of 100 percent RH, obtained with an aspirated wet-wick temperature or sling thermometer. *(Source: Adapted from ASHRAE, 1981, fig. 16, p. 8.21.)* The shaded area represents the ASHRAE comfort standard. The diamond-shaped area represents the modular comfort envelope (MCE) proposed by Rohles (1974).

Wet-Bulb Globe Temperature

This index (also called the WBGT index), used in enclosed environments, is a weighted average of dry-bulb temperature t_{db} (the usual measure of air temperature) and what is called the "naturally" convected wet-bulb temperature t_{nwb}. (Wet-bulb temperature takes into account the effects of humidity in combination with air temperature and air movement, and it is especially useful when the human body is near its upper limits of temperature regulation by sweating.) The WBGT index is derived with the following equation (ASHRAE *Handbook,* 1981, p. 8.17):

$$WBGT = 0.3t_{db} + 0.7t_{nwb}$$

(The WBGT index is measured with a wet-wick thermometer inside a copper ball painted black.)

One feature of this measure that makes it attractive is the fact that the air velocity need not be measured directly, since its value is reflected in the measurement of the natural wet-bulb temperature. The WBGT index is coming into more common use because it does take into account the combination of variables mentioned above.

Botsball

The Botsball (BB) index is a special type of thermometer that produces a single index of the combination of air temperature, humidity, wind speed, and radiation variables (Botsford, 1971). On the basis of an extensive experimental use of the Botsball index in many industrial work stations, Beshir, Ramsey, and Burford (1982) report that Botsball readings were highly correlated with WGBT values (the correlation being .96). Thus these two measures are virtually interchangeable. They can be converted from one to the other with the following equations (in degrees Celsius):

$$WGBT = 1.01BB + 2.6$$

$$BB = 0.905WGBT - 0.909$$

Beshir, Ramsey, and Burford point out that the BB index is a practical, simple device that can be used near workers, but without interfering with the workers. Thus it is very practical. However, it does have some operational limitations.

Heat Stress Index

The heat stress index (HSI) was developed by Belding and Hatch (1955). It is one of the most comprehensive indices available, but because of it complexity it is not in common use. Basically, the HSI is the ratio of the body's heat load from metabolism, convection, and radiation to the evaporative cooling capacity of the environment. It is predicated on the assumption that the heat load (from the sources mentioned above) must be dissipated through evaporation. Thus, the ratio of heat load to the evaporative cooling capacity of the environment indicates the relative ease or difficulty with which the heat load can be dissipated. The index takes into account environmental

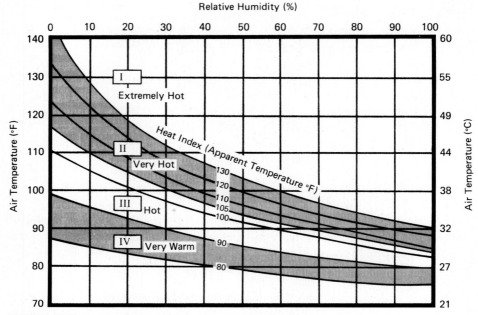

FIGURE 15–4
The heat index of the U.S. National Weather Service. *(Steadman, 1979.)*

factors such as temperature, humidity, and air movement, but in addition includes metabolic rate and clothing worn by individuals.

Heat Index

In 1985, the U.S. National Weather Service began using a *heat index* in certain of its forecasts—such as when the temperature is expected to reach 95°F (35°C). As an index of "midsummer misery" it is a measure of the contribution that heat makes with humidity in reducing the body's ability to cool itself. The heat index scale is shown in Figure 15–4. Each of the lines represents the combinations of air temperature and relative humidity that have the same general effect on people. Four categories of such effects on people in high-risk groups are given below:

	Category	Heat index	General effects
I	Extremely hot	130°F+ 55°C+	Heat/sunstroke highly likely with continued exposure
II	Very hot	105–130°F 41– 55°C	Sunstroke, heat cramps or heat exhaustion likely, and heatstroke possible with prolonged exposure and/or physical activity
III	Hot	90–105°F 32– 41°C	Sunstroke, heat cramps and heat exhaustion possible with prolonged exposure and/or physical activity
IV	Very warm	80– 90°F 27– 32°C	Fatigue possible with prolonged exposure and/or physical activity

The intent of the National Weather Service in using the heat index is to make people aware that the heat-humidity combination is dangerous and can lead to sunstroke or heat cramps.

Wind Chill Index

The indices discussed on page 434 are used largely under conditions of normal or high temperatures. The most commonly used index for cold conditions is the *wind chill index* (WCI) or, more typically, a derivative of it, the *equivalent wind chill temperature*. The original basis for the WCI was developed by Dr. Paul Siple (who, as an Eagle Scout, was in one of Admiral Byrd's expeditions to Antarctica). He developed the WCI for the Army in 1945 (Siple and Passel, 1945). The WCI is an index of the combined effects of air temperature and wind speed on subjective discomfort, and thus it is useful as an idea of the relative severity of cold environments, especially with wind velocities below 50 mi/h (80 km/h).

Actually, the WCI as such is not often used directly. Rather it is common practice to convert such values to corresponding equivalent wind chill temperatures. Any given equivalent wind chill temperature is the ambient temperature which, in a calm wind velocity, would yield the same WCI. The equivalent wind chill temperatures for selected air temperatures and wind speeds are given in Table 15-1. Any given value in the body of the table represents the air temperature (under calm conditions) that would produce the same "subjective discomfort" as the specific combination of air temperature and wind speed for that value. Thus, with an air temperature of 30°F (−1°C) and a wind speed of 40 mi/h (64 km/h), the discomfort would be the same as with an air temperature of −4°F (−22°C) with calm wind. Similarly, an air temperature of 0°F (−18°C) and a wind speed of 40 mi/h (64 km/h) would seem like a chilling −54°F (−47°C) with calm wind.

Discussion of Composite Indices

Certain of the composite indices discussed above (such as ET, ET*, and WBGT) are expressed in terms of temperature (degrees Fahrenheit or Celsius). In using these, however, keep in mind that they cannot be interpreted in terms of conventional dry-bulb temperatures, since they also take into account other variables (such as humidity, air flow, etc.).

CLO UNIT: A MEASURE OF INSULATION

Since there will be later reference to the clo unit, it is described here. The basic heat exchange process is, of course, influenced by the insulating effects of the clothing worn, varying from a bikini bathing suit to a parka as worn in the arctic. Such insulation is measured by the clo unit. The *clo unit* is a measure of the thermal insulation necessary to maintain in comfort a sitting, resting subject in a normally

TABLE 15-1
EQUIVALENT WIND CHILL TEMPERATURES FOR CERTAIN AIR TEMPERATURES AND WIND SPEEDS

Wind speed		Equivalent wind chill temperatures											
		Air temperature, °F						Air temperature, °C					
mi/h	km/h	30	20	10	0	-10	-20	-1	-7	-12	-18	-23	-29
Calm		30	20	10	0	-10	-20	-1	-7	-12	-18	-23	-29
5	8	27	16	7	-6	-15	-26	-3	-9	-14	-21	-26	-31
10	16	16	2	-9	-22	-31	-45	-9	-16	-23	-28	-36	-43
20	32	3	-9	-24	-40	-52	-68	-15	-23	-31	-39	-47	-55
30	48	-2	-18	-33	-49	-63	-78	-19	-28	-36	-45	-54	-61
40	64	-4	-22	-36	-54	-69	-87	-22	-30	-38	-47	-56	-65

Source: Environmental Science Services Administration.

ventilated room at 70°F (21°C) and 50 percent relative humidity. Since the typical individual in the nude is comfortable at about 86° F (30°C), one clo unit would be required to produce an equal sensation at about 70°F (21°F). A clo unit has very roughly the amount of insulation required to compensate for a drop of about 16°F and is approximately equivalent to the insulation of the clothing normally worn by men. To lend support to the old adage that there is nothing new under the sun, the Chinese have for years described the weather in terms of the number of suits required to keep warm, such as a *one-suit* day (reasonably comfortable), a *two-suit* day (a bit chilly), up to a limit of a *twelve-suit* day (which would be really bitter weather).

THERMAL COMFORT AND SENSATIONS

The concept of thermal comfort is admittedly elusive. First, there are distinct individual differences, which are in part the consequence of differences in metabolism. Otherwise, comfort can be influenced by the work being performed, the clothing worn, and even the season of the year. ASHRAE (1981) established a comfort standard that is represented in Figure 15-3 by the shaded area. In turn, Rohles (1974) carried out an extensive study of 1600 subjects that resulted in the development of a modular comfort envelope (MCE). This envelope falls between the 75 and 80°F (24 and 27°C) ET* lines, forming a diamond, which is also shown in Figure 15-3. Although these two areas do overlap somewhat, the MCE is generally higher than the ASHRAE standard. Rohles suggests that this difference arises from differences in the experimental conditions used in their development. The MCE study was carried out with subjects who wore clothing of 0.6 clo unit, whereas those used in the ASHRAE standard study wore clothing of 0.8 to 1.0 clo unit. In addition, the MCE applies to "sedentary" activity whereas the ASHRAE standard applies to office work, which is above the sedentary level. Rohles makes the point that the ASHRAE standard needs to be revised. However, since the two areas do overlap, he recommends that that area be considered as the "design conditions for comfort," which is specified as an ET* of 76°F (24°C). Rohles points out that the MCE applies only to situations in which dry-bulb temperature and humidity are varied, and not to conditions of increased air movement or higher activity levels or when different clothing is used.

Some indication of the influence of level of activity and of clothing on sensations of comfort is shown in Figure 15-5. This figure shows "comfort lines" for each of three levels of activity (sedentary, medium, and high) for persons with light and medium clothing. The differences in these lines (they are actually ET* lines) reflect the very definite effects on comfort of work activity and clothing.

Although much has been found out about the sensations of thermal comfort under various conditions, there are still many gaps in the available knowledge, especially with regard to the interactions among some of the variables that influence such sensations.

HEAT STRESS

Having discussed the heat exchange process and certain thermal indices, we now turn to the effects of heat stress.

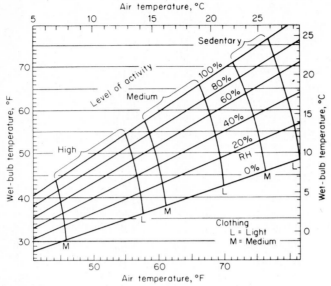

FIGURE 15–5
Comfort lines for persons engaged in three levels of work activity
with light clothing (0.5 clo) and medium clothing (1.0 clo). These
data are for a low level of air velocity. *(Source: Based on data
from Fanger, 1970; reprinted, with adaptations, with permission
from ASHRAE, 1981, chap. 8, fig. 18.)*

Physiological Effects of Heat Stress

One of the most direct effects of heat stress is on the temperature of the body.
However, there are several different measures of temperature, including oral temperature, rectal temperature, and skin temperature. The relationship among these is relatively limited except under conditions of high temperature and high humidity. Rectal
temperature is generally considered to be a representative temperature of the core.
Heat stress usually is accompanied by increases in rectal temperatures. This is illustrated in Figure 15-6, which shows the relation between ET and rectal temperature
for one individual engaged in three levels of physical activity and for the average of
three individuals at a single level of activity. In each instance the long, flat sections
of the curves represent what Lind (1963a) refers to as the *prescriptive zone* for whatever activity is involved. As a prelude to a later discussion of the effects of work
load, we can see the differential upswings for the three levels of work activity represented, the upswing starting at successively lower points with increasing work load.
The middle, light line (that represents a single individual working at a level of 300
kcal) parallels the heavy line (that represents the average for three individuals), a fact
which indicates considerable stability of the patterns of the curves.

Grandjean (1981) cites various studies as the basis for recommending that rectal
temperature should not be permitted to exceed 38°C (100°F). We can see from Figure
15-6 that such a level occurs with an effective temperature of about 30°C (86°F) for
persons performing physical work.

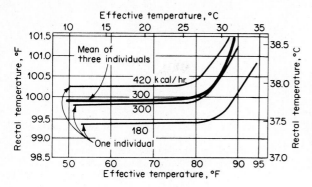

FIGURE 15–6
Rectal temperature as related to effective temperature for one individual working at three levels of work activity and for three individuals working at the same level (300 kcal/h). *(Source: Adapted from Lind, 1963a.)*

Heart rate also is affected by temperature. This was reflected, for example, by the result of a study by Meese et al. (1984) in which South African factory workers were given several tasks that simulated factory work under various thermal conditions (both hot and cold). The mean heart rates of the workers are shown in Figure 15-7. These rates increased over the range of temperatures for all four groups of workers.

Dehydration is another possible consequence of heat stress. In the case of 51 operating engineers working during mild summer weather in California's Central Valley, the average weight loss from water depletion per day was about 5 lb 2 oz (2.3/kg). Such a deficit, of course, needs to be made up during the evening and night.

Heat Stress and Performance

The relationship between heat stress and performance is admittedly complicated, the effects being related in part to the type of work activity. A few relevant studies are summarized to give some impression of the effects of heat stress on performance.

Physical Work   The effects of heat stress on performance of physical work are rather well documented. In the study by Meese et al. (1984), performance on the

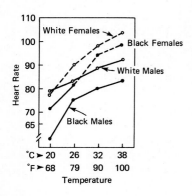

FIGURE 15–7
Mean heart rates of factory workers performing tasks that simulated factory work under various conditions of temperature. *(Adapted from Meese et al., 1984, fig. 1.4, p. 36.)*

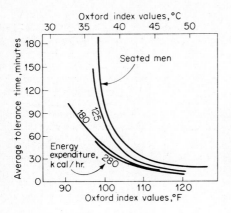

FIGURE 15–8
Average tolerance times of men seated, and of men working at three levels of energy expenditure, in relation to Oxford index values. Safe tolerance values should be taken to be no more than about 75 percent of the times shown. *(Source: Adapted from Lind, 1963b, as based on data from other studies.)*

simulated factory tasks generally deteriorated under the hottest conditions (100°F, 38°C), as contrasted with the temperatures of 90°F (32°C) and below. (The performance of white females on certain of the tasks departed somewhat from this pattern.) An additional example comes from Lind (1963b) as presented in Figure 15-8. This shows the tolerance times of men working at three levels of work expenditure in relation to variations in the Oxford index.

Attention-Demanding Tasks In many jobs the physical demands are relatively limited, but the attention demands may be substantial. On the basis of a review of research relating to performance in extreme heat conditions, Hancock (1981) synthesized the effects of heat on performance of people on three types of work: mental and cognitive work, tracking, and complex and dual tasks. His synthesis consisted in part of determining the tolerance time of people performing such activities under various thermal conditions (effective temperature).

The results of his synthesis are shown in Figure 15-9. Each curve is an "equal-tolerance" line in that it shows the various combinations of exposure time and effec-

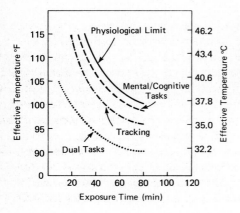

FIGURE 15-9
Equal-tolerance curves for three types of work activities along with a physiological tolerance curve. Based on a synthesis of data from various studies. *(Adapted from Hancock, 1981. Copyright by Human Factors Society, Inc. and reproduced by permission. The sources of data on which the curves are based are given in Hancock's paper.)*

tive temperatures that were equally tolerable to the subjects. In addition, a physiological exposure limit line is included. This figure shows quite clearly that there is, for all three types of work, a trade-off between exposure and ET (a high ET limits the tolerable exposure time). Also there are differences in the tolerance levels for the three types of activities: typical cognitive and mental activities can be performed closer to the "physiological" limit, with dual tasks (timesharing) having the lowest limits in terms of tolerance time and ET.

A further analysis by Hancock suggests that the impairment of performance on various types of jobs is associated with increases in core temperature. The following increases in core temperature were associated with reaching the tolerance levels in question:

	Increase
Physiological limit	3.0°F (1.7°C)
Mental and cognitive tasks	2.4°F (1.3°C)
Tracking	1.6°F (0.9°C)
Dual tasks	0.4°F (0.2°C)

The increase in core temperature can thus be viewed as an indication of the level of "tolerance" of the type of activity to heat stress: The lower the increase in core temperature, the lower the tolerance to heat stress during performance of the activity.

Note that Wing (1965) had developed a curve of the tolerance limits for performing mental activities similar to the curve for mental and cognitive tasks shown in Figure 15-9. However, Wing's curve was somewhat lower than Hancock's—actually a bit below the curve for tracking tasks. In discussing this difference Hancock argues that, on the basis of his synthesis, mental and cognitive activities can be performed at a level closer to the physiological limits than is reflected by Wing's presentation. The clarification of such a difference, however, probably is still dangling.

Discussion of Heat Stress and Performance A word of caution is in order concerning the evaluation of the results of much heat stress research relating to performance—that which has been based on laboratory studies. (This includes some of the research summarized above.) The performance of laboratory subjects is not necessarily representative of the performance of full-time workers because laboratory subjects may not be representative of workers, and because of possible differences in motivation and acclimatization.

Heat Exposure Limits

Individual differences in heat tolerance levels, the type of work, clothing worn, air circulation, and other variables preclude the establishment of absolute limits of tolerance for heat that differentiate "safe" and "unsafe" levels. With this caveat—that heat exposure limits must be considered as approximate—one set of recommended limits is shown in Figure 15-10 (ASHRAE, 1981). These recommendations were

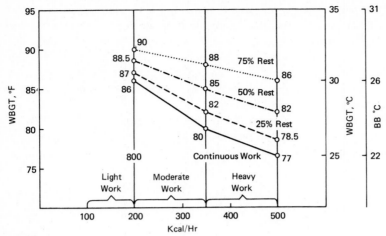

FIGURE 15–10
One set of recommended WBGT limits that would be permissible for various workloads for continuous work during a workday and for specified rest conditions. *(From ASHRAE, 1985, fig. 12, p. 8.16. Reprinted by permission from ASHRAE Handbook–1985 Fundamentals. Based on data from Dukes-Dubos and Henschel, 1971. Conversion to BB value based on Beshir, Ramsey, and Burford, 1982.)*

adopted by the American Conference of Governmental Industrial Hygienists (ACGIH) in 1973 and were accepted by the International Standards Organization (ISO) in 1981 in its draft standard. This set of recommendations specifies the permissible heat exposure threshold limit values in WBGT units for different workload levels and work-rest schedules. The limits depict the conditions under which 95 percent of the working population could be exposed without suffering adverse health effects. The lowest curve (for continuous work) represents the temperatures at which various workloads can be carried out over a workday without periodic rest. The other curves represent conditions under which workers should be given rest periods equal to 25, 50, or 75 percent of each hour during the workday. The conversion of WBGT values to BB equivalents is based on an equation from Beshir, Ramsey, and Burford (1982).

The WBGT limits for various levels of work as recommended by the Occupational Safety and Health Administration (OSHA) (1974) are virtually the same as those given in Figure 15-10. The OSHA recommendations are shown in Table 15-2, with limits given for both low and high air velocity.

Although there is reasonable agreement on the acceptable limits of exposure to heat for the conventional 8-h workday, there is considerable divergence of opinion regarding such limits for intermittent work in hot environments—when people are subject to occasional periods of work in hot environments. The problem is that the data about such exposure are quite limited. We do not really know the relative physiological impact of short and long exposures, the extent to which the temperature of rest areas affects strain, and whether short rest and short work periods produce the

TABLE 15-2
RECOMMENDED MAXIMUM WET-BULB GLOBE TEMPERATURES FOR VARIOUS
WORKLOADS AND AIR VELOCITIES

	Air velocity	
Workload	Low: below 300 ft/min (1.5 m/s)	High: 300 ft/min (1.5 m/s) or above
Light (200 kcal/h or below)	86°F (30.0°C)	90°F (32.2°C)
Moderate (201 to 300 kcal/h)	82°F (27.8°C)	87°F (30.6°C)
Heavy (above 300 kcal/h)	79°F (26.1°C)	84°F (28.9°C)

Source: Occupational Safety and Health Administration, 1974.

same strain as long rest and long work periods. In the absence of such data, one should be cautious in applying to intermittent-work situations the data and recommended exposure limits for an 8-h workday.

Reducing Heat Stress

Certain methods of reducing heat stress were implied in the previous discussion. These include reduction of air temperature, wall temperature, and humidity (when feasible); increasing air circulation (including fans); reduction of workload (in the case of physical tasks, if this is feasible); introduction of rest pauses; and use of lightweight apparel. Aside from such procedures, there have been some interesting developments in designing personal cooling systems for possible use in very hot environments such as steel mills and deep mines. These include air-cooled and water-cooled suits and helmets, frontal fans, jets of cooled air, and ice-pack suits (Zakay et al., 1983).

Because of the heat in deep mines in South Africa, Shvartz (1975) investigated the effects of body cooling with ice, with the ice being held in pockets made of polyvinyl chloride. He found that such cooling of the body reduced average heart rates from 158 to 108 beats per minute. Although the apparatus involved in such techniques can be complicated to use and may restrict work activity, there are some circumstances in which such procedures might well be justified.

Heat Reduction: Blast Furnaces Case Study Although most studies of heat stress have been carried out in laboratories, a major project was carried out in blast furnaces in Luxembourg and Belgium over a 4-year period to study how to minimize the effects of heat. Without going into details, at one stage certain modifications were made in a couple of blast furnaces to reduce heat stress. Figure 15-11 shows the comparison of the heat stress of these two as related to the heat stress of one of the blast furnaces before any changes were made. The changes in design of the two clearly reflect an appreciable reduction in heat stress.

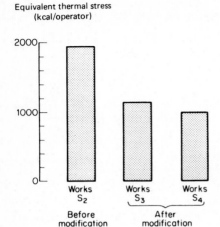

Equivalent thermal stress
(kcal/operator)

FIGURE 15–11
Heat stress of two blast furnaces that were partially redesigned (S_3 and S_4) as compared with that of a blast furnace before any modification. *(Source: Adapted from Vogt et al., 1977, fig. 10, p. 179.)*

COLD

Although civilization is generally reducing the requirement for many people to work in cold environments, there still are some circumstances where people must work, and live, in cold environments. These situations include outdoor work in winter, arctic locations (especially in military and exploration activities), cold-storage warehouses, and food lockers. As in the case of heat exposure, there are a number of complicated interrelationships between the physiological, performance, and sensory aspects of the effects of cold. In this section we discuss briefly a few of the effects of cold. In considering such effects, however, we need to keep in mind the interrelationships mentioned above as well as the admonition of Enander (1984) that an "integrated approach" is required to understand how people are affected by work in the cold. Such an integrated approach is still not entirely possible at this time.

Physiological Effects of Cold

As discussed, exposure to the cold triggers a restriction of blood to the surface of the body (to prevent body heat loss). This protective response, of course, causes a drop in skin temperature and may also cause shivering. Continued exposure can bring about frostbite, especially to the extremities. A fall in deep body (rectal) temperature below 95°F (35°C) usually threatens control of body temperature regulation, while 82°F (28°C) is considered critical for survival (ASHRAE, 1981, p. 8–15). (However, there have been cases of people who have survived extremely cold water or outdoor temperature for periods normally considered to be fatal.)

Subjective Sensations to Cold

We all have experienced sensations of discomfort in cold conditions. In part our sense of comfort or discomfort is related to skin temperature. This is shown in Table

TABLE 15-3
SUBJECTIVE SENSATIONS ASSOCIATED WITH SKIN
TEMPERATURES

Sensation	Mean skin temperature	Hand-skin temperature
Comfortable	92°F (33.3°C)	
Uncomfortably cold	88°F (31°C)	68°F (20°C)
Shivering cold	86°F (30°C)	
Extremely cold	84°F (29°C)	59°F (15°C)
Painful		41°F (5°C)

Source: ASHRAE 1981. pp. 8–15.

15-3, which shows the mean skin temperature (the mean of skin temperatures of various body areas) and hand-skin temperature for various sensations.

Performance under Cold Conditions

The effects of cold on performance of various types of human functions are not yet fully understood and, in any event, are rather intricate. Some of the interacting factors that complicate such understanding include the specific type of task or function; the interaction of air temperature, humidity, air flow, and radiation; duration of exposure; whether the body is being warmed or cooled; rate of cooling; differential exposure to different parts of the anatomy (as the body versus the hands); acclimatization; and individual differences.

Given this assortment of interacting variables, however, we can distill at least a few generalizations from the available research. A few such generalizations are given below, most of these being extracted from Enander's (1984) comprehensive review.

Manual Performance

• Manual performance is, without doubt, affected by the cold. The deterioration in manual performance is associated with (and presumably a function of) lower hand-skin temperature (HST). See Figure 15-12 on performance on two psychomotor tests as related to HST.

• Such performance is, in part, associated with the rate of cooling (with performance following slow cooling being affected more than that following fast cooling—probably because the deep temperature of the hands is lower).

• Movements involving the fingers are affected more than those of the hand.

• The lower bounds of unimpaired performance are still questionable. In terms of hand-skin temperatures, those from about 55 to 65°F (13 to 18°C) have been suggested as being such lower bounds, but there is still some quesiton about such limits. In terms of ambient temperature, Riley and Cochran (1984a) reported a moderate drop on several dexterity tests when the ambient temperature dropped from 75 to 55°F (24 to 13°C) but a precipitous drop between that level and 35°F (1.7°C); where the precipitous drop starts, however, is not known.

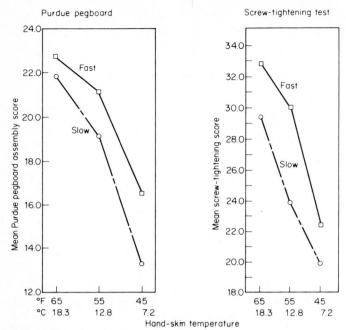

FIGURE 15–12
Deterioration in performance on two tasks as associated with
reduction in hand-skin temperature (HST) and with fast cooling of
the hands (within 5 min) or slow cooling (within 50 min). *(Source:
Adapted from Lockhart, Kiess, and Clegg, 1975, fig. 2, p. 111.
Copyright 1975 by the American Psychological Association.
Reprinted by permission.)*

Tracking Performance Tracking performance is adversely affected by the cold.
Although the lower limit of unimpaired performance is still uncertain, it probably is
between an ambient temperature of 39 and 55°F (4 and 13°C).

Reaction Time There is no systematic evidence that simple reaction time is
affected by cold. However, it seems that *errors* in a serial choice reaction time
(SCRT) task are increased (Ellis, 1982). Although the theoretical explanation for this
is not clear, the increased errors probably can be attributed more to cognitive factors
than to any effects on the simple physical responses required. In any event, Ellis
proposes that the test he used can be employed to measure the effects of cold stress
on performance.

Mental Activities Evidence about the effects of cold on mental activities is ambiguous. Some investigators have found virtually no effect, whereas at least one
investigator reports that the ''optimum'' air temperature for such activities is as low
as 59 to 63°F (15 to 17°C), implying some deterioration below that level.

Maintaining Performance and Comfort in the Cold

When people are in cold environments, for whatever reason (work, recreation, or otherwise), efforts should be made to optimize their comfort (and, when relevant, performance) in the conditions in question.

Thermal Effects of Clothing   Protection of the body with appropriate clothing is important in any event. The use of warm clothing can extend the tolerance of people for cold conditions, as illustrated in Figure 15-13. This shows the tolerable combination of exposure time and temperature for each of four insulation levels; the difference in tolerance between 1 and 4 clo units is upward of 60°F (16°C). This demonstrates the trade-off effects in terms of maintaining heat balance of exposure time and insulation.

In connection with the use of apparel in cold conditions, however, McIntyre and Griffiths (1975) report that adding garments does not compensate fully for the discomfort of cool conditions. Although their subjects did report increased feelings of warmth with long-sleeved woolen sweaters, the sweaters did not fully alleviate the feelings of discomfort. The authors attributed this to the fact that the subjects still reported that their feet felt cold. Thus, for adequate protection in the cold, attention should be given to providing footwear that is as warm as possible.

In view of current interests in outdoor activities such as skiing, backpacking, and camping, there is increased concern about thermal protection from various types of clothing, sleeping bags, etc. With many new types of such equipment having been developed in recent years—and with exaggerated claims by manufacturers of some such products—there is a need for some reliable method of measuring and comparing the heat transfer characteristics of outdoor garments and sleeping bags. Such a method is reported by McCullough and Rohles (1983), based on the experimental

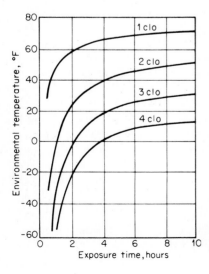

FIGURE 15–13
Exposure time and temperature that are tolerable for different levels of insulation (clo units). *(Source: Adapted from Burton and Edholm, 1955, as presented in Webb, 1964, p. 125.)*

testing of such equipment with electrically heated mannequins located in a climate-controlled chamber. Without going into details, they found that the method was satisfactory for comparing the thermal protection characteristics of such products, especially in terms of criteria of insulation and permeability.

Use of Gloves   When the work or other activities involve primarily the hands (rather than the fingers, as such), gloves usually can be worn. When finger dexterity is required, however, regular gloves may not be feasible. In such instances partial gloves may be useful. These are gloves with the extreme and middle phalanges (sections) of the fingers exposed. The protection of the rest of the hand, however, seems to provide at least some carry-over protection to the exposed portions of the fingers under at least cool (if not cold) conditions. In an investigation with such gloves, Riley and Cochran (1984b) had subjects perform several manipulative tasks with and without such gloves at 60, 50, and 40°F (16, 10, and 4°C). They reported that partial gloves can protect the hands and fingers with minimal detrimental effect on performance of the tasks in question, at least down to about 40°F (4°C). Some indication of the protection is shown by the following data on mean finger-skin temperature:

	Without gloves	With gloves
Males	49°F (9.4°C)	66°F (18.9°C)
Females	53°F (11.7°C)	71°F (21.7°C)

Use of Auxiliary Heaters   Most work in the cold involves some form of manual activity. In such cases efforts must be made to minimize reduction of hand-skin temperature. One such procedure is the use of infrared auxiliary heaters as proposed by Lockhart and Kiess (1971). In their investigation, one group used these heaters under 0°F (−18°C) conditions, another had no heaters under the same conditions, and a third control group performed the task indoors. The results for the Purdue Pegboard test are shown in Figure 15-14 and indicate that the auxiliary heating with the heaters was very effective in maintaining task performance. The use of such auxiliary heaters would not be feasible, of course, in some situations (such as in some construction or logging operations).

Use of Rewarming Facilities   There are times when exposure to cold exceeds reasonable tolerance levels of one or more relevant criteria (such as skin temperature, hand-skin temperature, core temperature, air temperature, work performance, etc.). When such tolerance levels are exceeded, individuals should go to a location (such as a warm room) where rewarming of the body or hands can take place. But there are still questions about how such rewarming should be carried out. In citing the results of various relevant investigations, for example, Enander (1984) reports that the effects of intermittent warming and cooling remain unclear. The ambiguities presumably revolve around the questions of optimum schedules for doing so, by considering such factors as the level of deep-body cooling that has taken place and the

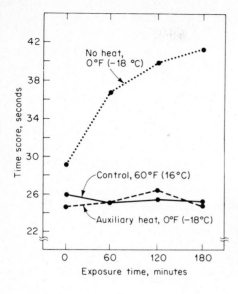

FIGURE 15–14
Time to perform the assembly part of the
Purdue pegboard test under air conditions of
0°F (− 18°C) with and without auxiliary warming
of the hands and for a control condition.
*(Source: Adapted from Lockhart and Kiess,
1971, fig. 1.)*

relative importance of hand cooling and body cooling in maintaining reasonable thermal balance.

AIR PRESSURE AND ALTITUDE

In the mundane lives of most mortals, the atmospheric variables that we complain about most are temperature and humidity. In less normal environments, however, other variables may play first fiddle. For people in high altitudes (e.g., mountainous areas and aircraft) and for people below sea level (e.g., diving and underwater construction work), air pressure and associated problems can be of paramount importance to well-being and to human performance.

The Atmosphere Around Us

The atmosphere of the earth consists primarily of oxygen (21 percent) and nitrogen (78 percent) but also includes a bit of carbon dioxide (0.03 percent) and other odds and ends. The density of the atmosphere, however, is reduced at higher altitudes; thus, at an altitude of, say, 20,000 ft (6100 m), there is less air in a given volume than at sea level. Further, because of the weight of the atmosphere, the pressure decreases with altitude. Air pressure is frequently measured in pounds per square inch (lb/in^2) or in millimeters of mercury (mmHg). At sea level the air pressure is 14.71 lb/in^2 (760 mmHg). Figure 15-15 shows the pressures at other altitudes. A reduction of pressure by some ratio would mean that the volume occupied by a given amount of air would be increased inversely (such as doubling the volume if the pressure is reduced by half).

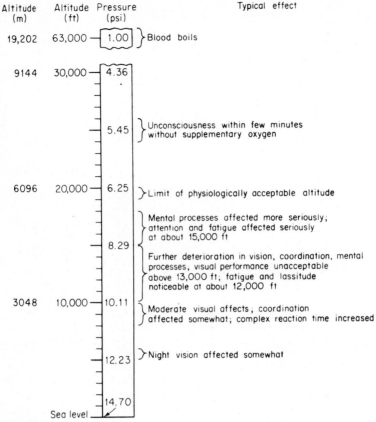

FIGURE 15–15
General effects of hypoxia at various altitude levels and equivalent pressure
levels. *(Source: Adapted from Roth, 1968, vol. 3, sec. 10; USAF, 1954,
1960; Tufts University, 1952, part 7, chap. 2, sec. 14, tables 4–1 to 4–4;
Balke, 1962.)*

Air Pressure and Oxygen Supply

A primary function of the respiratory system is transporting oxygen from the lungs
to the body tissue and picking up carbon dioxide on the return trip and carting it
back to the lungs, where it is exhaled. Under normal circumstances (including near-
sea-level pressure) the blood (actually the red blood cells) carries oxygen up to about
95 percent of the red blood cells' capacity. As air pressure is reduced, however, the
amount of oxygen that the blood will absorb is reduced. For example, at approximately 10 lb/in² (0.7kg/cm²) [equivalent to about a 10,000-ft (3048-m) altitude] the
blood will hold about 90 percent of its potential capacity; at about 7.3 lb/in² (0.51
kg/cm²) [18,000 ft (5486 m)] the percentage drops to about 70. At 1.0 lb/in² (0.07
kg/cm²) [63,000 ft (4429 m)] the pressure is so low that the blood actually boils, like
water in a teakettle.

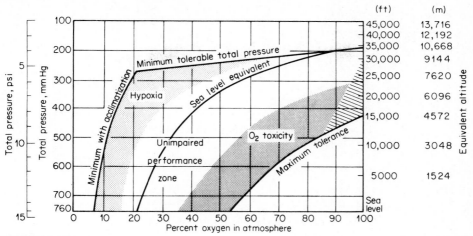

FIGURE 15–16
Physiological relationships between the percentage of oxygen in the atmosphere and pressure
(and equivalent altitude). The white band under the *sea-level equivalent* curve shows the
conditions in which human performance normally is unimpaired. *(Source: Adapted from Roth,
1968, fig. 1–2, based on data compiled by U. C. Luft and originally drawn by E. H. Green of
the Garrett Corporation.)*

Hypoxia

If the oxygen supply is reduced, a condition of *hypoxia* (also called *anoxia*) can
occur, the effects varying with the degree of reduction. Some indication of these is
given in Figure 15-16 as related to the amounts of oxygen that would be available at
various altitudes (or their equivalents). Generally speaking, the effects below about
8000 ft (2440 m) are fairly nominal, but above that level, or at least above about
10,000 ft (3050 m), the effects become progressively more serious, as shown in the
figure. In connection with the effects of hypoxia, however, two qualifications are in
order: There are marked individual differences, and acclimatization does increase
tolerance somewhat.

Use of Oxygen At altitudes where the hypoxia effects would normally be of
some consequence, the use of oxygen masks can stave off the onset of hypoxia or
minimize its degree. This is illustrated, for example, in Figure 15-17, which shows
the relationship between altitude and percentage of oxygen capacity for subjects
breathing air versus pure oxygen. This also shows the approximate degree of handicap for various percentage values. For example, while breathing pure air reduces the
percentage of capacity to about 70 at around 18,000 ft (5500 m), the use of oxygen
delays this amount of reduction up to about 42,000 ft (12,800 m).

Pressurization The ideal scheme to avoid hypoxia at high altitudes is to use a
pressurized cabin that maintains the atmospheric conditions of some lower altitude.
This is done in high-altitude civilian planes and in some military planes. The Air

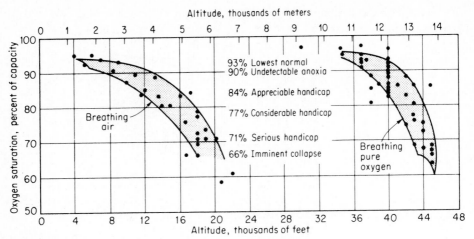

FIGURE 15–17
Relationship between altitude and oxygen saturation (percentage of capacity) for subjects breathing air and those breathing pure oxygen. *(Source: USAF, 1954.)*

Force generally requires the use of oxygen equipment for aircraft that will operate at altitudes of 10,000 ft (3050 m) or above or at 8000 ft (2440 m) on flights of 4 h or longer (*System Safety Design Handbook,* 1978), and at higher altitudes prescribes minimum differentials in pressure for different types of planes and flights (USAF, 1960). Pressure suits usually are prescribed for high-altitude aircraft, in part as protection against decompression.

Decompression

The volume of gas expands or contracts in proportion to the pressure applied to it (this is *Boyle's law*). The atmosphere within the human body is not immune from this law. Thus, as an individual changes from one pressure to another, the atmosphere in body tissues and cavities expands or contracts. In this connection, as the external air pressure is increased, that within the body follows suit fairly closely, and (aside from some discomfort here and there) there are no serious consequences. But when the external air pressure is reduced suddenly, there can be some unhappy consequences, generally referred to as *decompression sickness.* Actually, there are different kinds of physiological reactions. Among these is the formation of nitrogen bubbles in the blood, body tissue, and around the joints; the effect is somewhat like taking the top off a bottle of soda water. The manifestations of the physiological effects include various symptoms, such as those referred to as the *bends* and the *chokes,* and various skin manifestations. The *bends* consist of generalized pains in the joints and muscles. The *chokes* are characterized by breathing (choking) difficulty, coughing, respiratory distress, and accompanying chest pains; the skin sensations include hot and cold sensations and itching, and a mottling of the skin surface sometimes occurs. In extreme cases, the above symptoms become more severe, and in some cases shock, delirium, and coma occur; fatalities may occur in such cases.

Protection from Decompression   Decompression sickness occurs primarily in underwater construction work (in sealed caissons), in underwater diving operations, in submarines, and in aircraft [as in the case of rapid ascent to, say, 30,000 ft (9100 m) or more, or in rapid decompression from a rupture in a sealed cabin]. Where rapid decompression has occurred (say, by accident), it is usually the practice to subject the person to a higher pressure (near that to which he or she was originally exposed) and then gradually bring the pressure back to that of the earth.

THE AIR WE BREATHE

Human beings evolved in the "natural" atmosphere surrounding the earth. Aside from the aspects of heat and cold (previously discussed), we are pretty well adapted to that atmosphere. There are only a few other features that can have some impact on people.

Ions in the Air

One factor that can affect us is the level of negative and positive ions in the air. Such ions are molecules of the common atmospheric gases that have taken on negative or positive charges. High levels of positive ions are associated with the following: in nature, the arrival of warm, dry winds such as the Santa Anas of southern California; in human-created conditions, air conditioning, heat, and pollution (Tom et al., 1981). The following effects are alleged to be associated with positive ions: irritability, depression, tension, insomnia, lassitude, and general morbidity (Tom et al., 1981).

High levels of negative ions tend to occur after rainstorms or near bodies of water, such as waterfalls, through the breakup of droplets of water. In turn, the following effects are alleged to be associated with concentrations of negative ions: mental alertness, improved psychomotor performance, feelings of exhilaration, and increased energy.

In the late 1950s and early 1960s, there was quite a flurry of interest in the use of ion generators in homes, offices, stores, and factories. There were accompanying unsubstantiated huckster claims that such gadgets would improve the lot of human beings, the work performance of workers, and would even reduce shop lifting in department stores.

The research relating to ions has been rather meager and somewhat inconclusive. Much research still needs to be carried out, but an investigation by Tom et al. (1981) somewhat supported the claims of the beneficial effects of negative ions. These investigators found significant improvement in reaction time and self-reported feelings of ease of concentration and increased energy with high concentrations of negative ions (as contrasted with a "normal" environment). However, the investigators warn against unwarranted enthusiasm, pointing out that negative ions apparently do not affect performance on all types of tasks or create a euphoric syndrome on all types of psychological states. Further, some commercial ion generators may produce ozone (which is undesirable) in addition to negative ions.

Air Pollution

Science and technology have resulted in the creation of a whole host of atmospheric pollutants, including smoke, exhaust fumes, toxic vapors and gases, acid rain, insecticides, herbicides, and ionizing radiation. Some of these contaminants are the by-products of industrial processes and tend to be confined to their industrial environments, whereas others escape into the general atmosphere. And our automobiles, household furnaces, and other nonindustrial sources add their bit to the ever-increasing level of pollution in the air we breathe and the atmosphere that supports all forms of life, including our own. This is admittedly a problem of major proportions that faces civilization—one that has important human factors twists and implications. This problem is of such magnitude, however, that it is beyond the scope of this text.

REFERENCES

American Conference of Governmental Industrial Hygienists (ACGIH) (1973). Threshold limit values for chemical substances and physical agents in the workroom environment with intended changes for 1973. Cincinnati, OH.

American Society of Heating, Refrigerating, and Air-Conditioning Engineers (1985). *ASHRAE handbook. 1985 Fundamentals* New York.

Balke, B. (1962, April). *Human tolerances,* Rept. 62–6. Oklahoma City: Civil Aeromedical Research Institute, Federal Aviation Agency, Aeronautical Center.

Belding, H. S., and Hatch, T. F. (1955, August). Index for evaluating heat stress in terms of resulting physiological strains. *Heating, Piping and Air Conditioning,* pp. 129–136.

Beshir, M. Y., Ramsey, J. D., and Burford, C. L. (1982). Threshold values for the Botsball: A field study of occupational heat. *Ergonomics, 25*(3), 247–254.

Botsford, J. H. (1971). A wet globe thermometer for environmental heat measurement. *American Industrial Hygiene Association Journal, 32,* 1–10.

Burton, A. C., and Edholm, O. G. (1955). *Man in a cold environment: Physiological and pathological effects of exposure to low temperatures.* London: Edward Arnold.

Dukes-Dubos, F., and Henschel, A. (1971). *The modification of the WBGT index for establishing permissible heat exposure limits in occupational work* (TR-69). Washington: Public Health Service.

Ellis, H. D. (1982). The effects of cold on the performance of serial choice reaction time and various discrete tasks. *Human Factors, 24*(5), 589–598.

Enander, A. (1984). Performance and sensory aspects of work in cold conditions: A review. *Ergonomics, 27*(4), 365–378.

Fanger, P. O. (1970). *Thermal comfort, analysis and applications in environmental engineering.* Copenhagen: Danish Technical Press.

Grandjean, E. (1981). *Fitting the task to the man.* London: Taylor & Francis.

Hancock, P. A. (1981). The limitation of human performance in extreme heat conditions. *Proceedings of the Human Factors Society, 1981.* Santa Monica, CA: Human Factors Society. pp. 74–78.

Lind, A. R. (1963a). A physiological criterion for setting thermal environmental limits for everyday work. *Journal of Applied Physiology, 18,* 51–56.

Lind, A. R. (1963b). Tolerable limits for prolonged and intermittent exposures to heat. In J. D. Hardy (ed.), *Temperature—Its measurement and control in science and industry,* vol. 3, pt. 3, pp. 337–345. New York: Reinhold.

Lockhart, J. M., and Kiess, H. O. (1971). Auxiliary heating of the hands during cold exposure and manual performance. *Human Factors, 13*(6), 457–465.

Lockhart, J. M., Kiess, H. O., and Clegg, T. J. (1975). Effect of rate and level of lowered finger-surface temperature on manual performance. *Journal of Applied Psychology,* 60(1), 106–113.

McCullough, E. A., and Rohles, F. H. (1983). Quantifying the thermal protection characteristics of outdoor clothing. *Human Factors,* 25(2), 191–198.

McIntyre, D. A., and Griffiths, I. D. (1975). The effects of added clothing on warmth and comfort in cool conditions. *Ergonomics,* 18(2), 205–211.

Meese, G. B., Kok, R., Lewis, M. I., and Wyon, D. P. (1984). A laboratory study of the effects of moderate thermal stress on the performance of factory workers. *Ergonomics,* 27(1), 19–43.

Occupational Safety and Health Administration (OSHA) (1974, Jan. 9). *Recommendation for a standard for work in hot environments* (draft no. 5). Washington: Department of Labor.

Riley, M. W., and Cochran, D. J. (1984a). Dexterity performance and reduced ambient temperature. *Human Factors,* 26(2), 207–214.

Riley, M. W., and Cochran, D. J. (1984b). Partial gloves and reduced temperature. *Proceedings of the Human Factors Society, 1984.* Santa Monica, CA: Human Factors Society.

Rohles, F. H., Jr. (1974). The modal comfort envelope and its use in current standards. *Human Factors,* 64(3), 314-323.

Roth, E. M. (ed.) (1968, November). *Compendium of human responses to the aerospace environment* (vols. 1–4) (NASA CR-1205).

Shvartz, E. (1975). The application of conductive cooling to human operators. *Human Factors.* 17(5). 438–445.

Siple, P. A., and Passel, C. F. (1945). Movement of dry atmospheric cooling in subfreezing temperatures. *Proceedings of the American Philosophical Society,* 89, 177–199.

Steadman, R. C. (1979). The assessment of sultriness. *Journal of Applied Meteorology,* 18(7), 861–884.

Tom, G., Poole, M. F., Galla, J., and Berrier, J. (1981). The influence of negative ions on human performance and mood. *Human Factors,* 23(5), 633–636.

Tufts University (1952). *Handbook of human engineering data* (2nd ed.). Medford, Mass.

United States Air Force. (1954, July). *Flight surgeons manual* (USAF Manual 1960–5).

United States Air Force. (1960, Jan. 1). *Your body in flight* (USAF Pamphlet 160–10–3).

Vogt, J. J., Foehr, R., Kuntzinger, E., Seywert, L., Libert, J. P., Candas, V., and Van Peteghem, Th. (1977). Improvement of the working conditions at blast furnaces. *Ergonomics,* 20(2), 167–180.

Webb. P. (Ed.) (1964). *Bioastronautics data book.* (NASA SP–3006).

Wing, J. F. (1965, September). A review of the effects of high ambient temperature on mental performance. (TR 65–102). U.S. Air Force, AMRL. Wright Patterson Air Force Base, Ohio.

Winslow, C. E. A., and Herrington, L. P. (1949). *Temperature and human life.* Princeton, NJ: Princeton University Press.

Zakay, D., Shapiro, Y., Epstein, Y., and Bril, S. (1982). The effect of personal cooling systems on performance under heat stress. *Proceedings of the Human Factors Society.* Santa Monica, CA: Human Factors Society. pp. 128–131.

16

NOISE

Before the days of machines and mechanical transportation equipment, our noise environment consisted of noises such as those of household activities, animals, horse-drawn vehicles, hand tools, and nature. But human ingenuity changed all that by creating machines, motor vehicles, radios, guns, bombs, sirens, jet aircraft, and rock concerts. Noise has become such a pervasive aspect of working situations and community life that we refer to it as *noise pollution,* and some consider it to be a health hazard.

Although noise has commonly been referred to as *unwanted sound,* a somewhat more exact definition is the one proposed by Burrows (1960), in which noise is considered in an information-theory context, as follows: Noise is "that auditory stimulus or stimuli bearing no informational relationship to the presence or completion of the immediate task." This concept applies equally well to attributes of task-related sounds that are informationally useless as well as to sounds that are not task-related.

HOW LOUD IS IT?

We touched briefly in Chapter 6 on the fact that the human ear is not equally sensitive to all frequencies of sound. In general, we are less sensitive to low frequencies (below 1000 Hz) and more sensitive to higher frequencies. Thus a low-frequency tone will not sound as loud to us as a high-frequency tone of equal intensity (i.e., sound pressure). To put it another way, a low-frequency tone must have more intensity than a higher-frequency tone to be of equal loudness. This particular fact has led investigators to search for a metric to measure the subjective quality of sound. We discuss several basic measures here. When we discuss the annoyance quality of noise, we introduce several other measures that are derived from these basic measures.

Sound Level Meter Scales

As we indicated in Chapter 6, sound-pressure meters built to American National Standards Institute (ANSI) specifications contain frequency-response weighting networks (designated A, B, and C). Each network electronically attenuates sounds of certain frequencies and produces a weighted total sound-pressure level. Figure 16-1 shows the relative response curves of the A, B, and C scales and the response characteristics of the human ear at threshold. As can be seen, the C scale weights all frequencies almost equally. The B scale, originally intended to represent how people might respond to sounds of moderate intensity, is rarely (if ever) used. The most commonly used scale is the A scale. The Occupational Safety and Health Administration (OSHA) standards for daily occupational noise limits are specified in terms of this measure, and the Environmental Protection Agency (1974) has selected the A scale as the appropriate measure of environmental noise. As we will see, the many indices of loudness, noisiness, and annoyance are all based on the A scale (the unit is dBA). Of the three scales, the A scale comes closest to approximating the response characteristics of the human ear.

As if the A, B, and C scales were not enough, there exist on some meters D scales (there is more than one D scale). The scales were designed primarily to provide a measure of aircraft noise but have yet to gain complete universal acceptance,

FIGURE 16-1
Relative response characteristics of the A, B, and C sound-level meter scales and the human ear at threshold. *(Source: Jensen, Jokel, and Miller, 1978, fig. 2.4.)*

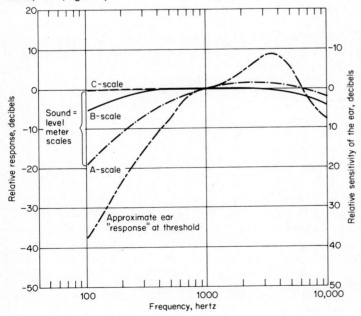

and currently they are used only rarely and for very specific measurement applications.

Psychophysical Indices

Loudness is a subjective or psychological experience related to both the intensity and the frequency of sound. Researchers have tried to develop scales or indices based on the *physical* properties of sound that will measure this *psychological* experience, hence the term *psychophysical*. Among the oldest and most widely recognized psychophysical indices of loudness are the *phon* and *sone*. The methodology used to define the phon and sone is common to other loudness indices and involves subjects matching comparison sounds to a reference sound in terms of subjective loudness (or, in one case, noisiness).

In the case of phons, Robinson and Dadson (1957) presented to subjects a 1000-Hz pure tone (the reference sound) at different sound-pressure levels and had the subjects adjust the intensity of various pure tones (the comparison sounds) until the comparison sound was judged to be of equal loudness to the reference sound. The decibel level of the comparison sound was recorded. From these data, equal-loudness curves were generated, and they are shown in Figure 16-2. Each curve shows the decibel intensity of different frequencies that were judged to be equal in loudness to

FIGURE 16-2
Equal-loudness curves of pure tones. Each curve represents intensity
levels of various frequencies that are judged to be equally loud. The
lowest curve shows the minimum intensities of various frequencies that
typically can be heard. *(Source: Robinson and Dadson, 1957. Crown
copyright reserved. Courtesy National Physical Laboratory, Teddington,
Middlesex, England.)*

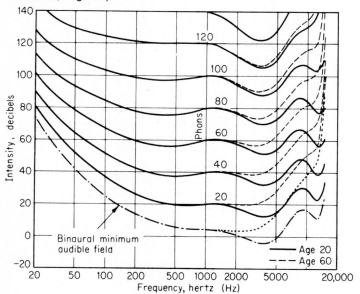

that of a 1000-Hz tone of a specified intensity level. To illustrate, one can see in Figure 16-2 that a 50-Hz tone of about 65 dB is judged equal in loudness to a 1000-Hz tone of only 40 dB. The unit *phon* was designated as the measure of loudness and was set equal to the decibel level of the 1000-Hz tone. All tones, for example, judged equal in loudness to a 60-dB, 1000-Hz tone are designated as having a loudness level of 60 phons. Our 65-dB, 50-Hz tone would, therefore, have a loudness level of 40 phons because it is equal in loudness to a 1000-Hz tone of 40 dB.

So far, so good. The phon tells us about the subjective *equality* of various sounds, but it tells us nothing about the *relative* loudness of different sounds. That is, we cannot say how many times louder a 40-phon sound is compared to a 20-phon sound. We know it is louder, but we do not know if it is, say, twice as loud or 4 times as loud. For such comparative judgments we need still another yardstick. Fletcher and Munson (1933) developed such a scale, and Stevens (1936) named it the *sone*. In developing this scale (as with the phon) a reference sound was used. One *sone* is defined as the loudness of a 1000-Hz tone of 40 dB (that is, 40 phons). A sound that is judged to be twice as loud as the reference sound has a loudness of 2 sones, a sound that is judged to be 3 times as loud as the reference sound has a loudness of 3 sones, etc. In turn, a sound that is judged to be half as loud as the reference sound has a loudness of ½ sone.

Manufacturers of stove-top exhaust fans in the United States list the loudness of their product in sones; unfortunately, most consumers have no idea what it means! To provide some basis for relating sones to our own experiences, consider the examples given in Table 16-1.

There is a relationship between phons and sones: 40 phons = 1 sone, and every 10 additional phons doubles the number of sones. For example, 50 phons = 2 sones, 60 phons = 4 sones, and 70 phons = 8 sones. In like manner 30 phons = 0.5 sone, and 20 phons = 0.25 sone. Given this, we can determine that our 40-phon sound is 4 times as loud as a 20-phon sound.

There is another set of indices used to measure loudness which was developed as a refinement to the original phon and sone. The two measures, analogous to the phon

TABLE 16-1
EXAMPLES OF LOUDNESS LEVELS

	Loudness	
Noise source	**Decibels**	**Sones**
Residential inside, quiet	42	1
Household ventilating fan	56	7
Automobile, 50 ft (15 m)	68	14
"Quiet" factory area	76	54
18-in (46-cm) automatic lathe	89	127
Punch press, 3 ft (1 m)	103	350
Nail-making machine, 6 ft (2 m)	111	800
Pneumatic riveter, 4 ft (1.2 m)	128	3000

Source: Bonvallet, 1952, p. 43.

and sone, are the *perceived level of noise* (PLdB) and the *Mark VII sone* (Stevens, 1972). There is also a set of indices used to measure noisiness which is not quite the same as loudness. The two measures are the *perceived noise level* (PNdB) and the *noy* (Kryter, 1970). These various indices differ from the original phon and sone in terms of the reference sound used and the relationship between the equality and relative measures. Table 16-2 summarizes these differences and lists references where computational procedures can be found.

Equivalent Sound Level

Noise often varies in intensity over time. Fortunately, a baby's scream will change to a whimper and with luck to a yawn. Over the years many single-number measures have been proposed for time-varying sound levels. We discuss a few when we talk about the annoyance of noise. The Environmental Protection Agency (1974), however, concluded that, insofar as cumulative noise effects are concerned, the long-term average sound level was the best measure for the magnitude of environmental noise. This long-term average is designated the *equivalent sound level* L_{eq} and is equal to the sound-pressure level (usually measured in dBA) of a *constant* noise that, over a given time, transmits to the receiver the same amount of acoustic energy as the actual time-varying sound.

L_{eq} depends on the time interval and acoustic events occurring during that period.

TABLE 16-2
SOME PSYCHOPHYSICAL INDICES OF SOUND LOUDNESS AND NOISINESS

	Psychophysical indices		
Characteristics	Phon and sone	Stevens, 1972	Kryter, 1970
Measure of:	Loudness	Loudness	Noisiness
Equality scale measured in:	Phons	Perceived level of noise (PLdB)	Perceived noise level (PNdB)
Reference sound for equality scale:	1000-Hz pure tone	1/3 octave band of noise centered at 3150 Hz	Octave band of random noise centered at 1000 Hz
Relative scale measured in:	Sones	Mark VII sones	Noys
Reference level for relative scale:	1 sone = 40 phon	1 Mark VII sone = 32 PLdB	1 noy = 40 PNdB
Relationship between equality and reference scales:	Each 10-phon increase doubles the number of sones	Each 9-PLdB increase doubles the number of Mark VII sones	Each 10-PNdB increase doubles the number of noys
Reference for computational procedures:	Peterson and Gross, 1978	Peterson and Gross, 1978	Williams, 1978

For example, if a 100-dBA noise occurred for 1 h, the L_{eq} for that hour would be 100 dBA. Consider, however, a situation in which it was quiet during the next 4 h. The L_{eq} for the total 5-h period would now be less than 100 dB; in fact, it would be 94 dBA. What this says is that 5 h of 94-dBA noise is equivalent in acoustic energy to 1 h of 100-dBA noise and 4 h of quiet.

Taylor and Lipscomb (1978) present a manual method for determining L_{eq} by using a sound-pressure meter. The decibel reading on the meter is noted every 5 or 10 s, and a worksheet is used to make tally marks adjacent to the appropriate decibel level. Fortunately today there are hand-held instruments that directly display L_{eq} values—another victory for technology.

NOISE AND LOSS OF HEARING

Of the different possible effects of noise, one of the most important and clearly established is hearing loss. There are really two primary types of deafness: *nerve* deafness and *conduction* deafness. Nerve deafness usually results from damage or degeneration of the hair cells of the organ of Corti in the cochlea of the ear (see Chapter 6). The hearing loss in nerve deafness is typically uneven, being greater in the higher frequencies than in the lower ones. Normal deterioration of hearing through aging is usually of the nerve type, and continuous exposure to high noise levels also typically results in nerve deafness. Once nerve degeneration has occurred, it can rarely be remedied.

Conduction deafness is caused by some condition of the outer or middle ear that affects the transmission of sound waves to the inner ear. It may be caused by different conditions, such as adhesions in the middle ear that prevent the vibration of the ossicles, infection of the middle ear, wax or some other substance in the outer ear, or scars resulting from a perforated eardrum. Conduction deafness is more even across frequencies and does not result in complete hearing loss. It is only a partial loss because airborne sound waves strike the skull and are transmitted to the inner ear by conduction through the bones of the skull. People with conductive deafness sometimes are able to hear reasonably well, even in noisy places, if the sounds to which they are listening (for example, conversation) are at intensities above the background noise. This type of deafness can sometimes be arrested or even improved. Hearing aids are more useful in this type of deafness than they are when deafness is caused by nerve damage.

Measuring Hearing

To review the effects of noise on hearing, we should first see how hearing (or hearing loss) is measured. There are two basic methods of measurement, namely, the use of simple tests of hearing and the use of an audiometer. Each method is discussed below.

Simple Hearing Tests   For some purposes simple hearing tests are used. These include a voice test, a whisper test, a coin-click test, and a watch-tick test. In the voice and whisper tests, for example, the tester (out of sight) speaks or whispers to the testee, and the testee is asked to repeat what was said. This may be done at different distances and with different voice intensities. The primary shortcoming of such tests is that they usually lack standardization. Even if reasonable standardization can be achieved (such as using a particular person's voice or a particular watch), such tests would still serve as only a gross test of hearing.

Audiometer Tests   *Audiometers* are of two types, the most common being an instrument that is used to measure hearing at various frequencies. It reproduces, through earphones, pure tones of different frequencies and intensities. As the intensity is increased or decreased, the people being tested are asked to indicate when they can just barely hear the tone or when it ceases to be audible. It is then possible to determine for each frequency tested the lowest intensity that can just barely be heard; this is the *threshold* for the frequency.

The second type of audiometer is a speech audiometer. Direct speech, or a recording of speech, is reproduced to earphones or to a loudspeaker, and intensity is controlled. Various types of speech intelligibility tests are discussed in Chapter 7.

Normal Hearing and Hearing Loss

Before we see what effect noise has on hearing, we should first see what normal hearing is like.

Surveys of Hearing Loss   Surveys have been made to determine the hearing abilities of people. In such surveys individuals are tested at various frequencies to determine their loss of hearing at each of the tested frequencies. The average hearing loss at each frequency is then determined.

Normal, nonoccupational, hearing loss is considered to be due to two sources: presbycusis and sociocusis. *Presbycusis* is hearing loss due to the normal process of aging. *Sociocusis* refers to hearing loss due to nonoccupational noise sources, such as household noises, television, radio, traffic, etc. Admittedly, what is considered nonoccupational to one person may be occupational to another. In essence, sociocusis excludes excess exposure to loud noises on a regular basis.

Kryter (1983a, 1983b) summarized various population surveys of hearing loss and derived idealized hearing loss curves due to presbycusis and sociocusis. Figure 16-3 shows the median hearing loss of males and females at various ages. [Figure 16-3 actually includes hearing loss due to pathological conditions (nosocusis), but Kryter indicates that this accounts for only a few decibels of the hearing loss shown.] It is clear that with age, hearing loss becomes increasingly severe at the higher frequencies. The difference between males and females is attributed to sociocusis; that is, males tend to be exposed to higher levels of nonoccupational noise than females. There are, of course, exceptions to this, as any mother who has taken care of two 3-year-olds can tell you. Based on Kryter's estimates, approximately 55 percent of normal hearing loss at higher frequencies is due to presbycusis, with sociocusis accounting for approximately 45 percent in industrialized countries.

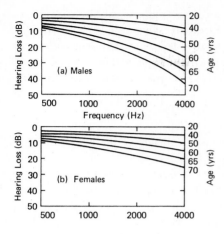

FIGURE 16-3
Idealized median (50th percentile) hearing loss due to presbycusis, sociocusis, and nosocusis for males and females as a function of age. *(Source: Kryter, 1983a, fig. 20.)*

Occupational Hearing Loss

Most hearing loss other than presbycusis and sociocusis can be considered occupationally related. Most such hearing loss is from continuous exposure over time, although exposure to noncontinuous noise (i.e., occasional or infrequent high levels of noise) also can take its toll.

After exposure to continuous noise of sufficient intensity there is some temporary hearing loss which is recovered a few hours or days after exposure. However, with additional exposure the amount of recovery gradually becomes less and less, and the individual is left with some permanent loss. The implications of permanent hearing loss are obvious. Temporary hearing loss, however, can also have serious consequences if a person depends on auditory information in the performance of a job or task.

Temporary Hearing Loss from Continuous Noise Since hearing generally recovers with time after exposure, the measurement of hearing loss must take place at a fixed time after exposure to be comparable. Traditionally, this has been done 2 min after the end of exposure. Any shift in threshold (from preexposure levels) is called the *temporary threshold shift at 2 min* (abbreviated TTS_2).

The relationship between TTS_2 and the sound level of the noise to which one is exposed is not a simple one. Some sound levels will not produce any measurable TTS_2 regardless of the duration of exposure. These sound-pressure levels define *effective quiet* where the hazardous effects of noise on hearing are concerned. This lower limit depends somewhat on frequency but is generally considered to be around 60 to 65 dBA. For exposures to noises of moderate intensity (80 to 105 dBA) TTS_2 increases in proportion to the logarithm of the sound-pressure level (SPL) of the exposure noise. Kryter (1985), after reviewing the literature, concluded that for exposures resulting in up to 10 dbA of TTS_2, $TTS_2 = 10 \log SPL + k$. For exposures resulting in 10 to 40 dBA of TTS_2, $TTS_2 = 20 \log SPL + k$ (where k is a constant that depends on the specifics of the exposure, such as intermittency, duration, etc.).

The growth, or acquisition, of TTS_2 is proportional to the logarithm of exposure time, building up quickly at first and then more slowly as exposure time increases. Recovery from temporary threshold shifts, when the level of TTS_2 is less than 40 dBA, also follows a logarithmic function and is proportional to the logarithm of recovery time. Although both the growth and recovery of temporary threshold shifts are proportional to the logarithm of time, recovery takes longer than acquisition. For example, it may take less than an hour to acquire a TTS_2 of 25 dBA, but complete recovery may take up to 16 h (Davies and Jones, 1982).

There is an interesting relationship between the frequency of the exposure noise and the TTS_2 produced. The maximum threshold shift is produced not at the frequency of the exposure noise, but at frequencies well *above* the exposure noise; e.g., exposure to a pure tone of 700 Hz will produce a maximum TTS_2 at 1000 Hz or higher.

There are, of course, marked individual differences in TTS_2 that result from a given noise. Some people may experience considerable TTS_2 while others may experience hardly any. Further, some people are more sensitive to high-frequency sounds, others to low-frequency sounds. These individual differences must be considered when criteria are set to protect ears from damage.

Permanent Hearing Loss from Continuous Noise   With repeated exposure to noise of sufficient intensity, a *permanent threshold shift* (PTS, or NIPTS for noise-induced permanent threshold shift) will gradually appear. Usually the PTS occurs first at 4000 Hz. As the number of years of noise exposure increases, the hearing loss around 4000 Hz becomes more pronounced, but is generally restricted to a frequency range of 3000 to 6000 Hz. With further noise exposure, the hearing loss at 4000 Hz continues and spreads over a wider frequency range (Melnick, 1979). (Note that 4000 Hz is in that region of frequencies to which the human ear is most sensitive.) This can be seen in Figure 16-4, which shows hearing loss curves for jute weavers exposed to wideband continuous noise with a noise spectrum that peaked in the octave bands centered at 1000 and 2000 Hz. The overall sound level varied from 99 to 102 dBA.

A committee under the aegis of the American Industrial Hygiene Association has teased out and condensed, from many sources, data that show the incidence of hearing impairment, in a consolidated figure, for several age groups of individuals who have been exposed (in their work) to noise intensities of different levels. This consolidation is shown in Figure 16-5. Actually this figure shows, for any group, the probabilities of individuals having hearing impairment (impairment being defined as an average hearing threshold shift in excess of 15 dB at 500, 1000, and 2000 Hz). Although the probabilities of impairment (so defined) are not much above those of the general population for individuals exposed to 85 dB, the curves shoot up sharply at higher levels, except for the youngest group (but their time will come!).

Must we wait 10 or 20 years to discover that a noise environment is potentially harmful? Perhaps not. It is widely accepted that the average TTS from an 8-h exposure to noise in young, normal ears is similar in magnitude to the average permanent threshold shift found in workers after 10 to 20 years' exposure to the same levels of

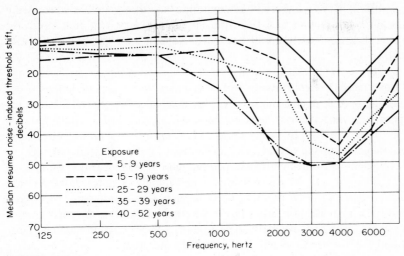

FIGURE 16-4
Median noise-induced permanent threshold shifts at various frequencies as a function of increasing exposure to noise among female jute weavers. The noise was wideband and continuous with a spectrum that peaked in the octave bands centered at 1000 and 2000 Hz, with overall sound level varying from 99 to 102 dBA. (*Source: Taylor, Pearson, Mair, and Burns, 1965, as presented by Melnick, 1979.*)

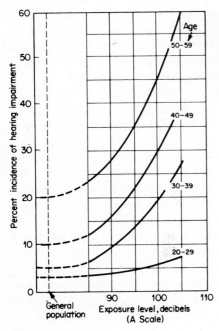

FIGURE 16-5
Incidence of hearing impairment in the general population and in selected populations by age group and occupational noise exposure; impairment is defined as a hearing threshold level in excess of an average of 15 dB at 500, 1000, and 2000 Hz. (*Source: Industrial Noise Manual, 1966, fig. 1, p. 420.*)

noise (Kryter, 1985). Kryter indicates that there are no data available to seriously challenge this basic assumption that TTS after an 8-h exposure is a good predictor of the permanent hearing loss that will occur with regular exposure 10 to 20 years down the road.

Hearing Loss from Noncontinuous Noise The gamut of noncontinuous noise includes intermittent (but steady) noise (such as machines that operate for short, interrupted periods), impact noise (such as that from a drop forge), and impulsive noise (such as from gunfire). In heavy doses, such noise levies its toll in hearing loss, but the combinations and permutations of intensity, noise spectrum, frequency, duration of exposure, and other parmeters preclude any simple, pat descriptions of the effects of such noise.[1] In the case of impact and impulsive noise, however, the toll sometimes is levied fairly promptly. For example, 35 drop-forge operators showed a noticeable increase in hearing threshold within as little as 2 years, and 45 gunnery instructors averaged 10 percent hearing loss over only 9 months, even though most had used hearing protection devices (Machle, 1945).

PHYSIOLOGICAL EFFECTS OF NOISE

Permanent hearing loss is, of course, the consequence of physiological damage to the mechanisms of the ear. Aside from possible damage to the ear itself, one would wonder whether continued exposure to noise might induce any other temporary or permanent physiological effects.

The onset of a loud noise will cause a *startle response,* characterized by muscle contractions, blink, and a head-jerk movement. In addition, larger and slower breathing movements, small changes in heart rate, and dilation of the pupils occur. There is also a moderate reduction in the diameter of blood vessels in the peripheral regions, particularly the skin (Burns, 1979). All these responses are relatively transient and settle back to normal or near normal levels very quickly. With repeated exposure to the noise, the magnitude of the initial response is also diminished.

Although there is considerable agreement on the startle-response reactions to the onset of noise, there is less agreement on the long-term physiological effects of repeated exposure to noise. There is considerable evidence that indicates that exposure to high noise levels (such as 95 dBA or more) is associated with generalized stress responses among those exposed to the noise (Jansen, 1969; Burns, 1979; Gulian, 1974). Kryter (1985), however, questions whether such effects are due to the direct overarousal of the autonomic nervous system by the noise itself or to psychological responses indirectly associated with the noise. In residential communities, for example, high noise levels (e.g., aircraft noise) disrupt sleep, cause annoyance, and may cause concern for one's safety. In industrial settings, high noise is often associated with moving and dangerous equipment; noise may mask sounds relevant to job performance, thus increasing the difficulty of the job; and workers may be concerned about the damaging effects of noise on hearing. These indirect consequences of noise may be more important than the actual noise itself as the cause of stress. Whether

[1]A thorough treatment of exposure to intermittent noise is presented by Kryter et al. (1966).

stress associated with noise is the consequence of direct overstimulation of the autonomic nervous sytem or some other indirect cause, one cannot deny that stress and high noise often go hand in hand. It should be pointed out and stressed (excuse the pun) that adverse health effects associated with noise (such as hypertension, gastrointestinal problems, electrocardiogram irregularites, and complaints of headaches) are usually evidenced only with relatively high levels of noise, often over 95 dBA.

As an example of the kinds of effects reported in studies of residential community noise, Meecham (1983) reported on an 8-year study conducted on 200,000 residents living near Los Angeles International Airport. He found death rates from heart attacks among those over 75 years of age and suicides among those 45 to 54 years old to be higher in an area exposed to jet aircraft landing noise than in a nearby matched "quiet" area. In the high-risk area, jets passing overhead generated noise levels of 115 dB, and during peak daytime hours they passed overhead every 2.5 min, enough to rattle anyone.

EFFECTS OF NOISE ON PERFORMANCE

The effects of noise on performance are not clear-cut. By choosing the right studies, one can show that noise produces a decrement, no effect, or even an improvement in performance. Gawron (1982), for example, reviewed 58 noise experiments and found 29 showed noise hindered performance, 22 showed no effect, and 7 showed noise facilitated performance. This apparent state of confusion is due, in part, to the wide variability of conditions tested in noise experiments. The noises tested were intermittent in some cases and continuous in others. Electronically generated white noise and pure tones; tape-recorded machine noise, office noise, and street sounds; bells, buzzers, and air horn blasts; tank, helicopter, airplane, and rocket noise; music; and even gibberish have been used at one time or another. The definition of "quiet" has ranged from 0 to 90 dB and often is not even specified by the experimenter. A variety of tasks have been studied, including tracking, reaction time, digit recall, arithmetic, letter cancellation, number checking, maze tracing, monitoring, distance judgment, hand steadiness, and accuracy of shooting. These tasks differ in difficulty and in the relative demands they place on the perceptual, cognitive, memory, and motor capabilities of the subjects. Even the experimental design used to test the effects of noise can influence the results. Adoption of a repeated-measures design (where the same subjects perform under both noise and quiet) may lead to different conclusions compared to a randomized-groups design (where different groups of subjects perform under noise and quiet) (Poulton, 1973; Poulton and Freeman, 1966).

General Conclusions Regarding Effects of Noise

Despite, or because of, this conglomeration of tasks and noise characteristics, we can make a few guarded conclusions about the effects of noise on performance: (1) With the possible exception of tasks that involve short-term memory, the level of noise required to obtain reliable performance effects is quite high, generally over 95 dBA. (2) Performance of simple, routine tasks may show no effect and often will even show an improvement as a result of noise. (3) The detrimental effects of noise are usually associated with tasks performed continuously without rest pauses between responses (Davies and Jones, 1982) and difficult tasks that place high demands on perceptual and/or information processing capacity (Eschenbrenner, 1971).

Specific Effects of Noise

In discussing the detrimental effects of noise, Broadbent (1976) identifies three effects which he has gleaned from the research available. First, he reports that certain decisions become more confident in the presence of noise. For example, if a person is presented a visual display and is asked to report signals that occur, that person will be more confident of the responses (whether correct or incorrect) when the task is performed in high levels of noise. Easily detected signals are usually not missed any more often in high-noise conditions than in quiet conditions. Doubtful and uncertain signals, however, may be missed more often in noisy conditions because the person sometimes is "confident" that they did not occur. In quiet conditions, the person may reconsider such signals and report their presence.

Broadbent also reports that there is a "funneling of attention" on the task. At high noise levels, a person typically focuses attention on the most important aspects of a task or on the most probable sources of information. If relevant task information is missed owing to this funneling phenomenon, then performance suffers. This phenomenon may explain why noise improves performance on simple, routine tasks. The person's attention is funneled onto the task rather than wandering because of possible boredom.

Broadbent also reports that continuous work in which there is no opportunity for relaxation will often show occasional moments of low performance and gaps in performance where no recorded response is made. The overall, average performance may not suffer, but the variability in performance increases.

These detrimental effects (increased confidence, funneling, and gaps), as we said, have typically been associated with high levels of noise. An apparent exception is performance on cognitive tasks (Hockey, 1978). Here we often find a disruptive effect of noise at levels of 70 or 80 dBA. This is especially true of tasks involving short-term memory requirements. Weinstein (1974, 1977), for example, reported that 68- to 70-dBA noise significantly impaired the detection of grammatical errors in a proofreading task (which requires short-term memory) but did not adversely affect the detection of spelling errors.

Basis for Specific Effects of Noise Although there is beginning to be some consensus on the effects of noise, there still remains considerable controversy over why, or how, noise has the effects it does. Poulton (1976, 1977, 1978) and Broadbent (1976, 1978) have carried on a lively "discussion" of the mechanisms underlying the effects of noise on performance. Poulton (1978) contends that all the known effects of noise on performance can be explained by four determinants which combine to affect performance: (1) masking of acoustic task-related cues and inner speech; (2) distraction; (3) a beneficial increase in arousal when noise is first introduced, which gradually lessens and falls below normal when the noise is first switched off; and (4) positive and negative transfer from performance in noise to performance in quiet.[2]

Broadbent (1976, 1978, 1979), however, rejects the notion that detrimental effects of noise are caused by masking of acoustic task cues or inner speech. Rather, Broadbent attributes most of the detrimental effects of noise to "overarousal." It is a long-standing observation that there exists an inverted U relationship between arousal and performance. Too little or too much arousal results in lower performance than does a moderate level of arousal. Poulton rejects the overstimulation hypothesis when it is applied to noise.

Whether masking plays a major role in determining the effects of noise remains to be seen. Probably, as is so often the case, both sides are right—to a degree. Undoubtedly if task-related acoustic cues are present and noise masks them, a decrement may well result. The question is whether noise can have a detrimental effect on a task in which no acoustic cue, inner speech, distraction, or transfer is present. Poulton would probably say no; Broadbent probably would contend that it would still be possible owing to overstimulation. Kryter (1985) seems to side with Poulton and, while acknowledging the role of arousal, believes much of the noise-induced performance decrements reported in the literature can be explained by masking inner speech or task-relevant auditory cues. In addition, Kryter also suggests that subjects in noise experiments construe various meanings to the noise which can affect motivation or cause stress. For example, subjects may believe that when a loud noise is present, it is a signal of importance of some kind, or they may believe that loud noise is supposed to affect them in some way. Some subjects may be concerned that the noise will be harmful to their hearing or that the experimental condition will become overbearing; such concerns may cause some physiological stress.

Discussion

Most of the evidence relating to either the degrading or the enhancing effects on performance is based on experimental studies, not on actual work situations. And extrapolation from such studies to actual work situations is probably a bit risky. Subjects used in most noise experiments are either students or military personnel under 30 years of age, and they are hardly representative of an industrial population. Subjects in experiments are usually highly motivated and will make an extra effort to maintian performance under adverse conditions. For simple or short-duration tasks, this may be adequate to mask any negative effects of noise. The long-term effects of noise on performance (i.e., those more representative of industrial exposure such as 5 to 7 h/d, 5 d/wk, for weeks on end) have been rarely studied in the laboratory. Further, tasks used in laboratory experiments often bear little resemblance to the types of tasks performed in noisy industrial environments, making generalization tenuous.

---

[2]Positive transfer results from the better learning of the task in noise under the influence of the increased arousal. Negative transfer results from techniques of performance in noise used to counteract the masking or distraction being used in quiet where they are not appropriate.

We reiterate one thing in closing: The level of noise required to exert measurable degrading effects on performance is, with the exception of short-term memory tasks, considerably higher than the highest levels that are acceptable by other criteria, such as hearing loss and effects on speech communications. Thus, if noise levels are kept within reasonable bounds in terms of, say, hearing loss considerations, the probabilities of serious effects on performance probably would be relatively nominal.

NOISE EXPOSURE LIMITS

In typical work situations, hearing loss is perhaps the prime criterion for acceptable noise levels. Standards that differentiate between continuous noise, impulse noise, infrasonic noise, and ultrasonic noise, have been set by various organizations. We do not attempt to review all the various standards; rather we concentrate on those promulgated by the Occupational Safety and Health Administration (OSHA) of the U.S. Department of Labor. Actually, other national and international standards follow the OSHA recommendations rather closely.

Continuous and Intermittent Noise

OSHA has established permissible noise exposures for persons working on jobs in industry (OSHA, 1983). The permissible levels depend on the duration of exposure and are shown in Table 16-3. A key concept in the OSHA requirements is *noise dose*. Exposure to any sound level at or above 80 dBA causes the listener to incur a *partial dose* of noise. (Exposures to sound levels less than 80 dBA are ignored in calculating doses.) A partial dose is calculated for each specified sound-pressure level above 80 dBA as follows:

$$\frac{\text{Time actually spent at sound level}}{\text{Maximum permissible time at sound level (see Table 16-3)}}$$

TABLE 16-3
PERMISSIBLE NOISE EXPOSURES ACCORDING TO OSHA

Sound level, dBA	Permissible time, h
80	32
85	16
90	8
95	4
100	2
105	1
110	0.5
115	0.25
120*	0.125*
125*	0.063*
130*	0.031*

*Exposures above 115 dBA are not permitted regardless of duration; but should they exist, they are to be included in computations of the noise dose.
Source: OSHA, 1983.

The total or daily noise dose is equal to the sum of the partial doses. The noise dose can then be converted to an *8-h time-weighted average (TWA) sound level* by using Table 16-4. The TWA is the sound level that would produce a given noise dose if an employee were exposed to that sound level continuously over an 8-h workday.

A noise dose of 50 percent (TWA = 85 dBA) is designated as the *action level,* or the point at which the employer must implement a continuing, effective hearing conservation program. The program must include exposure monitoring, audiometric testing, hearing protection, employee training, and record keeping. A noise dose of 100 percent (TWA = 90 dBA) is designated as the *permissible exposure level,* or the point at which the employee must use feasible engineering and administrative controls to reduce noise exposure. OSHA (1981) estimated that there are 2.9 million production workers in the United States with TWAs in excess of 90 dBA and an additional 2.3 million workers with TWAs exceeding 85 dBA.

The concept of noise dose produces a curious situation. Consider a worker who, during a workday, is exposed to the following noise levels:

95 dBA	for 3.5 h
105 dBA	for 0.5 h
85 dBA	for 4.0 h

Such exposure is within the *individual* OSHA permissible time limits set forth in Table 16-3, but the total noise dose is equal to 163.5 [that is, 100(3.5/4.0 + 0.5/1.0 + 4.0/16.0)]. This represents a TWA of approximately 93.5 dBA and thus exceeds the permissible exposure level.

TABLE 16–4
CONVERTING NOISE DOSE TO TWA

Noise dose	TWA,* dBA
10	73
25	80
50 (action level)	85
75	88
100 (permissible exposure level)	90
115	91
130	92
150	93
175	94
200	95
400	100

*Values are rounded to nearest decibel. The exact conversion from noise dose D to TWA is given by

$$\text{TWA} = 16.61 \log \frac{D}{100} + 90$$

Source: OSHA, 1983.

Impulse Noise

OSHA defines *impulse noise* as "a sound with a rise time of not more than 35 ms to peak intensity and a duration of not more than 500 ms to the time when the level is 20 dB below the peak" (OSHA, 1974). Table 16-5 lists the maximum number of permissible impulses for various peak intensity values. Note that OSHA (1983) is moving toward including impulse noise into the noise dose concept and doing away with regulating impulse noise by limiting the number of exposures.

Infrasonic Noise

Infrasonic noise is noise with frequencies below the audible range, typically less than 20 Hz. Currently, there are no national or international standards for permissible exposure limits to infrasonic noise. Von Gierke and Nixon (1976) present a review of the effects of infrasound and conclude that it is not subjectively perceived and has no effect on performance, comfort, or general well-being. To protect the auditory system, however, they recommend 8-h exposure limits ranging from 136 dB at 1 Hz to 123 dB at 20 Hz. If the level is increased 3 dB, then the permissible duration must be halved.

Ultrasonic Noise

Ultrasonic noise is noise with frequencies above the audible range, typically greater than 20,000 Hz. Action (1983) reviewed the topic and the various standards that do exist for ultrasonic exposure. The criteria are similar, typically limiting exposures to 110 dB for frequencies at and above 20,000 Hz. This translates to a one-third octave-band criterion of 75 dB at 20,000 Hz and 110 dB for bands at and above 25,000 Hz.

TABLE 16-5
MAXIMUM NUMBER OF PERMISSIBLE
IMPULSES IN AN 8-H DAY AS A FUNCTION
OF PEAK SOUND-PRESSURE LEVEL

Peak sound-pressure level, dB	Maximum number of impulses per 8h*
140	100
135	316
130	1,000
125	3,162
120	8,913
115	31,623
112.4	57,600‡

*Based on following formula: number $= 10^{16-P/10}$ (where P = peak decibels).
‡This would be considered continuous noise.
Source: Leavitt, Thompson, and Hodgson, 1982 based on OSHA, 1981. Reprinted with permission by American Industrial Hygiene Association Journal.

THE ANNOYANCE OF NOISE

No one need tell us that noise can be annoying; we have all had the experience at some time. Annoyance is not the same as loudness. Loud noises are usually more annoying than soft noises, but there are exceptions. Consider, for example, the slow, rhythmic drip of a water faucet versus the roar of the ocean surf—which would annoy you more? In general, more people are annoyed by aircraft and automobile noises than by loud neighbors, loud radios, or children (Kryter, 1985).

Annoyance is measured by having subjects rate noises on a verbal scale, such as noticeable–intrusive–annoying–very annoying–unbearable. There are a host of factors, both acoustic and nonacoustic, that influence the annoying quality of a noise. A list of some of these is presented in Table 16-6.

Measures of Noise Exposure

Considerable work has been done to develop a measure of noise exposure that would represent, in a single number, many of the important acoustic factors and some of the nonacoustic factors influencing the annoyance of noise. Sperry (1978) lists 13 different measures which are used around the world to assess community exposure to noise. Figure 16-6 lists these measures and shows the relationships among them. Most of these measures were developed in the context of community exposure to aircraft noise. From Figure 16-6 we see that all the measures are based on the A-weighted sound level (dBA). Equivalent sound level L_{eq} and perceived noise level (PNL), discussed previously, have formed the bases for a variety of other measures. The various measures make corrections for such factors as time of day, season of year, variability in the noise, and number of aircraft flyovers.

TABLE 16-6
SOME FACTORS THAT INFLUENCE THE ANNOYANCE
QUALITY OF NOISE

Acoustic factors	Sound level
	Frequency
	Duration
	Spectral complexity
	Fluctuations in sound level
	Fluctuations in frequency
	Risetime of the noise
Nonacoustic factors	Past experience with the noise
	Listener's activity
	Predictability of noise occurrence
	Necessity of the noise
	Listener's personality
	Attitudes toward the source of the noise
	Time of year
	Time of day
	Type of locale

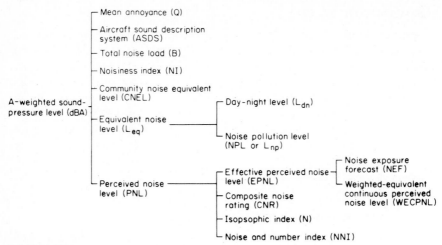

FIGURE 16-6
Various measures of exposure to noise. Many were developed in the context of community exposure to aircraft noise. Measures on the left of the figure are primary measures; those on the right are derived from, or based on, those to the left.

We do not take the time to describe each of the measures in Figure 16-6. However, one measure has been related to annoyance and community action and deserves further attention. The day-night level L_{dn} is used by the Environmental Protection Agency to rate community exposure to noise. The day-night level is the equivalent sound level L_{eq} for a 24-h period with a correction of 10 dB added to noise levels occurring in the nighttime (10 p.m. to 7 a.m.). Kryter (1985), however, recommends that 5dB be added to the noise levels occurring between 7 p.m. and 7 a.m. He proposes that the measure be called DENL (day-evening-night level).

It may be encouraging to note that many of the measures shown in Figure 16-6, when they are taken over a 24-h period, correlate almost perfectly with one another. For example, L_{eq}, L_{dn}, and CNEL rarely differ by more than ± 1 dB (Fidell, 1979). This should not be too surprising since they are all measuring the same noise environment.

Annoyance and Community Response

Whatever these indices of noise exposure measure, they are not highly correlated to whatever gives rise to community response (Fidell, 1979). From Table 16-6 we see that a host of nonacoustic factors influence the annoyance quality of noise. Most of these are not taken into consideration by the various noise exposure measures. Nevertheless, the Environmental Protection Agency, from a review of British and U.S. community noise surveys, found a linear relationship between L_{dn} and the percentage of people in the survey who were highly annoyed. This relationship is

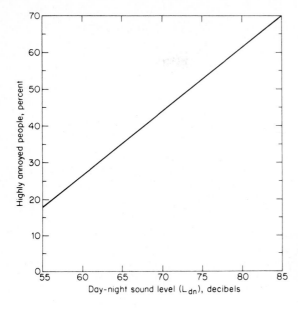

FIGURE 16-7
Relationship between noise exposure and percentage of community highly annoyed. *(Source: Environmental Protection Agency, 1974.)*

shown in Figure 16-7. Typically, such relationships show correlations on the order of .90.

Different noises having the same L_{dn}, as measured outdoors, however, are not necessarily equally annoying. For example, Kryter (1985) summarized several studies showing that people are more sensitive to aircraft noise than to ground-based vehicular noises, such as automobiles and trucks. Vehicular noises need to be about 10 dB higher than aircraft noises to be equally annoying. Kryter believes that this result is due to the difference in sound level, produced by these two types of noise, at the listeners' ears inside homes. More of the noise from aircraft is transmitted inside a house than is the case for vehicular noises. It is also possible that fears for one's safety are greater with respect to aircraft noise than with vehicular noise, which would add to the annoyance quality of the noise.

To take it a step farther, attempts have been made to relate noise exposure to community reactions such as complaints, threats, and legal action. There exists a giant step between being annoyed and taking action. A host of sociological factors, political factors, and psychological factors intervene in such a decision. Fidell indicates that the noise exposure itself usually does not account for even half the variance in community reactions.

The Environmental Protection Agency, using L_{dn}, however, has devised a procedure for predicting community reactions to noise. The procedure attempts to take into account some of the sociopolitical factors that influence community reactions. Table 16–7 lists corrections that are to be added to L_{dn} to obtain a *normalized* L_{dn}. The normalized L_{dn} is then used with Table 16–8 to predict community reactions.

Generally, a normalized L_{dn} of 55 dBA or lower will not result in complaints. It

TABLE 16–7
CORRECTION FACTORS TO BE ADDED TO THE MEASURED DAY-NIGHT LEVEL L_{dn}
TO OBTAIN NORMALIZED L_{dn}

Type of correction	Description	Correction added to measured L_{dn}, dB
Seasonal correction	Summer (or year-round operation)	0
	Winter only (or windows always closed)	−5
Correction for outdoor residual noise level	Quiet suburban or rural community (away from large cities, industrial activity, and trucking)	+10
	Normal suburban community (away from industrial activity)	+5
	Urban residential community (not near heavily traveled roads or industrial areas)	0
	Noisy urban residential community (near relatively busy roads or industrial areas)	−5
	Very noisy urban residential community	−10 +5
Correction for previous exposure and community attitudes	No prior experience with intruding noise	
	Community has had some exposure to introducing noise; little effort is being made to control noise. This correction may also be applied to a community which has not been exposed previously to noise, but the people are aware that bona fide efforts are being made to control it.	0
	Community has had considerable exposure to introducing noise; noisemaker's relations with community are good.	−5
	Community is aware that operation causing noise is necessary but will not continue indefinitely. This correction may be applied on a limited basis and under emergency conditions.	−10
Pure tone or impulse	No pure tone or impulsive character	0
	Pure tone or impulsive character present	+5

Source: Environmental Protection Agency, 1974.

TABLE 16–8
EXPECTED COMMUNITY RESPONSE BASED ON
NORMALIZED L_{dn}

Community response	Normalized L_{dn} dBA
No reaction or sporadic complaints	50–60
Widespread complaints	60–70
Severe threats of legal action or strong appeals to local officials	70–75
Vigorous action	75–80

Source: Environmental Protection Agency data as presented by R. Taylor, 1978, table 8–4.

must be remembered, however, that these predictions are not precise, but rather are only rough indications of the probable community reaction.

HANDLING NOISE PROBLEMS

When a noise problem is suspected or exists, there is no substitute for having good, solid information to bring to bear on the problem and for attacking the problem in a systematic manner. We provide an overview of some noise control techniques, but our discussion cannot substitute for the expertise of a qualified acoustic engineer.

Defining the Noise Problem

Defining a possible noise problem consists of essentially two phases. The first phase is the measurement of the noise itself. The overall sound-pressure level (e.g., dBA) gives a gross indication of a potential noise problem. An octave-band analysis of the noise, however, gives a more detailed and useful picture of the noise situation. The second phase is to determine what noise level would be acceptable, in terms of hearing loss, annoyance, communications, etc. Such limits normally would be adapted from relevant criteria such as discussed above. An example is shown in Figure 16–8. This figure shows the spectrum of the original noise of a foundry cleaning room and a tentative-design baseline that was derived from a set of relevant noise standards. Incidentally, part of the original spectrum was below this level, but the high frequencies were not; the difference represents the amount of reduction that should be achieved (in this case, the shaded area). The third line shows the noise level after abatement.

FIGURE 16-8
Spectrum of noise of foundry cleaning room before abatement, the baseline that represents a desired upper ceiling, and the spectrum after abatement. The shaded area represents the desired reduction. The abatement consisted primarily in spraying a heavy coat of deadener on the tumbling barrels and surfaces of tote boxes. *(Source: Foundry Noise Manual, 1966, p. 52.)*

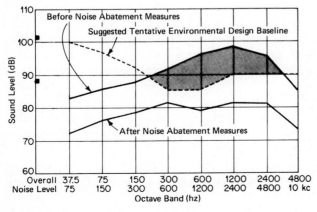

Noise Control

A noise problem can be controlled by attacking the noise at the source, along its path from the source to the receiver, and at the receiver. Often a combination of noise control techniques is required to achieve the desired level of abatement.

Control at the Source Noise is caused by vibration and can be reduced by decreasing either the amount of vibration or the surface area of the vibrating parts. Vibration can be reduced by proper design, maintenance, lubrication, and alignment of equipment. Isolating vibrating parts from other machine parts or structures by use of resilient materials such as rubber or elastomers reduces the number, and hence the surface area, of vibrating sources. This is illustrated in Figure 16–9 by the noise reduction accomplished by isolating the machine from the floor by resilient pads (treatment *a* in the figure). Adding damping materials to machine parts to increase their stiffness or mass can reduce the amplitude of vibrations as well.

Often, a potential noise problem can be averted by selecting quieter equipment initially. It is often more economical to pay extra for quieter equipment than to purchase noisier equipment that will require additional expenditures for noise control. Figure 16–10, for example, shows the noise spectra for two pneumatic screwdrivers and illustrates the considerable noise reduction that can be achieved simply by choosing the proper equipment. Low-frequency noise is less annoying and is tolerated better than high-frequency noise. Therefore, where possible, equipment that generates low-frequency noise should be selected over equipment that generates high-frequency noise. For example, use of a large, slow-speed blower would be preferred over a smaller, high-speed blower. Mufflers are effective in reducing exhaust noise, and reducing turbulences in pipes will decrease noise in fluid flows.

Control along the Path High-frequency noise is more directional than low-frequency noise and is more easily contained and deflected by barriers. Acoustic linings and materials can reduce noise, as illustrated in Figure 16–9. The noise reduction effect of acoustic material (treatment *b* in Figure 16–9) and rigid sealed enclosures (treatment *c*) is primarily in the high-frequency range. When full enclosures are designed, maintenance requirements of the enclosed equipment must be kept in mind. Although the noise from the machine in Figure 16–9 has been drastically reduced by enclosing it in an acoustically lined, sealed, double vault, pity the maintenance person who has to get inside to change the fan belt!

Full enclosures are not necessary to reduce high-frequency noise. A single wall, shield, or barrier placed between the source and the receiver will deflect much of this noise. Low-frequency noise, however, will not be reduced at all by such barriers, because such noise will easily go over or around the barrier. This is illustrated in Figure 16–11, which shows the before and after noise spectra achieved by installing a simple ¼-in (6-mm) thick safety glass in front of the operator of a punch press that used compressed air jets to blow foreign particles from the die. Notice that there is almost no reduction for frequencies below 1000 Hz.

Adding sound absorption materials to the walls, ceiling, and floors of a room can reduce noise levels by 3 to 7 dB under certain circumstances. The goal of using such materials is to reduce the noise caused by reverberation rather than to reduce the noise of the equipment per se. For this reason the overall reduction is rather limited.

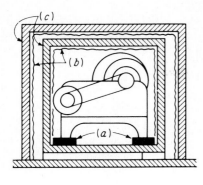

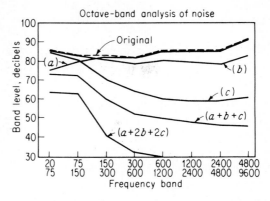

FIGURE 16-9
Illustrations of the possible effects of some noise control measures. The lines on the graph show the possible reductions in noise (from the original level) that might be expected by vibration insulation, *a;* an enclosure of acoustic absorbing material, *b;* a rigid, sealed enclosure, *c;* a single combined enclosure plus vibration insulation, *a + b + c;* and a double combined enclosure plus vibration insulation, *a + 2b + 2c. (Source: Adapted from Peterson and Gross, 1978.)*

Control at the Receiver Controlling noise at the receiver involves primarily the use of hearing protection but can include, secondarily, audiometric testing of exposed workers with job reassignment or reduced exposure times for those showing signs of hearing loss. OSHA, for example, requires employers to make available hearing protection devices to all employees whose noise dose exceeds 50 percent (TWA = 85 dBA), and workers must wear hearing protection if their noise dose is above 100 percent (TWA = 90 dBA).

Hearing protection devices come in two general types: insert type and muff type. Insert types can be premolded or custom-molded, can be made of an expandable foam or plastic, or can be a simple fiber plug. Muff types can be liquid-filled or foam-filled and are mounted on a headband or helmet. Lempert (1984) presented a compendium of noise attenuation data for virtually all hearing protection devices on the market at that time. The effectiveness of hearing protection devices varies widely

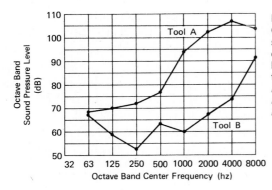

FIGURE 16-10
Octave-band analysis of two pneumatic screwdrivers illustrating the considerable noise reduction that can be achieved by choosing the proper tool. *(Source: American Industrial Hygiene Association, 1975, fig. 11.3. Reprinted with permission by American Industrial Hygiene Association.)*

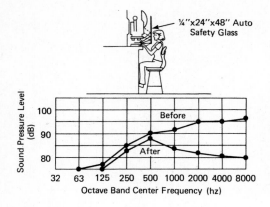

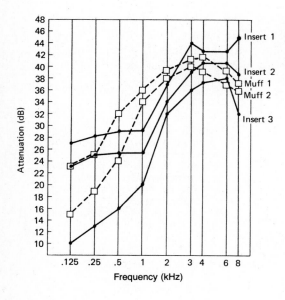

FIGURE 16-11
Use of a 1/4-in (6-mm) thick safety glass barrier to reduce high-frequency noise. The machine is a punch press which uses compressed air to blow foreign particles from the die. *(Source: American Industrial Hygiene Association, 1975, fig. 11.73. Reprinted with permission by American Industrial Hygiene Association.)*

from one type to another and even between brands within a specific type. Figure 16–12 illustrates the attenuation characteristics of several types of hearing protectors. In general, attenuation is greatest in the higher frequencies where the ear is most sensitive to sound. Figure 16–12 underscores the need to select hearing protection to match the characteristics of the intruding noise.

The Environmental Protection Agency (1979) requires manufacturers to compute and display the *noise reduction rating* (NRR) of their hearing protectors. The NRR is computed from an *octave band* analysis of the attenuation effectiveness of the device and is based on the attenuation value 2 standard deviations below the mean at each octave band. The NRR of a hearing protection device can be used to estimate the noise exposure of the wearer. If the sound-pressure level of the noise is measured

FIGURE 16-12
Noise attenuation of several insert- and muff-type hearing protection devices. *(Source: Bilsom International, Inc., 1984.)*

in decibels on the C-weighted scale, then the noise being experienced by the wearer is simply the sound level in dBC minus the NRR. If the sound-pressure level of the noise is measured in dBA, then the noise exposure is equal to the sound level (in dBA) plus 7 minus the NRR. The $+7$ is included to adjust for "spectral uncertainty" (OSHA, 1983).

Gasaway (1984), using data from Lempert (1984), plotted the NRR values of 79 insert-type and 163 muff-type protectors, as shown in Figure 16–13. In general, insert-type devices provided better protection than muff-type devices, with 43 percent of the inserts having NRR values above 25 compared to 23 percent of the muff types. The best insert types were the expandable foam inserts. These are small, disposable pieces of foam that are rolled (compressed) between the fingers to form a thin cylinder which is inserted in the ear canal. In a few seconds the foam cylinder begins to expand and fills the ear canal, providing an effective barrier to noise. Additional protection can be achieved in high-noise environments by combining an insert device with a muff. Berger (1984) reported that combining a low-NRR (17) insert device with a moderate-NRR (21) muff device yielded a relatively high level of protection (NRR = 29).

In using NRR data, one must be aware that the indicated attenuation, in the words of Gasaway (1984), may be "dangerously optimistic." NRR values are determined

FIGURE 16-13
Distribution of noise reduction ratings of insert- and muff-type hearing protection devices.
(Source: Gasaway, 1984, fig. 3. Reprinted by permission of the author from NATIONAL SAFETY NEWS, *November 1984.)*

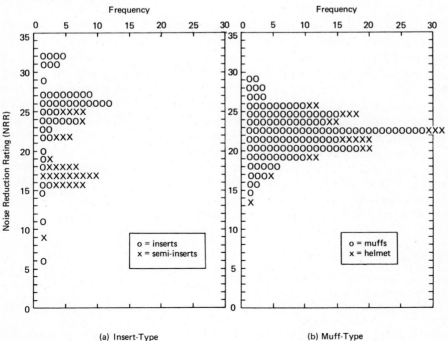

(a) Insert-Type   (b) Muff-Type

FIGURE 16-14
Effects of speaker and listener wearing hearing protection on speech intelligibility in a 92-dB noise environment. *(Source: Adapted from Hormann et al., 1984, fig. 5.)*

under ideal laboratory and fitting conditions. In real-world industrial settings, however, people often do not wear hearing protection devices properly; hair, beards, and eyeglass frames interfere with good fit, and insets are often not inserted properly. Berger (1983) suggests subtracting 10 from the reported NRR values to account for these realities. One final word on selecting hearing protection devices based on NRR values: The device with the highest NRR is not always the best choice. NRR is only one factor, although it is an important one, to consider. Other factors that should be considered include comfort, wearability, ease of use, sizing and cleaning requirements, durability, and compatibility with other safety gear and clothing. In many noise environments the highest-NRR devices are not really necessary and more comfortable; lower-NRR devices can be used.

Hearing Protection and Speech Communication One objection to wearing hearing protection is that it interferes with speech communication. Hormann et al. (1984) provided evidence that indeed speech communication is reduced in noise when hearing protection is worn. Figure 16–14 shows the percentage of correctly understood speech stimuli in a 92-dB noise environment when speaker and listener were wearing or not wearing hearing protection. When only the listener is wearing hearing protection, intelligibility declines about 33 percent. When the speaker also wears hearing protection, intelligibility declines an additional 30 percent. The reason for this latter finding is that speakers speak about 2 to 4 dB more softly and 20 percent faster, and their pauses are 25 percent shorter, when they wear noise protection in a noisy environment than when they do not (Hormann et al., 1984). Therefore, in a noise situation in which people are wearing hearing protection devices, extra effort must be made to speak loudly, slowly, and clearly.

REFERENCES

Action, W. (1983). Exposure to industrial ultrasound: Hazards, appraisal and control. *Journal of the Society of Occupational Medicine, 33,* 107–113.

American Industrial Hygiene Association (1975). *Industrial noise manual* (3d ed.). Akron, OH.

Berger, E. (1983). Using the NRR to estimate the real world performance of hearing protectors. *Sound and Vibration,* 18(5), 26–39.

Berger, E. (1984). *E.A.R.LOG 13: Attenuation of earplugs worn in combination with ear-muffs.* Indianapolis, IN: E.A.R. Division, Cabot Corporation.

Bilsom International, Inc. (1984). Attenuation data for Bilsom products. *Bilsom hearing protection catalog.* Reston, VA.

Bonvallet, S. (1952, February). *Noise.* Lectures presented at the inservice training course on the acoustical spectrum. Ann Arbor: University of Michigan Press.

Broadbent, D. (1976). Noise and the details of experiments: A reply to Poulton. *Applied Ergonomics, 7,* 231–235.

Broadbent, D. (1978). The current state of noise research: Reply to Poulton. *Psychological Bulletin, 85,* 1052–1067.

Broadbent, D. (1979). Human performance and noise. In C. Harris (ed.), *Handbook of noise control.* New York: McGraw-Hill.

Burns, W. (1979). Physiological effects of noise. In C. Harris (ed.), *Handbook of noise control.* New York: McGraw-Hill.

Burrows, A. A. (1960). Acoustic noise, an informational definition. *Human Factors, 2*(3), 163–168.

Davies, D., and Jones, D. (1982). Hearing and noise. In W. Singleton (ed.), *The body at work.* New York: Cambridge University Press.

Environmental Protection Agency. (1974). *Information on levels of environmental noise requisite to protect public health and welfare with an adequate margin of safety* (EPA 550/9-74-004). Washington.

Environmental Protection Agency (1979). Noise labeling requirements for hearing protectors. *Federal Register, 42,* 56139–56147.

Eschenbrenner, A. J., Jr. (1971). Effects of intermittent noise on the performance of a complex psychomotor task. *Human Factors, 13*(1), 59–63.

Fidell, S. (1979). Community response to noise. In C. Harris (ed.), *Handbook of noise control.* New York: McGraw-Hill.

Fletcher, H., and Munson, W. A. (1933). Loudness, its definition, measurement, and calculation. *Journal of the Acoustical Society of America, 5,* 82–108.

Foundry noise manual (2d ed.). (1966). Des Plaines, IL: American Foundryman's Society.

Gasaway, D. (1984, November). 1984 NIOSH compendium hearing protector attenuation. *National Safety News,* pp. 26–34.

Gawron, V. (1982). Performance effects of noise intensity, psychological set, and task type and complexity. *Human Factors, 24,* 225–243.

Gulian, E. (1974). *Noise as an occupational hazard: Effects on performance level and health. A Survey of the European literature.* Cincinnati, OH: National Institute for Occupational Safety and Health.

Hockey, G. (1978). Effects of noise on human work efficiency. In D. May (ed.), *Handbook of noise assessment.* New York: Van Nostrand Reinhold.

Hormann, H., Lazarus-Mainka, G., Schubeius, M., and Lazarus, H. (1984). The effect of noise and the wearing of ear protectors on verbal communication. *Noise Control Engineering Journal, 23*(2), 69–77.

Industrial noise manual (2d ed.). (1966). Detroit: American Industrial Hygiene Association.

Jansen, G. (1969, February). Effects of noise on physiological state. In W. D. Ward and J. E. Frick (eds.), *Noise as a public health hazard,* ASHA Rept. 4. Washington: American Speech and Hearing Association.

Jensen, P., Jokel, C., and Miller, L. (1978). *Industrial noise control manual* (ref. ed.). Cincinnati, OH: National Institute for Occupational Safety and Health.

Kryter, K. (1970). *The effects of noise on man.* New York: Academic.

Kryter, K. (1983a). Presbycusis, sociocusis and nosocusis. *Journal of Acoustical Society of America, 73*(6), 1897–1917.

Kryter, K. (1983b). Addendum and erratum: "Presbycusis, sociocusis, and nosocusis" [J. Acoust. Soc. Am. 73, 1897–1919 (1983)]. *Journal of Acoustical Society of America,* 74(6), 1907–1909.

Kryter, K. (1985). *The effects of noise on man* (2d ed.). Orlando, FL: Academic.

Kryter, K., Ward, W., Miller, J., and Eldredge, D. (1966). Hazardous exposure to intermittent and steady-state noise. *Journal of the Acoustical Society of America,* 39, 451–463.

Leavitt, R., Thompson, R., and Hodgson, W. (1982). Computation of permissible noise exposure. *American Industrial Hygiene Association Journal,* 43, 371–373.

Lempert, B. (1984). Compendium of hearing protection devices. *Sound and Vibration,* 18(5), 26–39.

Machle, W. (1945). The effect of gun blast on hearing. *Archives of Otolaryngology,* 42, 164–168.

Meecham, W. (1983, May 10). Paper delivered at Acoustical Society of America Meeting, Cincinnati, Ohio, as reported in *Science News,* 123, p. 294.

Melnick, W. (1979). Hearing loss from noise exposure. In C. Harris (ed.), *Handbook of noise control.* New York: McGraw-Hill.

Occupational Safety and Health Administration (1974, Oct. 24). Occupational noise exposure: Proposed requirements and procedures. *Federal Register,* 39, 37773–37777.

Occupational Safety and Health Administration (1981, Jan. 16). Occupational noise exposure: Hearing conservation amendment. *Federal Register,* 46, 4078–4179.

Occupational Safety and Health Administration (1983). Occupational noise exposure: Hearing conservation amendment. *Federal Register,* 48, 9738–9783.

Peterson, A., and Gross, E., Jr. (1978). *Handbook of noise measurement* (8th ed.). New Concord, MA: General Radio Co.

Poulton, E. (1973). Unwanted range effects from using within-subject experimental designs. *Psychological Bulletin,* 80, 113–121.

Poulton, E. (1976). Continuous noise interferes with work by masking auditory feedback and inner speech. *Applied Ergonomics,* 7, 79–84.

Poulton, E., (1977). Continuous intense noise masks auditory feedback and inner speech. *Psychological Bulletin,* 84, 977–1001.

Poulton, E. (1978). A new look at the effects of noise: A rejoinder. *Psychological Bulletin,* 85, 1068–1079.

Poulton, E., and Freeman, P. (1966). Unwanted asymmetrical transfer effects with balanced experimental designs. *Psychological Bulletin,* 66, 1–8.

Robinson, D. and Dadson, R. (1957). Threshold of hearing and equal-loudness relations for pure tones, and the loudness function. *Journal of the Acoustical Society of America,* 29(12), 1284–1288.

Sperry, W. (1978). Aircraft and airport noise. In D. Lipscomb and A. Taylor (eds.), *Noise control: Handbook of principles and practices.* New York: Van Nostrand Reinhold.

Stevens, S. S (1936). A scale for the measurement of a psychological magnitude: Loudness. *Psychological Review,* 43, 405–416.

Stevens, S. S. (1972). Perceived level of noise by Mark VII and decibels (E). *Journal of the Acoustical Society of America,* 51(2, pt. 2), 575–601.

Taylor, A., and Lipscomb, D. (1978). The use and measurement of equivalent sound level. In D. Lipscomb and A. Taylor (eds.), *Noise control: Handbook of principles and practices.* New York: Van Nostrand Reinhold.

Taylor, R. (1978). Exterior industrial and commercial noise. In D. May (ed.), *Handbook of noise assessment.* New York: Van Nostrand Reinhold.

Taylor, W., Pearson, J., Mair, A., and Burns, W. (1965). Study of noise and hearing in jute weavers. *Journal Acoustical Society of America,* 38, 113–120.

von Gierke, H., and Nixon, C. (1976). Effects of intense infrasound on man. In W. Tempest (ed.), *Infrasound and low frequency vibration*. New York: Academic, pp. 115–150.

Weinstein, N. (1974). Effects of noise on intellectual performance. *Journal of Applied Psychology,* 59(5), 548–554.

Weinstein, N. (1977). Noise and intellectual performance: A confirmation and extension. *Journal of Applied Psychology,* 62, 104–107.

Williams, K. (1978). An introduction to the assessment and measurement of sound. In D. Lipscomb and A. Taylor (eds.), *Noise control: Handbook of principles and practices,* New York: Van Nostrand Reinhold.

CHAPTER 17

MOTION

Technological ingenuity in recent times has resulted in the creation of methods of travel that our ancestors probably never dreamed about. These incude space shuttles, aircraft, zero-ground-pressure vehicles, rockets strapped to one's back, and, of course, various earthbound vehicles such as automobiles, buses, and trucks. Many of these make it possible for people to move at speeds and in environments never before experienced and to which they are not biologically adapted. The disparity between people's biological and physical nature, on the one hand, and the environmental factors imposed on them by their "exotic" modes of travel, on the other, defines a domain within which human factors can make a contribution.

The variables imposed by our increased mobility include vibration, acceleration and deceleration, weightlessness, and an assortment of more strictly psychological phenomena associated with these factors, such as disorientation and other illusions. All these involve, to one degree or another, the sense of motion and orientation. Therefore, before discussing these topics, we review briefly the sensory receptors associated with motion and body orientation.

MOTION AND ORIENTATION SENSES

The *five senses,* in the Aristotelian tradition (vision, audition, smell, taste, and touch), basically deal with stimuli external to the body. The sensory receptors involved (the eyes, ears, etc.) are referred to as *exteroceptors*. There are, however, a number of other sensory receptors related to motion and body orientation.

Proprioceptors

The *proprioceptors* are sensory receptors of various kinds that are embedded within the subcutaneous tissues, such as in the muscles and tendons, in the coverings of the bones, and in the musculature surrounding certain of the internal organs. These receptors are stimulated primarily by the actions of the body itself. A special class of these are the *kinesthetic receptors,* which are concentrated around the joints and which are primarily used to tell us where our limbs are at any given moment and to coordinate our movements.

Semicircular Canals

The three semicircular canals in each ear are interconnected, doughnut-shaped tubes that form, roughly, a three-coordinate system, as shown in Figure 17–1 (along with the vestibular sacs). With changes in acceleration or deceleration, the fluid shifts position in these tubes, which stimulates nerve endings that then transmit nerve impulses to the brain. Note that movement of the body at a constant rate does not cause any stimulation of these canals. Rather, they are sensitive only to a *change* in rate (acceleration or deceleration).

Vestibular Sacs: Utricle and Saccule

The *vestibular sacs* (also called the *otolith organs*) are two organs with interior hair cells that contain a gelatinous substance. The *utricle* is generally positioned in a horizontal plane, and the *saccule* is more in a vertical plane. As the body changes position, the gelatinous substance is affected by gravity, triggering nerve impulses via the hair cells. The utricle apparently is the more important of the two organs. The primary function of these organs is the sensing of body posture in relation to the vertical, so they serve as something of a gyroscope that helps to keep us on an even keel. Although their dominant role is that of aiding in sensing postural conditions of the body, they are also somewhat sensitive to acceleration and deceleration and presumably supplement the semicircular canals in sensing such changes.

Interdependence of Motion and Orientation Senses

In the maintenance of equilibrium and body orientation, or in reliably sensing motion and posture, all these senses play a role, along with the skin senses, vision, and

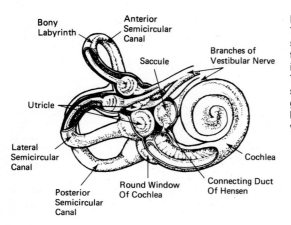

FIGURE 17-1
The body orientation organs. The semicircular canals form roughly a three-coordinate system that provides information about body movement. The vestibular sacs (the utricle and saccule) respond to the forces of gravity and provide information about body position in relation to the vertical. *(Source: Geldard, 1972.)*

sometimes audition. The importance of vision in orientation was illustrated in experiments with people in a "tilting room," in some cases with a chair that also could be tilted within the room (Witkin, 1959). Subjects seated in the room were tilted to various angles and asked to indicate what direction they considered vertical. When blindfolded, they were able to indicate the vertical more accurately than when they could see the inside of the room; when the subjects could see, they tended to indicate that the ceiling of the room was upright even if the room itself was tilted and the chair was actually vertical. The implication of such investigations is that misperceptions of the *true* upright direction may occur when there is a *conflict* between the sensations of gravity and visual perceptions; in such a case one's visual perceptions may dominate, even when they are erroneous.

WHOLE-BODY VIBRATION

"If it moves, it probably vibrates." This saying is especially true in the field of transportation; be it car, truck, train, airplane, or boat, all subject occupants to vibration to some degree. Vibration is not a new topic for us. In Chapter 16 we discussed noise, which is a form of vibration that is audible, and in Chapter 11 we discussed vibration induced from hand tools. Our concern in this chapter, however, is low-frequency (generally less than 100 Hz), whole-body vibration typically encountered in trucks, tractors, airplanes, etc.

Vibration Terminology

Vibration is primarily of two types: sinusoidal and random. *Sinusoidal vibration* may be a single sine wave of some particular frequency, or it may contain combinations of sine waves of different frequencies. Its principal characteristic is its regularity (that is, the waveform repeats itself at regular intervals). This type of vibration is most often encountered in laboratory studies. *Random vibration* is, as the name implies, irregular and unpredictable. This is the most common type of vibration encountered in the real world.

Vibration, be it sinusoidal or random, occurs in one or more directional planes. Table 17–1 lists these and indicates the terminology and symbols used to designate the direction of the vibration. The terminology relates to the direction of the vibration relative to the human body. Thus, a person standing on a platform which is vibrating

TABLE 17–1
VIBRATION TERMINOLOGY WITH RESPECT TO DIRECTION OF VIBRATION

Direction of motion	Heart motion	Other descripton	Symbol
Forward-backward	Spine-sternum-spine	Fore-aft	$\pm g_x$
Left right	Left-right-left	Side-to-side	$\pm g_y$
Headward-footward	Head-feet-head	Head-tail	$\pm g_z$

Source: Adapted from Hornick, 1973, table 7–1, p. 229.

up and down experiences head-tail vibration ($\pm g_z$). If, however, the person lies down on his or her back, the vibration is then considered fore-aft ($\pm g_x$).

In addition to type and direction, vibration is described in terms of *frequency* and *intensity*. Frequency is measured in hertz (i.e., cycles per second), as described in Chapter 16. Intensity is measured in a variety of ways such as peak or maximum: (1) *amplitude* (in or cm); (2) *displacement* (in or cm); (3) *velocity,* the first derivative of displacement (in/s or cm/s); (4) *acceleration,* the second derivative of displacement (in/s² or cm/s²) (sometimes acceleration is expressed in terms of numbers of gravities[1] and is labeled in terms of lowercase g units as $\pm g_x$, $\pm g_y$, or $\pm g_z$, depending on the direction of the oscillation as given in Table 17–1); or (5) *rate of change of acceleration,* also called *jerk,* the third derivative of displacement (in/s³ or cm/s³).

In the case of random vibration, maximum amplitude or acceleration is somewhat inappropriate because the oscillations vary randomly in terms of frequency and intensity. To handle this variability, the frequency spectrum is usually indicated by *mean-square spectral density* and expressed as *power spectral density* (PSD) in g^2 per hertz. The PSD defines the power at discrete frequencies within the selected bandwidth. A plot of PSD (g^2/Hz) versus frequency illustrates the power distribution of the vibration environment. Figure 17–2, for example, shows PSD plots for an automobile and an aircraft.

Intensity is usually expressed as a root-mean-square (rms) value of acceleration (in/s², cm/s², or g).[2] *Root-mean-square acceleration* defines the total energy across

---

[1]Note that $1g = 386$ in/s² $= 980$ cm/s².

[2]*Root mean square* (rms) is the square root of the arithmetic mean of instantaneous values (amplitude or acceleration) squared. In the case of a simple sine wave, rms $g = 0.707 \times$ peak g.

FIGURE 17-2
Power spectral density plots for an automobile and an aircraft in cruise. *(Source: Stephens, 1979, fig. 2.)*

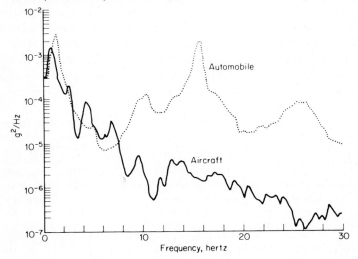

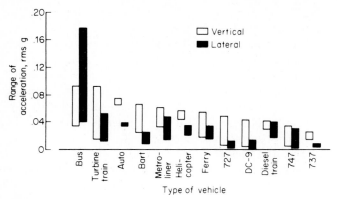

FIGURE 17-3
Root-mean-square acceleration ranges of various air and surface vehicles during cruise. The variability within a vehicle type is due to such factors as road quality, air turbulence, and age and general condition of the test vehicles. (*Source: Stephens, 1979, fig. 4b.*)

the entire frequency range. Figure 17–3 shows rms acceleration values (actually ranges of rms *g*) for various air and surface vehicles during cruise.

One last point should be made about vibration measurement: Acceleration, displacement, and frequency are all related. If, for example, acceleration is held constant and frequency is varied, displacement must also vary. It is impossible to hold two of the quantities constant and vary only the third. This makes it difficult to isolate the specific parameter that is affecting performance, subjective reactions, or physiological reactions.

Attenuation, Amplification, and Resonance

As vibration is transmitted to the body, it can be amplified or attenuated as a consequence of body posture (e.g., standing or sitting), the type of seating, and the frequency of the vibration. Every object (or *mass*) has a resonant frequency, in somewhat the same sense as a pendulum has a natural frequency. When an object is vibrated at its resonant frequency, the object will vibrate at a maximum amplitude which is larger than the amplitude of the original vibration. This phenomenon is called *resonance*.

The human body and the individual body members and organs have their own resonant frequencies. As the human body is made to vibrate (as in a vehicle), the effect on the body and on the body members and organs is the consequence of the interaction between the frequency of the vibrating source and the resonant frequencies of the individual *masses*. Since the body members and organs have different resonant frequencies, and since they are not attached rigidly to the body structure, they tend to vibrate at different frequencies, rather than in unison.

The degree to which a part of the body will amplify the input vibration is a function of the ratio of the input frequency to the resonant frequency of the body

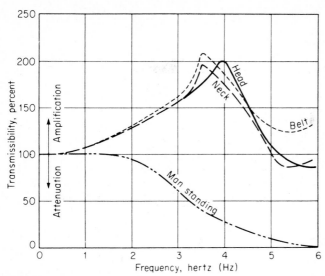

FIGURE 17-4
Mechanical response of person to vertical vibration, showing
amplification and attenuation, by frequency, for seated and
standing positions. *(Source: Radke, 1957.)*

part. *Amplification* will occur if this ratio is 1.414 or less (Radke, 1974) with, of
course, maximum amplification occurring at the resonant frequency (i.e., when the
ratio equals 1.0). When the ratio is greater than about 1.414, there is *attenuation*
(i.e., the displacement amplitude of the body part is less than that of the input vibration).

The amplification and attenuation that would result from the ratio mentioned
above, however, would be expected with no *damping*. In practice, some damping
usually would be provided, either by intent or otherwise. Examples of the actual
amplification and attenuation of the human body while standing and seated are illustrated in Figure 17-4. With the vibration of the vibrating tables used as a base
(expressed as 100 percent), the amplification, or attenuation, is measured by the use
of accelerometers positioned at different locations, such as at the belt, the neck, and
the head. For a person standing, there is a typical attenuation effect, because the legs
absorb the effects as the individual bends and straightens them in response to the
movement. In the case of a seated individual, however, there is an amplification
effect (at the head, neck, and belt) especially from about 3.5 to 4.5 Hz. Figure 17–5
shows the resonant frequency at the shoulder of a seated man in a relaxed posture
and with the trunk muscles tensed. The resonant frequency here is approximately 4.5
to 6.0 Hz depending on the state of the trunk muscles.

Damping can be controlled to some degree by appropriate seat design, spring
design, cushioning, etc. Figure 17–6 illustrates the amplification and attenuation effects of various seats. In this particular comparison, the suspension seat resulted in
substantial attentuation over the critical frequencies of 3 to 6 Hz.

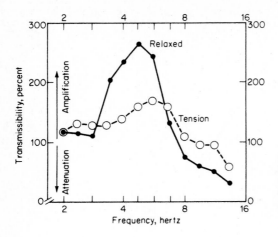

FIGURE 17-5
The effect of tensed and relaxed trunk muscles on the amplitude of vertical vibration of the shoulders of a seated man. *(Source: Adapted from Guignard, 1965, fig. 380.)*

Physiological Effects of Vibration

The evidence suggests that short-term exposure to vibration causes only very small physiological effects which are of little practical significance (Hornick, 1973). A slight degree of hyperventilation has been reported and increases in heart rate are found during the early periods of exposure. The elevated heart rate appears to be an anticipatory general stress response (Hornick). No significant changes are found in blood chemistry or endocrine chemical composition (Gell and Moeller, 1972).

The physiological effects of long-term exposure to whole-body vibration are not clear. Most of the evidence comes from epidemiological investigations of truck drivers and heavy-equipment operators (Troup, 1978). These workers have disproportional incidences of lumbar spinal disorders, hemorrhoids, hernias, and digestive and urinary problems (Wasserman, 1975). The problem with these sorts of studies is that it is difficult to separate the effects due to vibration from the effects due to sitting all day, or to manually loading and unloading a vehicle, all of which are part of operating trucks and heavy equipment. Carlsoo (1982) concluded that, although our knowledge about the link between vibration and changes in our joints is still very poor, it seems well established that vibration does increase muscle tension. This occurs because exposed persons tense their muscles to dampen the vibration.

Performance Effects of Vibration

Considerable research has demonstrated the effects of vibration on the tracking performance of seated subjects. Hornick (1973) has reviewed much of this body of literature and has gleaned a few generalizations from it. The effects of vibration are somewhat dependent on the difficulty of the tracking task, type of display, and type of controller used. For example, the use of side-stick and arm support can reduce vibration-induced error by as much as 50 percent compared with conventional center-mounted joysticks. Detrimental effects of vertical sinusoidal vibration generally occur in the 4- to 20-Hz range with accelerations exceeding 0.20g. These conditions can

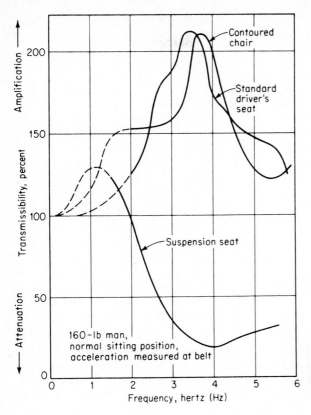

FIGURE 17-6
Mechanical response of a person's body to vibrations, when seated in different seats. *(Source: Simons, Radke, and Oswald, 1956.)*

produce tracking errors up to 40 percent greater than under nonvibratory control conditions. The effect of sinusoidal vibration, however, is not the same as the effect of random vibration. Random vibration does not seem to be as detrimental to tracking performance as sinusoidal vibration. Vertical vibration is generally more debilitating than lateral or fore-aft vibration. Further, the effects of vibration on tracking performance do not seem to end with the cessation of the vibration; residual effects may last up to ½ h after exposure.

Visual performance is generally impaired by vibration and is greatest in the range of 10 to 25 Hz. The amplitude of vibration seems to be a key factor. The effect, of course, depends greatly on the visual task and the display characteristics (e.g., size of letters, larger being better; or clutter, less being better).

Performance on tasks that require steadiness or precision of muscular control is likely to show decrements from vibration. Yet tasks that measure primarily central neural processes, such as reaction time, monitoring, and pattern recognition, appear to be highly resistant to degradation during vibration. In fact, Poulton (1978) points out that vibration between 3.5 and 6 Hz can have an alerting effect on subjects engaged in boring vigilance tasks. Figure 17–5 indicated that, within this frequency range, tensing the trunk muscles attenuates the amplitude of shoulder vibration. Tensing the muscles is a good method for maintaining alertness. Outside of the 3.5- to 6-Hz range the subject can attenuate shoulder vibration more by relaxing the trunk muscles—a good way to fall asleep.

Subjective Responses to Whole-Body Vibration

The subjective response most often assessed in vibration studies is comfort. *Comfort,* of course, is a state of feeling and so depends in part on the person experiencing the situation. We cannot know directly or by observation the level of comfort being experienced by another person; we must ask people to report to us how comfortable they are. This is usually done by using adjective phrases such as *mildly uncomfortable, annoying, very uncomfortable,* or *alarming.* Investigators have tried to link the physical characteristics of vibration, most notably frequency and acceleration, to subjective evaluations of comfort. Their studies usually result in *equal-comfort contours* for combinations of frequency and acceleration. Unfortunately, the resulting contours differ widely from study to study (e.g., see Oborne, 1976, for a review). This discrepancy is due to different methodologies, subject populations, vibration environments, and semantic comfort descriptors. It appears that individuals react in markedly different ways to whole-body vibration, and it is difficult to determine the causes of these differences (Oborne, Heath, and Boarer, 1981). In addition an individual's response to vibration is quite stable over time (Oborne, 1978). The issue of individual differences in response to vibration provides a challenge for researchers and designers alike. Oborne, Heath, and Boarer (1981), for example, determined equal-sensation contours for 100 subjects. They found that 70 percent of the subjects did not respond in the same way as depicted by the average contour. Designing for the average would, in this case, ''inconvenience'' more than half the population!

Probably the most comprehensive research on ride comfort (or more properly *discomfort*) has been conducted at the National Aeronautics and Space Administration (NASA), Langley Research Center, with over 2200 test subjects (Leatherwood, Dempsey, and Clevenson, 1980). To avoid semantic confusion, NASA developed a ratio scale of discomfort, measured in DISC units. The scale is anchored (1 DISC) at the discomfort threshold (i.e., where 50 percent of the passengers feel uncomfortable). A vibration rated 2 DISC is considered twice as uncomfortable as a vibration rated 1 DISC. The percentage of people feeling uncomfortable rises rapidly with each unit of DISC, as shown in Table 17–2.

TABLE 17–2
PASSENGERS FEELING UNCOMFORTABLE AT
VARIOUS LEVELS OF DISC

Discomfort scale, DISC	Percentage uncomfortable
1	50
2	90
3	100

Source: Leatherwood, Dempsey, and Clevenson, 1980.

The NASA experiments were carried out in a simulator with an interior configuration of a modern jet airliner (including actual airplane cushioned seats). Many studies were conducted to establish a model to predict subjective comfort ratings (in DISC units) for various complex ride environments. The methodology and resulting formulas and tables are too complex to present here. The approach, however, consists of three elements: (1) estimation of discomfort due to sinusoidal or random vibration, or both, within single axes; (2) estimation of the discomfort due to vibration in combined axes; and (3) application of corrections for the effects of interior noise and duration of exposure. Incidentally, perceived discomfort *decreased* with increasing exposure time.

Although we cannot present the details of the model, Figure 17–7 illustrates the equal-discomfort contours for seated subjects experiencing sinusoidal vertical vibration. As Leatherwood, Dempsey, and Clevenson (1980) point out, the model is useful for estimating the average level of discomfort experienced by a group of passengers in a combined noise and vibration environment, and not for predicting the discomfort of an individual passenger.

Another extensive study of whole-body vibration was carried out at the Institute of Sound and Vibration Research in Great Britain. [See Griffin, Parsons, and Whitham (1982) for a summary of the studies conducted.] In a series of studies researchers there determined equal-comfort contours for three linear directions of vibration (fore-aft, lateral, and vertical) and rotational vibration, each applied to three

FIGURE 17-7
Equal-discomfort curves for seated subjects experiencing sinusoidal
vertical vibration. Discomfort is expressed in DISC units (see text).
(Source: Leatherwood, Dempsey, and Clevenson, 1980, fig. 4a.)

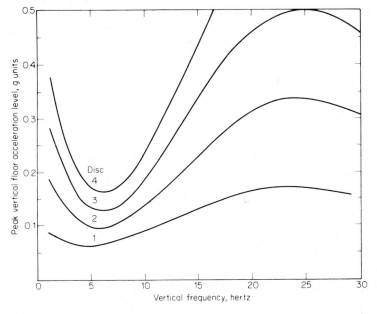

components of a seat (seat pan, backrest, and footrest). Subjects were seated on a hard seat in a laboratory setting. Figure 17–8 shows simplified equal-comfort contours for the three linear vibrations applied to the seat pan. These contours were based on the median contour values, but were simplified to facilitate application in practical situations. Although not shown, the shapes of the contours for backrest and footrest vibration were markedly different from those for seat pan vibration.

Comparison of Figures 17–8 and 17–7 must be approached with extreme caution. Figure 17–8 is based on data gathered in a laboratory setting by using a hard seat and measuring the intensity of the vibration at the seat pan in rms acceleration. In contrast, Figure 17–7 is based on data gathered in a simulated airplane by using soft airline passenger seats and measuring the intensity of the vibration at the floor in peak acceleration g units.

It is not easy to identify the bodily sensations that form the basis for comfort or discomfort judgments. Whitham and Griffin (1978) asked subjects to indicate which specific body locations were uncomfortable at various vibration frequencies (acceleration was held constant at 1.0 m/s^2 rms). In general, most responses of seated subjects implicated the lower abdomen at 2 Hz, moving up the body at 4 and 8 Hz, with most responses implicating the head at 16 Hz. At 32 Hz the responses were divided between the head and lower abdomen, while at 64 Hz they were mostly located near the principal vibration input site, i.e., the seat of the pants.

In actual vehicle environments, subjective evaluations of comfort depend on the expectations, anxiety, past experiences, and other psychological factors of the passengers. In addition to vibration, other physical environmental factors also affect comfort evaluations. Richards and Jacobson (1977), for example, found that among airline passengers the presence of smoke, lighting, and work space were important factors in their comfort ratings.

Limits for Exposure to Whole-Body Vibration

Specifying the limits for exposure to whole-body vibration depends on the criterion adopted. The criterion can be based on comfort, task performance, or physiological response. The equal-comfort contours discussed above are, in essence, limits based on the criterion of comfort.

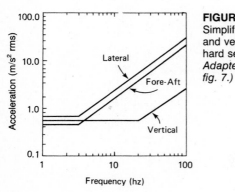

FIGURE 17-8
Simplified equal-comfort contours for fore-aft, lateral, and vertical vibration applied to the seat pan of a hard seat in a laboratory environment. *(Source: Adapted from Griffin, Parsons, and Whitham, 1982, fig. 7.)*

The International Organization for Standardization (ISO), after wrestling with the problem for 10 years, developed a standard for exposure to whole-body vibration (ISO, 1978). The standard specifies the limits in terms of acceleration, frequency, and exposure duration. Limits are specified for comfort, task proficiency, and physiological safety. Figure 17–9 illustrates the *fatigue-decreased proficiency* (FDP) boundaries for vertical vibration. The trough between 4 and 8 Hz corresponds to the resonance frequencies of the body (see Figures 17–4 and 17–5). Each line in Figure 17–9 represents an estimate of the upper limits that people generally can tolerate before fatigue effects catch up with them. Amplitudes about 10 dB less than those indicated tend to characterize the upper boundary of comfort, and the safe exposure limits are about 6 dB higher than the values shown. The standard allows one to raise or lower the FDP and comfort boundaries depending on the nature of the environment and task. The FDP boundaries can be altered by +3 to −12dB depending on the task; the comfort boundaries can be changed by +3 to −30 dB depending on environmental factors. The standard, however, does not specifically state what task or environmental factors are associated with these adjustments.

These standards, although heralded as a major contribution, have nonetheless been criticized (Morrissey and Bittner, 1975; Sandover, 1979, Oborne, 1983). Some of the criticisms are that (1) the comfort and FDP limits for short time exposures, less than 1 h, may be too high; (2) the limits appear to be based on mean results and do not take into account variability in the population (i.e., what percentage of the exposed population will experience fatigue? 50 percent? 100 percent?); (3) the proposals imply that the effects of combinations of single-axis vibration are additive, despite strong evidence to the contrary (e.g., Leatherwood, Dempsey, and Clevenson, 1980); (4) the use of the same shaped contours for the comfort, performance, and tolerance

FIGURE 17-9
The fatigue-decreased proficiency boundary for vertical vibration contained in ISO 2631. To obtain the boundary of reduced comfort, subtract 10 dB; to obtain boundary for safe exposure, add 6 dB. *(Source: International Organization for Standardization, 1974.)*

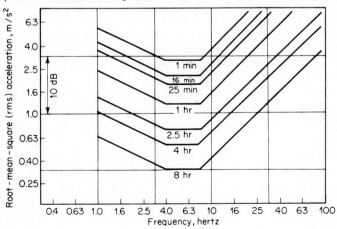

boundaries is an oversimplification; (5) little (if any) evidence is available to support the assumption that there exists a time-intensity trade-off for comfort or performance or for the specific function depicted by the standard (Kjellberg and Wikström, 1985); and (6) the entire standard was based on insufficient and inadequate experimental evidence, although it was the best information available at the time.

Given these criticisms, one may wonder what purpose the standard serves. Clearly the standard was a major step in the right direction, and it has stimulated discussion and research in an area that had been poorly investigated prior to its release. The standard represents a good first approximation of human response to whole-body vibration, but it will need additional modifications as new research findings accumulate. In fact, a new ISO study group has been formed to recommend what some people think might well be radical changes to the standard.

ACCELERATION

Acceleration of the human body is part and parcel of riding in any moving vehicle. In most vehicles, such as automobiles, trains, buses, or commercial aircraft, the levels of acceleration are moderate, and the effects are nominal. There are, however, vehicles that produce very high levels of acceleration for their occupants, such as high-performance jet aircraft and space rockets, with effects that can be of some consequence. Before talking about these, however, we should discuss certain basic concepts and terminology used to characterize acceleration.

Terminology

Acceleration is a rate of change of motion of an object having some mass. The rate of change of motion is expressed as feet per second per second (ft/s^2) or meters per second per second (m/s^2). The basic unit of acceleration, G, is derived from the force of gravity in our earthbound environment. The acceleration of a body in free fall is 32.2 ft/s^2 (9.81 m/s^2), this being $1G$. If a person is accelerated at $2G$, the body effectively weighs twice its normal amount.

Acceleration forces applied to a mass can be either linear or rotational. *Linear acceleration* is the rate of change of *velocity* of a mass, with the direction of movement kept constant. In turn, *rotational acceleration* is the rate of change of *direction* of a mass, whose velocity is kept constant. There are two forms of rotational acceleration. In one form, *radial acceleration,* the axis of rotation is external to the body (as in an aircraft turn). In the other form, commonly referred to as *angular* acceleration, the axis of rotation passes through the body (as when a ballet dancer is twirling on a toe). In some circumstances people are subjected to an admixture of these.

Although rotational acceleration is important in some circumstances (as in certain aircraft maneuvers), here we discuss only linear acceleration. Such acceleration can, of course, occur in any direction with respect to the human body. In this regard, if the human body is caused forcibly to change velocity, that is, to accelerate or decelerate, as in a vehicle that is changing velocity, then the change in velocity causes a physiological reactive force that is opposite to the direction of movement. This reactive force is manifested by a displacement of the heart and other organs, body

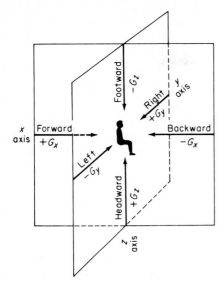

FIGURE 17-10
Illustration of three directions of linear acceleration.
The direction of displacement of the heart and
other organs is opposite to that of the motion of the
body.

tissues, and blood, since these body components are not rigid. Thus, as an automobile accelerates rapidly, the internal organs and the blood tend to "lag" the structure of the body as the body is propelled forward at increasing velocity.

The three possible directions of linear motion are depicted in Figure 17–10, these being called x (forward-backward), y (left-right), and z (upward-downward). Acceleration forces in these directions are expressed in terms of $\pm G$ units, as shown in column 4 of Table 17–3. Column 2 shows, for each direction of motion of the body, the (opposite) direction of motion of the heart and other internal organs. In turn, column 3 gives a vernacular label for each direction of motion as described by the sensation of movement of the eyeballs, such as "eyeballs in" or "eyeballs down."

TABLE 17–3
ACCELERATION TERMINOLOGY WITH RESPECT TO DIRECTION OF ACCELERATION

	Acceleration		
Direction of motion (1)	Heart motion toward (2)	Motion of eyeballs (3)	Symbol (4)
Forward	Spine	In	$+G_x$
Backward	Sternum	Out	$-G_x$
Right	Left	Left	$+G_y$
Left	Right	Right	$-G_y$
Headward	Feet	Down	$+G_z$
Footward	Head	Up	$-G_z$

Source: Adapted from Hornick, 1973, table 7–1, p. 229.

Rarely, if ever, is someone exposed to a simple unvarying linear acceleration in one direction. Instead, acceleration may vary in magnitude or direction and may be accompanied by complex oscillations and vibration. For purposes of summary, however, it is simpler to consider the effects of each direction of acceleration separately without adding to the complexity.

Effects of Headward (+ G$_z$) Acceleration

In headward acceleration, there is an increase in apparent body weight, a tendency for soft tissue to droop, and a tendency for blood to pool in the lower parts of the body. Figure 17–11 summarizes these effects as a function of acceleration magnitude. The predominant effects of headward acceleration are on gross body movement and vision.

Headward acceleration results in decreased ability to detect targets, especially in the peripheral visual field. Figure 17–12, for example, illustrates the change in threshold luminance (i.e., the minimum light intensity at which a stimulus can be perceived), for foveal and peripheral vision, as a function of $+G_z$ acceleration. Complete loss of peripheral vision generally occurs at about $+4.1G_z$, and loss of central vision occurs at about $+5.3G_z$ (Zarriello, Norsworthy, and Bower, 1958).

These effects of $+G_z$ acceleration have consequences on the performance of tasks which involve movement, vision, or both, including such tasks as visual reaction time, reading, tracking, and certain higher mental processes (Fraser, 1973).

In addition to these effects, various cardiovascular and respiratory effects of headward acceleration have been noted (Fraser, 1973). For example, Figure 17–13 illustrates some cardiovascular changes occurring with various levels of $+G_z$ acceleration. Cardiac output and stroke volume decrease, while heart rate, aortic pressure, and systematic vascular resistance increase (Wood et al., 1961).

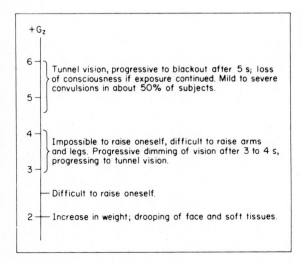

$+G_z$

6 — Tunnel vision, progressive to blackout after 5 s; loss of consciousness if exposure continued. Mild to severe convulsions in about 50% of subjects.

5 —

4 — Impossible to raise oneself, difficult to raise arms and legs. Progressive dimming of vision after 3 to 4 s, progressing to tunnel vision.

3 —

— Difficult to raise oneself.

2 — Increase in weight; drooping of face and soft tissues.

FIGURE 17-11
Summary of effects of headward
(+ G$_z$) acceleration. *(Source:
Fraser, 1973, p. 150.)*

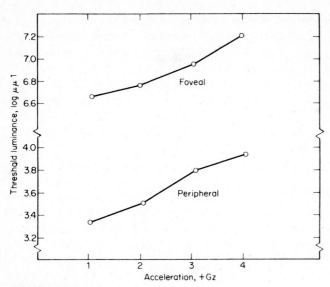

FIGURE 17-12
Threshold of foveal and peripheral vision under $+G_z$
acceleration. *(Source: White, 1960, as shown in Fraser, 1973,
figs. 4-5 and 4-6.)*

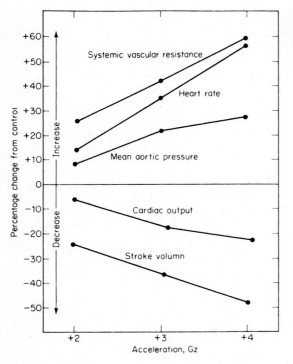

FIGURE 17-13
Cardiovascular response to
headward acceleration $(+G_z)$.
(Source: Wood et al., 1961.)

Effects of Footward ($-G_z$) Acceleration

Far less research has been conducted on the effects of footward acceleration, un-doubtedly because it is far less common than headward acceleration. In footward acceleration, the force is acting up toward the head.

At $-1G_z$ there is unpleasant, but tolerable, facial congestion. At $-2G_z$ to $-3G_z$, the facial congestion becomes severe. At these levels one experiences throbbing headaches, progressive blurring, graying, or occasionally reddening of vision after about 5 s. The limit of tolerance at $-5G_z$ is about 5 s, reached only by the most exceptional subjects (Fraser, 1973).

Effects of Forward ($+G_x$)Acceleration

Seated subjects experiencing forward ($+G_z$) acceleration are pushed back in their chairs. Figure 17–14 summarizes some of the known effects of forward acceleration. A more detailed portrayal of the effects of vehicular (forward) acceleration $+G_x$ on gross body movements is shown in Figure 17–15. This figure shows the typical levels of G that are near the threshold of ability to perform the acts indicated. Some impli-

FIGURE 17-14
Summary of the effects of forward $(+G_z)$ acceleration. *(Source: Fraser, 1973, p. 151.)*

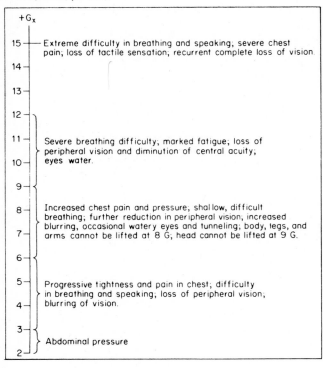

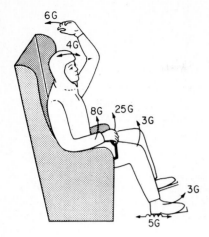

FIGURE 17-15
The forward $(+G_x)$ forces that are near the threshold of various body movements. For any given motion indicated, the movement is just possible at the G_x forces indicated; greater G_x forces usually would make it impossible to perform the act. *(Source: Chambers and Brown, 1959.)*

cations for design decisions are fairly obvious; it would seem unwise, for example, to place the ejection control lever of an airplane over the pilot's head.

In addition to the effects on gross body movement and vision, acceleration produces a marked effect on respiration. Figure 17–16 illustrates the effect of 120-s exposure to $+G_x$ acceleration on arterial oxygen saturation, i.e., the degree to which the blood leaves the lungs saturated with oxygen. At approximately $+6G_x$, one begins to experience sensory impairment, with mental impairment due to lack of oxygen occurring around $+8G_x$.

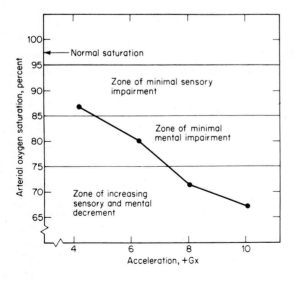

FIGURE 17-16
Arterial oxygen saturation after 120-s exposure to forward $(+G_x)$ acceleration while breathing air. *(Source: Adapted from Alexander, Sever, and Hoppin, 1965.)*

Effects of Backward (−G$_x$) Acceleration

The effects of backward acceleration are similar to those experienced in forward acceleration. Chest pressure is reversed, making breathing easier. There still exist, however, pain and discomfort from outward pressure against the restraint harness being worn (and one would not want to be in such an environment without some type of upper body restraint). The effects on vision are similar for both forward and backward acceleration, but the reduction in arterial oxygen saturation is less in backward acceleration than in forward acceleration (Fraser, 1973).

Effects of Lateral (±G$_z$) Acceleration

Little information is available on the effects of lateral acceleration (Fraser, 1973), but high levels of lateral acceleration are relatively rare in operational environments.

Voluntary Tolerance to Acceleration

The physiological symptoms brought about by acceleration, of course, have a pretty direct relationship to the levels that people are willing to tolerate voluntarily. The physiological effects of differential levels of linear acceleration in various directions, mentioned above, are paralleled by somewhat corresponding differences in average voluntary tolerance levels, such as shown in Figure 17–17. This figure shows the average levels for the various directions that can be tolerated for specific times. Tolerance to footward acceleration −G$_z$ is least, followed by headward acceleration, +G$_z$, with forward acceleration, +G$_z$ being the most tolerable. Individual differences are of considerable magnitude; trained and highly motivated personnel frequently can endure substantially higher levels than the average person.

Protection from Acceleration Effects

Although minor doses of acceleration (as in most land vehicles and most commercial airplanes) pose no serious problem either in safety and welfare or in performance, the effects of higher levels (especially with long exposures, as in space travel) usually necessitate some protective measures. One way to provide protection is to assume a posture which increases the tolerance to the direction of acceleration being experienced. Figure 17–18, for example, shows various body positions which resulted in the greatest tolerance to various directions of acceleration, as well as similar, but less advantageous, postures. Incidentally, in water immersion conditions, the same acceleration force affects both the water and the person immersed in it. This has the effect of equalizing the forces over the parts of the body. Individuals so immersed can even move their body members under acceleration and can tolerate more acceleration than under any other known condition, especially in a supine position with about a 35° angle of the trunk to the line of force.

Immersion, although providing considerable protection, is generally not considered feasible in space flight because of the problem of lugging around a private swimming pool. Aside from the use of optimum postures, the presently available protection schemes include restraining devices, anti-G suits, and special body supports such as contour and net couches.

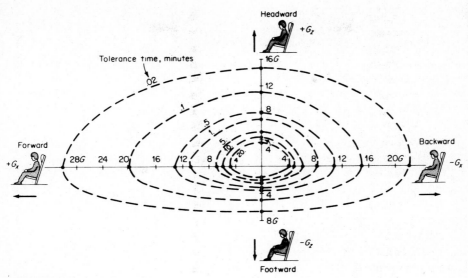

FIGURE 17-17
Average levels of linear acceleration, in different directions, that can be tolerated on a voluntary
basis for specified periods. Each curve shows the average G load that can be tolerated for the
time indicated. The data points obtained were actually those on the axes; the lines as such are
extrapolated from the data points to form the concentric figures. *(Source: Adapted from
Chambers, 1963, fig. 6.; pp. 193–320.)*

For high, sustained G loads in a forward direction, $+G_z$, the use of contour
couches (sometimes molded for the individual) is reasonably satisfactory. Such
couches, however, do not provide adequate support for the rearward direction $-G_x$.
For such purposes additional contraptions are required to provide adequate restraint,
such as a restraint helmet and supporting face and chin pieces, and frontal supports
(such as nylon netting) for the chest, torso, and body members; to further hem in the
individual, an anti-G suit may be in order. Anti-G suits, however, generally have
their greatest utility in connection with headward ($+G_z$) acceleration, with some
designs being more effective than others (Nicholson and Franks, 1966). Anti-G suits
typically apply pressure to the lower abdominal area and legs, thus reducing the
tendency of the blood to drain to those areas during headward ($+G_z$) acceleration.
There is no particularly effective scheme for providing protection against footward
acceleration ($-G_z$); so if you insist on being shot out of a cannon, have them shoot
you out headward, and don't forget your anti-G suit!

Deceleration and Impact

The normal deceleration of a vehicle imposes forces upon people that are essentially
the same as those of actual acceleration, but in reverse. Acceleration usually is a
gradual affair (except in unusual instances, such as with astronauts during blast-off),
but deceleration can be extremely abrupt, especially in the case of vehicle accidents.
When a vehicle hits a solid object or another vehicle head on, an unrestrained occu-

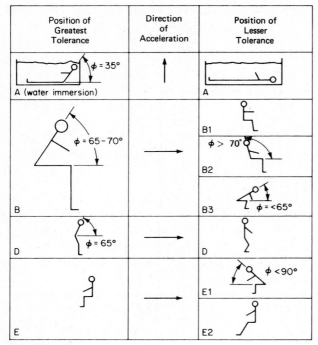

FIGURE 17-18
Comparison of various body postures which influence tolerance
to acceleration. *(Source: Bondurant et al., 1958, as adapted
by Fraser, 1973, fig. 4-30.)*

pant will continue forward at the initial velocity until the body strikes some part of
the interior of the vehicle or is thrown clear from the vehicle. This is sometimes
called the *second collision*. The deceleration of the occupant is a function of the
deformation of the part of the interior which is hit by the occupant's body, if any.
To be protected from this second collision, the occupant must be "tied" to the vehicle (as with restraining devices) and decelerate with the vehicle (which may occur
through a distance of a couple of feet or more) or must be provided with some
energy-absorbing object that results in more gradual deceleration by increasing the
travel distance of the occupant during deceleration (air bags, collapsible steering
wheels, etc.).

Restraining Devices

An occupant can be tied to a vehicle by quite a few types of restraining devices; in
automobiles and commercial aircraft these are nearly always seat (lap) belts. In some
automobiles shoulder harnesses are used alone or in combination with seat belts; in
military aircraft more complex restraints are used, individually or in combination. A
major deficiency of automobile seat belts is, of course, the possibility of head damage caused by the torso and head being propelled forward with the body hinged at

the hips. Considerably greater protection is provided by shoulder-lap belts, which aid the occupant in "riding down" the vehicle as it crashes. A general comparison of some of these effects is shown in Figure 17–19. (However, there is some suspicion that, under impact conditions, shoulder belts might apply excessive force on the spine.) Some indication of the protection provided by seat belts comes from a number of papers presented at the 1984 SAE International Congress and Exposition (SAE, 1984). Seat belt use has been associated with reducing fatalities by 50 percent or more (Hartemann et al., 1984), largely because seat belts prevent the ejection of the individual from the vehicle. Seat belts reduce minor and moderate injuries, by about 25 percent (Norin, Carlsson, and Korner, 1984). In addition, injuries tend to be less severe when belts were worn.

Another method of tying passengers to a vehicle is through the use of rear-facing seats. It has been estimated that properly designed aft-facing seats could protect occupants against as much as 40G impacts with little or no injury (Eiband, 1959). Such seats would be particularly feasible in aircraft, trains, etc., but are not completely unrealistic for use in automobiles (if we could adapt ourselves to them). In the case of vehicles with forward-facing seats involved in rear-end crashes, the use of a full back support with headrest minimizes the possibility of whiplash.

Energy-Absorbing Objects

The air bag may be destined to replace seat belts as the primary device for protection from vehicular head-on accidents. The air bag system is designed so that when the vehicle hits the back of another vehicle, the air bag is immediately inflated in front of the occupant and is then deflated. The effect is to absorb the energy of the occupant being catapulted forward on impact. In other words, the deceleration takes place

FIGURE 17-19
Velocity of vehicle occupants following impact with lap belt and shoulder-lap belt. The shoulder-lap restraint causes the upper torso and head to follow more closely the vehicle velocity change than does the seat belt alone. On impact, an unrestrained occupant would be propelled forward at the impact velocity until he or she hit some part of the interior. *(Source: Cichowski and Silver, 1968, fig. 1.)*

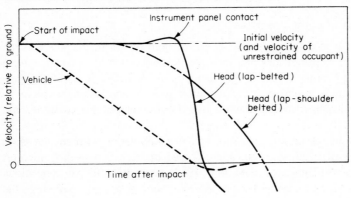

over a longer travel distance than if the occupant were propelled forward at the vehicle velocity at the time of impact until she or he struck a rigid object such as the windshield or the steering wheel.

WEIGHTLESSNESS

Since human beings have evolved in an earthbound environment, the force of gravity has been our constant companion. Its constant presence has influenced our physiological makeup and is basic to all our activities. Except for the case of a few roller coasters, it is only in space flight and in certain aspects of aircraft flight that the "natural" phenomenon of weightlessness, or reduced gravity, is experienced. For those few who do venture into outer space, two aspects of the weightlessness, or reduced-weight, state are particularly important (Berry, 1973). The first is the absence of weight itself; the removal from the normal gravitational environment could be expected to have an impact on the human organism, such as on it physiological functioning and on perceptual-motor performance and sensory performance. Second, the weightless condition is accompanied by a tractionless condition when the person is moving and working.

Since the weightless condition does not exist in our earthbound environment, various schemes have been developed for creating such a condition, or simulating it, in order to study its effects. True weightlessness can be achieved within the earth's gravitational field by the use of aircraft in a parabolic flight maneuver. During the maneuver (up and down) a weightless condition can be achieved for upward of 1 min (Moran, 1969).

Various forms of partial simulation of the weightless state include cable suspension rigs (which reduce the friction between persons and the surface on which they are operating), gimbals, air bearings, and water immersion (to achieve neutral buoyancy) (Deutsch, 1969). Although some such methods have provided some inklings about the effects of weightlessness, the most reliable basis for estimating the long-term effects is undoubtedly actual experience in orbital space flights. We are accumulating a wealth of such information, since U.S. and Russian space travelers have already spent tens of thousands of hours in space. To date, the longest flight has been 237 days in the Soviet Salyut 7 space station.

Physiological Effects of Weightlessness

In summarizing the early space flight experience of both the United States and the Soviet Union, Berry (1973) indicates that although some physiological changes have been consistently noted, none have been permanently debilitating. Some of the temporary effects that have been observed include aberrations in cardiac electrical activity, changes in the number of red and white blood cells, loss of muscle tone, and loss of weight.

More recently, space sickness (also called *space adaptation syndrome*) has been reported by roughly half of all space travelers during the first 2 to 4 days in orbit (Oman, 1984; Oman, Lichtenberg, and Money, 1984). Symptoms include sensations of tumbling, overall discomfort, prolonged nausea, and vomiting. Although the symptoms often disappear after a few days in orbit, task performance can be adversely affected during that time.

Thornton (1978) reports various anthropometric changes that occur in a weightless environment. Space travelers grow in height by approximately 3 percent [approximately 2 in (5 cm)]. This has implications for, among other things, pressure-suit design and control stations with critical eye-level requirements. Even more significant is the natural relaxed posture assumed in a weightless environment, as shown in Fig. 17–20. Notice the lower line of sight under zero-*g* conditions, the angled foot, and the height of the arms. Seats, work stations, etc., must be designed to accommodate this unique posture. Thornton reports that the body "rebels" with fatigue and discomfort against any attempt to force it into a more normal earthbound posture.

Performance Effects of Weightlessness

Locomotion within a spacecraft adds a third dimension to our usual two-dimensional movements. Apparently no serious problems have occurred in normal locomotion within spacecraft (Berry, 1973). Although serious consideration has been given to

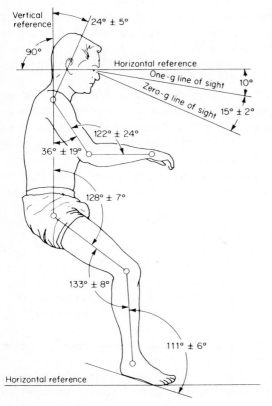

FIGURE 17-20
Typical relaxed posture assumed by
people in weightless conditions.
(Source: Thornton, 1978, fig. 16.)

generating artificial gravity in spacecraft (by rotating the entire space vehicle or station or by having an on-board centrifuge), the experience of U.S. astronauts so far has suggested that this would be unnecessary.

Performance outside a spacecraft in an extravehicular activity (EVA) is a bit of a different story, since some astronauts have experienced considerable exhaustion in such activities. As Berry points out, this probably argues for careful consideration of workload in planning EVA missions.

ILLUSIONS DURING MOTION

When humans are in motion, they receive cues regarding their whereabouts and motion from sense organs, especially the semicircular canals, the vestibular sacs (the otolith organs), the eyes, the kinesthetic receptors, and the cutaneous senses. These sense organs and the intricate interactions among them were specifically designed for a self-propelled terrestrial animal. They were not intended to cope with the unusual and prolonged forces encountered in three-dimensional flight. Such forces can push our sense organs beyond their capabilities to function accurately and cause them to signal erroneous information concerning the position and motion of the body.

Pilots in airplanes sometimes experience *disorientation* regarding their position or motion with respect to the earth. Actually numerous phenomena cause disorientation; some give rise to sensations that the person is spinning (true vertigo), and others cause dizziness (a feeling of movement within the head). Indeed, the three terms *disorientation, vertigo,* and *dizziness* are usually, though inaccurately, used interchangeably to describe a variety of symptoms such as false sensations of turning, linear velocity, or tilt (Kirkham et al., 1978).

Disorientation can be the basis for fatalities in an aircraft environment. Kirkham et al. (1978), for example, report that disorientation was the cause of 16 percent of all fatal general aviation aircraft accidents in the United States. Reason (1974) reports that virtually all pilots experience some form of disorientation at least once in their flying careers—for some, it is their last.

There are a multitude of disorienting effects, and we cannot hope to present or discuss them all. To understand some of them requires a knowledge of physics and physiology that is more than we wish to tackle here.

Reason (1974) distinguishes two types of disorientation effects: those arising from false sensations and those arising from misperception.

Disorientation from False Sensations

The mental confusion or misjudgment resulting from false sensations is due to inaccurate or inappropriate sensory information coming from the semicircular canals, otolith organs, or both. That which results from false sensations from the semicircular canals involves angular acceleration cues. One such situation occurs when a pilot executes a roll at an angular acceleration that is below the threshold of perception of the semicircular canals. The pilot is not aware of the full extent of the roll and so overcorrects when attempting to reestablish straight, level flight. The aircraft ends up in a bank turned in the opposite direction, but the pilot thinks it is flying straight and level.

Another example occurs after recovery from a spin; a pilot will often think the aircraft is actually spinning in the opposite direction. Attempting to recover from this false spin will often send the airplane into another real spin. Pilots call this the *graveyard spin,* and for good reason.

Another form of disorientation arising from inaccurate semicircular canal information is called the *Coriolis illusion,* or *cross-coupling effect.* This can occur when the head is tilted during a long-established turning or circling maneuver. The head must be tilted in a different plane from the plane of the aircraft's turn. The experience is an illusion of roll, often accompanied by dizziness.

False sensations from the otolith organs often give rise to false sensations of tilt. The *oculogravic illusion* is one example. Whenever an aircraft's forward speed is suddenly increased (i.e., accelerated), the acceleration force vector combines with the gravity force vector to fool the otoliths into signaling that the body (and hence the aircraft) is tilted in a ''nose-up'' attitude. The eyes also roll back, thus adding to the impression of tilt. Attempting to level the aircraft (when, in fact, it is already level) can send it into the ground rather abruptly. Incidentally, decelerating often gives rise to the sensation of tilting nose-down.

These false-sensation forms of disorientation tend to occur under conditions of poor visibility, e.g., at night, in clouds, or in fog. It is for this reason that pilots are trained to trust their instruments rather than their senses. Usually a visual reference can reduce or eliminate disorientation, but not always.

Disorientation Resulting from Misperception

These forms of disorientation arise because the brain misinterprets or misclassifies perfectly accurate sensory information, usually provided by the visual sense. One form is *autokinesis,* in which a fixed light appears to move against its dark background. It has been reported that pilots have attempted to ''join up'' in formations with stars, buoys, and street lights which appear to be moving (Clark and Graybiel, 1955). Another common visual misperception occurs under conditions of limited visibility wherein a pilot will accept a sloping cloud bank as an indication of the horizontal and, as a result, align the wings with the cloud bank and fly along at a ''rakish'' tilt.

Discussion

The reduction of disorientation and illusions generally depends more on procedural practices and training than on the engineering design of aircraft. Among such practices are the following (Clark and Graybiel, 1955): understanding the nature of various illusions and the circumstances under which they tend to occur, maintaining either instrument *or* contact flight, avoiding night aerobatics, shifting attention to different features of the environment, learning to depend on the correct cue to orientation (such as visual cues or instruments), avoiding sudden accelerations and decelerations at night, and avoiding prolonged constant-speed turns at night.

MOTION SICKNESS

We would be remiss if we failed to include motion sickness in a chapter on humans in motion.[3] Although motion sickness may never actually kill us, there are times when we wish it would! Motion sickness, of course, is not really a sickness in the pathological sense at all; it only makes us "feel sick." Motion sickness is associated with most forms of travel—cars, boats, trains, and even camels, but not horses. (We explain the camel-horse paradox shortly.) People differ in their susceptibility to motion sickness, but that susceptibility is a relatively stable and enduring characteristic of the individual (Reason, 1974). Further, a person susceptible to one type of motion sickness is probably susceptible to all types except, for some unexplained reason, space sickness.

The symptoms of motion sickness are familiar to us all. Reason (1974) puts them into two classes: head and gut. *Head symptoms* include drowsiness and a general apathy, together called the *sopite syndrome* by Graybiel and Krepton (1976). Headaches are also experienced. *Gut symptoms* range from a disconcerting awareness of the stomach region to acute nausea and vomiting (or, more politely, *emesis*).

The most widely accepted theory of motion sickness, called *sensory rearrangement theory* (Reason and Brand, 1975), considers motion sickness as a consequence of incongruities among the spatial senses, i.e., the organs of balance, the eyes, and the nonvestibular position senses (joints, muscles, and tendons). The incongruity among the senses is incompatible with what we have come to expect on the basis of our past experience. Reason (1974, 1978) points out that the vestibular system, i.e., the semicircular canals and otolith organs, must be implicated for motion sickness to be an outcome. And, since the vestibular system responds only to accelerations (linear and angular), acceleration (or apparent acceleration) must always be involved in motion sickness.

Reason (1974, 1978) identifies two classes of sensory rearrangement: *visual-inertia rearrangements* and *canal-otolith rearrangements*. Further, these rearrangements, or incongruities, are of two types: (1) both systems (e.g., canal and otolith) simultaneously signal contradictory information, and (2) one system signals in the absence of an expected signal from the other. For example, some people can get motion sickness from watching a movie filmed inside a roller coaster. In this case, the visual system senses accelerations, but the semicircular canals and otoliths sense no motion.

Low-frequency oscillations of less than 1 Hz, as in the up-and-down motion of a ship, usually cause motion sickness. Reason and Brand (1975) believe this is because the otolith signals are out of phase with the semicircular canal signals; however, the exact mechanism is unclear. McCauley and Kennedy (1976) summarized the results of exposing over 500 subjects to 2 h of vertical sinusoidal vibration of various combinations of frequency (<1 Hz) and acceleration. The dependent variable was the percentage of subjects who vomited—not the kind of research you dream about doing after graduate school! Figure 17–21 shows the percentage of subjects vomiting at various frequency-acceleration combinations. Vibration from 0.15 to 0.25 Hz (or 9

---

[3]See Kennedy and Frank (1984) and Money (1970) for reviews of the motion sickness literature.

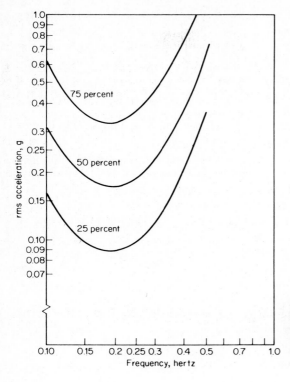

FIGURE 17-21
Equal motion sickness contours, based on percentage of emesis within 2 h for 500 male subjects exposed to vertical sinusoidal motion. *(Source: McCauley and Kennedy, 1976, fig. 4.)*

to 15 cycles per minute) is, for some reason, especially effective for inducing gut symptoms of motion sickness. Incidentally, McCauley and Kennedy indicate that those who are going to get sick generally do so within the first hour when they are exposed to very regular (sinusoidal) motion. There is some evidence (Guignard and McCauley, 1982), however, that the response to low-frequency motion depicted in Figure 17–21 may not strictly hold in real-world environments with waveforms that are more complex than single sine waves.

Were you wondering about the camel-horse paradox? It has nothing to do with whether horses smell better than camels; actually, the answer lies in Figure 17–21. The characteristic frequency of a camel's gait happens to be in the 0.2Hz range, while a horse tends to oscillate at higher frequencies.

You will recall that frequency, amplitude, and acceleration are all related in sinusoidal vibration. For very low-frequency vibration, relatively large amplitudes are required to achieve relatively low levels of acceleration. For example, a 0.2-Hz vibration, having a 0.09 rms *g*, would have a peak-to-peak displacement of 5.2 ft (1.58 m).[4]

Now that we know what causes motion sickness, what can we do to reduce its effects? Reason (1974) lists several practical suggestions, including getting off the

---

[4]Peak-to-peak displacement (in) for a sine wave $= (27.675 \times$ rms $g)/[$frequency $(Hz)^2]$.

boat. In the case of seasickness, lying down on your back will often help; holding your head still will also reduce the symptoms. Several drugs are available which seem to be effective against motion sickness. Most people will eventually adapt to the motion environment and "get their sea legs" in a few days. NASA uses adaptation schedules in which space crews are exposed to sickness-inducing stimulation in graded doses, the idea being to build up tolerance (Reason, 1974). We can all add a few home-grown preventive measures to the list.

In addition to the remedies listed above, Bittner and Guignard (1985) discuss several work-space design approaches for minimizing seasickness based on human factors principles. Included are these suggestions: (1) Locate the work station near the ship's effective center of rotation to reduce rotational heave; (2) group displays and controls to minimize the need for large-angle head turning; (3) place tools and equipment within close reach so that bending and twisting are not required to retrieve them; (4) align the operator with the longitudinal axis of the ship; and (5) provide an external visual frame of reference.

One last point worth mentioning involves motion sickness and ride quality. Obviously a major source of ride discomfort is motion sickness, yet it is rarely addressed in the "ride quality" literature. Look back at the equal-discomfort curves in Figures 17–7 and 17–8 and the ISO recommended criteria shown in Figure 17–9. In all cases the curves start at 1 Hz with no mention of discomfort below 1 Hz. Future research in ride quality will have to address the region below 1 Hz to provide comprehensive data on ride comfort.

DISCUSSION

The common denominator of the exotic experiences of astronauts in outer space, of the more mundane processes of moving around in automobiles or other vehicles, and of the operation of tractors and other related items of equipment is the fact that the human body is being moved somehow, e.g., accelerated, decelerated, transported, jostled, or shaken. Although some forms and degree of such shuffling are inconsequential, other forms or degrees obviously have potentially adverse effects in terms of physiological, performance, or subjective criteria. The challenge to the human factors disciplines is to design the physical systems or protective devices in question so as to minimize such effects and, where they cannot be eliminated, to design the system to match the reduced capabilities of the operator and to make whatever motion is involved more tolerable for passengers.

REFERENCES

Alexander, W., Sever, R., and Hoppin, F. (1965). *Hypoxemia induced by sustained forward acceleration in pilots breathing pure oxygen in a five psi absolute environment* (NASA-TX-X-51649). Washington.

Berry, C. A. (1973). Weightlessness. In *Bioastronautics data book* (2d ed.) (NASA SP-3006). Washington.

Bittner, A., and Guignard, J. (1985). Human factors engineering principles for minimizing adverse ship motion effects: Theory and practice. *Naval Engineers Journal, 97*(4), 205–213.

Bondurant, S., Clark, N. P., Blanchard, W. G., et al. (1958). *Human tolerance to some of the accelerations anticipated in space flight* (TR 58-156). Ohio: Wright-Patterson Air Force Base, WADC.

Carlsoo, S. (1982). The effect of vibration on the skeleton, joints and muscles: A review of the literature. *Applied Ergonomics,* 13, 251–258.

Chambers, R. M. (1963). Operator performance in acceleration environments. In N. M. Burns, R. M. Chambers, and E. Hendler (eds.), *Unusual environments and human behavior*. New York: Free Press.

Chambers, R. M., and Brown, J. L. (1959). *Acceleration*. Paper presented at Symposium on Environmental Stress and Human Performance. American Psychological Association.

Cichowski, W. G., and Silver, J. N. (1968). *Effective use of restraint systems in passenger cars* (SAE 680032). Detroit: Paper presented at Automotive Engineering Congress, Society of Automotive Engineers.

Clark, B., and Graybiel, A. (1955). *Disorientation: A cause of pilot error,* Res. Project NM 001 110 100.39. U. S. Navy School of Aviation Medicine.

Deutsch, S. (1969). Preface to special issue on reduced-gravity simulation. *Human Factors,* 5(11), 415–418.

Eiband, A. M. (1959). *Human tolerance to rapidly applied acceleration,* NASA Memo 5-19-59E. Washington.

Fraser, T. (1973). Sustained linear acceleration. In *Bioastronautics data book* (2d. ed.) (NASA-SP-3006). Washington.

Geldard, F. (1972). *The human senses* (2d ed.). New York: Wiley.

Gell, C., and Moeller, G. (1972). The biodynamics aspects of low altitude high speed flight. *Ergonomics,* 15, 655–670.

Graybiel, A., and Krepton, J. (1976). Sopite syndrome: A sometimes sole manifestation of motion sickness. *Aviation, Space, and Environmental Medicine,* 47, 873–882.

Griffin, M., Parsons, K., and Whitham, E. (1982). Vibration and comfort. IV. Application of experimental results. *Ergonomics,* 25, 721–739.

Guignard, J. (1965). Vibration. In J. Gillies (ed.), *A textbook of aviation physiology*. Oxford, England: Pergamon.

Guignard, J., and McCauley, M. (1982). Motion sickness incidence induced by complex periodic waveforms. *Aviation, Space, and Environmental Medicine,* 53, 554–563.

Hartemann, F., Henry, C., Faverjon, G., Tarriere, C., Got, C., and Patel, A. (1984). Ten years of safety due to the three-point seat belt. In SAE (ed.), *Advances in belt restraint systems: Design, performance and usage* (SAE/P-84/141). Warrendale, PA: Society of Automotive Engineers, pp. 7–14.

Hornick, R. (1973). Vibration. In *Bioastronautics data book* (2d ed.) (NASA SP-3006). Washington.

International Organization for Standardization. (1978). *Guide for the evaluation of human exposure to whole-body vibration* (ISO 2631).

Kennedy, R., and Frank, L. (1984, Mar. 9). *A review of motion sickness with special reference to simulator sickness* (rev.). Prepared for distribution at the National Academy of Sciences/National Research Council Committee on Human Factors, Workshop on Simulator Sickness, September 26–28, 1983, held at Naval Postgraduate School, Monterey CA.

Kirkham, W., Collins, W., Grape, P., Simpson, J., and Wallace, T. (1978). Spatial disorientation in general aviation accidents. *Aviation, Space, and Environmental Medicine,* 49, 1080–1086.

Kjellberg, A., and Wikström, B. (1985). Whole-body vibration: Exposure time and acute effects—a review. *Ergonomics,* 28(3), 535–544.

Leatherwood, J., Dempsey, T., and Clevenson, S. (1980). A design tool for estimating passenger ride discomfort within complex ride environments. *Human Factors,* 22, 291–312.

McCauley, M., and Kennedy, R. (1976). *Recommended human exposure limits for very low frequency vibration* [TP-76-36(U)]. Point Mugu, CA: Pacific Missile Test Center.

Money, K. (1970). Motion sickness. *Physiological Reviews,* 50(1), 1–38.

Moran, M. J. (1969). Reduced-gravity human factors research with aircraft. *Human Factors,* 11(5), 463–472.

Morrissey, S., and Bittner, A. (1975). *Effects of vibration on humans: Performance decrements and limits* (TP-75-37U). Point Mugu, CA: Pacific Missile Test Center.

Nicholson, A. N., and Franks, W. R. (1966). Devices for protection against positive (long axis) acceleration. In P. I. Altman and D. S. Dittmer (eds.), *Environmental biology,* Tech. Rept. 66–194. Aerospace Medical Research Laboratory, pp. 259–260.

Norin, H., Carlsson, G., and Korner, J. (1984). Seat belt usage in Sweden and its injury reducing effect. In SAE (ed.), *Advances in belt restraint systems: Design, performance and usage* (SAE/P-84/141). Warrendale, PA: Society of Automotive Engineers, pp. 15–28.

Osborne, D. (1976). A critical assessment of studies relating whole-body vibration to passenger comfort. *Ergonomics,* 19, 751–774.

Osborne, D. (1978). The stability of equal sensation contours for whole-body vibration. *Ergonomics,* 21, 651–658.

Osborne, D. (1983). Whole-body vibration and international standard ISO 2631: A critique. *Human factors,* 25, 55–69.

Osborne, D., Heath, T., and Boarer, P. (1981). Variation in human response to whole-body vibration. *Ergonomics,* 24, 301–313.

Oman, C. (1984, August). Why do astronauts suffer space sickness? *New Scientist,* 23, 10–13.

Oman, C., Lichtenberg, B., and Money, K. (1984, May 3). Space motion sickness monitoring experiment: Spacelab 1, Paper 35. NATO-AGARD Aerospace Medical Panel Symposium: *Motion sickness; Mechanisms, prediction, prevention and treatment.* Williamsburg, VA.

Poulton, E. (1978). Increased vigilance with vertical vibration at 5 Hz: An alerting mechanism. *Applied Ergonomics,* 9, 73–76.

Radke, A. O. (1974, Dec. 3). *Vehicle vibration: Man's new environment,* Paper 57-A-54. American Society of Mechanical Engineers.

Reason, J. (1974). *Man in motion: The psychology of travel.* London: Weidenfeld and Nicolson.

Reason, J. (1978). Motion sickness: Some theoretical and practical considerations. *Applied Ergonomics,* 9, 163–167.

Reason, J., and Brand, J. (1975). *Motion sickness.* London: Academic.

Richards, L., and Jacobson, I. (1977). Ride quality assessment III: Questionnaire results of a second flight programme. *Ergonomics,* 20, 499–519.

SAE (1984). *Advances in belt restraint systems: Design, performance and usage* (SAE/P-84/141). Warrendale, PA.

Sandover, J. (1979). A standard on human response to vibration—One of a new breed? *Applied Ergonomics,* 10, 33–37.

Simons, A. K., Radke, A. O., and Oswald, W. C. (1956, Mar. 11). *A study of truck ride characteristics in military vehicles,* Rep. 118. Milwaukee: Bostrom Research Laboratories.

Stephens, D. (1979). Developments in ride quality criteria. *Noise Control Engineering,* 12, 6–14.

Thornton, W. (1978). Anthropometric changes in weightlessness. In Anthropology research staff (eds.), *Anthropometric source book,* vol. I: *Anthropometry for designers* (NASA RP-1024). Houston.

Troup, J. (1978). Driver's back pain and its prevention. *Applied Ergonomics,* 9, 207–214.

Wasserman, D. (1975). Bumps, grind take toll on bones, muscles, mind. *Health and Safety,* 45, 19–21.

White, W. (1960). *Variations in absolute visual threshold during acceleration stress,* WADC Tech. Rept. 60–34. Ohio: Wright-Patterson Air Force Base.

Whitham, E., and Griffin, M. (1978). The effects of vibration frequency and direction on the location of areas of discomfort caused by whole-body vibration. *Applied Ergonomics,* 9, 231–239.

Witkin, H. A. (1959, February). The perception of the upright. *Scientific American.*

Wood, E., Sutterer, W., Marshall, H., Lindberg, E., and Headley, R. (1961). *Effects of headward and forward accelerations on the cardiovascular system,* WADC Tech. Rept. 60–634. Ohio: Wright-Patterson Air Force Base.

Zarriello, J., Norsworthy, M., and Bower, H. (1958). *A study of early grayout thresholds as an indicator of human tolerance to positive radial acceleratory force* (Project NM–11–02–11). Pensacola, FL: Naval School of Aviation Medicine.

PART SIX

HUMAN FACTORS: SELECTED TOPICS

HUMAN FACTORS APPLICATIONS IN SYSTEM DESIGN

Some previous chapters have dealt, in part, with the human factors aspects of the design of certain types of things people use (displays, controls, hand tools and devices, etc.) and of the environments in which people work and live. This chapter deals with some procedures and considerations involved in the application of human factors principles and data to the design of systems. We review the system design process and discuss the role played by human factors in the various stages. In the second half of the chapter we switch gears somewhat and discuss a specific area of application of human factors, namely, product liability litigation and the use of warnings to enhance the safety of products.

SYSTEM DESIGN PROCESS

In Chapter 1 we discussed briefly some characteristics of systems. In the design of such systems (including equipment, facilities, and other physical items that people use) typically certain basic stages or processes have to be carried out.

There are various ways to depict the major stages in the system design process. One rather straightforward representation is shown in Figure 18-1. This representation (and, in fact, much of the following discussion of the system design process) applies to the design of very complex systems that often include major hardware components. The comprehensive, systematic approach to the design process discussed here has been developed and used primarily in the design of major military and space systems. Although the "systematic" features of such an approach have not been widely used in the design of other types of systems and products, often the basic stages and considerations involved need to be addressed somehow, even if in a nonsystematic fashion. In the development of some systems or products, however, some aspects of the design process may simply be irrelevant; this may be particularly

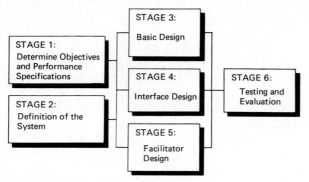

FIGURE 18-1
Major stages in the system design process. These stages are
carried out in iterative fashion as the system develops. Later
stages may modify decisions made in earlier stages. For some
simple systems, these stages are carried out informally and, in
some cases, not at all. *(Source: Adapted from Bailey, 1982.)*

true with simple systems or products. Elaborate design processes may not be warranted in many circumstances, but the basic point is this: If a particular aspect of the
design process *is* relevant, that aspect should be given appropriate attention.

Characteristics of the System Design Process

Although representations of the system design process such as the one shown in
Figure 18-1 imply an orderly, structured process, in practice the stages often overlap
and are performed in iterative fashion. In the design of a system, decisions in later
stages often necessitate the modification and refinement of information and decisions
from earlier stages. Meister (1985) lists the following characteristics of the process:

1 *Molecularization.* The process works from broad molar functions to more molecular tasks and subtasks.
2 *Requirements are forcing functions.* Design options are developed to satisfy
system requirements. (For this reason formal behavioral requirements must be included in the initial design specifications.)
3 *System development is discovery.* Initially there are many unknowns about the
system, and during the design process these unknowns are clarified and addressed.
4 *System development involves transformation.* There is a transformation from
physical requirements to behavioral implications of those requirements, and from
behavior implications to the physical mechanisms (e.g., controls, displays) required
for implementing the behavior implications.
5 *Time.* There is never enough time for practitioners to perform the analyses,
studies, and tests that they would like to do.
6 *Cost.* There is usually not enough money to support the design effort, and
behavioral recommendations, if too costly, will be rejected automatically.

7 *Iteration*. Activities are repeated as more detailed information about the system becomes available.

8 *Design competition*. Large systems are designed by groups or teams of specialists, each with their own concerns and criteria. Less influential groups (often including human factors people) must function under constraints established by the dominant groups.

9 *Relevance*. Design relevance as perceived by engineers is critical for the acceptance and judged value of behavioral inputs.

We now turn our attention to the design stages depicted in Figure 18-1, and we discuss some of the human factors activities, concerns, and tools associated with each.

STAGE 1: DETERMINE OBJECTIVES AND PERFORMANCE SPECIFICATIONS

Before a system can be designed, it must have a purpose or reason for being. The purposes of a system are stated as objectives and usually in rather general terms. For example, the objective of a system might be to provide communication links between offices in different cities. System performance specifications detail what the system must do to meet its objectives. Examples might include transmitting both voice and visual information across the country, linking up to 100 offices, providing data security for transmitting sensitive information, etc. System performance specifications should reflect the context in which the system will operate, including consideration of the skills available in the existing or projected user population, organizational constraints, etc. System requirements often have behavioral implications that will have an impact on later stages of development. For example, a system that must be able to operate in arctic conditions will impose constraints on personnel and will require special clothing that could have an impact on the design of controls and work spaces.

Bailey (1982) identified two human factors activities appropriate to this stage of the design process: Identify the intended users of the system, and identify and define the activity-related needs of users which the system will address. Identification of user needs is done through observation, interviews, and questionnaires. Usually a system is designed to improve upon some already existing system or set of systems. By observing how activities are carried out with the existing system and by interviewing and surveying users of the existing system, information can be gathered to ensure that the new system's specifications meet the needs of existing and potential users. This information can also be valuable in later stages of the system's development.

STAGE 2: DEFINITION OF THE SYSTEM

The principal activity during stage 2 is the definition of the functions which the system has to perform to meet its objectives and performance specifications. In theory, no attempt is made during this stage to assign functions to hardware, software, or humans. Functions can be instantaneous (e.g., to power up equipment) or prolonged (e.g., to store incoming messages) and simple (e.g., to determine message destination) or complex (e.g., to prioritize incoming messages). In general, functions are relatively molar behaviors; the individual tasks and behaviors needed to implement or carry out the functions are more detailed. There are, however, no rules for deciding whether a set of behaviors should be called a function or a task (Meister, 1985). To aid in the conceptualization of system functions, *functional flow diagrams* are often developed; an example is shown in Figure 18-2. Such diagrams depict the interrelationships among the system functions, with each box representing one function. Often subsidiary flow diagrams are developed to analyze the subfunctions necessary to carry out each function identified in the higher-order diagrams. In the case of an urban transportation system, some functions that would need to be performed include the sale of tickets, collection of tickets, movement of passengers to loading platforms, opening and closing of vehicle doors to permit entrance and exit of passengers (possibly to take into account handicapped passengers), vehicle control, and movement of passengers to exits of stations, along with providing relevant information and directions for passengers and a communications system for use in the control and spacing of vehicles. This function analysis initially should be concerned with *what* functions need to be performed to fulfill the objectives, and not with the *way* in which the functions are to be performed (such as whether they are to be performed by individuals or by machine components).

During this stage, human factors specialists help insure that the functions identified match the needs of the intended users. Further, these specialists may collect

FIGURE 18-2
An example of a functional flow diagram used to graphically depict the functions of a system.
(Source: Geer, 1981.)

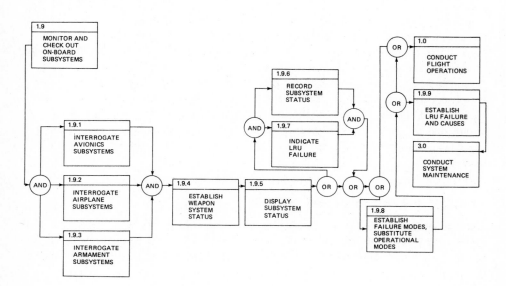

more detailed information regarding the characteristics, capabilities, and limitations of the intended user population which will be used in the next stage of development.

STAGE 3: BASIC DESIGN

It is in this stage of system development that the system begins to take shape. The principal human factors activities are (1) allocation of function to humans, hardware, and software; (2) specifications of human performance requirements; (3) task analysis; and (4) job design. During this stage usually numerous changes are made to the system which necessitate revisions and updating of the human factors activities. Rarely are the four activities listed above done only once; rather, they are performed in iterative fashion as the design matures.

Allocation of Functions

Given functions that have to be performed, in some instances there may be an option as to whether any particular function should be allocated to a human being or to some physical (machine) component(s). In this process, the allocation of certain functions to human beings, and of others to physical components, is virtually predetermined by certain manifest considerations, such as obvious superiority of one over the other or economic considerations. Between these two extremes, however, may lie a range of functions that are within the reasonable repertoire of both human beings and physical components. (The control of vehicles in an urban transportation system, for example, could be achieved by individual operators or by a central computer system.)

Because of the importance of these decisions for some systems, one would hope for some guidelines to aid the system designer in allocating specific functions to human beings versus physical machine components. The most common types of guidelines previously proposed consist of general statements about the kinds of things humans can do better than machines, and vice versa. As Chapanis (1965) points out, however, such comparisons serve a useful function in only the most elementary way. Such generalizations are useful primarily in directing one's thinking toward human-machine problems and in reminding one of some general characteristics that humans and machines have as system components. Aside from the potential utility of such comparisons for use in the allocation process, however, Jordan (1963) and others raise the more basic question of the appropriate role of human beings in systems, especially given the trend toward automation. Before discussing such views, however, we present a set of such generalizations, both to demonstrate what such "lists" are like and to serve what Chapanis refers to as the "elementary" purposes for which these lists may have some utility.

Relative Capabilities of Human Beings and of Machines The following generalizations about the relative capabilities of human beings and of machine components are drawn from various sources (Chapanis, 1960, p. 543; Fitts, 1951, 1962; Meister, 1971; Meister and Rabideau, 1965; and others), plus some additional items and variations on previously expressed themes.

Humans are generally better in their abilities to

• Sense very low levels of certain kinds of stimuli: visual, auditory, tactual, olfactory, and taste
• Detect stimuli against high-noise-level background, such as blips on CRT displays with poor reception
• Recognize patterns of complex stimuli which may vary from situation to situation, such as objects in aerial photographs and speech sounds
• Sense unusual and unexpected events in the environment
• Store (remember) large amounts of information over long periods (better for remembering principles and strategies than masses of detailed information)
• Retrieve pertinent information from storage (recall), frequently retrieving many related items of information (but reliability of recall is low)
• Draw upon varied experience in making decisions; adapt decisions to situational requirements; act in emergencies (do not require previous ''programming'' for all situations)
• Select alternative modes of operation if certain modes fail
• Reason inductively, generalizing from observations
• Apply principles to solutions of varied problems
• Make subjective estimates and evaluations
• Develop entirely new solutions
• Concentrate on most important activities when overload conditions require
• Adapt physical response (within reason) to variations in operational requirements

Machines are generally better in their abilities to

• Sense stimuli outside the normal range of human sensitivity, such as x-rays, radar wavelengths, and ultrasonic vibrations
• Apply deductive reasoning, such as recognizing stimuli as belonging to a general class (when the characteristics of the class need to be specified)
• Monitor for prespecified events, especially infrequent ones (but machines cannot improvise in case of unanticipated types of events)
• Store coded information quickly and in substantial quantity (for example, large sets of numerical values can be stored very quickly)
• Retrieve coded information quickly and accurately when specifically requested (although specific instructions need to be provided on the type of information to be recalled)
• Process quantitative information following specified programs
• Make rapid and consistent responses to input signals
• Perform repetitive activities reliably
• Exert considerable physical force in a highly controlled manner
• Maintain performance over extended periods (machines typically do not ''fatigue'' as rapidly as humans)
• Count or measure physical quantities
• Perform several programmed activities simultaneously

- Maintain efficient operations under conditions of heavy load (humans have relatively limited channel capacity)
- Maintain efficient operations under distractions

In discussing such lists of relative advantages of humans and machines, Jordan (1963) boils the assortment to a nub, as follows: "Men are flexible but cannot be depended upon to perform in a consistent manner, whereas machines can be depended upon to perform consistently but have no flexibility whatsoever."

Limitations of Human-Machine Comparisons The previously implied limitations regarding the practical utility of general comparisons of human and machine capabilities stem from various factors. Some have been pointed up by Chapanis (1965), Price (1985), and Corkindale (1967) and are discussed in the following (with a fair portion of editorial license):

1 General human-machine comparisons are not always applicable. Given some general superiority of humans or machines, in some circumstances it would be inappropriate to apply the dictates of that generality. For example, the amazing computational abilities of computers do not imply that one should use a computer whenever computations are required.

2 There is a lack of adequate data on which to base function allocations.

3 Relative comparisons are subject to continual change. For example, machines are not very effective in pattern recognition, but this may change.

4 It is not always important to provide for the "best" performance. For example, although human beings who serve as toll collectors on superhighways offer some advantage over mechanical collectors, the mechanized devices do the job well enough to be acceptable.

5 Function performance is not the only criterion. One also has to consider the trade-off values of other criteria such as availability, cost, weight, power, reliability, and cost of maintenance. At present, there are very few systematic guidelines to follow in figuring out the trade-off values of various criteria.

6 Function allocation should reflect social and related values. The process of allocation of functions to humans versus machine components directly predetermines the role of human beings in systems and thereby raises important questions of a social, cultural, economic, and even political nature. The basic roles of human beings in the production of the goods and services of the economy have a direct bearing on such factors as job satisfaction, human motivation, and the value systems of individuals and of the culture. Since our culture places a premium on certain human values, the system should not require human work activities that are incompatible with such values. In this vein, Jordan postulates the premise that humans and machines should be considered not comparable, but rather complementary. Whether one would agree with his conclusion that the allocation concept becomes entirely meaningless, the fact does remain that decisions (if not allocations) need to be made concerning the relative roles of humans and machines. In this context and given that the objective of most systems is not the entertainment of the operators (pinball machines and gambling devices excepted), it would seem that, within reason, the human work activities generated by a system should provide the opportunity for reasonable intrinsic satisfaction to those who perform them. That is, we must design jobs that people can do and want to do.

A Strategy for Allocating Functions The discussion above suggests that functions cannot be allocated by formula (Price and Pulliam, 1983). The allocation process must rely on expert judgment as the final means for making meaningful decisions; allocation of function is as much an art as a science. Although there are no clear-cut guidelines for making specific allocation decisions, Price (1985) suggests four rules for developing an allocation strategy:

1 *Mandatory allocation.* Some functions or portions of functions may have to be allocated to a human or machine because of system requirements (e.g., a human must maintain abort control), hostile environments, safety considerations, or legal or labor constraints. Mandatory allocations should be identified and made first.

2 *Balance of value.* The relative goodness of a human and that of the to-be-available machine (including software) technology for performing the function are estimated and represented as a point in the decision space of Figure 18–3. If the point is in the unacceptable *(Uhm)* region (i.e., it cannot be satisfactorily performed by either human or machine), then either the function must be redefined or the system requirements and constraints must be modified. If the point is in region *Uh* (unacceptable for humans) or region *Um* (unacceptable for machines), then the function should be treated as a mandatory allocation. If the function is in the *Ph* (better performed by humans) or *Pm* (better performed by machines) region, it can be tentatively allocated to the preferred alternative, but that allocation may be changed by the next two rules. If the function is in the *Pmh* region (no significant difference between human or machine performance), proceed to the next two rules.

3 *Utilitarian and cost-based allocation.* In utilitarian allocation a function may be allocated to humans simply because human beings are present and there is no compelling reason why they should not perform the work. The relative costs of human and machine performance must be considered, and allocations can be made on the basis of least cost.

4 *Affective and cognitive support allocations.* The final rule recognizes the unique needs of humans. Allocation decisions may have to be revised to provide affective and cognitive support for the humans in the system. *Affective support* refers to the emotional requirements of humans, such as their need to do challenging work, to know their work has value, to feel personally secure, and to be in control. This has a bearing on issues of job satisfaction and motivation. *Cognitive support* refers to the human need for information so as to be ready for action or to make decisions that may be required. The human must maintain an adequate "mental model" of the system and its condition in order to take control in an emergency. Another consideration in cognitive support is that the human be given sufficient activity to ensure alertness.

The allocation of functions can best be accomplished by a team of individuals, including engineers and human factors specialists. The rules outlined above should

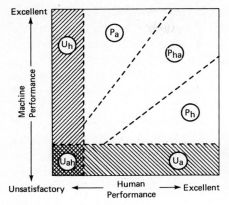

FIGURE 18-3
"Decision space" for representing relative machine and human performance as an aid in making allocation of function decisions. *(Source: Price, Maisano, and Van Cott, 1982.)*

be viewed as a reasonable starting point with the understanding that the detailed decisions still depend on the judgments of experts.

Human Performance Requirements

After the design team identifies the functions to be carried out by humans, the next step is to determine the human performance requirements for those functions. Human performance requirements are the performance characteristics which must be met in order for the system to meet its requirements. Addressed are such things as required accuracy, speed, time necessary to develop performance proficiency, and user satisfaction. Sometimes the human performance requirements are beyond the anticipated capacity of the user population. In such cases it is necessary to reallocate functions or even to redefine the system requirements developed in stage 1.

Task Description and Analysis

Meister (1985) makes a distinction between task description and task analysis, although other authors combine the two processes and speak only of task analysis. Even Meister, however, admits that the dividing line between the two is unclear. Yet both must be carried out during the system design process. It is not appropriate here to include an extensive discussion of various task description and analysis procedures. [The interested reader can consult McCormick (1979) or Meister (1985) for more information.] Our intent here is to present the general nature of the process, the types of information collected, and the purposes to which it can be applied.

Nature of the Process The general procedure for developing a task description and analysis (hereafter referred to as simply task analysis) is to list in sequence all the tasks that must be performed to accomplish the functions in which a human plays a part. Each task is then further broken down into the steps required to carry out the task. Each step, then, is usually analyzed to determine such things as the stimuli that initiate the step; decisions the human must make in performing the step; actions required during the step; information necessary to carry out the step; feedback information resulting from performing the step; potential sources of error or stress; and the criterion for successful performance. In addition, determinations of task criticality and difficulty can be made, the number and skill level of people needed to operate the system can be estimated, and projections regarding training requirements can be outlined.

Often one product of a task analysis is an *operational-sequence diagram* (OSD). An OSD is a graphical depiction of the interactions between people and equipment in the performance of a task over time. An example of an OSD is shown in Figure 18–4. Figure 18–5 summarizes the standard symbols used to construct such diagrams. Although these diagrams are developed during the design of complex systems, often the diagrams themselves have limited utility. Their value lies in the fact that, to construct them, the task analyst must gather a great deal of important information and in so doing develops a detailed understanding of the tasks (Geer, 1981). In this case it may be the process rather than the product that is important.

Purposes for Task Analysis A thorough task analysis is essential to insure that the system will be operable and maintainable in a safe and efficient manner. Task analysis is the basis for designing human-machine interfaces (i.e., the controls, displays, work stations, etc.) and instruction manuals and job aids; determining person-

FIGURE 18-4
An example of an operational-sequence diagram for a task involved in railroad operations. The symbols used to construct the diagram are explained in Figure 18-5. *(Source: Sanders, Jankovich, and Goodpaster, 1974.)*

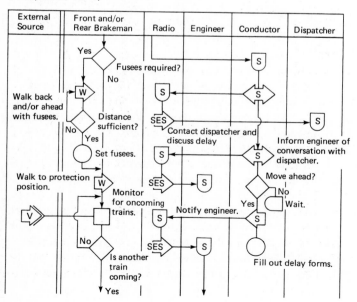

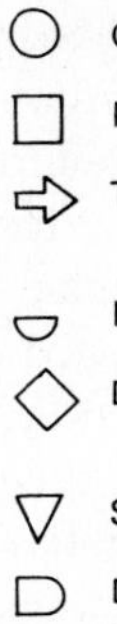

○   OPERATE   — An action function to accomplish or continue a process.

▢   INSPECT   — To monitor or verify quantity or quality.

⇨   TRANSMIT[1]— To pass information without changing its form. Also used to denote travel between two locations.

▽   RECEIPT[1] — To receive information or objects.

◇   DECISION   — To evaluate and select a course of action or inaction based on receipt of information.

▽   STORAGE   — To retain information or objects.

▭   DELAY   — Period of inactivity.

---

1. Mode of transmission and receipt is indicated by a code letter within the TRANSMIT and RECEIPT symbols:

V — Visual	T — Touch
E — Electrical/Electronic	M — Mechanically
S — Sound (Verbal)	W — Walking
	H — Hand deliver

FIGURE 18-5
Standard symbols and notation used to construct operational-sequence diagrams.

nel requirements; developing training programs; and designing the evaluation of the system. In essence, task analysis is at the heart of the human factors contributions to system design.

Computer Aids for Basic Design

Over the years various computerized design aids have been developed to help the designer make some of the basic design decisions discussed above. Although several of these design aids have been developed [see, for example, Geer (1981) or Meister (1985)], we present a brief overview of only one.

CAFES (computer-aided function allocation evaluation system), probably the most comprehensive and sophisticated of the computerized design aids, contains several modules that are useful throughout the system design process (Parks and Springer, 1975). The most complicated of the CAFES modules is the function allocation model (FAM). The purpose of FAM is to guide the designer in allocating functions and tasks. To use it requires a great deal of input data, such as task start and stop times, task attributes (e.g., criticality), functional flow diagrams, mission time lines, candidate allocations, and various weighting criteria. The designer inputs the data, and the computer calculates such useful information as the probabilities of successful task and mission performance, gross workload for each person in the system, a rank ordering of the various allocation strategies, a time line showing who in the system is doing what and when (useful for developing operational sequence diagrams), and the percentage of tasks completed, interrupted, or being performed simultaneously. With such information the designer can assess potential problems, alter the allocation of functions, and rerun the program until a satisfactory allocation strategy has been developed.

Another CAFES module is the workload assessment model (WAM). The designer inputs information regarding the task sequences and durations as well as the sensory and motor channels used to perform each task. WAM compares the time required to perform a task sequence with the time available for performance. This is done for eyes, hands, feet, and auditory, verbal, and cognitive channels. The program identifies time segments and tasks in which some criterion level of workload is exceeded.

The results from programs such as FAM and WAM should not be considered the final word on function allocation. The data they generate can be valuable, but such programs, as yet, do not take into account aspects such as the affective and cognitive support requirements of function allocation. In the end experts will allocate functions based on their judgment and, perhaps, computer analyses similar to those of FAM and WAM. In this regard, Meister (1985) points out that despite their potential, programs such as FAM and WAM are used much less frequently than the traditional manual methods.

Job Design

To a very considerable degree the design of the equipment and other facilities that people use in their jobs predetermines the nature of the jobs they perform. Thus, the designers of some types of equipment are, in effect, designing the jobs of people who use the equipment. They may, or may not, be aware of this, but they certainly should be.

Human Values in Job Design   As discussed above in the context of function allocation, there are certain philosophical considerations in the design of jobs. In this regard there has been considerable interest in the notion of job enlargement and job enrichment, this concern being based on the assumption that enlarged or enriched jobs generally bring about higher levels of motivation and job satisfaction. There are a number of variations on this theme, such as increasing the number of activities to be performed, giving the worker responsibility for inspection of her or his work, delegating responsibility for a complete unit (rather than for a specific part), providing opportunity to the worker to select the work methods to be used, rotating jobs, and placing greater responsibility on work groups for production processes.

The human factors approach has placed primary emphasis on efficiency or productivity, and then it has tended toward the creation of jobs that are more specialized and that require less skill than jobs that are enlarged. In discussing these somewhat disparate objectives, McCormick and Ilgen (1985) emphasize the fact of individual differences in value systems, indicating that some people do not like "enlarged" or "enriched" jobs.

The degree of possible conflict between these two approaches is not yet clear, and there are as yet no clear-cut guidelines for designing jobs that would be compatible with the dual objectives of work efficiency and providing the opportunity for motivation and job satisfaction on the part of workers. Although we cannot offer any pat solutions to this problem, it is to the credit of the human factors clan that increased attention is being given to this issue.

STAGE 4: INTERFACE DESIGN

After the basic design of the system is defined and the functions and tasks allocated to humans are delineated, attention can be given to the characteristics of the human-machine and human-software interfaces. These include work spaces, displays, controls, consoles, computer dialogs, etc. Many of the earlier chapters have dealt with the design and human factors considerations involved in such interfaces. Although the actual design of the physical components is predominantly an engineering chore, this stage represents a time when human factors inputs are of considerable moment. The specific nature of the design decisions made during this phase can (if inappropriate in terms of human considerations) forever plague the user and cause decrement in system performance or, conversely (if appropriately designed), facilitate use of the equipment and bring about better system performance.

Human Factors Data for Interface Design

A major and continuing effort of the human factors members of a design team during this stage of system development lies in tracking down and interpreting existing human factors data potentially relevant to the design problem. The data relevant to various human factors design problems cover a wide range (with many gaps yet to be filled) and exist in many forms, such as the following:

- Common sense and experience (such as the designer has in memory, some of which may be valid and some not)
- Comparative quantitative data (such as relative accuracy in reading two types of visual instruments, as illustrated in Chapter 5)
- Sets of quantitative data (such as anthropometric measures of samples of people, as given in Chapter 12, and error rates in performing various tasks)
- Principles (based on substantial experience and research, which provide guidelines for design, such as the principle of avoiding or minimizing glare when possible)
- Mathematical functions and equations (which describe certain basic relationships with human performance, such as in certain types of simulation models)
- Graphical representations (nomographs or other representations, such as tolerance to acceleration of various intensities and durations)
- Judgment of experts
- Design standards or criteria (consisting of specifications for specific areas of application, such as displays, controls, work areas, and noise levels)

Much of the research in human factors is directed toward the development of design standards or criteria (sometimes called *engineering standards)* that can be used directly in the design of relevant items. The recommendations regarding the design of control devices in Table B-2 represent examples of such standards. Such design specifications sometimes are incorporated into the form of checklists; in the case of military services, they are incorporated into military specifications called MIL-SPECS. The principal human factors military design standard is MIL-STD-1472C (1981) which specifies both general and detailed design criteria for a wide range of interface problems. Another source of human factors design guidance is found in MIL-HDBK-759A (1981). Although the military specifications were developed for use in designing military systems, they also have substantial value in designing other types of systems. Appendix D lists major human factors books which can also be of value in interface design.

Problems in Applying Human Factors Data

It would be very satisfying to say that available human factors data are being extensively applied to the design of the things people use in their work and daily living. But such is not the case. Meister (1971), for example, reported that (unfortunately) many design engineers did *not* consider human factors in their design procedures. And Rogers and Armstrong (1977), on the basis of a study dealing with the use of human factors design standards in engineering design, came to the dismal conclusion that such standards have little, if any, effect on product design. They suggested that the apparent reasons for this are varied and complex, involving resistance on the part of designers and managers, education of human factors specialists and designers, the standards themselves, and interdisciplinary communication.

The human factors research community probably has contributed to this state of affairs in at least a couple of ways, in particular in failing to communicate adequately with people in other disciplines (such as engineering, architecture, and industrial design) and in failing to present human factors data in a reasonably useful form.

Presentation of Human Factors Data   Blanchard (1975) proposes a set of guidelines for the development of human factors data banks that would better contribute to the use of such data. In connection with the development of human factors design standards, Rogers and Armstrong (1977) state that such standards would be enhanced if the following suggestions were incorporated:

1 Eliminate general or ambiguous terms such as *proper feel* or *high torque*.
2 Present quantitative data in a manner consistent with designer preference, i.e., graphical or pictorial first, followed by tabulations.
3 Eliminate the use of narrative statements when data can be presented quantitatively.
4 Eliminate inconsistencies in data within and among standards.
5 Provide revisions and updating of standards on a more timely basis.

Rogers and Pegden (1977) elaborate on some of these recommendations and add certain others, such as providing a quick access system, including relevant definitions, and providing adequate cross-referencing.

Considerations in Applying Human Factors Data

The application of relevant human factors data to some design problems is a fairly straightforward proposition. More generally, however, it is necessary to use judgment in this process—in particular, in evaluating the applicability of potentially relevant data to specific design problems. In making such evaluations at least four considerations should be taken into account.

Practical Significance   One consideration is the practical significance of the application of some relevant human factors information. For example, although the time required to use control device A might be significantly less (statistically) than that for device B, the difference might be so slight, and of such nominal utility, that it would not be worthwhile to use device A if other factors (such as cost) argued against it.

Extrapolation to Different Situations   Much of the available human factors data has been based on research findings or experience in certain specific settings. The second consideration in the application of human factors data deals with possible extrapolation of such data to other settings. For example, could one assume that the reaction time of astronauts in a space capsule would be the same as in an earthbound environment? Referring specifically to laboratory studies, Chapanis (1967) urges extreme caution in applying the results of such studies to real-world problems.

Perhaps three points might be made about these sobering reflections. First, despite Chapanis's words of caution, there are some design problems for which available research data or experience is fully adequate to resolve the design problem. Second, in certain areas of investigation it may be possible to simulate the real world with sufficient fidelity to derive research findings that can be used with reasonable confidence. Third, there are undoubtedly circumstances in which one might extrapolate from data that are not uniquely appropriate to the design problem at hand, on the expectation that this would increase the odds of achieving a better design than if such data were not used. Clearly this strategy involves risks and should be followed only in the case of noncritical design problems.

Consideration of Risks   A third consideration in applying human factors data is associated with the seriousness of the risks incurred in making bad guesses. Bad guesses in designing a shuttle for shipping an astronaut into space obviously would be a matter of greater concern than those in designing a hat rack. The more serious the risks, the greater the need for relevant, high-quality data.

Consideration of Trade-off Function   A fourth consideration is trade-off values. Since frequently it is not possible to achieve an optimum in all possible criteria for a given design of a human-machine system, some give and take must be accepted. The possible payoff of one feature (as suggested by the results of research) may have to be sacrificed, at least in part, for some other more desirable payoff.

Tools for Interface Design

In addition to the application of relevant human factors data, human factors specialists use other tools to assist in the interface design. Mock-ups of proposed equipment layouts can be constructed, and people representative of the user population can be placed in the mock-up and walked through the planned tasks. Clearances, reach envelopes, visibility envelopes, etc., can be systematically assessed by using people of different sizes.

Construction and modification of mock-ups to reflect alternative designs and the evaluation of each version by using human subjects are both expensive and time-consuming. In practice, only a few alternative designs may be considered. An alternative to constructing mock-ups is to use computer simulation with its graphics capabilities to compare alternatives early in the design process.

Several computer-based work station design and evaluation aids are available, including such acronym-based programs as HECAD (Topmiller and Aume, 1978), COMBIMAN (Evans, 1978), and SAMMIE (Bonney and Case, 1976). To illustrate the nature of such design aids, we briefly discuss two programs that are part of CHESS (crew human engineering software system) (Jones, Jonsen, and Van, 1982). The first program is called IRA (instrument readability analysis). IRA calculates for any control station component the physical height required of a marking (e.g., a label or number on a display) to make it subtend a given visual angle at the eye of the observer. The program simulates realistic head motion (Line, 1979).

The second program, called CAR (crew station assessment of reach) (Harris, Bennett, and Dow, 1980), was developed for the U.S. Navy to evaluate aircraft cockpit accommodations; but its use is not limited to that application. Using anthropometric data, CAR "builds" a representative population of people in the form of three-dimensional "stick" or "link" figures, as shown in Figure 18–6a. (The outline of a body is drawn around the stick figure to clarify the link relationships.) Some CAR-based programs "enflesh" the basic link figures; an example is shown in Figure 18–6b. CAR analyzes the ability of each "person" in its sample to reach (by using any of several types of reach) a particular control. The CAR user can specify the type of clothing and type of seat belt constraint being worn. The results include data on head clearance, the percentage of the population achieving a satisfactory visual envelope, and the percentage capable of reaching each control. The report also indicates the distance and direction of control-location change needed to accommodate additional members of the population. Data comparing CAR-generated reach envelopes with envelopes generated by real people have shown very good agreement, especially in the areas to the front and side of the person (Harris, Bennett, and Stokes, 1982).

STAGE 5: FACILITATOR DESIGN

The main focus of this stage of the system design process is to plan for materials that will promote acceptable human performance. Included are instruction manuals, performance aids, and training devices and programs. Although not substitutes for good design, these support materials can have an impact on the decisions made during the interface design stage. Most, if not all, complex systems require some training and properly designed documentation to aid system users. It is important that these aspects of the system's design be given the same consideration and time as the design of the human-hardware and human-software interfaces.

In some systems a concept referred to as *embedded training* is utilized. Embedded training means that training programs are "built into" the system so that the operational equipment can be switched over to a training mode during periods when it is not needed for operational use.

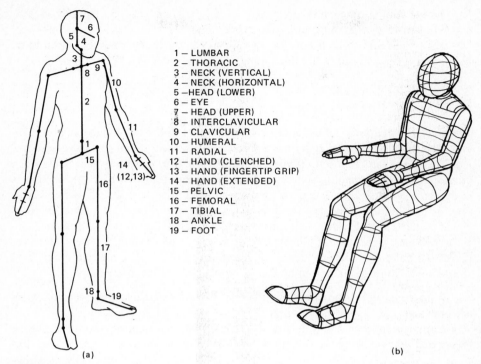

1 — LUMBAR
2 — THORACIC
3 — NECK (VERTICAL)
4 — NECK (HORIZONTAL)
5 — HEAD (LOWER)
6 — EYE
7 — HEAD (UPPER)
8 — INTERCLAVICULAR
9 — CLAVICULAR
10 — HUMERAL
11 — RADIAL
12 — HAND (CLENCHED)
13 — HAND (FINGERTIP GRIP)
14 — HAND (EXTENDED)
15 — PELVIC
16 — FEMORAL
17 — TIBIAL
18 — ANKLE
19 — FOOT

(a)

(b)

FIGURE 18-6
Representations used by computerized anthropometric programs. *(a)* CAR link-person model with body outline drawn to clarify relationships; *(b)* an example of an enfleshed CAR link model. *(Source: Stone and McCauley, 1984.)*

One form of system documentation that specifies how to operate and, perhaps, maintain the system is instruction manuals. Instruction manuals can be elaborate books or a simple piece of paper packed with a product. We take a moment to discuss a few issues involved in the design of instruction manuals.

Instruction Manuals

The design of effective instruction manuals is more an art than a science. Although we cannot discuss the many human factors issues involved in the design of documentation, listing a few will illustrate the complexity of the task. One needs to consider such things as whether the documentation will be "hard copy" or online (i.e., part of the software) or some combination; how to utilize illustrations; how to combine text and illustrations; how much detail to present; and how to index, tab, and organize the material. In regard to some of these issues Coskuntuna and Mauro (1980) offer these overly simplified rules of thumb:

- Less is more: Avoid informational overload.
- Avoid abstract information; use only concrete information.

- Forget why's; concentrate on how's.
- Remember that learning will come from doing.
- Forget the *hype*. The users already have the product, and what they want to do now is set it up or use it with minimal hassle.

As an aside, they urge putting important material at the front or in a prominent position.

Use of Illustrations   Illustrations have their place either in connection with manuals or instructions or simply by themselves to give instructions or to serve as warnings. In this regard Johnson (1980) differentiates between pictures and symbols; A *picture* is a realistic photograph or drawing or an object about which information is to be conveyed; a *symbol* is a photograph or drawing that represents something *else,* such as the Red Cross symbol. In turn he describes a *pictogram* as a series of associated pictures intended to give information about performing a series of actions.

In discussing the use of illustrations to convey information to users, Johnson makes the point that concepts about which information is to be conveyed range from simple to complex, as indicated by the following ends of the continuum:

- Simple concept: Instructions are given to perform a single action (why and when are obvious).
- Complex concept: Multiple actions are required, and multiple results can ensue (when and why an action is to be performed may not be obvious).

He also makes the point that the concepts in question can range from concrete (such as an emergency door) to abstract (with no physical referent, such as the general notion of danger). The various combinations of these distinctions, along with proposed methods of conveying information about them, are given in Table 18–1.

STAGE 6: TESTING AND EVALUATION[1]

Evaluation in the context of system development has been defined as the measurement of system development products (hardware, procedures, and personnel) to verify that they will do what they are supposed to do (Meister and Rabideau, 1965, p. 13). In turn, *human factors evaluation* is the examination of these products to ensure the adequacy of attributes that have implications for human performance. Actually, almost every decision during the design of a system includes some evaluation, such as deciding whether to use a visual signal or an auditory signal. Although many such evaluative decisions need to be made as part of the ongoing development cycle, for most items developed for human use there should be some systematic evaluation upon the completion of the development stage and before the item goes into production. Although such evaluation should be carried out, unfortunately many things people use in their work and everyday lives do not benefit from such evaluation. Thus many things do not serve their purposes effectively or safely and in some instances have to be recalled for expensive retrofitting or replacement.

[1]For a thorough treatment of human factors evaluation in system development, see Meister and Rabideau (1965).

TABLE 18–1
PROPOSED METHODS FOR CONVEYING INSTRUCTIONAL
INFORMATION

	Concrete data	Abstract data
Simple	use pictures/words	use symbols/words
Complex	use pictograms/words	use words

Source: Johnson, 1980

The Nature of Human Factors Evaluation

The evaluation of some items in terms of human factors considerations should be carried out systematically, in much the same manner as an experiment. In many instances, however, probably the evaluation procedures leave much to be desired in terms of acceptable procedures.

In connection with experimental procedures, Meister (1978) differentiated between what he called controlled experimentation (CE) and personnel subsystem measurement (PSM). The CE experimenter deals largely with the *explanation* of relationships and phenomena, such as explaining why reaction time is a function of the numbers of stimuli and the possible responses to them. The PSM investigator, however, must be concerned with the measurement of human performance (and related criteria) *in operational terms,* that is, in terms relevant to the system, subsystem, or item in question, specifically to the functions of the personnel involved in the operation and maintenance of the system.

Special Problems in Human Factors Evaluation

The above comments may give some inklings about the problems of human factors evaluation. We can touch here on only a few problems that haunt those responsible for human factors evaluation in actual operational situations or in circumstances that approximate operational conditions. Although much of the research on which human factors data and principles are developed is carried out in laboratory settings, DeGreene (1977) makes the point that the real laboratory for human factors research should be the real world. This is indeed so in the final evaluation and testing of things to be used by people.

The overview of research methods discussed in Chapter 2 is as applicable to the final evaluation of the things people are to use as it is to the research aimed toward developing human factors data and principles. However, the problems involved in evaluation in the real world usually are much more complex than those in the laboratory. In the design of evaluation procedures, consideration must be given to three factors: subjects, criteria, and experimental procedures.

Subjects The evaluation should be carried out with the same types of individuals who are expected to use the item in question, such as homemakers, trained pilots, persons with driver's licenses, factory assemblers, handicapped persons, or the public at large.

Criteria   The criteria (the dependent variables) should be in some manner relevant to the operational use of the item, and depending on the item they can include work performance, physiological effects, accidents, effects on health, learning time, job satisfaction, attitudes and opinions, and economic considerations, to mention a few. The problems of measuring some of these can drive an experimenter crazy. The use of multiple criteria is characteristic of most evaluation research. Rarely, if ever, is a single criterion sufficient basis for an evaluation.

Experimental Procedures and Controls   The problem of carrying out appropriate experimental procedures, including adequate controls (involving in some instances control groups), is much greater in the real world than in a laboratory. We cannot discuss this problem in detail here, but it should be recognized by the investigator.

Johnson and Baker (1974) point out that in many field studies it is not possible to replicate the study, control groups cannot always be used, and the physical environment cannot be completely controlled. Recognizing these and other limitations of evaluation in real-world circumstances, the investigator should make every reasonable effort to follow appropriate experimental procedures.

DISCUSSION OF HUMAN FACTORS AND SYSTEM DESIGN

Clearly the application of human factors data to design processes does not (at least yet) lend itself to the formulation of a completely routine, objective set of procedures and solutions. However, a systematic consideration of the human factors aspects of a system at least focuses attention on features that should be designed with human beings in mind. In this connection, it might be useful to list at least some reminders that are appropriate when one is approaching a design problem. These reminders are presented in the form of a series of questions (with occasional supplementary comments). Two points should be made about these questions. First, some of these would not be pertinent in the design of particular items; in turn, this is not intended as an all-inclusive list of questions. Second, (and as indicated frequently before), the fulfillment of one objective may of necessity be at the cost of another. Nevertheless, this list of questions should serve as a good start in the design process.

1 What functions need to be carried out to fulfill the system objective?

2 If there are any reasonable options available, which should be performed by human beings?

3 For a given function, what information external to the individual is required? Of such information, what information can be adequately received directly from the environment, and what information should be presented through the use of displays?

4 For information to be presented by displays, what sensory modality should be used? Consideration should be given to the relative advantages and disadvantages of the various sensory modalities for receiving the type of information in question.

5 For a given type of information, what type of display should be used? The display generally should provide the information when and where it is needed. These considerations may reflect the general type of display, the stimulus dimension and codes to be used, and the specific features of the display. The display should provide for adequate sensory discrimination of the minimum differences that are required.

6 Are the various visual displays arranged for optimum use?

7 Are the information inputs collectively within reasonable bounds of human information-receiving capacities?

8 Do the various information sources avoid excessive timesharing?

9 Are the decision-making and adaptive abilities of human beings appropriately utilized?

10 Are the decisions to be made at any given time within the reasonable capability limits of human beings?

11 In the case of automated systems or components, do the individuals have basic *control,* so that they do not feel that their behavior is being controlled by the system?

12 When physical control is to be exercised by an individual, what type of control device should be used?

13 Is each control device easily identifiable?

14 Are the controls properly designed in terms of shape, size, and other relevant considerations?

15 Are the operational requirements of any given control (as well as of the controls generally) within reasonable bounds? The requirements for force, speed, precision, etc., should be within limits of virtually all persons who are to use the system. The human-machine dynamics should capitalize on human abilities so that, in operation, the devices meet the specified system requirements.

16 Is the operation of each control device compatible with any corresponding display and with common human response tendencies?

17 Are the control devices arranged conveniently and for reasonably optimum use?

18 Is the work space suitable for the range of individuals who will use the facility?

19 Are the various components and other features of the facility arranged in a satisfactory manner for ease of use and safety?

20 When relevant, is the visibility from the work station satisfactory?

21 If there is a communication network, will the communication flow avoid overburdening the individuals involved?

22 Are the various tasks to be done grouped appropriately into jobs?

23 Do the tasks which require timesharing avoid overburdening any individual or the system? Particular attention needs to be given to the possibility of overburdening in emergencies.

24 Is there provision for adequate redundancy in the system, especially of critical functions? Redundancy can be provided in the form of backup or parallel components (either persons or machines).

25 Are the jobs such that personnel can be trained to do them?

26 If so, is the training period expected to be within reasonable time limits?

27 Do the work aids and training complement each other?

28 If training simulators are used, do they achieve a reasonable balance between transfer of training and costs?

29 Is the system or item adequately designed for convenient maintenance and repair, including any individual components? For example, is there adequate clearance space for reaching parts that need to be repaired or replaced? Can individual parts be repaired or replaced easily? Are proper tools and adequate troubleshooting aids available? Are there adequate instructions for maintenance and repair?

30 Do the environmental conditions (temperature, illumination, noise, etc.) permit satisfactory levels of human performance and provide for the physical well-being of individuals?

31 In any evaluation or test of the system (or components), does the system performance meet the desired performance requirements?

32 Does the system in its entirety provide reasonable opportunity for the individuals involved to experience some form and degree of self-fulfillment and to fulfill some of the human values that we should all like to have the opportunity to fulfill in our daily lives?

33 Does the system in its entirety contribute generally to the fulfillment of reasonable human values? In systems with identifiable outputs of goods and services, this consideration would apply to those goods and services. In the case of systems that relate to our life space and everyday living, this consideration would apply to the potential fulfillment of those human values that are within the reasonable bounds of our civilization.

In the resolution of these and other kinds of human factors considerations, one should draw upon whatever relevant information is available. This information can be of different types, including principles that have been developed through experience or research, sets of normative data (such as frequency distributions of, say, body size), sets of factual data of a probability nature (such as percentage of signals detected under specified conditions), mathematical formulas, tentative theories of behavior, hypotheses suggested by research investigations, and even the general knowledge acquired through everyday experience.

With respect to information that would have to be generated through research (as opposed to experience), while very comprehensive information is available in certain areas of knowledge, in others it is pretty skimpy, and there are some areas in which one draws a complete blank. Where adequate information is not available, perhaps considered judgments based on partial information will, in the long run, result in better design decisions than those that are arbitrary. But let us reinforce the point that such judgments usually should be made by those whose professional training and experience put them in the class of experts, whether in the field of night vision, physical anthropometry, hearing disorders, perception, heat stress, acceleration, learning, or decision making.

Here, again, we mention the almost inevitability of having to trade certain advantages for others. The balancing of advantages and disadvantages generally needs to take into account various types of considerations—engineering feasibility, human considerations, economic considerations, and others. Given that there probably are few guidelines to follow, the general objective of this horse trading is fairly clear.

This goal basically reflects the stated or implied system objectives and the accompanying performance requirements. In other words, any trade-offs should be made on the basis of the considerations of their relative effects in terms of system objectives.

PRODUCTS LIABILITY

The production and the sale of products (be they hair dryers or nuclear power plants) are not necessarily the end of the line for the producer or seller. For certain items the producer or seller maintains relationships with the buyers for service, maintenance, or repair, for example. In recent years in the United States, the matter of products liability has taken on increasing importance, to the point that producers and sellers must give advance attention to such liability in designing, producing, and selling products. The liability issue applies equally to products that are used in industry (machines, tools, etc.) and to those purchased by consumers (household appliances, utensils, etc.).

Products liability is the legal term used to describe an action in which an injured party (the plaintiff) seeks to recover damages for personal injury or loss of property from a manufacturer or seller (the defendant) because the plaintiff believes that the injuries or damages resulted from a defective product. Today, more than ever, people look to the courts for redress when they suffer injury or damage. This attitude has been acutely felt in the area of products liability. Each year more and more cases are brought to court for adjudication. This growth in products liability cases has created a greater demand for human factors experts, both in the initial design to make products safer and in the courtroom as expert witnesses.

Human factors specialists have served as expert witnesses in cases involving such products as automobiles, forklift trucks, medical equipment, boats, conveyers, industrial machinery, railroad crossings, CB antennas, and even stairs. The issues that human factors experts have addressed in such cases include hearing discrimination, visual acuity, visual illusions, eye-hand coordination, reaction time, temperature sensitivity, biomechanics of lifting, reading speed, control and display design, and the adequacy and effectiveness of warnings and instructions. Several cases have been decided principally on the testimony of human factors experts (Wallace and Key, 1984).

In this section we present some basic concepts in products liability that are relevant to human factors people. We do not dwell on the innumerable nuances of the law involved in products liability cases; our intention here is not to make the reader a lawyer. (God knows we have enough lawyers already). Products liability is case law; that is, each new court decision adds, changes, clarifies, or sometimes clouds the accumulated legal precedents. Each state has different legal precedents; hence, cases may be tried and decided differently in different states. Even as you read this, a case may be decided that drastically alters the nature and course of products liability cases. In addition, Congress has been considering a comprehensive products liability bill that would alter the scope and nature of products liability litigations. The fate of the bill, however, is uncertain.

Products liability cases are usually tried under one of the following bodies of law: (1) *negligence,* which tests the conduct of the defendant; (2) *strict liability,* which tests the quality of the product; (3) *implied warranty,* which also tests the quality of the product; and (4) *express warranty* and *misrepresentation,* which tests the performance of a product against the explicit representations made about it by the manufacturer or sellers.

Products liability cases typically involve three types of defects: manufacturing defects, design defects, and warning defects. The Interagency Task Force on Products Liability (1977) reported that 35 percent of all products liability cases involve manufacturing defects, 37 percent involve design defects, and 18 percent involve warning defects. The absence of a proper warning can easily be seen as a kind of design defect. Weinstein et al. (1978), for example, repeatedly point out the trade-off between providing warning to decrease the danger of a product and designing to reduce the danger. They rightly stress the desirability of designing dangers out of the product rather than warning of their presence.

Making a Case

Certain assertions must be established to make a products liability case, regardless of the body of law under which the case is tried (Weinstein et al., 1978). It must be established that the product was defective in manufacture or design. This requires a definition of when a product is defective. We discuss this further in the next section. Suffice it to say, however, that a product can be dangerous without being defective; a knife is a good example. It must be established that the product was defective at the time the product left the defendant's hands. Products are often abused by users, and through abuse the product becomes defective. One might think that in such a case the manufacturer would not be held liable; actually the manufacturer could still be held liable if the abuse was foreseeable. The mere presence of a defect in a product at the time of injury is not enough to make a case. It must be established that the defect was involved in the injury. The injury, after all, may have had no relationship whatever to the defect. A closely related, and more difficult, question that must also be addressed is whether the defect actually caused the harm. Here the *"but for"* test is applied: "But for the presence of the defect, product failure, or malfunction, would the injury have occurred?" Imagine, for example, that a car is sold with defective brakes. The plaintiff was driving down a street, encountered a patch of ice, slammed on the brakes, which failed, and hit the car in front. Is this a products liability case? Not necessarily; the court could find that even with no defect, slamming on the brakes would not have stopped the car under the icy conditions.

Let us now discuss how to determine when a product is defective. The rules here have changed over the years and will undoubtedly continue to do so. However, a few common threads emerging from recent decisions probably will set the trend for the near future at least.

When Is a Product Defective?

We said that a product can be dangerous without being defective in design. In fact, for many years the legal precedent, established in *Campo v. Scofield,*[2] held that a

---

[2]*Campo v. Scofield,* New York, Vol. 301, pp. 468, Northeastern 2d, Vol. 95, pp. 802 (1950).

manufacturer was not responsible for dangers in a product that were open and obvious. This came to be called the *patent-danger rule,* i.e., a patently obvious danger. It was not until 26 years later that the patent-danger rule was rejected in the landmark case of *Micallef v. Miehle Company.*[3] The plaintiff was employed as an operator on a huge photo-offset printing press. One day he discovered a foreign object on the printing plate, called a *hickie* in the trade, which causes a blemish on the printed page. To correct the situation, the plaintiff informed his supervisor that he intended to "chase the hickie," a common practice wherein a piece of plastic is inserted against the plate which is wrapped around a cylinder that spins at high speed. The plastic caught and drew the plaintiff's hand into the unit between the cylinder and an ink roller. The machine had no safety guards to prevent such an occurrence, and the plaintiff was unable to stop the machine quickly because the shutoff button was distant from his position at the machine.

The plaintiff was fully aware of the obvious danger, but it was the custom to "chase hickies on the run" because once the machine was stopped, it required 3 h to start it up again. The court, if it maintained the patent-danger rule, would have been forced to deny payment to the plaintiff. The court, however, rejected the rule and stated that it would judge a product for *reasonableness.* Therefore, a product is defective if it presents an *unreasonable danger* to the user.

The question, then, is, What is unreasonable? or What is reasonable? First, most products present some risks to the user. These risks, however, must be balanced against the functions the product performs and the cost of providing for greater safety. The California Supreme Court, in *Barker v. Lull Engineering Company,*[4] went even further in specifying the conditions under which a product would be found defective in design. The court set out a two-pronged test. A product would be found defective (1) if it failed to perform safely as an ordinary user would expect when it was used in an intended or *reasonably foreseeable manner* (emphasis added) or (2) if the risks inherent in the design outweighed the benefits of that design. In determining whether the benefits of the design outweigh such risks the jury may consider, among other things, the gravity of the danger posed by the design, the likelihood that such danger would cause damage, the feasibility of a safer alternate design at the time of manufacture, the financial cost of an improved design, and the adverse consequences to the product and the user that would result from an alternate design.

The idea that a manufacturer can be held liable even if a user abuses or misuses a product, as long as the abuse or misuse was reasonably foreseeable, has considerable implications for human factors. Human factors deals with human behavior and how people respond in certain situations. The courts have now extended their concern to include consideration of what unusual, perhaps even harebrained, things people might do with a product for which it was not intended. Weinstein et al. (1978) present the example of a hair dryer. Is it reasonable to expect that a hair dryer might be used to defrost a freezer, or thaw frozen water pipes, as well as dry things other

[3]*Micallef v. Miehle Company,* New York 2d, Vol. 39, pp. 376, Northeastern 2d, Vol. 348, pp. 571, New York Supplement 2d, Vol. 384, pp. 115 (1976).

[4]*Barker v. Lull Engineering Company,* California 3d, Vol. 20, pp. 413, California P. 2d, Vol. 573, pp. 431, California Reporter Vol. 143, pp. 225 (1978).

than hair? It will be used in humid environments and near water. Even in drying hair, it may be reasonable to foresee the user sitting in a bathtub when he or she turns on the unit. Surely it is foreseeable that users will be male and female, young and old, and have a broad range of manual dexterity and levels of understanding and awareness.

Ritter v. Narragansett Electric Company[5] is a case that hinged on the issue of foreseeable use. The defendant manufactured a small freestanding 30-in (76-cm) gas range. The plaintiff, a 4-year-old girl, opened the oven door and used it as step stool to look into a pot on top of the stove. The stove tipped forward, seriously injuring the plaintiff. Expert testimony concluded that the oven door could not hold a weight of 30 lb (14 kg) without tipping. The issue was whether the use of the open door as a step stool was so unforeseeable that the manufacturer should not be held liable. Note, however, that had the stove tipped because a homemaker used the open door as a shelf for a heavy turkey, there probably would have been no question of the manufacturer's liability since it could be argued that one of the intended uses of the door was as a shelf for checking food during preparation. But, as a step stool, was that foreseeable? The jury said yes and attached liability to the manufacturer.

One last point before we leave the question of when a product is defective. Designing a product to meet government or industry standards does not guarantee that it will not be found unreasonably dangerous. Weinstein et al. (1978) point out that traditionally courts have taken the position that all standards provide, *at best,* lower limits for product acceptability. Consider a case in point, *Berkebile v. Brantly Helicopter Corporation.*[6] The plaintiff took off in a helicopter with a nearly empty gas tank. Shortly after takeoff the helicopter crashed. The charge of defect was that the helicopter system of autorotation (wherein the aircraft "glides" to the ground) required that the pilot be able to throw the helicopter into autorotation within 1 s. The plaintiff contended that this was too short a time and that this was the cause of the crash. The defendant argued that the 1-s time met Federal Aviation Administration regulations. The court agreed with the defendant, but the appeals court stated that "such compliance (with regulations) does not prevent a finding of negligence where a reasonable man [and we assume woman] would take additional precautions."

Designing a Reasonably Safe Product

Products must be designed for reasonably foreseeable use, not solely intended use. This requires an analysis be made to determine the types of use and misuse a product could be subjected to; this may require a survey of users of similar products or potential users of a new product. Laboratory simulation tests might also be conducted on product prototypes to gain insight into user behaviors and product performance. Weinstein et al. (1978) list seven steps that should be included in the design process of a product to ensure that a reasonably safe product evolves:

1 Delineate the scope of product uses.
2 Identify the environments within which the product will be used.

[5]*Ritter v. Narragansett Electric Company,* Rhode Island, Vol. 109, pp. 176, Atlantic 2d, Vol. 285, pp. 255 (1971).
[6]*Berkebile v. Brantly Helicopter Corporation,* Pennsylvania Superior, Vol. 219, pp. 479, Atlantic 2d, Vol. 281, pp. 707 (1971).

3 Describe the user population.

4 Postulate all possible hazards, including estimates of probability of occurrence and seriousness of resulting harm.

5 Delineate alternative design features or production techniques, including warnings and instructions, that can be expected to effectively mitigate or eliminate the hazards.

6 Evaluate such alternatives relative to the expected performance standards of the product, including the following:

 a Other hazards that may be introduced by the alternatives

 b Their effect on the subsequent usefulness of the product

 c Their effect on the ultimate cost of the product

 d A comparison to similar products

7 Decide which features to include in the final design.

WARNINGS

Closely allied to the issue of whether a product is unreasonably dangerous are the warnings and instructions that accompany it. A distinction can be made between warnings and instructions, although it is rather fuzzy. *Warnings* inform the user of the dangers of improper use and, if possible, tell how to guard against those dangers. *Instructions* tell the user how to use the product effectively. In addition, instructions themselves may, and often do, contain warnings. We concentrate on warnings and indicate some of the human factors and legal issues involved in their use.

There are three basic approaches for making a product safer:

- Design the dangerous feature out of the product.
- Protect against the hazard by guarding or shielding.
- Provide adequate warnings and instructions for proper use and reasonable foreseeable misuse.

In general, designing the dangerous feature out of a product is the most effective means of making the product safer. Often, however, it is not possible or economically feasible to do that. In such cases guarding and shielding should be implemented where possible. Warnings should be considered after the first two design methods have been applied and where unreasonable dangers still exist.

Purposes of Warnings

We can identify four principal purposes for warnings: (1) Inform the users or potential users of a hazard or danger, of which they may not be aware, that is inherent in the use or reasonably foreseeable misuse of the product. (2) Provide users or potential users with information regarding the likelihood and/or severity of injury from the use or reasonably foreseeable misuse of the product. (3) Inform users or potential users how to reduce the likelihood and/or severity of injury. (4) Remind users of a danger at the time and place where the danger is most likely to be encountered.

Designing Warnings

Designing appropriate warnings is a complex task, and where the risks of serious injury or death are involved, warnings should be tested for effectiveness by using people representative of the foreseeable user population. We cannot hope to discuss in detail all the various aspects of warning design; however, we outline the major sorts of considerations involved.

The ultimate aim of a warning is to alter behavior by encouraging the user either to not engage in a particular act or to change the manner in which the act is performed. For a warning to change behavior, the warning must be sensed (seen or heard), received (read or listened to), understood, and finally heeded. Each of these steps involve human factors design considerations.

Sensing a Warning   The warning must catch the attention of the consumer under the circumstances in which the product will be used. The following considerations are important: size; shape; color; graphical design; contrast; placement; use of "active" attention getters such as bells, waving flags, or blinking lights; and the physical durability of the warning itself to such things as weather and physical abuse. In addition, in Chapter 6 we discuss some design considerations for auditory warnings.

Receiving a Warning   Sensing the presence of a warning does not ensure that it will be read or listened to. With visual warnings, the length of the warning may influence the willingness of a user to read it. Probably, the longer and more complicated the warning, the less likely it is that people will invest the time and effort required to read it. The perceived level of hazard inferred by the warning may also influence whether it is read. The warning should, of course, be commensurate with the dangers inherent in using the product. Traditionally three levels of hazard have been differentiated:

• *Danger* is used where there is an immediate hazard which, if encountered, will result in severe personal injury or death.

• *Warning* is the signal word for hazards or unsafe practices which *could* result in severe personal injury or death.

• *Caution* is for hazards or unsafe practices which could usually result in minor personal injury, product damage, or property damage.

Presumably a person would be more likely to read a warning that started out "DANGER—THIS PRODUCT CAN KILL YOU!" than one that said "NOTE—IF THIS PRODUCT IS USED IN AN IMPROPER MANNER, IT IS POSSIBLE THAT INJURY COULD OCCUR."

The degree of familiarity with a product and its perceived hazard potential may also influence whether people read warnings (Godfrey and Laughery, 1984). Generally, the more familiar people are with a product, the more confident they are of their ability to use the product safely; and the less dangerous the product appears, the less likely it is that they will read permanently affixed warnings. There is some evidence, however, that people will read a temporarily posted warning that informs them of the development of a hazard in a familiar product or situation (Godfrey, Rothstein, and Laughery, 1985).

Understanding the Warning   The words and/or graphics of a warning must be understood by the user population. If words are used, they must be chosen carefully and tested for comprehension. Do not use vague, ambiguous, or ill-defined terms; highly technical words or phrases; double negatives; or long, grammatically complex phrasing. Consideration must also be given to the primary languages of the user population. Symbols must be carefully chosen to convey the intended message, and this can be more difficult than one may think. Collins (1983), for example, tested comprehension of standard hazard symbols among mining industry personnel. In several cases respondents interpreted the symbol to mean the opposite of its intended meaning. The symbol for corrosive hazard, for example, was thought by 21 percent of the people to denote "emergency hand-wash location!" The symbol for emergency eye-wash location was interpreted by 24 percent of the people to mean "eye irritant located here"!

At a minimum, a warning should contain the following fundamental elements:

• *Signal word*—to convey the gravity of the risk, for example, *danger, warning, caution*.
• *Hazard*—the nature of the hazard.
• *Consequences*—what is likely to happen if the warning is not heeded.
• *Instructions*—appropriate behavior to reduce or eliminate the hazard.

An example of a minimum warning would be (Wogalter, Desaulniers, and Godfrey, 1985)

DANGER
HIGH VOLTAGE WIRES
CAN KILL
STAY AWAY

How specific a warning must be in spelling out hazards is a difficult question. Courts have often opted for a great deal of specificity in warnings. For example, the following warning was found to be inadequate in a court case:[7]

CAUTION: FLAMMABLE MIXTURE. DO NOT USE NEAR FIRE OR FLAME. CAUTION! WARNING! EXTREMELY FLAMMABLE! TOXIC! CONTAINS NAPHTHA, ACETONE, AND METHYL ETHYL KETONE. ALTHOUGH THIS ADHESIVE IS NO MORE HAZARDOUS THAN DRYCLEANING FLUIDS OR GASOLINE, PRECAUTIONS MUST BE TAKEN. USE WITH ADEQUATE VENTILATION. KEEP AWAY FROM HEAT, SPARKS, AND OPEN FLAME. AVOID PROLONGED CONTACT WITH SKIN AND BREATHING OF VAPORS. KEEP CONTAINER TIGHTLY CLOSED WHEN NOT IN USE.

It was found to be inadequate because it did not mention the danger of using the product in the vicinity of closed and concealed pilot lights (Kreifeldt and Alpert, 1985).

---

[7]*Florentino v. A.E. Dtaley Mfg. Co.* Massachusetts 416 N.E. 2d 998.

There is a problem, however, in warning about all possible hazards inherent in the use and reasonable foreseeable misuse of a product. The problem is *warning overload*. That is, if there is a warning against every conceivable hazard, the effect of all the warnings may be so diluted as to make any one of them ineffective. Take, for example, a large industrial machine with electrical hazards, flammable liquids, high-pressure lines, sharp edges, pinch points, slippery surfaces, etc. Warning against all such hazards and, in addition, reasonable foreseeable misuses could result in virtually the entire machine being plastered with warning labels. It is doubtful that people working in and around such a machine would pay much attention to so many warnings.

Heeding the Warning   Just because a warning is sensed, received, and understood does not ensure that a person will heed or comply with the warning. These are some factors that may influence whether people heed a warning: whether the user is capable of performing the action required; whether the warning is remembered at the time and place where the action is required; whether heeding the warning will result in additional time, cost, inconvenience, or discomfort to users; and the riskiness and general carelessness of the individual.

Effectiveness of Warnings

The ultimate measure of effectiveness is whether the warning was heeded by the user. Unfortunately, there has not been a great deal of research assessing the effectiveness of warnings by using behavioral measures. Most research on warning effectiveness assesses the perceived effectiveness by asking subjects which of several warnings they *think* would be most effective. There are, however, a few behavioral studies which may shed some light on the issue of effectiveness.

Warnings in Instructions   Warnings properly placed in instruction manuals that are used by people to operate, assemble, or repair a product can be quite effective. Wogalter, Fontenelle, and Laughery (1985), for example, found that placing a warning ("Wear rubber gloves and mask") at the beginning of an instruction sheet used to perform a chemistry experiment resulted in 87 percent of the subjects complying with the warning. When the warning was placed at the end of the instructions, however, only 44 percent complied. Using a survey technique, the Consumer Product Safety Commission (1980) reported that 69 percent of people who felt they had been exposed to warning labels or hazard-avoidance instructions related to installing CB antennas near power lines said that the warnings had caused them to consider the promixity of the power lines when selecting an installation site. There was also evidence that CB antennas installed *before* warning labels and hazard-avoidance instructions were made mandatory were mounted *closer* to power lines than were CB antennas installed *after* the warnings and instructions became mandatory.

Warnings within instructions, however, may be effective only when the instructions are being used. If the product is used again without rereading of the instructions, there is a good probability that the warning may be forgotten. For this reason, a warning is often placed on the product itself.

On-Product Warnings The literature addressing the effectiveness of on-product warnings indicates that people often do not heed such warnings. Dorris and Purswell (1977) reported that of 100 subjects, not one even noticed a warning placed on the handle of a hammer they were asked to use to perform a task. The widely recognized lack of effectiveness of the warning on cigarette packages[8] ("Warning: The Surgeon General has determined that cigarette smoking is dangerous to your health.") has often been cited as evidence for the general lack of effectiveness of product warnings (McCarthy et al. 1984). To be fair, however, there are features of cigarette smoking that are not at all comparable to other on-product applications of warnings. Cigarette smoking, for example, may be addictive, it is pleasurable (at least to those who do the smoking), and considerable advertising is aimed at promoting smoking. In how many other situations is noncompliance with a warning addictive, pleasurable, and encouraged in the media? The lack of effectiveness of cigarette warnings, therefore, may be a special case rather than a general indictment of on-product warnings.

In a revealing study of on-product warnings Strawbridge (1985) had subjects scan a label on the back of a bottle of liquid adhesive to determine whether the adhesive could be used to glue two specific materials and if so, to do so. The label contained the following warning:

DANGER: Contains Acid. To avoid severe burns, shake well before opening.

Various conditions were tested including placement of the warning at the top, middle, or bottom of the label; highlighting with inverse type (white letters on a black background); and embedding the warning in a paragraph of text or starting the paragraph with the warning. Surprisingly, the addition of highlighting and the position of the warning (top-middle-bottom) had little effect on compliance. Embedding the warning, however, did reduce compliance. For the nonembedded conditions, 94 percent of the subjects noticed the warning, but only 81 percent read it. Even more surprising, although 81 percent of the subjects read the warning, only 47 percent complied with it. The reason given by the subjects who read the warning but did not comply was that they simply forgot; yet the time between reading the warning and opening the bottle was no more than 10 s!

More encouraging results came from a study reported by Godfrey, Rothstein, and Laughery (1985). They found that placing temporary warning signs on a copy machine ("*Caution:* Machine does not work. May cause delay. Use another machine."), a telephone ("*Caution:* Telephone is out of order. Money will be lost. Use other telephone."), and a drinking fountain ("*Warning:* Bad filter caused water contamination. Do not drink water.") were effective in modifying behavior in the desired manner. Compliance rates were 73 percent, 100 percent, and 67 percent for the copier, telephone, and water fountain, respectively.

Whether people will comply with a warning depends in part on the cost of compliance. Godfrey, Rothstein, and Laughery (1985) in another part of their study posted warning signs indicating that a door was broken and that to avoid injury people should (1) use an adjacent door, (2) use another exit 50 ft (15 m) away, or

---

[8]New warnings have been mandated, but their effectiveness has not, as yet, been assessed.

(3) use another exit 200 ft (61 m) away. The percentage of people complying with the warnings in the three situations were 94 percent, 6 percent, and 0 percent, respectively.

DISCUSSION

If has not been the intent of this book to compile and organize a tremendous amount of available human factors data or to provide how-to-do-it procedures and guidelines for applying such data to the design of the things people use. Rather, one objective has been to present at least some information about certain human characteristics (such as sensory and motor processes) that might contribute to greater understanding of human performance and behavior and, in turn, have some relevance for those concerned with designing things for people to use. In addition, it has been the intent to include discussions of at least certain methods and techniques relevant to the human factors data and principles to practical design problems.

But perhaps the most important underlying objective has been to develop an increased sensitivity to, or awareness of, the many human aspects of the systems and situations that abound in our current civilization. Such awareness is at least the first prerequisite to the subsequent processes of creating those systems and situations in which human talents can be most effectively utilized in the furtherance of human welfare.

To date, the systematic consideration of human factors in the design of things people use has not taken on epidemic proportions, but rather has been limited to certain types and classes of systems and facilities. In this regard, however, we would like to propose that virtually any kind of equipment, facility, gadget, "thing," process, service, or environment that is created or influenced by humans—presently in existence or to be created—be viewed as a prime candidate for human factors attention. Such systematic concern should, of course, be directed toward the identification of any existing or potential human factor problems, with the objective of designing the product or conditions for more effective human use and the improvement of the quality of life.

The remaining chapters of this book deal with examples of human factors research, and in certain instances of the application of human factors data and principles, as related to the following areas: the "built" environment (buildings, communities, etc.), transportation and related facilities, and certain work-related situations that have not been dealt with extensively in other chapters. In presenting such examples it should not be assumed that they all represent exceptionally good examples of research or applications. Rather, they are intended to reflect some types of human factors design problems that people have worked on.

REFERENCES

Bailey, R. (1982). *Human performance engineering*. New York: Prentice-Hall.
Blanchard, R. E. (1975). Human performance and personnel resource data store design guidelines. *Human Factors*. 17(1), 25–34.

Bonney, M., and Case, K. (1976). The development of SAMMIE for computer aided work place and work task design. *Proceedings, 6th Congress of the International Ergonomics Association and Technical Program for the 20th Annual Meeting of the Human Factors Society*. Santa Monica, CA: Human Factors Society. pp. 340–348.

Chapanis, A. (1960). Human engineering. In C. D. Flagle, W. H. Huggins, and R. H. Roy (eds.), *Operations research and systems engineering*. Baltimore, MD: Johns Hopkins.

Chapanis, A. (1965). On the allocation of functions between men and machines. *Occupational Psychology*, 39, 1–11.

Chapanis, A. (1967). The relevance of laboratory studies to practical situations, *Ergonomics*, 10(5), 557–577.

Collins, B. (1983). *Use of hazard pictorials/symbols in the minerals industry* (NBSIR 83-2732). Washington: Department of Commerce, National Bureau of Standards.

Consumer Product Safety Commission (1980, December). *Evaluation of the impact of the communications antenna labeling rule*. Washington.

Corkindale, K. G. (1967). Man-machine allocation in military systems. *Ergonomics*, 10(2), 161–166.

Coskuntuna, S., and Mauro, C. L. (1980). Human factors and industrial design in consumer products. *Proceedings of the Symposium: Human Factors and Industrial Design in Consumer Products*. Medford, MA: Tufts University, pp. 300–313.

DeGreene, K. B. (1977). Has human factors come of age? *Proceedings of the Human Factors Society*. Santa Monica, CA: Human Factors Society. pp. 457–461.

Dorris, A., and Purswell, J. (1977). Warnings and human behavior: Implications for the design of product warnings. *Journal of Products Liability*, 1, 255–264.

Evans, S. (1978). *Updated user's guide for the COMBIMAN*, Rept. AMRL-TR-78-31. Wright-Patterson Air Force Base, OH: Aerospace Medical Research Laboratory.

Fitts, P. M. (ed.) (1951). *Human engineering for an effective air-navigation and traffic-control system*. Washington: NRC.

Fitts, P. M. (1962). Functions of men in complex systems. *Aerospace Engineering*, 21(1), 34–39.

Geer, C. (1981). *Human engineering procedures guide*. (AFAMRL-TR-81-35). Wright-Patterson Air Force Base, OH: Aerospace Medical Research Laboratory.

Godfrey, S. and Laughery, K. (1984). The biasing effects of product familiarity on consumers' awareness of hazard. *Proceedings of the Human Factors Society 28th Annual Meeting*. Santa Monica, CA: Human Factors Society.

Godfrey, S., Rothstein, P., and Laughery, K. (1985). Warnings: Do they make a difference? *Proceedings of the Human Factors Society 29th Annual Meeting*. Santa Monica, CA: Human Factors Society. pp. 669–673.

Harris, R., Bennett, J., and Dow, L. (1980). *CAR-II-A revised model for crewstation assessment of reach*, Tech. Rept. 1400.06b. Willow Grove, PA: Analytics, Inc.

Harris, R., Bennett, J., and Stokes, J. (1982). Validating CAR: A comparison study of experimentally-derived and computer-generated reach envelopes. *Proceedings of the Human Factors Society 26th Annual Meeting*. Santa Monica, CA: Human Factors Society. pp. 969–973.

Interagency Task Force on Products Liability (1977). Final Rept. II-54. Washington: Department of Commerce.

Johnson, D. A. (1980). The design of effective safety information displays. *Proceedings of the Symposium: Human Factors and Industrial Design in Consumer Products*. Medford, MA: Tufts University, pp. 314–328.

Johnson, E. M., and Baker, J. D. (1974). Field testing: The delicate compromise. *Human Factors*, 16(3), 203–214.

Jones, R., Jonsen, G., and Van, C. (1982). Evaluation of control station design: The crew human engineering software system. *Proceedings of the Human Factors Society 26th Annual Meeting.* Santa Monica, CA: Human Factors Society. pp. 40–43.

Jordan, N. (1963). Allocation of functions between man and machines in automated systems. *Journal of Applied Psychology,* 47(3), 161–165.

Kreifeldt, J., and Alpert, M. (1985). Use, misuse, warnings: A guide for design and the law. In T. Kvalseth (ed.), *Interface 85: Fourth Symposium on Human Factors and Industrial Design in Consumer Products,* Santa Monica, CA: Consumer Products Technical Group of The Human Factors Society. pp. 77–82.

Line, C. (1979). *Instruments readability analysis users guide,* Memo G-2574-CCL-099. Seattle: Boeing Computer Services Company.

McCarthy, R., Finnegan, J., Krumm-Scott, S., and McCarthy, G. (1984). Product information presentation, user behavior, and safety. *Proceedings of the Human Factors Society 28th Annual Meeting.* Santa Monica, CA: Human Factors Society. pp. 81–85.

McCormick, E. J. (1979). *Job analysis: Methods and applications.* New York: Amacom.

McCormick, E. J., and Ilgen, D. R. (1985). *Industrial psychology* (8th ed.). Englewood Cliffs, NJ: Prentice-Hall.

Meister, D. (1971). *Human factors: Theory and practice.* New York: Wiley.

Meister, D. (1978). A theoretical structure for personnel subsystem management. *Proceedings of the Human Factors Society.* Santa Monica, CA: Human Factors Society. pp. 474–478.

Meister, D. (1985). *Behavioral analysis and measurement methods.* New York: Wiley.

Meister, D. and Rabideau, G. F. (1965). *Human factors evaluation in system development.* New York: Wiley.

MIL-HDBK-759A (1981, June). *Military handbook: Human factors engineering design for army materiel.*

MIL-STD-1472C (1981, May). *Human engineering design criteria for military systems, equipment, and facilities.*

Parks, D., and Springer, W. (1975). *Human factors engineering analytic process definition and criterion development for CAFES,* Rept. D180-18750-1. Seattle: Boeing Aerospace Company.

Price, H. (1985). The allocation of functions in systems. *Human Factors.* 27(1), 33–45.

Price, H., Maisano, R., and Van Cott, H. (1982). *The allocation of functions in man-machine systems: A perspective and literature review.* (NUREG-CR-2623). Oak Ridge, TN: Oak Ridge National Laboratories.

Price, H., and Pulliam, R. (1983). Control room function allocation—A need for man-computer symbiosis. *Proceedings of the 1982 IEEE Computer Forum.* Denver, CO: Institute of Electrical and Electronics Engineers.

Rogers, J. G., and Armstrong, R. (1977). Use of human engineering standards in design. *Human Factors,* 19(1), 15–23.

Rogers, J. G., and Pegden, C. D. (1977). Formatting and organization of a human engineering standard. *Human Factors,* 19(1), 55–61.

Sanders, M. S., Jankovich, J., and Goodpaster, P. (1974). *Task analysis for the jobs of freight train conductor and brakeman.* (RDTR 263). Crane, IN: Applied Sciences Department, Naval Ammunition Depot.

Stone, G., and McCauley, H. (1984). *Flight deck design methodology using computerized anthropometric models,* Douglas Paper 7508. Long Beach, CA: Douglas Aircraft Co.

Strawbridge, J. (1985). The influence of highlighting, imbedding, and positioning on warning effectiveness. Unpublished master's thesis. Northridge: California State University.

Topmiller, D., and Aume, N. (1978, January). Computer graphic design for human performance. *Proceedings of the 1978 Annual Reliability and Maintainability Symposium.* Los Angeles, pp. 385–388.

Wallace, W., and Key, J. (1984, December). Human factors experts: Their use and admissibility in modern litigation. *For the Defense,* pp. 16–24.

Weinstein, A., Twerski, A., Piehler, H., and Donaher, W. (1978). *Products liability and the reasonably safe product.* New York: Wiley.

Wogalter, M., Desaulniers, D., and Godfrey, S. (1985). Perceived effectiveness of environmental warnings. *Proceedings of the Human Factors Society 29th Annual Meeting.* Santa Monica, CA: Human Factors Society, pp. 664–668.

Wogalter, M., Fontenelle, G., and Laughery, K. (1985). Behavioral effectiveness of warnings. *Proceedings of the Human Factors Society 29th Annual Meeting.* Santa Monica, CA: Human Factors Society, pp. 679–683.

CHAPTER 19

THE BUILT ENVIRONMENT

Although the human factors discipline has traditionally been primarily concerned with the design of equipment, facilities, and environments as related to work activities, the basic approach of human factors is equally applicable to a wide spectrum of other areas, including virtually all our total life space. Our life space can be viewed as encompassing our cities and communities (including their physical and cultural aspects), our neighborhoods, the buildings in which we live and work and that we use for other purposes, and the natural outdoor environment. This chapter deals largely with some of the human factors aspects of what has been called the *built environment*, particularly the buildings and related facilities that people use for various purposes.

In previous chapters we touched on topics related to the built environment, such as illumination in Chapter 14 and arrangement of components within a physical space in Chapter 13. In this chapter, however, we highlight what might be called the architectural aspects of the built environment. To do this, we will focus on two environments with which we are familiar: offices and places where we live. Our purpose is to illustrate some of the human factors considerations relevant to such environments, and some relevant research findings, rather than presenting an exhaustive review of the literature.

EVALUATING THE BUILT ENVIRONMENT

The varied features of the built environment can have profound effects on people. In addition to the effects of the physical features of such environments, people are influenced by such nonphysical features as social, cultural, technological, economic, and political factors characteristic of the environment. The built environment and its associated nonphysical factors affect a wide gamut of human experience, including performance in work and other aspects of life, social behavior, attitudes, satisfaction and dissatisfaction, mental health, quality of life, and physical health and well-being.

Both field and laboratory studies are used to assess the effects on people of various aspects of the built environment. A common laboratory methodology is to show to subjects drawings, pictures, or scale models of built environments, such as a living room or an office, and ask them to rate it along various dimensions, such as friendliness, warmth, complexity, etc. Features of the environment, such as placement of windows, ceiling height, or color of walls, are varied in order to assess whether ratings change when the feature changes.

We discuss some of these studies later; however, such studies have some potentially serious limitations. Looking at a drawing or scale model of an environment can be a vastly different experience from being in the actual full-scale version of it. Further, opinions and attitudes about an environment formed from a one- or two-minute exposure may be quite different from the opinions and attitudes about that environment developed after living or working in it for several weeks or months. In laboratory studies either subjects are not instructed about what activities will take place in the space or they are asked to imagine the environment will be used for some purpose. The purpose or functions carried out in an environment can affect people's opinions and attitudes about the environment. We may, for example, prefer rooms with lower ceilings when we eat than when we are watching television.

Field studies and surveys, of course, have the advantage of involving prolonged exposure to the environment under actual "real-life" conditions. But it is often difficult to isolate the specific aspect(s) of the environment contributing to the attitudes and behaviors being measured. As in other areas of human factors, a combination of laboratory and field studies probably is the best method of unraveling the varied effects of the built environment on people's attitudes and behaviors.

Discussion

The design of buildings and other features of the built environment is a different process because good design should be directed toward satisfying several goals (i.e., criteria) simultaneously. However, it usually is not feasible to fulfill all the desired objectives with a single design. In this regard, Bennett (1977) makes the point that various design objectives are not equal and tend to form a hierarchy. In planning a private home, for example, a person might consider such factors as location, convenience to transportation, type of neighborhood, size and number of rooms, style of architecture, and type of building materials, along with cost. The relative values the individual places on these and other features usually would place the values in some rough hierarchy; so if not all the desirable features can be fulfilled, those low in the hierarchy usually would be dropped or modified. The decisions made under such circumstances represent the trade-offs that plague human factors designers in the design of many items.

THE OFFICE AS A BUILT ENVIRONMENT

To many people the office environment is like a second home where they spend almost half their waking hours. In the United States there were more people working in offices in 1980 than in factories. There were 44.8 million white-collar workers compared to only 29.8 million blue-collar workers (Kleeman, 1981). With all those people spending all that time in offices, it is no wonder that office planning and design has become a big business.

In previous chapters we discussed aspects of office design such as illumination, speech communications, noise, and work station design. These are areas traditionally addressed by human factors specialists and other professionals. In recent years, our attention has widened to consider other, more global aspects of the office environment including overall size, arrangement of furnishings, and even the value of windows. We have come to recognize that such aspects of the office environment can affect the way people work and their attitudes toward their jobs.

To provide some context for our discussion of office design, we briefly discuss the nature of office work. Our discussion is not exhaustive, but does provide some information that can influence the design of the office.

Office Activities

As with any design effort, it is essential that we understand the functions and tasks to be performed by the people using whatever it is we are designing. This is equally true for designing offices. There have been several attempts to develop taxonomies of office tasks. Table 19-1 is representative and illustrates the range of tasks typically performed in offices.

One major activity performed in offices is information development and gathering; one criterion for good office design is to facilitate that function. To gain some in-

TABLE 19-1
A TAXONOMY OF OFFICE TASKS

Cognitive	Physical
Information development and gathering	Filing and retrieval
Information storage and retrieval	Writing
Reading and proofreading	Mail handling
Data analysis and calculating	Traveling
Planning and scheduling	Copying and reproducing
Decision making	Collating and sorting
Social	Pickup and delivery
	Typing and keying
Telephoning	Keeping calendars
Dictating	Using equipment
Conferring	
Meeting	
Procedural	
Completing forms	
Checking documents	

Source: Galitz, 1980, fig. 4-1

TABLE 19-2
WHERE OFFICE WORKERS GO FIRST FOR INFORMATION

Situation	Source	Frequency, %
No search required	Recall	19.0
	Received with task	10.5
	Asked a colleague	14.5
	Searched own collection	13.0
	Internal company consultant	9.5
Local work environment	Departmental files	5.5
	Assigned to subordinate	4.5
	Respondent's own action	2.5
	Asked supervisor	1.0
	Library or librarian	10.0
	Manufacturer or supplier	6.0
External to work environment	Customer	2.0
	External consultant	1.0
	DOD information systems	1.0

Source: Hodges and Angalet (1968). Copyright by the Human Factors Society, Inc. and reproduced by permission.

sight, Hodges and Angalet (1968) interviewed 1500 engineers, scientists, and technical personnel employed by 84 companies, institutes, and universities. Table 19-2 presents the findings concerning where information was first sought when it was needed. About 30 percent of the respondents' informational needs were satisfied without search, and 50 percent were sought within the local work environment. This suggests that informal and personal information sources be documented and incorporated into the office design and that formal information sources (personal files, department files, etc.) be close to people so they can use them more productively (Galitz, 1980).

The percentage of time spent on office tasks varies with the position in the organization. Table 19-3, for example, presents some data from a study of 1200 white-collar workers from 11 companies at 40 different locations (Knopf, 1982). As one moves down the organizational ladder from executive to secretary, the percentage of

TABLE 19-3
PERCENTAGES OF MAJOR TYPES OF COMMUNICATIONS IN OFFICES AS A
FUNCTION OF POSITION IN THE ORGANIZATION

Type of communication	Executive	Manager	Knowledge worker	Secretary
Face-to-face	53	47	23	Negligible
Document	27	29	42	55
Telephone	16	9	17	20
Other	4	15	18	25

Source: Knopf (1982).

time spent in face-to-face communications decreases, while time spent in document-based communications increases. The same study reported that executives spend 47 percent of their time in meetings, both formal and informal. An office with executives should be designed, therefore, to provide space where meetings can be held without distraction and with adequate privacy.

What Kind of Office?

Arguments have been made for the pros and cons of various types of offices—large and small, landscaped and conventional. A few studies have dealt with this matter, although the results have been somewhat ambiguous, if not conflicting. Part of the problem lies in the proliferation of terms used to describe various office designs. At the risk of stepping on a few toes, let us suggest the following distinctions. A *large* office refers to size rather than to how the space is partitioned. A large office may be an *open-plan* or a *cellular* office (those divided into many smaller offices for individuals or small groups). An open-plan office is not absolutely open, but it can be subdivided by movable furniture and screens. One type of open-plan office is the "bull pen" office with desks arranged in neat rows, often as far as the eye can see. Another type of open-plan office is the landscaped, or *Bürolandschaft*, office which originated in Germany (Brookes, 1972). Such a office consists of one large, open—but "landscaped"—area which is planned and designed about the organizational processes to take place within it. The people who work together are physically located together, the geometry of the layout reflecting the pattern of the work groups. The areas of the various work groups are separated by plants; low, movable screens; cabinets; shelves; etc., as shown in Figure 19-1.

In a nationwide survey of 1047 office workers (Steelcase, 1978) almost one-third reported they were working in an open-plan office other than a bull pen office. Of the respondents only 15 percent reported working in bull pen offices.

Flexibility is one of the major advantages cited for open-plan offices. Space can be reconfigured easily, and work stations can be rearranged at little cost to meet changing needs or work patterns. In addition, more of the available space is utilized in an open-plan office because less space is devoted to hallways and walls. We briefly discuss a few studies that have compared open-plan offices (usually the term as used in such studies does not include bull pen offices) with traditional cellular offices to illustrate other possible benefits and some potential limitations of open-plan offices.

Social Behavior in Offices   There are some indications that small-office environments are more conducive to the development of social affinity for others than large offices. This was reflected, for example, by the results of a sociometric study by Wells (1965) in which office personnel in large and small offices were each asked to indicate their choices of individuals beside whom they would like to work. The data summarized in Table 19-4 (page 562) show the percentages of the choices made by individuals (in open and small areas) of other persons in their own section or department. Although there was greater internal cohesion among personnel working in the

FIGURE 19-1
Example of a landscaped office in which individual offices and work groups are separated by
plants, screens, cabinets, and shelves. This office is in the Administration Services Building at
Purdue University.

smaller areas than in the open areas, also there were more isolates (individuals not
chosen by anyone).

Disturbances and Distractions in Offices   There appear to be indications that
open-plan offices contribute to an increased number of distractions and disturbances,
some work-related, others of a social nature. Mercer (1979), for example, observed
samples of workers in three office environments: traditional cellular offices, open-
plan offices, and what was called an *action office* (apparently similar to landscaped
offices). Some of the results related to the number of disturbances and distractions
occurring per hour in the three environments are presented in Table 19-5. Disturbances tended to be higher in the action and open-plan offices than in the traditional
ones. Work-related distractions were considerably higher in the open-plan than in the
other two types of offices. What is also somewhat amazing is that in all three offices
there was an average of one disturbance or distraction every 2 min! No wonder so
many office workers are needed to get the work done. Complaints about disturbances
and distractions are fairly common in open-plan offices, as Nemecek and Grandjean

TABLE 19-4
SOCIOMETRIC CHOICES OF OFFICE WORKERS IN OPEN
AND SMALL OFFICE AREAS

	Choices of members of own section, %	Reciprocal choices within own section,%
Open areas	64	38
Small areas	81	66

Source: Wells, 1965.

TABLE 19-5
NUMBER OF DISTURBANCES AND DISTRACTIONS OCCURRING PER HOUR IN THREE
TYPES OF OFFICE PLANS

	Type of office		
	Action	Open plan	Traditional
Disturbances			
Work-related	11.4	8.1	7.1
Social	5.5	5.3	4.5
Total	16.9	13.4	11.6
Distractions			
Work-related	8.9	23.4	10.3
Social	2.4	2.3	4.8
Total	11.3	25.7	15.1
Total	28.2	39.1	26.7

Source: Adapted from Mercer (1979), table 2.

(1973) found when they surveyed 519 workers in 15 open-plan offices in Switzerland. They found that 69 percent of the respondents complained about disturbances in concentration and 11 percent mentioned that confidential conversations were impossible. To reinforce the importance of these sorts of findings, Steelcase (1978) found that 41 percent of their nationwide sample of office workers indicated that the characteristic most important in helping to get the work done well was the ability to concentrate without noise or other distractions.

Health in Office Environments Hedge (1984) reported that complaints of frequent headaches were almost twice as prevalent among open-plan office employees as among traditional-plan office employees. Problems of eye irritation, coughs, colds, and sore throats, however, appeared more related to whether the office was air-conditioned than to the type of office plan. All these symptoms were more prevalent in air-conditioned offices.

Designing Open-Plan Offices   A key factor in the success of an open-plan office is whether the layout is based on actual communication patterns and work group needs. Surveys should be conducted to assess who talks to whom, who works with whom, what type of interactions take place, privacy needs, and what facilities (equipment, storage, etc.) are needed by each person to do the work. Given such information, an intelligent office layout can be formulated.

Wichman (1984) offers the following specific design recommendations to enhance the utility of an open-plan office:

• *Use sound-absorbing materials on all major surfaces wherever possible.* Noise is often more of a problem than expected.

• *Leave some elements of design for the work station user.* Work stations are usually overly designed and inflexible. People need to have control over their environments; so leave some opportunities for changing or rearranging things.

• *Provide both vertical and horizontal surfaces for the display of personal belongings.* People like to personalize their work stations.

• *Install telephones that ring "silently."* In small work stations phones that flash a light for the first two "rings" before emitting an auditory signal dramatically reduce disturbances.

• *Provide all private work areas with a system to signal the willingness of the occupant to be disturbed.* There is a need to express changing needs for privacy and community.

• *Provide several easily accessible islands of privacy.* This would include small rooms with full walls and doors that can be used for conferences and for private or long-distance telephone calls.

• *Have clearly marked flow paths for visitors.* For example, hang signs from the ceiling showing where secretaries and department boundaries are located. The visual stimulus configuration of an open-plan office is often too complex for visitors to quickly develop a cognitive map.

• *Design work stations so it is easy for drop-in visitors to sit down while speaking.* This will tend to reduce disturbances to other workers.

• *Plan for ventilation air flow.* Most traditional offices have ventilation ducting. This is usually not the case with open-plan cubicles, so they become dead-air cul-de-sacs that are extremely resistant to post hoc resolution.

• *Overplan for storage space.* Open-plan systems with their emphasis on tidiness seem to chronically underserve the storage needs of people.

Discussion   The attitudes and opinions of people about landscaped offices are something of a mixed bag. People who work in such offices have been reported by certain investigators (Brookes, 1972; Brookes and Kaplan, 1972; McCarrey et al., 1974) to complain of noise, visual bustle, lack of privacy and confidentiality of communications, and lack of "territory definition." However, such offices are generally viewed as developing greater solidarity and offering greater personal contact. In addition, most people react favorably to the aesthetic aspects of such offices. The implications regarding productive efficiency are ambiguous. Brookes (1972), for example, states, "It looks better but it works worse," whereas McCarrey et al. (1974) report opinions tending to imply increased productivity.

At present it is probably not feasible to say that the reported advantages of landscaped offices, do, or do not, outweigh the disadvantages. However, probably landscaped offices would be preferred over large, open, bull pen types.

Windows or No Windows?

Because of factors such as cost and energy conservation, there has been renewed interest in windowless buildings. Windows are no longer necessary to provide light and ventilation, but there is some concern that they may be of some value in fulfilling what Manning (1965) refers to as a "psychological need" for some "contact with the outside world." The use of deep buildings, where only the outer ring of offices has windows, and underground workplaces (Wise and Wise, 1984) have further focused attention on the psychological value of windows.

In general, office workers seem to prefer working in building with windows. Without windows, workers often complain about lack of daylight, poor ventilation, not knowing the weather outside, and not having a view (Collins, 1975). Ruys (1970), for example, in a survey of workers in small windowless offices found that 87 percent expressed dissatisfaction with the lack of windows.

Although many office workers, when asked, express dissatisfaction with a lack of windows, there is no evidence of reduced work performance or of workers being "driven crazy" because of a lack of windows. Also studies of windowless schools and factories where large groups of people interact in the windowless space have not always revealed negative attitudes. In the case of schools, for example, windows are often seen as a source of distraction, and removing them is often seen as benefiting the educational process (Chambers, 1963; Tikkanen, 1970). As Boyce (1981) points out, there is probably no single, general conclusion about attitudes toward windowless buildings. Factors such as the size of the windowless space, the level of activity within the space, and workers' expectations all mediate attitudes toward such environments. One thing seems clear, however: Where windows are available, it is usually the executives who have them. And this probably says more about the value of windows than all the surveys that have been done to date.

Office Furnishings and Arrangements

Some fairly definite indications of the subjective reactions of people to office furnishings and arrangements can be reported. For one thing, people tend to report feelings of spaciousness being greatest in offices that are moderately furnished (in terms of number of chairs, desks, etc.), as contrasted with those that are more empty or over furnished (Imamoglu, 1973).

In addition, people react more favorably to offices that have living things (as plants) and aesthetic objects (as posters) and are tidy, than to those that lack these features (Campbell, 1979). These features make visitors to feel more welcome and comfortable, as reflected by the responses of subjects presented with slides of offices with various combinations of features (see Table 19-6).

TABLE 19-6
RATINGS OF RESPONDENTS REGARDING SLIDES OF OFFICES WITH VARIOUS FEATURES

	Mean ratings (9 = favorable; 1 = unfavorable)					
Quality	**Plants**	**No plants**	**Posters**	**No posters**	**Tidy**	**Messy**
Comfort of visitor	6.0	4.6	5.7	4.9	6.1	4.5
Invitingness of office	5.9	4.3	5.6	4.7	6.1	4.1
How welcome visitor feels	6.0	4.7	5.8	5.0	6.2	4.5

Source: Campbell, 1979.

Another feature of offices that influences the reaction of visitors and interpersonal relationships is the arrangement of chairs for visitors. When a visitor is seated opposite the desk of the person being visited, the desk tends to serve as a barrier. In general, visitors have a greater sense of friendliness when both chairs are on the same side of the desk, as illustrated in Figure 19-2a as contrasted with 19-2b (Campbell, 1979; Zweigenhaft, 1976).

Office Automation

We would be remiss if we did not at least mention office automation and its possible impact on office design. The computer and visual display terminal (VDT) are probably the most visible aspects of office automation, but there are other aspects as well.

FIGURE 19-2
Examples of the drawings of rooms used in a study relating to the perceived dimensions of rooms. These examples are of rooms that were appraised to be most friendly and least friendly. *(Source: Adapted from Wools and Canter, 1970, figs. 4 and 5.)*

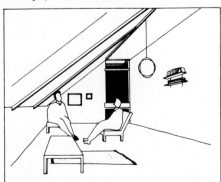
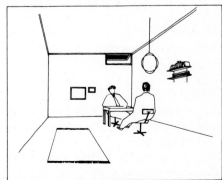

(*a*) Most friendly room   (*b*) Least friendly room

Helander (1985) presents a taxonomy of office automation tools based on the type of interaction and office task being automated as shown in Figure 19-3. Helander points out that, at present, there are no automated aids for the face-to-face social skills of supervising, counseling, persuading, and negotiating. Automated aids are making their biggest impact in the areas of perceptual motor skills, rule-based decision making, and analysis and problem solving.

The implications for office design of various automated tools such as voice mail, teleconferencing, and electronic filing are difficult to foresee at this time. Undoubtedly, the boundaries of the office will be expanded, sources of information and methods of gathering information will change, and interaction patterns will broaden. With easier access to more sources of information, however, the danger of information overload will probably increase.

Discussion

Large or small offices? Landscaped offices? Windows or no windows? Although research data about these and other aspects of offices are still quite skimpy, the research that is available (such as that discussed above) gives an impression of ambiguity, inconsistency, and lack of support for at least certain expectations or hypotheses. In reflecting about this disturbing state of affairs, we need to keep in mind the fact that in part the measures of the effects of some design features consist of subjective reactions of people, such as preferences, attitudes, and aesthetic impressions. Overt manifestations in terms of behavioral criteria such as work performance are difficult to document, but at the same time the favorable disposition of workers to certain types of working situations indicates that they probably have some long-range hidden values. Further, although people might prefer a particular environment situation, probably they have a fair quota of resiliency or adaptability that makes it possible to adjust to a variety of circumstances.

DWELLING UNITS

There are many aspects of dwelling units that have human factors implications. There has been a considerable body of literature to address such issues as space requirements in the kitchen and bathroom and indoor climate [see, for example, Grandjean (1973)]. As in our discussion of the office environment, we focus on architectural and interior design features. Even within this limited domain, we touch only briefly on a few such aspects.

Room Usage in Dwellings

One human factors aspect of dwelling units relates to the usage of rooms, including indices of the spatial adequacy of dwelling units in relation to the number of occupants. One index of spatial adequacy is the number of *persons per room* (PPR), which is simply the total number of occupants divided by the number of rooms. As pointed out by Black (1968, p. 58), the upper limits of what are usually considered acceptable values of PPR are somewhere around 1.00 or 1.20, with a national aver-

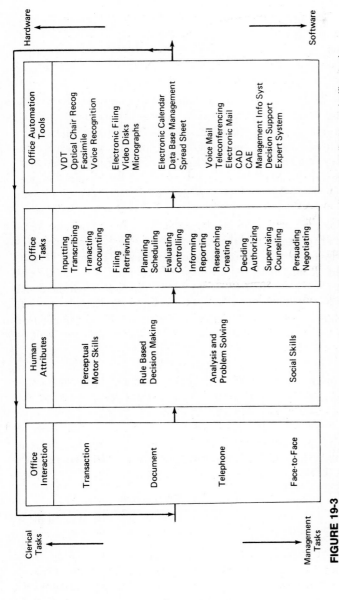

FIGURE 19-3

Taxonomy of office automation tools based on the type of office interaction, human attribute, and office tasks involved. (*Source: Helander, 1985, fig. 1. Copyright by the Human Factors Society, Inc. and reproduced by permission.*)

TABLE 19-7
RESULTS OF SURVEY OF ADEQUACY OF THREE
CATEGORIES OF SFPP OF LIVING AREA

SFPP	Category	Salt Lake City survey, %
< 130	Poor housing	2
131–215	Adequate	21
> 215	Very good	77

Source: Black, 1968, pp. 62–63.

age of about 0.69. Another index that is sometimes used is the *square feet per person* (SFPP). As pointed out by Black, there are no U.S. norms,, or standards, for the SFPP, although Chombart de Lauwe (as cited in Black, 1968) has proposed the categories given in Table 19-7; the last column shows the percentage in each category resulting from a survey of 121 houses in Salt Lake City. It might be added that, of the home owners surveyed, 7 out of 10 were satisfied with their present houses and in the case of those who were not, house size had no apparent relation to their dissatisfaction.

Besides the derivation of gross indices of spatial adequacy, concern has also been paid to how rooms in a dwelling are used. Grandjean (1973), for example, presents the results of a survey of households in Switzerland carried out by Bächtold (1964). Table 19-8 presents results dealing with the use made of kitchens as a function of their size. The larger the kitchen, the more likely the meals would be eaten there and the children would play there, (hopefully not at the same time). Such data on room usage can be valuable for designing a dwelling to be more useful to the occupants.

Room Arrangement

Dwelling units (be they houses or apartments) come in many arrangements and sizes. In this regard an extensive survey by Becker (1974) of apartment occupants of public

TABLE 19-8
USE MADE OF KITCHENS IN RELATION TO THEIR SIZE, GIVEN AS PERCENTAGE OF
THOSE PARTICIPATING IN SURVEY

Activity	Kitchen area		
	Up to 6 m^2	7 m^2	8 m^2 or more
Breakfast	49	66	85
Midday meal	42	48	71
Evening meal	40	53	74
Washing, ironing	42	74	78
Children playing and working	2	8	25
Spending leisure time	11	15	26

Source: Bächtold (1964) as presented in Grandjean (1973).

housing developments included data obtained by interviews with 257 residents and questionnaire checklists from 591 residents. Certain questions dealt with room size and arrangement of the living-dining-eating spaces. The responses to these questions are summarized:

• Many variations of room size and arrangement were considered to be equally satisfactory by the residents, but (as would be expected) satisfaction tended to be greater with larger rooms.

• More residents preferred a separate dining area (39 percent) or separate dining area and large eat-in kitchen (28 percent) to combined living-dining area (19 percent) or living room with large eat-in kitchen (18 percent). A couple of these arrangements are illustrated in Figure 19-4 along with proposed designs that would permit several options for arrangement of eating and living space. Although most residents preferred a separate dining area, 83 percent said they would not be willing to give up any of their living room space for a separate dining area or an all-purpose room. In other words, residents presumably found living room space so essential (whatever the size) that they could not conceive of reducing it.

FIGURE 19-4
Order of preference of residents of a couple of multifamily housing units regarding dining and eating arrangements, and a couple of proposed designs, one of which *(d)* would permit several options for eating and arrangement of living space (this one having a movable wall unit which could be placed in various locations, such as at *A* or *B*). *(Source: Adapted from Becker, 1974, figs. 5a, 5b, 5c, and 5d.)*

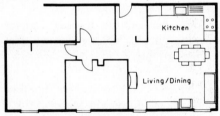

(*a*) One of least preferred arrangements (in convenience of carrying food, food odors in living area, limited hobby space)

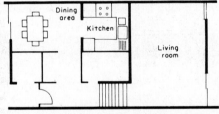

(*b*) More preferred arrangement (particularly because of separated eating and living areas)

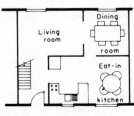

(*c*) Proposed desirable arrangement

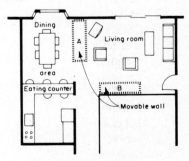

(*d*) More flexible proposed arrangement (with movable "wall" unit to permit flexibility)

In addition to dining and living areas, Grandjean (1973) lists the following guidelines for the placement of bedrooms within a dwelling:

- Bedrooms should be separated from the living part of the dwelling (i.e., living room, dining room, kitchen).
- Bedrooms should be near a bathroom.
- Bathrooms should be accessible from children's rooms without having to pass through the living room.
- To make it easier to watch small children, at least one child's bedroom should be close to that of the parents.
- Because the parents' bedroom is used so little during the day, it may face north.
- Since children's rooms are used so much in the daytime, they should face south.

Interior Design Features

Numerous aspects of interior design can influence people's perceptions of a dwelling or living space. For illustrative purposes we briefly discuss a few.

Ceiling Height and Slope. The usual height for rooms in dwelling units is 8 ft (2.4 m). In a study by Baird, Cassidy, and Kurr (1978), subjects were placed in a room that had an adjustable ceiling. After the ceiling was set at a given height, the subject was asked to give a preference rating on a scale ranging from -10 to $+10$, these ratings being given in two hypothetical frames of reference—one in which the subject would be involved in no activty and another in which the subject would be dining. The mean ratings, shown in Figure 19-5, indicate that the preferred heights were about 2 ft (6 m) above the conventional height.

In another phase of this study the authors found that subjects tended to prefer sloping ceilings (in particular those with $3/12$ and $5/12$ pitches) and wall corners that were greater than 90° rather than less than 90°.

Windows People seem to prefer spaces with windows to those without, but how do windows affect our subjective perceptions of a room? Kaye and Murray (1982) asked 176 students to rate watercolor perspective drawings of a living room on 30 adjectives, such as *cluttered, interesting, confined, gloomy,* and *colorful.* Using a statistical technique called *factor analysis,* the authors identified four underlying factors that seemed to account for the ratings on the individual adjectives. The four factors are listed in Table 19-9 along with some of the adjectives most associated with each. The presence or absence of a window in the picture affected the social-aesthetic, mood, and size factors. The presence of a window made the rooms appear more friendly, inviting, and exciting (social-aesthetic). Windowed rooms were also seen as less drab, gloomy, and closed (mood) and larger and more spacious (Size).

Color There have been many speculations about human reactions to color; however, the topic is dominated more by opinions of people than by supporting research evidence. People differ markedly in their preferences for color, but generally blue, green, and red hues seem to be most commonly preferred. However, the context in

—see Rm. Dens. on Journ Art.

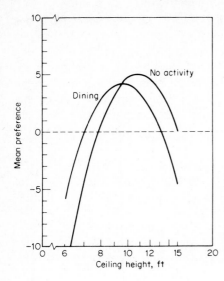

FIGURE 19-5
Mean preferences for ceiling height of rooms for "no activity" and for dining. Note that the ceiling height is shown on a logarithmic scale. *(Source: Adapted from Baird, Cassidy, and Kurr, 1978, fig. 3, p. 725. Copyright 1978 by the American Psychological Association. Reprinted by permission.)*

which colors are used can have a bearing on preferences. Aside from hue, Bennett (1977) reports research by others indicating that people tend to prefer light to dark colors and saturated colors to unsaturated colors. When objects are placed against a background, people tend to prefer high-contrast combinations, i.e., light-colored objects with dark background colors or vice versa.

In all this, however, we need to remember that the aesthetics of interior design, with various combinations of colors in rooms and furnishings, are essentially based on subjective preferences, and that one person's meat is another person's poison.

Kunishima and Yanase (1985) presented to subjects color slides of empty living rooms with different wall colors. A total of 60 slides varying in hue, brightness, and saturation were presented. Subjects rated each room on 16 bipolar-adjective scales

TABLE 19-9
FOUR FACTORS UNDERLYING RATINGS FOR VARIOUS ROOM ARRANGEMENTS USING AN ADJECTIVE RATING FORM

Factor	Representative adjectives
Social-aesthetic	Bright, happy, inviting, friendly, exciting
Physical organization	Cluttered, organized, accidental
Mood	Unattractive, drab, dead, closed, gloomy
Size	Large, spacious

Source: Kaye and Murray (1982).

(e.g., spacious-crowded, calm-restless). Through factor analysis, three underlying factors were identified that seemed to account for the ratings on the individual bipolar scales: activity (gay, fresh, happy); evaluation (elegant, refined, comfortable, good); and warmth (warm, stuffy). The authors found that hue had its greatest effect on the warmth factor with reds rated high, greens neutral, and blues low. Increased brightness resulted in higher ratings on the activity factor. Saturation affected the evaluation factor and to a lesser degree the activity factor. As saturation increased, the rating on the evaluation factor decreased sharply while the rating on the activity factor increased somewhat.

The relationship between hue and perceived warmth has been reported by many investigators. Greene and Bell (1980), however, carried the idea a step further. They assessed whether different colors actually affected people's perception of thermal comfort. After all, if people felt warmer in the presence of "warm" colors (reds and oranges), why not lower the thermostat and still achieve the same level of thermal comfort? Unfortunately, being in a room of a given temperature, whether it was painted red (warm color) or blue (cool color), resulted in the same degree of thermal comfort.

In assessing the influence of color in our lives, one has the impression of being caught between Scylla (accepting the many common beliefs and pronouncements about the effects of color) and Charybdis (rejecting such beliefs and notions, assuming that color is of only nominal consequence in human life—except possibly for the fate of paint stores). An in-between frame of reference actually seems to be warranted, but hard data about the various aspects of color are still limited.

Stairs and Ramps

Accidents in the United States involving stairs, both in and out of the home, annually result in about 800,000 injuries that require emergency-room treatment and are responsible for 1.8 to 2.6 million injuries of a more minor and temporary nature (Consumer Product Safety Commission, 1982). Most stair accidents occur in the home where they often rank as the number 1 type of accident.

Alessi and Brill and Associates (1978) described various accident scenarios that accounted for about 80 percent of the stair accidents investigated. A review of the scenarios highlights the following contributing factors:

- Being in a hurry.
- Poor eyesight.
- Being under the influence of alcohol or medication.
- Carrying children, groceries, or other objects.
- Darkness.
- Wearing high-heeled shoes.
- Presence of loose carpeting or stair covering.
- Wet or icy exterior stairs.
- Misplaced or missing handrails.
- An object, view, or activity that causes the person to be distracted.
- Stair treads too shallow to accommodate the length of a foot or footwear.

• Irregular riser heights or tread depths. A dimensional irregularity of as little as ¼ in (6 mm) between adjacent risers or treads can cause an accident.
• Objects on the stairs (e.g., leaves, trash, bicycles, shoes).

There has been extensive research dealing with stair accidents and design features contributing to stair safety (Pauls, 1984). In general, recommended are riser heights of 4 to 7 in (10 to 18 cm), tread depths of at least 11 in (28 cm), provision and proper placement of handrails, use of nonslip surfaces on tread surfaces, and dimensional nonuniformities (such as differences in the heights of risers) between adjacent steps of not more than ³⁄₁₆ in (5 mm).

In some situations ramps can be used instead of steps. Unfortunately, there is no simple answer to the question of whether ramps or stairs should be used in specific situations. Certainly ramps are preferable to stairs for use with wheelchairs. In connection with circumstances in which there might be an option, Corlett et al. (1972) carried out an experiment in which the subjects climbed and descended a height of 704 in (1788 cm), using at various times steps with risers of 4 in (10 cm) and 6 in (15 cm) with various tread lengths that resulted in the slopes shown in Table 19-10. The ramp conditions corresponded to these slopes. The eight subjects were young males, four being short and lightweight and the other four being tall and heavyweight. The results are a bit complicating and are summarized only briefly, as follows:

• Where joint rotation and muscle strength are not limiting factors, stairs are more efficient than ramps from a physiological point of view (in terms of oxygen consumption and heart rate).
• The higher 6-in (15-cm) step resulted in less physiological cost than the lower 4-in (10-cm) step.
• Where knee angle (and probably ankle angle) is important (as with old or lame people), a ramp would be easier to negotiate for any given slope, although the maximum ramp angle requires further study to specify.

SPECIAL-PURPOSE DWELLINGS

We briefly discuss two types of special-purpose dwellings: college dormitories and housing for the elderly and handicapped. College dormitories are an example of how the design of buildings can influence the social behavior of people. For the handi-

TABLE 19-10
SLOPES OF STAIR RISERS USED IN STUDY
BY CORLETT ET AL.

Riser	Slope		
4 in (10 cm)	10.5°	13.6°	19.5°
6 in (15 cm)	15.8°	20.0°	30°

Source: Corlett et al., 1972.

capped and many elderly people, circumstances argue for special attention to housing features. These circumstances include physical limitations and health conditions, the fact that such persons spend more time in their housing units, and the additional cost constraints realized for many such people.

College Dormitories

Let us consider the effects of dormitory design on social behavior. As one phase of a study reported by Baum and Valins (1977), the investigators compared residents of suite-type and corridor-type dormitories in terms of the residential locations of their friends. (The individual room in both types of dormitories were nearly the same size, and the average square footage per person, including lounges and bathrooms, was about the same.) The comparison showed quite clearly that more of the friends of residents of suite-type dormitories lived within the same dormitory (an average of 61 percent) as contrasted with friends of residents of corridor-type dormitories (an average of 27 percent). These results strongly suggest that suite-type dormitories are more conducive to the formation of friendships.

Another phase of this study dealt with residents of long versus short corridors. In general, those living on short corridors had a sense of greater privacy and of less crowding, and they experienced less aggressiveness on the part of other residents than did those living on long corridors. These and other results from this and other studies rather consistently reflect differences in the social behavior and attitudes of residents of different type of dormitories.

Housing for the Handicapped and Elderly

It has been estimated that by the year 2010 almost one-quarter of the U.S. population will be 55 years old or older and 40 percent of the elderly population will be 75 years of age or older (Newman, Zais, and Struyk, 1984). Roughly 45 percent of the current elderly population reported some sort of health limitation which, to some degree, restricted their activities (U.S. Public Health Service, 1981). The types of disabilities that are receptive to design of facilities and products for the elderly and handicapped include difficulty in interpreting information; loss of sight; loss of hearing; poor balance; lack of coordination; limited stamina; difficulty in moving the head; loss of upper-extremity skills; difficulty with handling and fingering; difficulty bending, kneeling, etc.; reliance on walking aids; and inability to use lower extremities.

Numerous publications discuss designing for the elderly and handicapped (see, for example, Altman, Lawton, and Wohlwill, 1984; Hale, 1979; Howell, 1980; Koncelik, 1982; Parsons, 1981; and Steinfeld, 1979a, 1979b). And it is not feasible here to discuss the many facets of the problem. Rather, we simply highlight a few of the major aspects that have important human factors implications.

Some of the specific features of housing facilities cited as being especially relevant for the aged are the following (Browning and Saran, 1978; Lang, 1978; Parsons, 1978; Rohles, 1978): safety and security; bathing facilities; handles and railings; warning devices; wide doorways and halls (for possible wheelchairs); adequate provision for mobility in general; furniture; facilities for recreation and entertainment (including TV, hobby facilities, etc.). Adequate provision for such features would apply equally to private homes and to various types of facilities for the elderly and retired, such as retirement housing developments, retirement homes, public housing, and convalescent homes.

BEYOND THE DWELLING UNIT

Multifamily housing developments involve human factors considerations over and above those associated with the individual dwelling units comprising them. These factors are associated with such features as the height, size, and arrangement of apartment buildings and with certain associated outdoor features. Becker (1974), for example, in a survey of housing developments in urban and suburban areas throughout New York State, found that the exterior appearance of a development was ''very important'' to most residents (67 percent). Variations in the shape, pattern, and form of buildings that increased their individuality were much appreciated; straight, rectilinear, and symmetric forms were strongly disliked. In addition, the barrier around a development can influence the affective response to the development, as illustrated in Figure 19-6. The wire fence at the left gives an institutional impression; yet probably it would not serve as an impenetrable barrier for someone intent on breaching it. The hedgelike fence at the right, however, creates a psychological boundary and a feeling of enclosure and so, while not communicating rejection to nonresidents, would tend to discourage strangers from entering. The burgeoning problems of urban centers are undoubtedly one of the major challenges of current life. The many facets of these problems leave few inhabitants unscathed. A partial inventory would include

FIGURE 19-6
Illustration of two types of barriers for a multifamily housing development. The one at the left gives an institutional feeling, whereas the one at the right has a more pleasing appearance yet still serves as a psychological boundary and gives a feeling of enclosure. *(Source: Adapted from Becker, 1974, figs. 7 and 8.)*

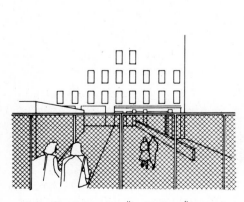

Chain-like fence gives "institutional" impression.

Hedge-like fence creates psychological boundary, with "soft" territorial definition.

problems associated with health, recreation, mobility, segregation, education, congestion, physical housing, crime, and loss of individuality. The current manifestations of these problems lend some validity to the forebodings of Ralph Waldo Emerson and Henry Thoreau, who viewed with deep misgivings the encroachment of civilization on human life, especially in the form of large population centers. The tremendous population growth, however, makes it inevitable that many people must live in close proximity to others (and thus requires the existence of urban centers). Given this inevitability, however, we propose to operate on the hypothesis that, by proper design, urban centers can be created which might make it possible to achieve the fulfillment of a wide spectrum of reasonable human values—perhaps even those that Emerson and Thoreau and maybe those that we might esteem.

As Proshansky, Ittelson, and Rivlin (1976) point out, cities are not just social, political, economic, and cultural systems, but geographical and physical systems as well. The interaction of these various features of cities with the inhabitants clearly influences the quality of life, but these influences are, of course, not the same for all inhabitants. The crowding and density in slum areas are, of course, in marked contrast to those features in spacious luxury apartments or in posh suburban areas.

It is not feasible within the limits of these pages to discuss the problems of urban communities and to speculate about their possible solution. The sociologists, social psychologists, and clinical psychologists clearly could contribute through research to the understanding of the impact of such communities on the lives of the inhabitants. In addition, however, it is believed that the human factors disciplines have much to contribute to the solution of some of the human problems of urban communities.

DISCUSSION

It is probable that utopia in the built environment that forms our living spaces is a will-o'-the-wisp, an unattainable goal. In other words, it is doubtful whether human beings' living space in the buildings and other structures they use and in the communities they build could ever provide for the broad-scale fulfillment of the various criteria relevant to each of us—physical and mental health and welfare, aesthetic values, opportunity for social interchange or privacy, recreation, entertainment, culture, convenience, mobility, safety and security, psychological identity, optimum facilities for contribution to efficient production of goods and services, or whatever. Although perfection in our built environment is not realistic, we should never stop trying to achieve it in rehabilitating our existing built environment and ensuring that newly constructed components (buildings, cities, etc.) are designed to be reasonably optimum in terms of fulfilling human needs.

REFERENCES

Alessi, D., and M. Brill and Associates (1978). *Home safety guidelines for architects and builders* (NBS-GCR 78–156). Washington: National Bureau of Standards.

Altman, I., Lawton, M., and Wohlwill, J. (eds.) (1984). *Elderly people and the environment.* New York: Plenum.

Bächtold, R. (1964) Der moderne Wohnungs—und Siedlungsbau als soziologisches Problem. Dissertation, University of Fribourg, Switzerland.

Baird, J., Cassidy, B., and Kurr, J. (1978). Room preference as a function of architectural features and user activities. *Journal of Applied Psychology,* 63(6), 719–727.

Baum, A., and Valins, S. (1977). *Architecture and social behavior.* Hillsdale, NJ: Lawrence Erlbaum Associates.

Becker, F. D. (1974). *Design for living: The resident's view of multi-family living.* Ithaca, NY: Center for Urban Development Research, Cornell University.

Bennett, C. (1977). *Spaces for people: Human factors in design.* Englewood Cliffs, NJ: Prentice-Hall.

Black, J. C. (1968). Uses made of spaces in owner-occupied houses. Ph.D. thesis. Salt Lake City: University of Utah.

Boyce, P. (1981). *Human factors in lighting.* New York: Macmillan.

Brookes, M. J. (1972). Office landscape: Does it work? *Applied Ergonomics.* 3(4), 224–236.

Brookes, M. J.,and Kaplan, A. (1972). The office environment: Space planning and affective behavior. *Human Factors*, 14(5), 373–391.

Browning, H. W., and Saran, C. (1978). A proposed plan of home safety for the older adult. *Proceedings of the Human Factors Society.* Santa Monica, CA: Human Factors Society, pp. 592–596.

Campbell, D. E. (1979). Interior office design and visitor response. *Journal of Applied Psychology*, 64(6), 648–653.

Chambers, J. (1963). A study of attitudes and feelings towards windowless classrooms. Ed.D. dissertation, University of Tennessee.

Collins, B. (1975). *Windows and people: A literature survey.* NBS building science series. Washington: National Bureau of Standards.

Consumer Product Safety Commission (1982, April–June). *NEISS data highlights.* Washington.

Corlett, E., Hutcheson, C., DeLugan, M., and Rogozenski, J. (1972). Ramps or stairs. *Applied Ergonomics*, 3(4), 195–201.

Galitz, W. (1980). *Human factors in office automation.* Atlanta: Life Office Management Association.

Grandjean, E. (1973). *Ergonomics of the home.* London: Taylor & Francis.

Greene, T., and Bell, P. (1980). Additional considerations concerning the effects of "warm" and "cool" colours on energy conservation. *Ergonomics.* 23, 949–954.

Hale, G. (1979). *The source book for the disabled.* London: Imprint Books.

Hedge, A. (1984). Ill health among office workers: An examination of the relationship between office design and employee well-being. In E. Grandjean (ed.), *Ergonomics and health in modern offices.* London: Taylor & Francis.

Helander, M. (1985). Emerging office automation systems. *Human Factors*, 27(1), 3–20.

Hodges, J., and Angalet, B. (1968). The prime technical information source—The local work environment. *Human Factors*, 10(4), 425–430.

Howell, S. (1980). *Designing for aging: Patterns of use.* Cambridge, MA: M.I.T. Press.

Imamoglu, V. (1973). The effect of furniture density on the subjective evaluation of spaciousness and estimation of size of rooms. In R. Küller (ed.) *Architectural Psychology: Proceedings of the Lund Conference.* Stroudsburg, PA: Dowden, Hutchinson & Ross, pp. 341–352.

Kaye, S., and Murray, M. (1982). Evaluation of an architectural space as a function of variations in furniture arrangement, furniture density, and windows. *Human Factors*, 24(5), 609–618.

Kleeman, W., Jr. (1981). *The challenge of interior design*. Boston: CBI Publishing.

Knopf, C. (1982). *Proceedings of Probe Research Seminar on Voice Technology*. New Brunswick, NJ: Probe Research, Inc.

Koncelik, J. (1982). *Aging and the product environment*. Stroudsburg, PA: Dowden, Hutchinson & Ross.

Kunishima, M., and Yanase, T. (1985). Visual effects of wall color in living rooms. *Ergonomics*, 28(6), 869–882.

Lang, C. (1978). Seniors plan senior housing. *Proceedings of the Human Factors Society*. Santa Monica, CA: Human Factors Society. pp. 545–549.

McCarrey, M. W., Peterson, L., Edwards, S, and Von Kulmiz, P. (1974). Landscape office attitudes. *Journal of Applied Psychology*, 59(3), 401–403.

Manning, P. (ed.). (1965). *Office design: A study of environment* [SfB (92): UDC 725.23]. Liverpool, England: Pilkington Research Unit, Department of Building Science, University of Liverpool.

Mercer, A. (1979). Office environments and clerical behavior. *Environment and Planning B*. 6, 29–39.

Nemecek, J., and Grandjean, E. (1973). Results of an ergonomic investigation of large-space offices. *Human Factors*, 15(2), 111–124.

Newman, S., Zais, J., and Struyk, R. (1984). Housing older America. In I. Altman, M. Lawton, and J. Wohlwill (eds.), *Elderly people and the environment*. New York: Plenum.

Parsons, H. (1978). Bedrooms for the aged. *Proceedings of the Human Factors Society*. Santa Monica, CA: Human Factors Society. pp. 550–557.

Parsons, H. (1981). Residential design for the aging (for example, the bedroom). *Human Factors*, 23(1), 39–58.

Pauls, J. (1984). Stair safety: Review of research. *Proceedings of the 1984 International Conference on Occupational Ergonomics*, vol. 1: *Reviews*. Rexdale, Ontario Canada: Human Factors Association of Canada. pp. 171–180.

Proshansky, H. M., Ittelson, W. H., and Rivlin, L. G. (1976). *Environmental psychology*. New York: Holt.

Rohles, F. H. Jr. (1978). Habitability of the elderly in public housing. *Proceedings of the Human Factors Society*. Santa Monica, CA: Human Factors Society. pp. 693–697.

Ruys, T. (1970). Windowless offices. Master's thesis, University of Washington.

Steelcase (1978). *The Steelcase national study of office environments: Do they work?* Steelcase, Inc.

Steinfeld, E. (1979a). *Access to the built environment: A review of literature* (HUD-PDR-405). Washington: Department of Housing and Urban Development.

Steinfeld, E. (1979b). *Accessible buildings for people with walking and reaching limitations* (HUD-PDR-397). Washington: Department of Housing and Urban Development.

Tikkanen, K. (1970). *Significance of windows in classrooms*. Master's thesis, University of California at Berkeley.

U.S. Public Health Service (1981). Current estimates from the national health interview survey, United States, 1980. *Vital and Health Statistics*, ser. 10, no. 139. Washington: National Center for Health Statistics.

Wells, B. W. P. (1965). The psycho-social influence of building environment: Sociometric findings in large and small office spaces [SfB (92): UDC 301.15]. *Building Science*, 1, 153–175.

Wichman, H. (1984). Shifting from traditional to open offices: Problems suggested design principles. In H. Hendrick and O. Brown, Jr. (eds.), *Human factors in organizational design and management*. London: Elsevier.

Wise, J., and Wise, B. (1984). Humanizing the underground workplace: Environmental problems and design solutions. In H. Hendrick and O. Brown, Jr. (eds.), *Human factors in organizational design and management*. London: Elsevier.

Wools, R., and Canter, D. (1970). The effect of the meaning of buildings on behavior. *Applied Ergonomics*, 1(3), 144–150.

Zweigenhaft, R. L. (1976). Personal space in the faculty office: Desk placement and the student-faculty interaction. *Journal of Applied Psychology*, 61(4), 524–532.

CHAPTER 20

HIGHWAY TRANSPORTATION AND RELATED FACILITIES

One of the features that characterizes the present-day world is the use of transportation—transportation of people and physical materials. The various forms of transportation include those that go by land (automobiles, trucks, buses, trains, etc.), air, and sea. All these—and their related facilities such as highways—encompass a whole host of human factors problems, some of which we have touched on in previous chapters. Since we can touch on only certain illustrative human factors aspects, and since highway vehicles (especially automobiles) and public transportation systems represent the most common mode of transportation, we deal particularly with them.

VEHICULAR ACCIDENTS

The dominant problem with highway vehicles is accidents and resulting personal injuries and deaths. As Näätänen and Summala (1976) point out, this is a worldwide problem. Their summary of deaths and injuries for 19 countries, over more than 40 years (1930 to 1973) reflects different patterns for various countries, those patterns being related to the *level of motorization* of the countries. Although the ratio of number of deaths to the number of motor vehicles has tended to go down, the problem is still of such magnitude that it should be given greater attention. Rumar (1982), for example, estimated that traffic accidents account for approximately 250,000 deaths and 10 million injuries in the world every year.

Terminology dealing with accidents is used rather loosely, with such terms as *accidents, collisions,* and *crashes* sometimes being used interchangeably. For our purpose we consider accidents to be those traffic mishaps that involve some property damage, personal injury or death, or both.

Causes of Accidents

The causes of accidents can be characterized in various ways and at different levels of specificity. At a very general level, accidents can be attributed to human behavior, the environment, and the vehicle, usually with some interaction among these. In discussing accident causation, Older and Spicer (1976) suggest that accidents can be the consequence of conflict situations involving the driver and the environment (and presumably the vehicle) that lead to evasive actions on the part of the driver. Such evasive actions may or may not result in what they call a collision (or, in our terms, an accident).

In connection with accident causation, Shinar (1978) presents data on the percentage of accidents from two samples that were attributed to human, vehicle, and environmental causes and their combinations. One sample of 2258 accidents was investigated on site, and the other sample of 420 was investigated more thoroughly, in depth. The results, given in Figure 20-1, show that human behavior was clearly the dominant cause.

We discuss accident causation in terms of those three categories, and then we elaborate on certain more specific variables associated with accidents.

FIGURE 20-1
(a) The percentage of accidents caused by human, environmental, and vehicular causes and *(b)* the relative proportion of combinations of these causes. Because many accidents have multiple causes, the sum of the percentages in *(a)* exceeds 100 percent, while the sum in *(b)* is 100 percent. *(Source: Shinar, 1978, fig. 5.2, p. 111.)*

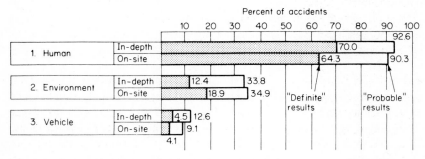

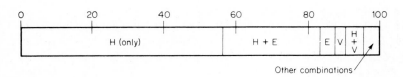

PERSONAL FACTORS IN ACCIDENTS

Discussion of personal factors in accident occurrence frequently involves reference to the notion of *accident proneness*. As usually interpreted, this term refers to some *attribute* that certain people have which causes them to be accident repeaters. We are disinclined to accept this interpretation, and we prefer to use this term in a statistical sense—to characterize those individuals who, for whatever reason, tend to have more accidents over time than can be attributed to chance. Note that the mere fact that an individual has a high accident rate for a while is *not* necessarily indicative of this statistical concept of accident proneness—a person who gets a bridge hand with 10 spades should not be called *spades-prone*. It is probable that there are some accident repeaters in driving, although collectively such individuals do not account for a very large proportion of all accidents. Further, there are inklings that different types of personal factors in certain instances, are related to driving behavior. These can be put into the following categories:

- Personal characteristics
- Driver behavior patterns
- Driver-related capabilities (experience, training, etc.)
- Temporary impairments (Shinar, 1978, refers particularly to fatigue, alcohol, and drugs)

For illustrative purposes we discuss a few factors which in specific circumstances, have been found to be related to driving behavior.

Personal characteristics

Various personal characteristics have been investigated as possible contributors to driving behavior; these include personality factors, vision, perception, physical abilities, reaction time, intelligence, age, and sex. Although we discuss certain of these briefly, probably most have significant impact on driving behavior to only a limited degree and with respect to only certain individuals.

Personality   There is an increasing body of evidence that manifestations of certain personality characteristics are associated with accident behavior. In summarizing the results of various studies, for example, McGuire (1976) concludes that some highway accidents are just another correlate of being emotionally unstable, unhappy, asocial, antisocial, impulsive, or under stress—or a host of similar conditions referred to by other labels. Such attributes presumably cause some people to be less cautious, less attentive, less responsible, less caring, less knowledgeable, or less capable of driving—and thus increase the risks of accident.

Vision   The vision test usually required to obtain a driver's license is a test of *static* visual acuity; it is administered when the visual stimuli are stationary and under normal daylight conditions. Although poor static visual acuity has been found to contribute somewhat to accident frequency. Booher (1978) and Shinar (1978) argue for the use of tests that measure more specific, driving-related visual skills. Henderson and Burg (1974), for example, identified the following visual abilities as those one would expect to be important for driving: (1) ability to perceive the presence, rate, and direction of relative angular movement between the observer and objects in the environment; (2) ability to detect changes in the size of an image on the eye as an indication of movement in depth; (3) peripheral vision; (4) static acuity; (5) ability to make visual discriminations when a visual target or the observer is moving (i.e., dynamic acuity); and (6) ability to visually fixate a moving object. Henderson and Burg, however, were unable to find any relationships between these abilities and accidents. Boyce (1981) believes that such weak correlations between visual abilities and accident records may be due, in part, to drivers with poor abilities being aware of their limitations and compensating by altering their driving behavior.

Shinar (1978) urges the use of measures of visual acuity under low levels of illumination similar to those encountered in nighttime driving. This argument is supported by data that indicate that nighttime accident involvement is related to poor visual acuity under nighttime levels of illumination, but unrelated to acuity under high (daytime) levels of illumination.

Field Dependence   The visual input to drivers is influenced both by the characteristics of the drivers and by the nature and location of objects within their environments. One personal characteristic that seems relevant is a perceptual style called *field dependence–field independence*. Field-independent people are better at distinguishing relevant from irrelevant cues in their environment than are those who are field-dependent. Various tests are used for measuring field dependence. Some are described by Goodenough (1976). A few studies have indicated that persons who are field-dependent may be more likely to have accidents than those who are field-independent. One such indication comes from a study by Barrett and Thornton (1968) using an automobile simulator. Figure 20-2 shows the relationship between field dependence-field independence and deceleration rate of subjects reacting to an emergency in the simulator, specifically the figure of a child appearing on the "road" ahead.

Goodenough summarizes the results of a few investigations that tend to confirm the implications of this perceptual style contributing to accidents. A study by Harano (1970) revealed that drivers with accident records (three or more accidents in 3 years) tended to be more field-dependent than did accident-free drivers.

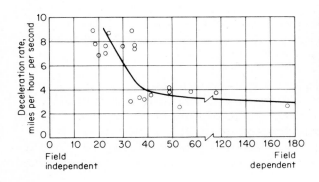

FIGURE 20-2
Relationship between field dependence–field independence style and deceleration rate of subjects reacting to an emergency situation in vehicle simulator. *(Source: Barrett and Thornton, 1968, fig. 3.)*

Although Goodenough points out that the reasons for the relationship between this type of perceptual style and driving behavior are not yet clear, he refers to evidence suggesting that field-dependent drivers do not quickly recognize developing hazards, are slower in responding to embedded road signs (those surrounded by many other stimuli), have difficulty in learning to control a skidding vehicle, and fail to drive defensively in high-speed traffic. In further efforts to "explain" the relationship, Shinar et al. (1978) report data that show a relationship between field dependence and eye-movement behavior. The more field-dependent the driver, the longer his or her eye-fixation duration, thus the longer it takes to pick up relevant information. Dewar (1984) indicates that field-dependent individuals tend to concentrate their visual attention within a narrow field of view, thus reducing peripheral vision.

Although it has been suggested that training might help field-dependent individuals to become more field-independent, there is as yet no strong evidence to indicate that this would be possible.

Age and Sex of Drivers   Age has been found, quite consistently, to be one of the higher correlates of accidents. The rate is highest for teenagers, decreases sharply in the twenties and increases for older drivers. This basic pattern is shown in Figure 20-3, this being based on data for males in Great Britain for a 2-year period. These data are for fatal and serious injury accidents, but the pattern for slight injury accidents is somewhat similar. Although some differences have been reported by sex, McGuire (1976) states that gender, per se, is probably not an important correlate of

FIGURE 20-3
Accident rate of males by age for fatal and serious injury accidents in
Great Britain for a 2-year period. *(Source: Jones, 1976, fig. 3.4A, p. 73.)*

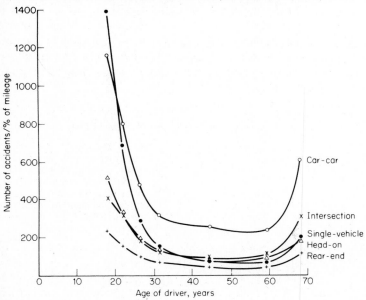

accidents, because the differences tend to be very small when age, exposure, etc. are controlled.

Behavior Patterns of Drivers

The behavior of people when driving presumably has its roots in their personal characteristics and experiences. Regardless of the origins, however, various aspects of driver behavior have been investigated to determine what relationships there are, if any, between such behaviors and accident occurrence. Such *behaviors* have been characterized in various ways and at various levels of specificity, such as inattention, improper lookout, risk taking, judgment (as of gaps between vehicles and of car-following distances), steering behavior, reaction time, etc. Here we cannot deal extensively with all possible variables, but we discuss a few for illustrative purposes.

As something of an overview of human behaviors associated with accident occurrence, some data from a study by Fell (1976) are summarized in Figure 20-4. These data are based on analyses of "definite involvement" and "definite or probable involvement" of four "direct-cause" categories. Of these it is clear that decision errors and recognition errors are dominant.

Sensory and Perceptual Input in Driving   Boyce (1981) presents data from on-the-spot investigations of 2036 traffic accidents in a rural area carried out by Sabey and Staughton (1975). Perceptual errors were a contributing factor in 44 percent of the accidents. A more detailed analysis of perceptual errors revealed the following contributions:

- Looked but failed to see—17 percent
- Distraction—16 percent

FIGURE 20-4
Summary of the causal accident involvement of four human direct-cause categories, expressed as percentages of accidents in which they were implicated. *(Source: Fell, 1976, fig. 2, p. 92.)*

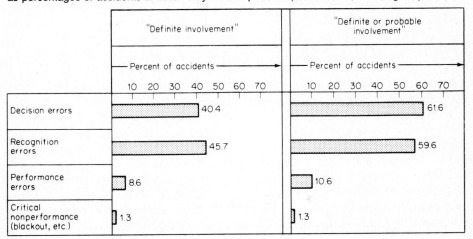

- Lack of attention—6 percent
- Incorrect interpretation—5 percent
- Misjudged speed and distance—5 percent

One aspect of sensory and perceptual input is the visual and scan patterns drivers use. Such patterns typically have to be studied under somewhat controlled conditions. In one such study, Mourant and Rockwell (1970) used an eye camera to record the eye movements of subjects driving on an expressway at 50 mi/h (80 km/h). One implication of the study was that route familiarity plays a role in such patterns. Over unfamiliar routes the drivers typically "sampled" a wide area in front of them, but with increasing familiarity their eye movement tended to be confined to a smaller area. Other studies of driver eye-scanning behavior have shown that novice drivers sample the roadway environment more narrowly than experienced drivers; thus novices tend to receive less information from the periphery of the visual field (Mourant and Rockwell, 1972). While following a car, drivers tend to focus on the center of the road, closer to the car ahead of them, and spend less time looking at traffic control devices. Alcohol and fatigue also reduce visual scanning of the environment. In addition, increasing speed is accompanied by a narrowing of visual attention. These sorts of results have implications for placement and design of road signs and road-edge markers.

The perceptions of individuals of what they sense (as by vision or hearing) serve as the basis for their judgments about things in their environment—what Shinar calls *perceptual judgments*. An example or two will illustrate this point. One example involves judgments of safe distances in following other vehicles. In this regard a 2-s rule of thumb has been proposed by a number of officials as a safe practice in following other cars. Thus at the speeds of the cars in question, it would take 2 s for the following car to reach the momentary position of the car ahead. Colbourn, Brown, and Copeman (1978) found in a driving experiment that most drivers did allow this distance; however the authors point out that the 2-s rule of thumb is inadequate under conditions of poor visibility, bad weather, etc.

On the other side of the coin, Harte and Harte (1976) summarize studies indicating that there is a common tendency in actual driving situations (as contrasted with controlled experiments) for many drivers to underestimate following distances. Such underestimates, of course, can be serious when a driver decides to overtake the car ahead. Harte and Harte also report that such judgments are influenced somewhat by the type of road cues. They found that underestimation of distances was greater on a segmented-line road (which is a bit of a surprise) than on unlined roads, and greater on unlined roads then on solid-line roads.

In connection with passing other cars, there are inklings that people may tend to underestimate the time required to pass. To add to this potential danger in passing, there is a further tendency on the part of some drivers to make less realistic judgments when passing vehicles traveling at high speeds than at low speeds (Gordon and Mast, 1970). One cannot provide protection from all the harum-scarum drivers of the world. But drivers who are aware of the human tendency to underestimate the time required for passing can increase their own survival probabilities by *not* passing other vehicles if there is the slightest doubt about the time factor.

An another factor that influences a driver's perceptual judgments of speed is the phenomenon of *adaptation*. In driving, adaptation is reflected by the tendency to perceive a given speed to be *less* when a person has previously adapted to a *higher* speed and to be *higher* when a person has previously adapted to a *lower* speed. Various investigators have confirmed this idea (Denton, 1976; Mathews, 1978). This effect is especially noticeable in changing from a high speed to a lower speed—people tend to think they are going slower than they really are (Schmidt and Tiffin, 1969). Further, the longer the exposure to the high speed, the greater the underestimation of the lower speed. Thus people who slow down (as on an exit ramp of a superhighway) would tend to drive at a higher speed than they think they are driving. In this and other driving circumstances, a person might be driving at a higher speed than is warranted, thereby increasing the liklihood of accidents.

Decision Processes   A driver makes many decisions in the course of a trip and executes these in driving behavior by controlling the steering wheel, accelerator, brake, and sometimes the horn. As an example of the decision processes, we mention briefly the matter of risk taking. Although the genesis of variations in risk taking on the part of different individuals is shrouded, there are indeed differences in the extent to which people are inclined to take risks. And even the same individual varies from time to time in the risks she or he is apparently willing to take.

While granting the fact of rather marked individual differences in risk taking, Näätänen and Summala (1976) remind us of a fairly universal tendency for people to view themselves as relatively immune to accidents ("accidents happen to other people—not to me") and they discuss the factors that tend to decrease the subjective danger of traffic for themselves as road users. Some involve:

• Deluding perceptual and cognitive processes (as sensory adaptation to speed discussed above, underestimation of speed, underestimation of the physical forces at work in a traffic accident, and the interpretation of threatening situations and accidents with only slight consequences)

• Learning (the consequences of observations of many situations that were potentially dangerous, but in which no unfortunate consequences occurred; this is also something of an adaptation process)

• Subjective easiness in driving (the feeling that driving, most of the time, is quite an easy task)

• The driver as the operator of the vehicle (the driver may feel that the risk of future traffic situations can be reduced because he or she can have some influence on the situation)

• Expectancy (most people have little basis for estimating the probabilities of accidents)

• Little traffic supervision (the risk of apprehension for traffic violation is limited)

• Examples, norms, and images (these are viewed by many individual drivers as being "for other people," not themselves)

The risk-taking tendencies of people are intricately intertwined with people's expectations—that is, the probability of some unhappy consequence resulting from given behaviors. And since accidents are relatively rare (and the odds of accidents, therefore, are low), the taking of limited risks tends to lend a sense of false immunity.

Reaction Time   One factor that separates the quick from the dead in traffic is reaction time in emergency conditions. Reaction time is generally slower when one is surprised than when one is prepared to make a response. Reaction time also increases as the stimulus event and required response becomes more complex. For these reasons, reaction time must be measured under realistic conditions by using representative drivers who are not anticipating the need to make a response. Summala (1981) summarized a series of studies in which unobtrusive measurements were made of the steering response of drivers to roadside stimuli. In one daytime situation, the door of a car parked on the side of the road was opened as cars approached, and the evasive steering response (turning toward the center of the road) was measured as a function of time from onset of the stimulus (i.e., the opening of the door). In another study performed at night, a light was turned on by the side of a dark road as cars approached, and again steering response was measured. In these studies drivers did not know they were participating in an experiment, so most likely they were responding in a natural manner.

Some of the results from these studies are shown in Figure 20-5. The evasive response started in no case until more than 1 s had passed from the onset of the stimulus. The halfway point of the steering response was reached at about 2.5 s, and maximum steering deflection occurred between 3 and 4 s. Also, the response latencies were very similar in both daytime and nighttime situations.

The U.S. standard for perceptual reaction time used by traffic engineers to design highways is 2.5 s (Oglesby, 1975), which corresponds well with the halfway point for evasive steering responses reported by Summala. Hooper and McGee (1983), however, argue, from a review of the literature, that the standard should be 3.2 s to be appropriate for the 85th percentile driver.

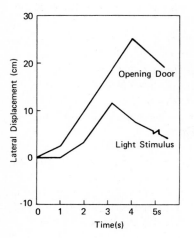

FIGURE 20-5
Average lateral displacement of cars in response to the opening of a car door at the side of the road during the day and to a light appearing at the side of the road at night as a function of the time available from the onset of the stimulus. *(Source: Adapted from Summala, 1981, fig. 1. Copyright by The Human Factors Society, Inc. and reproduced by permission.)*

Driver-Related Capabilities

Methods of driver improvement include driver training programs, licensing of drivers, enforcement practices, and driver improvement programs (especially for those involved in violations or accidents). At various times questions have been raised about the adequacy of all these programs—in some instances with good cause. Although such programs are relevant to vehicle operation, we cannot deal with them in any comprehensive manner here. So we simply state that such programs, when properly established and administered, undoubtedly can contribute to the improvement of driving and the reduction of accidents.

Temporary Impairment

The primary factors that contribute to temporary impairment of driving are fatigue and alcohol and drugs.

Fatigue   The primary culprit in generating fatigue is travel time and associated loss of sleep. The degradation in driving performance because of fatigue is reported by Shinar (1978) to account for a small, but significant, percent of highway accidents. However, among long-haul commercial drivers, who often must drive long hours, fatigue can be a significant cause of accidents. The Bureau of Motor Carrier Safety, for example, investigated 286 commercial vehicle accidents and found that 38 percent of them were categorized as attributable to the driver's being either asleep at the wheel or inattentive (Harris and Mackie, 1972).

Some drivers are often aware of the early stages of fatigue and can take steps to minimize it, as by stopping for a snooze or a cup of coffee. Hulbert (1972) lists other behaviors as preceding the final stage of actually falling asleep.

- Longer and delayed deceleration reactions in response to changing road demands
  - Fewer steering corrections
  - Reduced galvanic skin response (GSR) to emerging traffic events
  - More body movement such as rubbing the face, closing the eyes, and stretching

Presently the primary basis for reducing the risk from fatigue lies with the driver, who must be sensitive to the signs of the onset of fatigue and must stop driving. On the freeways in certain locations *Botts dots* are used to demarcate the lanes. These are small, raised bumps, and they are arranged so that if a driver changes lanes, the tires roll over the lanes with a ''roar.'' The dots are quite effective in alerting drowsy or inattentive drivers (They cannot be used in snow areas since they interfere with the use of snow removal equipment.)

Alcohol and Drugs   Alcohol and drugs contribute to a major proportion of all traffic accidents. It is commonly accepted that alcohol and/or drugs are a contributing factor in over 50 percent of all fatal traffic accidents. Alcohol slows reaction time, makes it more difficult to do several things at once, and reduces visual scanning of the environment. This is so well documented that we need not discuss it further.

Discussion

The extent to which corrective action can be taken about personal factors associated with accidents varies markedly with the factors in question. Nonetheless, let us mention a few illustrative factors along with possible corrective actions. For example, some cases of poor vision can be assisted with properly prescribed glasses, and proper road markings and signs could be useful. To help people who tend to be field-dependent maximize their perception, road signs might be placed in locations that are not cluttered, so they are clearly visible. The high incidence of accidents among teenagers might be reduced by more adequate driver training programs and more careful screening in issuing drivers' licenses.

The major problem associated with alcohol and drugs probably needs to be attacked through drivers' training and education programs and enforcement practices. In connection with strictly experimental innovative approaches to the problem of driving while under the influence of alcohol, some exploratory work has been carried out in the development of a computerized digit memory test. A set of digits would be presented to a person before she or he starts the car, and the person would be required to punch them into a keyboard correctly; otherwise, the starting mechanism of the car would not operate. The theory behind this is as follows: A person under the influence of alcohol would not be able to remember the digits and so could not start the car. It has been suggested that persons convicted of driving while under the influence should be required to install such a device in their car.

DRIVING ENVIRONMENT

The driving environment includes the roads and highways, street and highway lighting, road markings, road signs, and traffic as well as the natural features of the ambient environment such as rain and snow.

Road Characteristics

There are various indications regarding the effects of road characteristics on driving performance. One of the most clear-cut factors involves the use of superhighways versus conventional two-way roads. The accident rate on superhighways is clearly lower, this lower incidence being attributed to the reduced number of circumstances in which any given vehicle encounters other vehicles moving in other directions.

An example of a different kind of "effect" of road design characteristics is reported by Rutley and Mace (1972), who rigged electrodes to record the heart rates of a few drivers approaching the London airport. An example of these recordings for one driver is shown in Figure 20-6. This figure shows the heart rate in relation to the road being traversed and the "events" that occurred in driving over the section of the road in question. One can consider such physiological responses on the part of the driver (in this instance, changes in heart rate) as indices of strain that reflect varying levels of stress generated by the driving situation.

Road Markings

Everyday experience in driving clearly demonstrates the value of road markings, such as edge markings and center lines, especially in night driving. Aside from such mark-

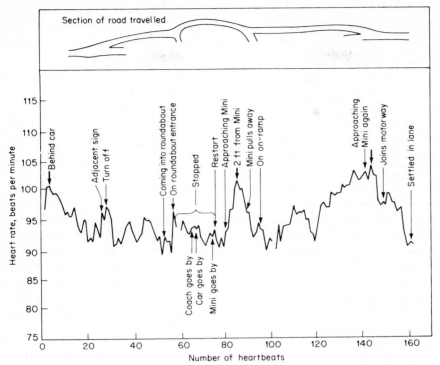

FIGURE 20-6
Heart rate of one driver on a section of a road near the London airport. The horizontal
scale represents the time during which 160 heartbeats were recorded. The features of
the road are sketched at the top, and the written entries indicate the "events" occurring
during the time period. *(Source: Rutley and Mace, 1972, fig. 3.)*

ings, however, there are some interesting developments in the value of special markings in specific situations. In Great Britain, for example, Rutley (1975) experimented
with the use of yellow transverse lines on the road starting 0.4 km (about ¼ mi)
ahead of traffic circles (called *roundabouts* in Great Britain). The spacing of these
lines was reduced as they got closer to the traffic circle. The driver, first becoming
"adapted" to lines being crossed at a given (slow) rate, then becomes aware of lines
being crossed at ever-increasing rates. This creates an illusion of speed (like telephone poles being passed at a high rate of speed) that typically causes the driver to
slow down as he or she approaches the traffic circle. Rutley presents the data shown
in Table 20-1 on the mean approach speed of vehicles at one traffic circle. Mean
speed was reduced immediately after the markings were painted on the road, but
after 1 year, the mean speed had returned to about the same level as before the
markings were applied. In addition, at this circle there had been 14 accidents the
year before the stripes were painted on the road, and there was only one accident
during the year after. This strongly suggests that such stripes can influence driver
behavior at least during the year following their application.

TABLE 20-1
MEAN SPEEDS OF VEHICLES BEFORE
AND AFTER PAINTING OF SPACED
TRANSVERSE LINES ON ROADS

	Mean speed	
	km/h	mi/h
Before	57.0	35.4
Immediately after	44.1	27.3
1 year after	52.4	32.5

Source: Rutley, 1975.

Another road-marking scheme was used by Shinar, Rockwell, and Malecki (1980) for marking a road curve. The markings were based on an illusion effect discovered many years ago by a German psychologist, Wundt, who found that V-shaped lines between parallel lines make the parallel lines appear closer together at the center than they really are, as illustrated in Figure 20-7. Shinar, Rockwell and Malecki (1980) painted V-shaped lines in a herringbone pattern on a road starting 97.5 m (about 320 ft) ahead of an obscured curve and on the curve itself. They hypothesized that the herringbone pattern of lines would create the illusion of narrowing of the road and cause drivers to slow down.

The herringbone pattern presumably did have the intended effect. Before the painting of the pattern, the mean reduction in speed of vehicles approaching the curve was 4.0 mi/h (6.4 m/h). After the pattern was painted, the mean speed reduction was 7.3 mi/h (11.7 km/h).

Road Signs

Considerable research has been directed toward improving the effectiveness of road signs. In Chapter 4 we discussed several factors that influence the effectiveness of signs and made some recommendations for their design. We cannot hope to cover, even briefly, all the aspects of roadway sign design. We do, however, briefly discuss driver's response to signs and legibility of signs at night.

Response to Road Signs   Several studies over the years have attempted to measure drivers' recall of road signs previously encountered on a stretch of road. Typically, a sign is placed on the road, then around a curve cars are stopped at random (often by a police officer), and the drivers are asked to recall the sign that was on the road. Generally, the percentage of drivers who can recall a sign is small, usually less than 50 percent and often less than 20 percent. Recall, however, is directly related to the degree of urgency of the information contained in the sign; the greater the urgency, the higher the probability of recall (Johansson and Backlund, 1970). The generally poor recall performance may, in part, be due to the emotional upset caused by being stopped on the road.

FIGURE 20-7
The Wundt illusion causes parallel lines to appear closer together
than they are at the center. This illusion was used by Shinar,
Rockwell, and Malecki (1980) as the basis for drawing a
herringbone pattern of lines approaching, and on, a curve in a road
to cause drivers to slow down on the curve.

Drory and Shinar (1982) suggest that in daylight driving, on an open road, drivers can obtain most of the information contained in warning signs from their view of the road ahead. Therefore, only the information they consider relevant and that cannot be obtained directly from the roadway is registered and recalled. This would explain why signs indicating police patrol and change in speed limits are recalled better than general warning and pedestrian-crossing signs (Johansson and Rumar, 1966). Shinar and Drory (1983) tested this hypothesis by comparing sign recall during the day and at night. Mean percentage recall was higher during the night (20 percent) than during the day (6 percent). This finding, then, tends to support the notion that signs are recalled better if they present information that is relevant and not otherwise easily obtainable.

Recall of signs is important; however, failure to recall does not necessarily mean that a sign was not perceived. Summala and Hietamaki (1984), for example, unobtrusively measured speed changes of 2185 drivers at a place in the road where an experimental sign (''*Danger*: Children'' or ''Speed Limit 30 km/h'') suddenly came into view. The change (decrease) in speed was dependent on the significance of the sign; however, all the drivers responded to each sign by releasing the accelerator pedal, indicating that the sign was perceived.

Legibility of Signs at Night   The legibility of road signs at night depends on several factors, such as the size of symbols, the colors of sign legends and of their backgrounds, the luminance of the legends and backgrounds, the type of material used, any ambient illumination, and the lights from approaching cars.

An illustration of the effect of sign brightness on legibility of numerals on signs of high and of low reflectance is based on a controlled experiment by Hicks (1976). In this study, the drivers used both low-beam and high-beam headlights on different nights. Legibility was measured in terms of the distance at which the numerals could be read by subjects when sober and when alcohol-impaired. For one group of subjects, the mean legibility distances under the different experimental conditions are shown in Figure 20-8. Both reflectance of signs and the position of the headlights influenced legibility. Alcohol impaired the performance of the drivers, and high sign reflectance and high-beam headlights made the signs more legible than low sign reflectance and low-beam headlights.

Sivak, Olson, and Pastalan (1981) measured the nighttime legibility distance for groups of drivers under 25 and over 61 years of age with equal visual acuity as measured under high luminance. The legibility distances for the older subjects were

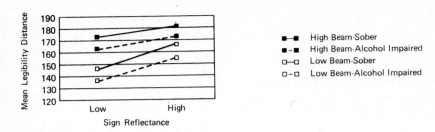

FIGURE 20-8
Legibility of numerals by sober and alcohol-impaired drivers under specified reflectance and headlight conditions. *(Source: Adapted from Hicks, 1976.)*

65 to 77 percent of those for the younger subjects. This implies that older drivers are likely to have less distance, and hence less time, in which to act on information contained in road signs.

Roadway Illumination

Only about one-quarter of all road miles are driven during darkness, yet the number of road fatalities at night often is equal to or greater than the number occurring during the day (Boyce, 1981). Roadway illumination, or lack of it, undoubtedly contributes to the higher accident rates at night. In general, fixed overhead roadway luminaires are superior to vehicle headlights for target detection at night; under some circumstances overhead lights can be 300 times more effective (Staplin, 1985). The before-and-after accident records of streets and highways which have been illuminated provide extremely persuasive evidence that this pays off. Table 20-2 summarizes the data compiled by the Street and Highway Lighting Bureau of Cleveland for 31 thoroughfare locations throughout the country, showing traffic deaths for the year before they were illuminated and for the year after. Such evidence is fairly persuasive and argues for the expansion of highway illumination programs.

Factors of Roadway Lighting   The CIE (1977) has published a set of recommendations for roadway lighting. Criteria are given for good roadway lighting in

TABLE 20-2
TRAFFIC DEATHS BEFORE AND AFTER ILLUMINATION OF
31 THOROUGHFARE LOCATIONS

		Reduction	
Year before illumination	Year after	Number	Percentage
556	202	354	64

Source: Public lighting needs, 1966.

terms of average road surface luminance, uniformity of luminance, extent of discomfort and disability glare, and directional guidance provided to the driver. We do not discuss these criteria, but we briefly discuss a few aspects of roadway lighting that affect visibility. One is transition lighting. Where roadway luminaires are to be used (such as at an intersection), it is desirable to provide some "transition" on the approaches and exits by graduating the size of lamps used. This helps facilitate visual adaptation to and from the highly illuminated area. Another factor is the *system geometry*, which refers in particular to the mounting height of luminaires; in some installations the mounting height is 35 ft (10.7 m) or more, and in some European systems installations at heights of 40 ft (12 m) or more are used. Such heights (while requiring larger lamps) help to increase the *cutoff* distance, that is, the distance from the light source at which the top of the windshield cuts off the view of the luminaire from the driver's eyes. This has an impact on visual comfort and visibility because overhead lamps can be a source of glare. Under average conditions this cutoff distance is about 3.5 times the mounting height of the luminaires.

In addition, the reflectivity of roadway surfaces varies substantially, a factor which affects visibility. For example, asphalt pavement surface (about 6 years old) reflects approximately 8 percent of the light, whereas concrete pavement has a surface reflectance of about 20 percent. It has been estimated that the amount of illumination required for equal brightness of asphalt pavements, compared with concrete, is roughly 2:1 (Blackwell, Prichard, and Schwab, 1960). Boyce (1981) also points out that more work is needed to assess the influence of wet roads on the effectiveness and design of roadway and vehicle lighting. We know that night visibility is reduced when the roads are wet; perhaps, this is due in part to the glare due to lights reflecting off the surface water on the roadway.

Vehicle Headlights

Closely related to roadway illumination is the illumination from vehicle headlights. A primary problem arises from glare produced by the headlights of oncoming vehicles. The effects of glare on visibility are shown by the results of a study by Mortimer and Becker (1973) as illustrated in Figure 20-9. This shows the *visibility distance* of targets ahead of drivers (on the left side of the driving lane) as the drivers approached oncoming cars with high-beam or low-beam lights. Glare increases and visibility decreases as the oncoming headlights approached; but as the headlights of the oncoming vehicle start to move away from the center of the visual field (when the two cars are nearing each other), visibility starts to increase. High-beam headlights of oncoming cars clearly reduce visibility more than low-beam headlights, except when the two cars are fairly close to each other.

Making the point that driver behavior is difficult to change, Shinar (1978, p. 146) suggests the use of polarized headlight beams and the possible addition of a third light beam (the *midbeam*) as means of reducing glare from oncoming headlights.

Traffic Events

Although it is not feasible to discuss many implications of the traffic environment in driving behavior, we mention one interesting procedure used by Helander (1978). He

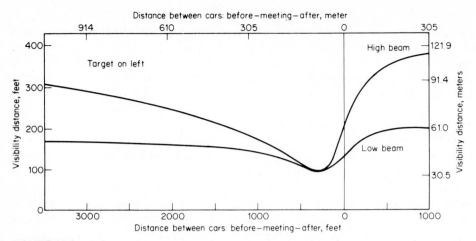

FIGURE 20-9
The effects of glare from oncoming headlights on visibility at varying distances between approaching cars. *(Source: Adapted from Mortimer and Becker, 1973.)*

placed electrodes on 60 subjects driving a rural test route and obtained measures of brake pressure. The electrodes recorded heart rate; electrodermal response (EDR), which is the same as galvanic skin response (GSR); and two electromyograms. The various measures were then related to 15 types of "traffic events" such as "other car passes own car" and "meeting other car." Brake pressure was considered as a criterion of the *mental difficulty* of the various traffic events. The event that was most difficult was "cycle or pedestrian meeting other car," and the next one was "other car merges in front of own car."

VEHICLE CHARACTERISTICS

Some of the previous chapters have dealt with topics that have certain implications relating to vehicle design features, such as the discussion of displays, control processes and control mechanisms, seating and space considerations, and vibration and motion. In this chapter we deal with a few vehicle characteristics that affect driving performance. Specifically we discuss manual and automatic transmissions and vehicle conspicuity.

Manual versus Automatic Transmissions

The use of manual transmissions involves more control actions than the use of automatic transmissions. In addition, the demand for executing more control actions might add at least some mental stress, especially in heavy traffic situations. Zeier (1979) carried out a study in Zurich, Switzerland, with drivers using both types of transmissions in a 14-km (about 9-mi) route, taking certain physiological measures during and after the driving task (we do not go into the details). He found significant differences between the two types of transmission users in rate of adrenaline excretion, skin conductance activity (SCR), heart rate, and heart rate variability, and he concluded that driving with the manual transmission produces greater activation of the sympathetic nervous system, in effect reflecting a higher level of stress. He suggests that the stress reduction due to the use of an automatic transmission may constitute a valuable measure for improved health and safety since it enables a driver to concentrate more on traffic events.

Vehicle Conspicuity

It stands to reason that the more adequately drivers are able to identify the presence and movements of vehicles in their traffic area, the more likely they are to avoid accidents. One of the most important factors influencing the detectability of an object in the environment is the contrast between the object and the background. Although the colors of vehicles as contrasted with their backgrounds can be useful in this regard, the use of lights on vehicles can also provide desired contrast. In fact, Hörberg and Rumar (1979) cite evidence that the conspicuity of any car with low-beam running lights is equal or superior to that of a car having the best color contrast for that background. Some years ago in the United States there was a flurry of interest in the use of a single front running light on automobiles, but this flurry (along with the sale of such lights) died out after a year or two. However, the demise of that practice may have been premature. Hörberg and Rumar summarize the results of a few studies that indicate that the use of daytime running lights raises the probability of detection of vehicles by drivers and a reduction of accidents. They report that because of such evidence, daylight running lights are compulsory in Sweden, and, to some extent, in Finland and Norway. (In these countries low-beam headlights are used as the running lights, instead of having special running lights.) Although vehicle conspicuity is affected by certain related variables such as viewing angle and intensity, we do not report the details of such effects. Rather, we want to emphasize the basic point that the conspicuity of vehicles is enhanced by the use of running lights, especially during the twilight hours, and that their use seems to contribute to the reduction of accidents.

Another aspect of vehicle conspicuity is the brake light system used on automobiles. Some years ago Malone et al. (1978) found that rear-end accident rates of taxicabs were significantly reduced by adding a center, high-mounted brake light to the conventional rear light system. Sivak et al. (1980) measured reaction times of unsuspecting drivers to conventional brake light systems with and without a single high-mounted (on the top of the trunk) brake light. Drivers following the experimental car applied their brakes in response to the onset of the experimental car's brake lights 31 percent of the time when the experimental car was equipped with only conventional rear brake lights. When the single high-mounted brake light was added to the experimental car, 55 percent of the time the following car responded. On those trials where the following car responded, however, the mean reaction times (1.39 s) were the same for both light configurations. In part as a result of these studies and others, all new cars sold in the United States are now required to have, in addition to the conventional brake lights, a single, center, high-mounted brake light on the rear of the car.

There are many other aspects of vehicle lighting that offer promise of reducing vehicle accidents, but these examples point up the fact that the use of appropriately designed vehicle lights can aid in driving safety by providing relevant information to drivers about the presence and movements of other vehicles.

PUBLIC TRANSPORTATION SYSTEMS

The previous discussion has dealt largely with the operation of vehicles and some of the individual and situational factors that influence such operation. Let us now consider some human factors aspects of public transportation systems. The various forms of public transportation include buses, subways and elevated systems, trolleys, and a few new forms such as monorails. The particular forms, of course, vary with distances to be traveled and the countries in question. With the higher costs of automobile transportation, there has been some shift to the use of public transportation systems, and this trend undoubtedly will continue, especially in connection with urban transportation systems.

Human Factors Aspects of Public Transportation

The human factors overtones of public transportation systems are manifest. From the point of view of the users (or potential users), some of the human factors aspects fall into the following categories:

- Station or terminal facilities and features
- Convenience and mobility of vehicles
- Vehicle interior (convenient entrance and exit, space and seating comfort, environmental factors such as minimum noise and vibration, adequate illumination, air quality, etc.)
- Safety and security
- Social factors (an adequate personal space, minimum crowding, etc.)
- Aesthetics and other psychological factors

It is, of course, not feasible here to deal extensively with the many specific human factors aspects of transportation systems, so we discuss only a few, especially those related to urban transportation systems.

Importance of Design Features to Users

The relative importance of specific design features of transportation systems varies from individual to individual, but some impression of the judgments of importance across a sample of people comes from a survey by Golob et al. (1970), based on interviews with, and questionnaires from, 786 individuals. The questionnaires include pairs of items such as the following:

A. Making a trip without changing vehicles
B. Easier entry and exit from the vehicle

The items generally depicted something about the service that might be provided by a transportation system. All together there were 32 items. The items were divided into several "blocks," and those within each block were paired. Respondents then indicated a preference for one of the items in each pair. On the basis of these preferences a scale value was derived for each respondent for each item in the block of items in the questionnaire. The mean scale value was then derived for each item, based on the responses of all who had that item in their questionnaires. These scale values are given for all 32 items in Figure 20-10.

FIGURE 20-10
Scale values of preferences for 32 characteristics of public transportation systems based on the paired-comparison responses of 786 subjects. The different symbols represent three different groups of characteristics. *(Source: Adapted from Golob et al., 1970, figs. 6, 8a, 8b, and 8c.)*

A word of caution is in order in evaluating the scale values, since the items were paired within each of nine blocks. In a strict sense, the scale values can be compared with each other only within each such block. Since one item ("lower fare") was common to all nine blocks, it did serve as a common link; but even so, some caution in interpretation is in order. Given this possible distortion of the scale values, the figure shows that "arriving when planned" comes out with the highest scale value, followed by "having a seat" and "no-transfer trip." These in turn are followed by a cluster of nine characteristics concerned mainly with the customers' time, fare, and shelters. Lower on the scale is a large cluster of 18 characteristics concerned primarily with interior design, aesthetic aspects of the trip, and passenger convenience. The items at the bottom of the scale are "coffee, newspapers, etc." on the vehicle and "stylish vehicle exterior."

As an additional analysis, the characteristics were separated into three subgroups: levels of service, convenience factors, and vehicle design characteristics. Figure 20-10 represents these groups by different symbols (a solid dot, a triangle, and a square).

Functional Aspects of Transportation Systems

A number of functional aspects of transportation systems can influence overall effectiveness, functions such as ticket selling, baggage handling, and loading and unloading of passengers. The time for loading and unloading of passengers, for example, influences the *dwell time* of a system, that is, the time a vehicle (such as a bus) would have to stop at stations or locations, thus, of course, affecting total trip times. The dwell time itself is also influenced by the physical features of the doors, platform, and mode of collection of fares (if collected on the vehicle, such as a bus). The unloading times for various types of vehicles as reported by Bauer (1980) range from 0.8 passenger per second (on a curb-level platform bus in Toronto) to 1.6 passengers per second on Toronto subways.

In connection with the functional aspects of transportation systems, greater attention is now being paid to features that facilitate the use of systems by the handicapped. The Urban Mass Transportation Administration (UMTA) (1978) estimated that there are 7.4 million transportation-handicapped individuals in urban areas of the United States. To address the needs of this population, the Department of Transportation has developed a procedure for evaluating the suitability of vehicles for the handicapped. In particular, this procedure provides for the calculation of the percentages of handicapped who would be able to use a particular transportation mode with the removal of various combinations of barriers. In the case of buses and trolleys, these barriers are sudden movement of the vehicle; riding while standing; rapid self-locomotion; movement in crowds; wait while standing; taking short steps; rising from a seat; aisle width (restricted); and walking long distances. If these barriers were removed, 99 percent of the handicapped could use the transportation mode in question, whereas if only the first four were removed, only 35 percent could use it. Although the removal of such barriers could pose a serious design problem, solutions or partial solutions for at least some have been proposed (*Travel barriers*, 1970).

In addition to designing vehicles for the physically handicapped, Richards, Hoel, and Foulke (1980) stress the need to design transit stations for the visually impaired as well. UMTA (1978) estimated that over 1.5 million persons in urban areas alone have difficulty using public transportation because of visual impairment. Richards, Hoel, and Foulke list several means for improving station accessibility for the visually impaired:

- *Eliminate barriers and obstacles*. Especially important are suspended fixtures that hang in the path of a pedestrian but cannot be detected by using a long cane.
- *Provide adequate signals and warnings*. For example, the Washington, D.C., subway system uses a texture change in the floor near the platform (for the blind) and a row of inlaid floor lights that flash as the train approaches (for the deaf).
- *Provide guidance for movement through space*. Floor texture patterns could create paths for blind users to follow through the station. Brightly colored paths could aid both sighted and partially sighted travelers. Distinct sounds or audio cues could be developed to identify specific locations within the station.
- *Provide knowledge of the environment*. Examples are tactile maps and familiarization training.
- *Provide special services*. Examples include the use of trained attendants, special passes to eliminate problems with fare card machines, and public awareness campaigns to increase aid and cooperation.

Clearly the implementation of recommendations such as those listed above would improve accessibility for unimpaired transit patrons as well as the impaired.

Effects of Commuting

Commuting by train into urban centers is virtually a way of life for millions of people. A rather interesting survey of the subjective feelings of commuters is reported by Singer, Lundberg, and Frankenhaeuser (1974). As one part of their survey they asked a sample of volunteer passengers to rate their subjective comfort, boredom, unpleasantness, and crowdedness after boarding the train at the suburb where they had started and after stopping at six stations before arriving at Stockholm, the final destination.

The results, summarized in Figure 20-11, show a marked decrease in comfort and increased boredom, unpleasantness, and crowdedness during the commuting trip.

Crowding seems to be a particularly important negative aspect of public vehicles. Kogi (1979), for example, reports changes in certain physiological measures (heart rate and palmar skin resistance) with increases in passenger density in an electric railcar in Japan. Although commuting probably never can be an enjoyable experience for many riders, ultimately vehicle design, with minimum crowding, might reduce the unpleasant features as well as minimizing physiological and psychological stress.

DISCUSSION

The sampling of human factors aspects of automobile driving and public transportation systems discussed in this chapter barely touches the surface of even these aspects of transportation. A potpourri of other aspects of transportation with numerous human factors implications could include ship design, motorcycle safety, seat belts for adults and children, airport design, public transport station design, airport control-

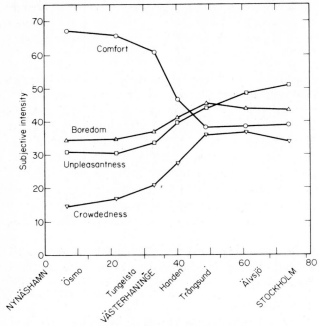

FIGURE 20-11
Results of a survey of commuters in Sweden of their subjective
reactions during a commuting trip. *(Source: Singer, Lundberg,
and Frankenhaeuser, 1974, fig. 3, p. 9.)*

tower operations, moving walkways for pedestrians, pedestrian safety, hand signals
of bicycle riders, and powerboat design and safety. The list could go on, but we
hope that the principal aim of this chapter has been achieved, namely that transportation and related facilities are simply permeated with human factors implications,
with the hope that such awareness may contribute to their ultimate improvement.

REFERENCES

Barrett, G. V., and Thornton, C. L. (1968). Relationship between perceptual style and driver
reaction to an emergency situation. *Journal of Applied Psychology*, 52(2), 169–176.
Bauer, H. J. (1970, April). *Public transportation and human factors engineering*, Res. Publ.
GMR-9982. Warren, MI: Research Laboratories, General Motors Corporation.
Blackwell, H., Prichard, B., and Schwab, R. (1960). *Illumination requirements for roadway
visual tasks*. Highway Research Board Bull. 255, Publ. 764.
Booher, H. (1978). Effects of visual and auditory impairment in driving performance. *Human
Factors*, 20, 307–320.
Boyce, P. (1981). *Human factors in lighting*. New York: Macmillan.
Colbourn, C., Brown, I., and Copeman, A. (1978). Drivers' judgments of safe distances in
vehicle following. *Human Factors*, 20, 1–11.

Commission Internationale de l'Eclairage (1977). *International recommendations for the lighting of roads for motorized traffic*, CIE Publ. 12/2.

Denton, G. (1976). The influence of adaptation on subjective velocity for an observer in simulated rectilinear motion. *Ergonomics*, 19, 409–430.

Dewar, R. (1984). Ergonomics in highway transportation. In M. Matthews and R. Webb (eds.), *Proceedings of the 1984 International Conference on Occupational Ergonomics*, vol. 1: *Reviews*. Rexdale, Ontario: Human Factors Association of Canada.

Drory, A., and Shinar, D. (1982). The effect of roadway environment and fatigue on sign perception. *Journal of Safety Research*, 13, 25–32.

Fell, J. C. (1976). A motor vehicle accident causal system: the human element. *Human Factors*, 18(1), 85– 94.

Golob, T. F., Canty, E. T., Gustafson, R. L., and Vitt, J. E. (1970, October) Research Publ. GMR-1037. Warren, MI: Research Laboratories, General Motors Corporation.

Goodenough, D. (1976). A review of individual differences in field dependence as a factor in auto safety. *Human Factors*, 18(1), 53–62.

Gordon, D. A., and Mast, J. M. (1970). Drivers' judgments in overtaking and passing. *Human Factors*, 12(3), 341–346.

Harano, R. M. (1970). Relationship of field dependence and motor-vehicle accident involvement. *Perceptual and Motor Skills*, 31, 372–374.

Harris, W., and Mackie, R. (1972, November). A study of the relationships among fatigue, hours of service, and safety of operations of truck and bus drivers, Rept. BMCS-RD-71-2. Washington: Bureau of Motor Carrier Safety.

Harte, D. B., and Harte, M. R. (1976). Estimates of car-following distances on three types of two-laned roads. *Human Factors*, 18(4), 393–396.

Helander, M. (1978). Applicability of drivers' electrodermal response to the design of the traffic environment. *Journal of Applied Psychology*, 63(4), 481–488.

Henderson, R., and Burg, A. (1974). *Vision and audition in driving* (TM(L)-5297/000/00). Santa Monica, CA: System Development Corp.

Hicks, J., III. (1976). The evaluation of the effect of sign brightness on the sign-reading behavior of alcohol-impaired drivers. *Human Factors*, 18, 45–52.

Hooper, K., and McGee, H. (1983). Driver perception-reaction time: Are revisions to current specification values in order? *Transportation Research Board*, 904, 21–30.

Horberg, V., and Rumar, K. (1979). The effect of running lights in vehicle conspicuity in daylight and twilight. *Ergonomics*, 22, 165–173.

Hulbert, S. (1972). Effect of driver fatigue. In J. Forbes (ed.), *Human factors in highway traffic research*. New York: Wiley.

Johansson, G., and Backlund, F. (1970). Drivers and road signs. *Ergonomics*. 13, 749–759.

Johansson, G., and Rumar, K. (1966). Drivers and road signs: A preliminary investigation of the capacity of car drivers to get information from road signs. *Ergonomics*, 9, 57–62.

Jones, I. S. (1976). *The effect of vehicle characteristics on road accidents*. New York: Pergamon.

Kogi, K. (1979). Passenger requirements and ergonomics in public transport. *Ergonomics*, 22(6), 631–639.

McGuire, F. L. (1976). Personality factors in highway accidents. *Human Factors*, 18(5), 433–442.

Malone, T., Kirkpatrick, M., Kohl, J., and Baker, C. (1978). *Field test evaluation of rear lighting systems*. Alexandria, VA: Essex Corp.

Mathews, M. (1978). A field study of the effects of drivers' adaptation to automobile velocity. *Human Factors*, 20, 709–716.

Mortimer, R. G., and Becker, J. M. (1973) *Development of a computer simulation to predict the visibility distance provided by headlamp beams*, Rept. UM-HSRI-HF-73-15. Ann Arbor: Highway Safety Research Institute, University of Michigan.

Mourant, R. R., and Rockwell, T. H. (1970). Mapping eye-movement patterns to the visual scene in driving: An exploratory study. *Human Factors*. 12(1), 81–87.

Mourant, R., and Rockwell, T. (1972). Strategies of visual search by novice and experienced drivers. *Human Factors*, 14, 325–335.

Näätänen, R., and Summala, H. (1976). *Road-user behavior and traffic accidents*. Amsterdam: North-Holland.

Oglesby, C. (1975). *Highway engineering*. New York: Wiley.

Older, S. J., and Spicer, B. R. (1976). Traffic conflicts: A development in accident research. *Human Factors*. 18(4), 335–350.

Public lighting needs. (1966). *Illuminating Engineering*, 61(9), 585–602.

Richards, L., Hoel, L., and Foulke, E. (1980). Improving access to public transportation for persons with sensory impairments. In D. Oborne and J. Lewis (eds.), *Human factors in transport research* (vol. 1). New York: Academic.

Rumar, K. (1982). The human factor in road safety. *Australian Road Research Board Proceedings*, 2: 66–80.

Rutley, K. S. (1975). Control of drivers' speed by means other than enforcement. *Ergonomics*, 18(1), 89–100.

Rutley, K. S., and Mace, D. G. W. (1972). Heart rate as a measure in road layout design. *Ergonomics*. 15(2), 165–173.

Sabey, B., and Staughton, E. (1975). Interacting roles of road environment, vehicle and road user in accidents. *5th International Conference of the International Association for Accident and Traffic Medicine*. London: International Association for Accident and Traffic Medicine.

Schmidt, F., and Tiffin, J. (1969). Distortion of drivers' estimates of automobile speed as a function of speed adaptation. *Journal of Applied Psychology*, 53, 536–539.

Shinar, D. (1978). *Psychology on the road: The human factor in traffic safety*. New York: Wiley.

Shinar, D., and Drory, A. (1983). Sign registration in daytime and nighttime driving. *Human Factors*, 25, 117–122.

Shinar, D., McDowell, E. D., Rackoff, N. J., and Rockwell, T. H. (1978). Field dependence and driver visual search behavior. *Human Factors*, 20(5), 553–559.

Shinar, D., Rockwell, T. H., and Malecki, J. A. (1980). The effects of changes in driver perception on road curve negotiation. *Ergonomics*, 23(3), 263–275.

Singer, J. E., Lundberg, U., and Frankenhaeuser, M. (1974, December). *Stress on the train: A study of urban commuting*, Rept. 425. Stockholm: Psychological Laboratories, University of Stockholm.

Sivak, M., Olson, P., and Pastalan, L. (1981). Effect of driver's age on nighttime legibility of highway signs. *Human Factors*, 23, 59–64.

Sivak, M., Post, D., Olson, P., and Donohue, R. (1980). Brake responses of unsuspecting drivers to high-mounted brake lights. *Proceedings of the 24th Annual Meeting of the Human Factors Society*. Santa Monica, CA: Human Factors Society. pp. 139–142.

Staplin, L. (1985). Nighttime hazard detection on freeways under alternative reduced lighting conditions. *Proceedings of the 29th Annual Meeting of the Human Factors Society*. Santa Monica, CA: Human Factors Society. pp. 725–729.

Summala, H. (1981). Latencies in vehicle steering: It is possible to measure drivers' response latencies and attention unobtrusively on the road. *Proceedings of the 25th Annual Meeting of the Human Factors Society*. Santa Monica, CA: Human Factors Society. pp. 711–715.

Summala, H., and Hietamaki, J. (1984). Drivers' immediate responses to traffic signs. *Ergonomics*, 27, 205–216.

Travel barriers (1970). U.S. Department of Transportation, National Information Service (PB187-237).

Urban Mass Transportation Administration (1978). *Summary report of data from national survey of transportation-handicapped people*. New York: Grey Advertising.

Zeier, H. (1979). Concurrent physiological activity of driver and passenger when driving with and without automatic transmission in heavy city traffic. *Ergonomics*, 20, 799–810.

CHAPTER 21

HUMAN ERROR AND
WORK-RELATED TOPICS

The major thrust of this text has been to emphasize the importance of human factors in designing equipment, facilities, procedures, and environments for people involved in the production of goods and services, as in manufacturing industries, service industries, military services, government, and other types of organizations. In large part, human factors efforts are directed toward designing things people use in order to enhance performance and minimize errors. (This takes us back to the matter of *human reliability* discussed in Chapter 2; such reliability, of course, being the complement of error.)

This chapter deals with certain human factors topics that have some rather special relevance to human errors and that, in part, share the common denominator of requiring continual or substantial attention on the part of people. These topics are monitoring activities, inspection processes, and accidents and safety. Before discussing these topics, however, we briefly describe the nature and classification of human errors.

HUMAN ERROR

To some people the term human error has a connotation of blame or cause. A much more productive approach, however, is to consider human error simply as an event whose cause can be investigated. Numerous definitions have been proposed for human error, but the following embodies the essence of most of them: *Human error is an inappropriate or undesirable human decision or behavior that reduces, or has the potential for reducing, effectiveness, safety, or system performance.* Two things should be noted about this definition. First, an error is defined in terms of its undesirable effect or potential effect, on system criteria or on people. Forgetting to pack cookies in a lunch would not be considered a human error in the context of a construction crew building a bridge, but forgetting to take safety shoes and glasses to the work site would be. Second, an action does not have to result in degraded system performance or an undesirable effect on people to be considered an error. An error that is corrected before it can cause damage is an error nonetheless. The important point is that an action must have the potential for adversely affecting system or human criteria.

Although there is a tendency among some to view errors as those of "operators," other people involved in the design and operation of systems also can make errors, such as equipment designers, managers, supervisors, and maintenance personnel. Therefore, in talking about human error, we should consider the entire system and not focus on only the operator.

Human-Error Classification Schemes

Various error classifications schemes have been developed over the years. An effective classification scheme can be of value in organizing data on human errors and for giving useful insights into the ways in which errors are caused and how they might be prevented.

Discrete-Action Classifications   One of the simplest classification schemes for individual, discrete actions is that used by Swain and Guttman (1983):

- Errors of omission
- Errors of commission
- Sequence errors
- Timing errors

Errors of omission involve failure to do something. For example, an electrician was electrocuted while attempting to position himself on the steel framework of an electrical substation. There were several points to disconnect in order to shut off power completely to the substation, and he apparently forgot to disconnect one of them.

Errors of commission involve performing an act incorrectly. For example, a mechanic sitting on a conveyer belt called for his partner to lightly hit the start button to jog the belt forward a few inches. The helper lost his balance momentarily and hit the button hard enough to actually start the belt moving at full speed, rather than just jogging it forward. The mechanic was pulled between the belt and a steel support 9 in (23 cm) above it.

A *sequence error* (really a subclass of errors of commission) occurs when a person performs some task, or step in a task, out of sequence. An example occurred in the case of a crane operator who was lifting a 24-ton block of stone. Rather than lifting the boom and then rotating it 90 degrees, he rotated the near-flat extended boom first, and before he could lift it, the crane overturned.

A *timing error* (also a subclass of errors of commission) occurs when a person fails to perform an action within the allotted time, either performing too fast or too slowly. Taking too long to remove one's hand from a workpiece in a drill press, for example, is a timing error that can result in a nasty injury.

Information Processing Classifications   Several authors have used an information-processing model to classify human errors. Usually the models classify errors as input, mediation, or output errors. DeGreen (1972), for example, classifies errors on the basis of where in the flow of information processing the error occurred. Table 21-1 lists his classification scheme. As Meister (1966) points out, human error occurs when any element in this chain of events is broken, such as failure to perceive a stimulus, inability to discriminate among various stimuli, misinterpretation of the meaning of stimuli, not knowing what response to make to a particular stimulus, physical inability to make a required response, and responding out of sequence. Identifying the *source* of errors in terms of input-mediation-output behaviors is a first step in developing ideas about how to reduce the likelihood of errors.

Rook (1962) combined a discrete-action classification with an information processing classification and added a dimension of intent. Table 21-2 shows the scheme proposed. Errors of commission are classified as intentional or unintentional. In this formulation, *intentional* refers not to intentional errors but rather to acts that an individual performed intentionally, thinking it was the correct action when in fact it was not. The crane operator who overturned the crane, for example, probably committed an intentional act.

Discussion   Although error classification schemes abound in the literature, no scheme has really been particularly useful. Part of the problem is that human error is complex, and simple classification schemes such as those described above do not capture that complexity. Often information is not available to make a classification decision; this is especially true with information processing classifications that require one to "get inside" the head of the person making the error. Consider the crane operator. Did he forget to raise the boom (error of omission), or did he intend to raise the boom after swinging it to the side (sequence error)? Tragically, we will never know; he was killed in the accident. Finally, the classification schemes are

TABLE 21-1
AN INFORMATION PROCESSING MODEL OF HUMAN ERROR

Inputs	Sensing
	Detecting
	Identifying
	Coding
	Classifying
Decisions	Estimating
	Logical manipulation
	Problem solving
Outputs	Chaining
	Omissions
	Insertions
	Misordering

Source: DeGreen, 1972.

TABLE 21-2
ROOK'S SYSTEM OF CLASSIFICATION OF ERRORS

Conscious level or intent in performing act	Behavior component		
	Input I	Mediation M	Output O
A Intentional	AI	AM	AO
B Unintentional	BI	BM	BO
C Omission	CI	CM	CO

Source: Rook, 1962.

deceptively simple. In practice, experts often disagree as to how to classify an error because rarely are the categories unambiguously defined.

Dealing with Human Error

It is inevitable that humans will err. There are numerous specific strategies for reducing the likelihood or negative consequences of human errors, but we do not try to enumerate them here. However, a brief discussion of generic approaches might be useful. In general, the likelihood or consequences of errors can be reduced by personnel selection and training and by design of the equipment, procedures, and environment.

Selection Selecting people with the capabilities and skills required to perform a job will result in fewer errors being made. Such things as perceptual, intellectual, and motor skills should be considered. The limitations with this approach are that (1) it is not always easy to determine what skills and abilities are required, (2) reliable and valid tests do not always exist for measuring the required skills and abilities, and (3) there may not be an adequate supply of qualified people.

Training Errors can be reduced by proper training of personnel. Unfortunately, people do not always perform as they were trained. They can forget or revert to old habits acquired before training. Training can also be expensive because it must be given to each person and, in critical situations, should include refresher training as well. We will have a little more to say about training later.

Design One of the important themes of this book is that the design of equipment, procedures, and environments can improve the performance of people, including reducing the likelihood and consequences of errors. There are three generic design approaches for dealing with human error:

- *Exclusion designs*: The design of things makes it impossible to commit the error.
- *Prevention designs*: The design of things makes it difficult, but not impossible, to commit the error.

• *Fail-safe designs*: The design of things reduces the consequences of errors without necessarily reducing the likelihood of errors.

One of the authors was an expert witness in a product liability case involving a roller press in which exclusion, prevention, and fail-safe designs were issues. In this case a worker was seriously injured because the safety guards were rendered inoperative due to incorrect wiring of a three-phase motor used to operate the press. The very nature of three-phase motors makes it a 50-50 chance that the wiring will be done incorrectly. An exclusion design would have made it impossible to start the machine if the motor were wired incorrectly. Prevention designs would have included such things as instructions on how to test if the motor was wired correctly or a warning system that signaled when the motor was incorrectly wired. A fail-safe design would dictate a system in which the safety guards would remain operable no matter which way the motor was wired. The particular roller press involved in this case, however, provided none of these designs.

Designing to reduce errors or their consequences can often be the most cost-effective approach to the problem of human error. A system need be designed only once, while selection and training must be repeated as new people become part of a system. In essence, it is easier to bend metal than to twist arms.

MONITORING TASKS

Certain types of work consist basically of monitoring or vigilance functions. In such tasks the monitor's function is that of giving attention to an operation or process to identify circumstances or events that require some action or response. A primary requirement of the monitor is the correct identification of all, or most, events that should require action. The input relating to these events may be presented to the monitor by various displays (such as dials, gauges, cathode ray tubes or other visual displays, or auditory signals of some kind), or it may be observed or detected directly (such as by noticing a change in the sound or the output of a machine or by observing boxes moving off a conveyer after they have been labeled). The indication of the event—whatever it may be—usually is called a *signal*; when a signal occurs, the monitor usually is supposed to take some action. Performance on monitoring tasks—whether in actual job situations or in laboratory studies—usually is measured with such criteria as (1) failure to detect relevant stimuli or signals, (2) false detection (*false alarms*), and (3) response lag.

Over the years, there have been many research studies dealing with vigilance. Davies and Parasuraman (1982), for example, cite over 700 references in their book on vigilance. And where there is research, there is theory! As pointed out by Loeb and Alluisi (1980), however, theories of vigilance are really not theories specific to vigilance, but rather are general theories applied to the vigilance situation. Examples include inhibition theory, expectancy theory, filter theory, arousal theory, habituation theory, and signal detection theory. Unfortunately, all describe and predict data almost equally well, and no one of them explains all vigilance phenomena.

Despite all the vigilance research conducted, some people still question the operational relevance of much of it. Mackie (1984), for example, reviewed 86 vigilance studies and found that the experimental conditions were so dissimilar to most operational vigilance situations (e.g., radar and sonar operations, process control operations, etc.) that he questioned whether the conclusions reached in the experiments would generalize to the "real world." Nevertheless, no one denies that performance of real-world vigilance tasks is often inadequate and usually needs improvement. There are, of course, individual differences in performance of vigilance or monitoring tasks, but a primary concern is how the situational and task-related aspects of such tasks affect performance.

Duration of Monitoring Periods

A very significant task-related variable is the duration of the monitoring periods. For tasks that consist exclusively of monitoring for signals, the evidence is quite clearcut that performance deteriorates with time (especially in terms of failure to detect signals that occur). This is illustrated by some of the results of an analysis of 37 studies by Teichner (1974), as shown in Figure 21-1. This particular figure shows summarized data for 12 monitoring tasks for which the initial detection rate was between 90 and 100 percent. For tasks with lower initial detection rates (i.e., the signals being more difficult to detect), the curves start lower and also show deterioration with time.

Some monitoring studies have shown a reasonable level of detection for perhaps ½ h or so before deterioration sets in. (Such a pattern is not revealed in Figure 21-1, perhaps because it summarizes data from several studies.) Because of such a pattern it has become the practice in some circumstances (as in radar watches in the military services) to limit monitoring periods to, say, ½ h. One of the clearest implications

FIGURE 21-1
Percentage of signals detected in monitoring tasks as a function of monitoring time, based on data from 12 studies. (*Source: Adapted from Teichner, 1974, fig. 10, p. 349.*)

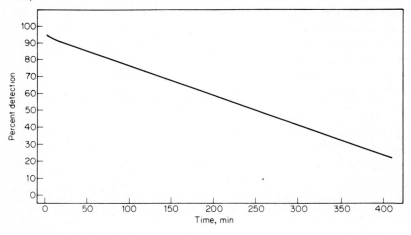

of monitoring research is that rest periods should be provided for continuous, demanding monitoring tasks.

Frequency Rate of Signals and Other Stimuli

Another factor that sometimes influences performance on monitoring tasks is the frequency rate of signals. The *percentage* of detection of signals that have a very *low* frequency rate tends to be *lower* than if the signal rate is somewhat *higher*. This tendency has been explained by various theories. The arousal level of subjects may be higher with higher signal rates, higher signal rates may promote better learning of the signal characteristics, or people may be more willing to call a stimulus a signal when the signal rate is high than when it is low.

This effect tends to occur with all-or-nothing displays in which the signals to be detected are the only stimuli presented. A somewhat different pattern occurs if the signals to be detected must be differentiated from other somewhat similar nonsignal stimuli. This effect was reflected by the results of a study by Wiener (1977) in which 32 signals were presented in a 48-min vigil, along with other interspersed nonsignal stimuli at a "regular" rate of one stimulus per second or at a "slow" rate of one stimulus every 5 s. (The signals to be detected were pairs of dots 13.3 cm apart on a cathode ray tube, and the nonsignals were 10.8 cm apart.) The results showed that performance (percentage of actual signals detected) was poorer when the nonsignal stimuli were presented under the regular rate conditions (approximately 60 percent detected) than under the slow rate conditions (approximately 80 percent detected). The poorer performance under the regular rate probably implies that the requirement to differentiate between the two types of stimuli once every second was so demanding that detection of the actual signals was adversely affected.

Discussion

Other variables can also affect performance on monitoring tasks, such as the regularity or irregularity of the signals (regularly occurring signals are more likely to be detected), the intensity or clarity of the signals to be detected, feedback (knowledge of results), and the complexity or clutter of the visual background of the signals. Although we do not review all the relevant research here, Bergum and Klein (1961), on the basis of a review of available research, suggested certain principles for more effective design of people-monitored systems; these are given below, with minor adaptations:

1 Visual signals should be as large as is reasonably possible. This includes size, intensity, and duration.

2 Visual signals should persist until they are seen (or otherwise detected) or as long as is reasonably possible.

3 In the case of visual signals, the area in which a signal can appear should be as restricted as possible.

4 Although "real" signal frequency often cannot be controlled, where possible it is desirable to maintain signal frequency at a minimum of 20 signals per hour. If necessary, this should be accomplished by introducing artificial (noncritical) signals to which the operator must respond.

5 Where possible, the operator should be provided with anticipatory information. For example, a buzzer might indicate the subsequent appearance of a critical signal.

6 Whenever possible and however possible, the monitor should be given knowledge of results.

7 Noise, temperature, humidity, illumination, and other environmental factors should be maintained at optimal levels.

Aside from these principles, it is important that adequate training be provided to those who are to serve as monitors. This training should make clear the nature of the signals that are to be identified and the response that is to be made to each.

INSPECTION PROCESSES

The inspection of parts and products in manufacturing processes is in certain respects similar to monitoring tasks: the inspector (usually making visual inspections) must continually be on the lookout for some condition that deviates from the norm—in particular, defective items. One difference is that inspection tasks are usually more discrete (e.g., looking at individual parts or products on a conveyer belt) than monitoring tasks, which tend to be more continuous (e.g., monitoring a nuclear power plant control board). Despite such differences, much of what has been said about monitoring tasks also applies to inspection tasks.

Inspection Periods

As an example of the similarity, there are cues that long periods of inspection without breaks usually result in reduced proficiency in terms of defects detected. Purswell and Hoag (1974), for example, compared the proficiency of inspection in an actual production facility under a "traditional" work schedule (a regular 8-h day with a mid-morning and mid-afternoon 10-min break and time out for lunch) as compared with a "new" schedule with a 5-min break every 30 min. Under the traditional schedule, there was a total of 450 min of working time. For 120 min, detection proficiency was maintained at 85 percent, but for the remaining 330 min of working time proficiency dropped to about 60 percent. Under the new schedule, however, although the total working time was only 395 min (because of the break periods), proficiency was maintained at 85 percent for the entire time. Thus, despite the reduced working time, the overall inspection proficiency of the new schedule was superior to that of the traditional schedule.

Additional evidence of the improved inspection with frequent breaks comes from an earlier study in Great Britain by Colquhoun (1959), as shown in Figure 21-2. In this case the subjects inspected defective items for 1 h in a simulated inspection. One group was given a 5-min break after 30 min; the other group had to carry on the inspection for a full hour. The performance of the "no rest" group was clearly worse.

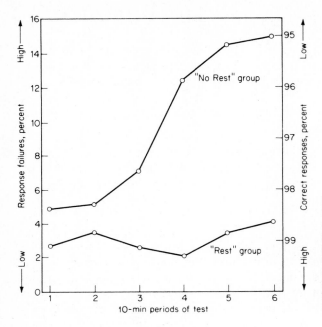

FIGURE 21-2
Percentage of failure to detect and correct detections in a simulated inspection task for two groups of inspectors. One group was given a 5-min break midway during the hour, and the other group worked continuously for the full hour. (*Source: Colquhoun, 1959, p. 370.*)

Detectability of Defects

Another factor that can affect the performance of inspection is the difficulty of making the visual and perceptual distinctions involved. Difficulty can be influenced by various factors relating to the nature and size of the defects, the total number of stimuli to be scanned, and the illumination available, along with others. Such effects are illustrated by the results of a study by G. L. Smith (1972), who varied the detectability of defects in a simulated inspection task by varying (1) the illumination, (2) the magnification used (that influenced the size of the defects), and (3) the total number of stimuli (the total number of stimuli to be scanned at a given time in order to identify the defective stimuli). The stimuli were circles, and the "defects" were circles with a slight break that made them look like the letter C (actually they were Landholdt rings described in Chapter 4). Four combinations of the three variables were ordered in terms of difficulty level of detectability, and the performance of the inspectors for these four conditions is shown in Figure 21-3. That figure shows the performance under two sets of instructions, namely "to avoid misses" and "to avoid false alarms," although the patterns under these two instructions are quite similar. The pattern shows quite clearly the deterioration in inspection performance with increasing level of difficulty of detectability.

Somewhat confirming evidence regarding the effects of the number of stimuli to be scanned comes from a study by Purswell, Greenhaw, and Oats (1972) in which the "defects" (an open circle and a black square) were included in either a 5 × 5 grid (with 25 open and black geometric figures) or a 7 × 7 grid (with 49 figures). The results are summarized in Table 21-3.

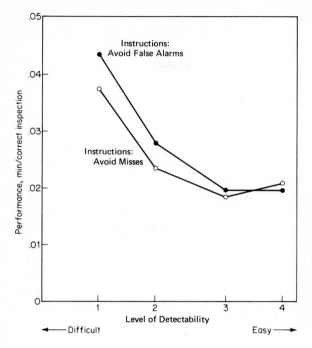

FIGURE 21-3
Inspection performance as related to the difficulty of detectability of defects. Level of detectability was varied by combination of illumination, the magnification used (that affected the apparent size of the defects), and the total number of stimuli to be scanned when identifying defects. (*Source: G. L. Smith, 1972, fig. 2, p. 289.*)

The effect of size of defects on inspection was demonstrated by Drury (1975) in the inspection of glass sheets for a single flaw, the flaws varying from 0.19 to 10 mm^2. In inspecting the sheets with a special shadowgraph screen, the time required to identify the flaws varied markedly by size of flaw, with some flaws taking several times as long to detect as others, and one flaw (the smallest) taking over 12 s as contrasted with less than 1 s in the case of the largest flaws.

It is, of course, frequently not possible to enhance the detectability of defects; but when this is possible, it should be done, by illumination, magnification, minimizing the number of stimuli to be scanned, and otherwise.

TABLE 21-3
ERROR IN IDENTIFYING DEFECTS AS RELATED TO TOTAL NUMBER OF STIMULI TO BE SCANNED

Grid	Total stimuli to be scanned	Error in identifying defects, %
5 × 5	25	4.7
7 × 7	49	11.2

Source: Purswell, Greenhaw, and Oats, 1972.

Rate of Presentation of Stimuli

Next in the inventory of variables that can influence inspection efficiency is the rate at which the stimuli to be scanned are presented, such as when they are presented at a controlled rate on a conveyer belt. Some indication of this effect comes from the study by Purswell, Greenhaw, and Oats (1972), in which the target presentation rate was varied by the velocity of the conveyer and the spacing of the stimuli. These variations were called the *target throughputs* in terms of targets per second. The total errors for the 5 × 5 and 7 × 7 grids are given in Figure 21-4.

It is admittedly somewhat risky to extrapolate directly from simulated laboratory studies to actual inspection processes. But such studies as this illustrate how the rate of presentation of items to be inspected can influence the efficiency of inspection processes. If the operation requires that the rate of presentation in actual inspection situations be controlled, two suggestions are offered: (1) The effect of rate of presentation of items on inspection efficiency should be determined empirically for the particular items or parts in question. (2) Usually some judgment needs to be made about the trade-off between speed of inspection and the effects of errors in inspection (either failure to identify defects or false alarms); if failure to identify a defect could have serious consequences, certainly every effort should be made to identify all defects. When feasible, however, it is preferable for the inspectors to control the inspection rate.

Use of Single and Multiple Inspectors

Data on the relative effectiveness of single versus multiple inspectors come primarily from simulated inspection studies, but indicate that having two or more inspectors tends to increase the effectiveness of inspection processes. Some such evidence comes from Waikar (1973, as reported by Purswell and Hoag, 1974) and is illustrated in Figure 21-5. The results of this study indicated improvement in the percentage of defective items rejected. These results indicate improvement from one to two, and from two to three, inspectors, but no further improvement with four. However, if all inspectors must recognize a defective item before it is rejected, the number of defective items rejected decreases as the number of inspectors increases. This implies that each inspector should be able individually to reject an item.

Lion et al. (1975), using a simulated inspection task, also found that inspection was significantly better when pairs of inspectors worked together than when each

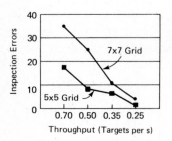

FIGURE 21-4
Inspection errors as related to target presentation rate (throughput) and spacing of target stimuli. (*Source: Adapted from Purswell, Greehaw, and Oats, 1972.*)

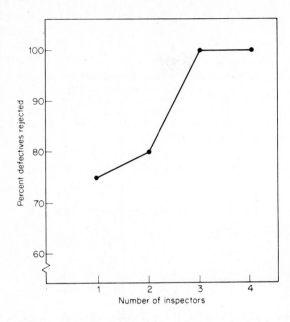

FIGURE 21-5
Percentage of defective items rejected when one, two, three, or four inspectors made individual decisions about items. (*Source: Waikar, 1973, as presented by Purswell and Hoag, 1974, fig. 2, p. 400.*)

inspector working singly. Paired inspectors viewed disks being moved along six parallel lines, while with single inspectors each inspected those on only three lines. The rate of movement with the single-inspector procedure, however, was twice as fast as when both inspectors scanned all six lines. The advantage of the paired-inspector operation was attributed in part to the stimulus of doing a job in unison, the feeling of competition, and possible reduced boredom because of the opportunity to talk.

Granting that the use of two or more inspectors can increase inspection proficiency, the question arises again as to the trade-off values of additional inspection costs versus the benefits from improved detection of defects.

Number of Fault Types

In industrial inspection tasks usually inspectors are looking for several possible defects or faults in a product. For example, a cabinet inspector might have to check for chipped paint, proper color, placement of label, and rough surfaces. As one might expect, as the number of different things an inspector is required to check increases, performance decreases. This was illustrated in a study by Ainsworth (1982) in which subjects inspected plastic parts moving along a conveyer system. Although other variables were also investigated, Ainsworth found that increasing the number of faults for which subjects had to search seriously degraded detection performance, as shown in Figure 21-6. Detection performance was measured in terms of detectability d' as defined in signal detection theory, which we discussed in Chapter 3. These results suggest that where inspection performance is critical, increasing the number of inspectors (so that each inspector looks for only a limited number of fault types) might be cost-effective.

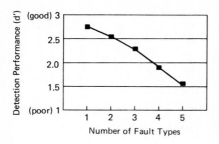

FIGURE 21-6
The effect of number of fault types being inspected on detection performance as measured by *d′*, a sensitivity measure measure based on signal detection theory. (*Source: Ainsworth, 1982, fig. 1. Copyright by the Human Factors Society, Inc. and reproduced by permission.*)

Aids in Inspection Processes

The effectiveness of many inspection tasks leaves much to be desired. Harris and Chaney (1969, p. 169), for example, report that the percentages of defects identified by 26 experienced inspectors ranged from 20 to 40, with a mean of about 40 percent. Certain methods of improvement of inspection operations have been discussed or implied, such as providing rest periods, arranging for inspection of stationary items rather than moving items when feasible, providing adequate illumination, minimizing (when feasible) the total number of items to be scanned simultaneously, using moderate rates of presentation for moving items, and using multiple inspectors when feasible. Certain other methods may work in certain circumstances. A few are mentioned here.

Visual Aids One scheme proposed by Chaney and Teel (1967), for example, used visual aids. In their application of this technique, they developed aids for use in inspection of machined parts for such defects as mislocated holes, improper dimensions, and lack of parallelism and concentricity. These aids consisted of simple drawings of the sample parts. The dimensions and tolerances for each feature to be inspected were placed on the drawings to minimize the need for calculation or reference to other materials. The inspection aids were introduced as a part of the regular inspection process in such a way that it was possible to compare the before and after inspection accuracy over 6 months, along with the effects of a specially designed training program. The results of the comparison, shown in Figure 21-7, clearly demonstrated the improvement associated with the visual aids as well as with the training. Needless to say, visual aids would be suitable only in certain special inspection processes.

Limit Samples The psychological literature is sprinkled with references to the fact that people are generally better at making relative judgments than absolute judgments. To take an oversimplified case, it is easier to say that one person is taller than another when they are seen side by side than to judge the actual height of either in absolute terms. As another example, most people can make "absolute" identifications of perhaps a couple of dozen colors varying in hue, saturation, and brightness, but they can differentiate pairs of many thousands if each pair is presented as a unit.

Limit samples are based on this principle. A *limit sample* as used in inspection operations is an example of a product or part that is just barely acceptable in terms

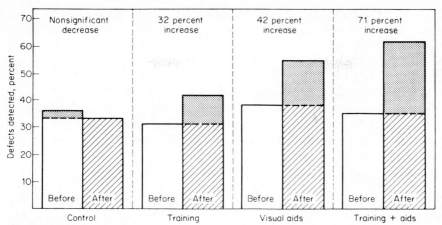

FIGURE 21-7
Percentage of improvement in detecting specified defects in machined parts following introduction of visual inspection aids, special training, and their combinations. (*Source: Chaney and Teel, 1967, as presented by Harris and Chaney, 1969, fig. 9.20. Copyright 1967 by the American Psychological Association. Reprinted by permission.*)

of some particular feature, such as a minor defect that still is acceptable. In operational use the limit sample is placed before each inspector with instructions to reject any item that is "worse" than the limit sample.

An example of this approach, as reported by Kelly (1955), involved the inspection of glass panels for the front of TV picture tubes. As a prelude to selecting the limit sample, she had a sample of 10 panels rated by the paired-comparison method (in which each item is judged, in comparison with every other). The results of this consisted of an ordering of the 10 panels from good to poor in judged appearance. In turn, management representatives made a determination of the panel that they considered to be just barely acceptable, and that panel was then the limit sample used in later actual inspection. It was estimated that inspection accuracy after the introduction of the limit sample improved over 75 percent.

There have been reports that limit samples do not serve a useful purpose in certain inspection processes. However, this may have been due to the fact that the inspectors actually were not using them. Additional training and supervision might help to ensure their continued use. Photographs can also be used in much the same manner as actual limit samples.

Stationary versus Moving Inspection Items

Some inspection operations are carried out when the items or parts to be inspected are stationary (static), whereas in other instances the items are moving, such as the inspection of fruit or glassware on a conveyer.

It is generally assumed that inspection is more efficient in detection of defects if the items are stationary rather than moving, and Williges and Streeter (1972) report their earlier research that tended to confirm this assumption. However, the nature of certain inspection operations virtually requires inspection of items while they are moving on a conveyer belt. In searching for ways and means to improve inspection of moving items, Williges and Streeter performed a study in which the inspectors carried out inspection processes under various combinations of stationary and moving "modes."

The details of their experiment and of the results are a bit complex and are not given here. But in general terms they found that inspectors whose training trials were with moving displays presumably learned visual scanning strategies that had a beneficial effect when they were later involved in the inspection of stationary items.

Discussion

The development of inspection processes involves a couple of major decisions. The first relates to the development of the procedures and techniques that are reasonably optimum from a human factors frame of reference and thus result in the most efficient method of inspection for the items in question. The second decision required in some circumstances relates to the question of trade-offs that may be necessary between the costs of the "best" inspection method and the consequences of less effective inspection, such as the criticality of the consequences from failure to detect defects in the inspection. The consequences of a defect in a Ping-Pong paddle would not be nearly as catastrophic as those of a defect in a nuclear power plant.

One by-product of the computer revolution is the increased use of sophisticated computer-based inspection devices to replace human inspectors. Such devices, at present, appear best suited for inspection tasks that require measurement of physical variables, such as tolerances, voltages, or weights. In turn, inspection tasks that require assessment of attributes, such as surface defects or missing components, may be performed better by a human than by a machine. Drury and Sinclair (1983), for example, compared human inspectors and a prototype optical scanning inspection device in ability to detect surface defects (i.e., nicks, dents, tool marks, scratches) in small steel cylinders. In general, they found human performance to be better than machine performance. The problem with the machine was that to increase its detection accuracy (i.e., rejection of bad parts), the decision threshold had to be changed, and that resulted in high false-alarm rates (i.e., rejecting good parts). Human inspectors, however, maintained good detection accuracy at relatively low false-alarm rates. These sorts of results underscore the need to assess empirically the relative performance of human and automated inspection and to consider both detection accuracy and false-alarm rates in the evaluation.

ACCIDENTS AND SAFETY

As indicated early on, one objective of human factors is to maintain or enhance certain human values, one of these being safety and physical well-being. In this regard the previous pages are sprinkled with references to accidents and injuries, these frequently being used as criteria in human factors research, as to the design of displays, controls, tools, work areas, and vehicles. To a very considerable extent many of the design features discussed in the various chapters (and scads of others not specifically referred to) are directed toward the creation of safer conditions for people in their work and everyday lives.

Since much of the previous material in this text has direct or indirect implications for safety, it is not intended here to refer further to such examples or to bring in other illustrations of human factors research or applications to specific design problems. Rather, here we discuss accidents in a general frame of reference both to illustrate certain techniques and methods that have been developed for use in the analysis of accidents and to mention certain methods for reducing accidents and injuries. In this regard some of the discussion of vehicular accidents in Chapter 20 will be relevant.

Human Error and Accidents

What percentage of accidents is caused by human error? This is a question that has vexed researchers for years. One often hears that approximately 85 percent of accidents are due to human error. This figure came from an analysis of insurance company records carried out by Heinrich (1959) and probably should not be taken too seriously. The percentage of accidents one attributes to human error depends on how one defines human error, the data source used to compile the statistics, and the alternative causes (other than human error) included in the tabulation.

Some studies of accidents have classified the cause of an accident as due to either unsafe acts (i.e., human error) or unsafe conditions (i.e., faulty equipment). The relative importance of these two factors varies greatly with the work situation. Conway, Muckler, and Peay (1980), for example, on the basis of a review of 12 studies, found the percentages of accidents due to human error ranged from 4 to 90 with the median being 50 percent.

For analytic purposes, however, such a broad dichotomy (i.e., unsafe acts versus unsafe conditions) has its limitations, and one would need to pinpoint the specific human variables (behaviors, characteristics, etc.) and the specific situational conditions associated with accident occurrence. For such purposes certain "models" of accident-related behavior have been proposed.

Models of Accident-Related Behavior

Various models of accident-related behavior have been proposed, some having at least certain common denominators. A model proposed by Drury and Brill (1980), for example, poses the view that accidents occur when the momentary demands of the task to be performed exceed the momentary abilities of the individual in question. In accident analysis they propose a form of task analysis in which the task demands are expressed in the same terms as information about what humans *can* do on the task. Thus the task demands might be specified in terms of the force that must be applied and the direction and distance of the movement involved, and human capabilities in turn could be specified in terms of the proportion of the population that is capable of applying such force. This type of comparison provides a very useful frame of reference in viewing accident occurrence.

A sequential model of the stages leading up to accidents is proposed by Ramsey (1978) as represented in Figure 21-8. Although his model focused specifically on potentially hazardous consumer products, it is equally relevant to work situations and other circumstances.

His model is intended to trace sequentially the activities that take place within the individual in potentially hazardous circumstances. The first stages deal with the *perception* and the *cognition* (i.e., the recognition) of the hazard. If the hazard is not perceived or recognized as such, the likelihood of an accident occurring is, of course, increased.

If the hazard is perceived and recognized as such, the next stage is *decision making*. A decision to avoid or not to avoid the hazard is, of course, influenced by the individual's attitudes and previously acquired behavior patterns. A person's decision not to avoid the hazard would be based on his or her risk-taking proclivities, some people being more willing to take risks than others.

If the individual decides to avoid the hazard, the next stage depends on the person's *ability* to do so, which is based on such factors as anthropometry, biomechanics, and motor skills, along with other human characteristics and skills such as experience, training, and reaction time.

The fulfillment of the sequence of behaviors involved in this model (including the exercise of the abilities to avoid a hazard) does not ensure avoidance of accidents, but it certainly reduces such possibilities. Even with the best of intentions and the presence of requisite abilities, chance factors bring accidents into human affairs. But

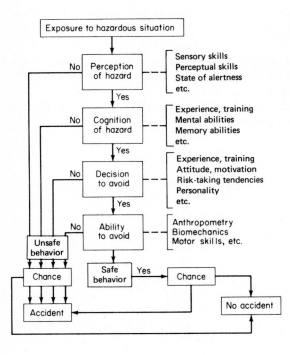

FIGURE 21-8
Accident sequence model representing various stages in the occurrence or avoidance of accidents in a potentially hazardous situation. (*Source: Adapted from Ramsey, 1978.*)

the failure to fulfill the indicated sequence certainly increases the chances of accidents.

A model such as this, which crystallizes the behavioral components involved in accidents, can be useful in focusing attention on such human factors aspects as the adequate presentation of relevant information relating to hazards (as in displays), the design of situational features that can be useful in avoiding accidents (control devices, work space, etc.), and the training of personnel in taking appropriate action in the face of hazards.

An example of a study that used a behavioral model was carried out by Lawrence (1974) in his investigation of 405 South African gold mining accidents. The percentages of accidents attributed to the various stages of the model were as follows:

Failed to perceive hazard	36%
Underestimated hazard	25%
Failed to respond to a recognized hazard	17%
Responded to hazard, but ineffectively	14%

Collection and Analysis of Accident and Injury Data

Actions that would be useful for reducing accidents and injuries should be based on the collection and analysis of relevant data. In this regard a number of techniques can be used, although each usually has some advantages and limitations.

Data Reporting Procedures   Data on accidents and injuries are compiled as a matter of routine by various organizations, including insurance companies, the Occupational Safety and Health Administration (OSHA), the Bureau of Labor Statistics (BLS), the National Safety Council, and many trade associations. This is not the place to present the classification systems used by various organizations or the masses of data based on them. However, it may be relevant to illustrate certain categories included in some classification systems. In discussing the identification of *causal* factors included in various classification systems, for example, Ramsey (1973) refers to the following (along with a few examples):

- Nature of injury (amputation, burn, fracture)
- Part of body (head, back, elbow)
- Type of accident (struck by, fall, caught in)
- Unsafe condition (inadequate guarding, placement hazard, defect of agency or equipment involved)
- Source of injury (body motion, air pressure, machine, hand tool, vehicle)
- Activity of injured (using hand tool, body movement, handling, operating power equipment)
- Unsafe act (operating or working at unsafe speed, using unsafe equipment, taking unsafe position or posture, failure to wear protective gear)

Certain sets of injury data involve frequencies and frequency rates of injuries as related to the above types of categories and their specific subcategories.

Accident data bases such as those maintained by OSHA or BLS serve a useful purpose in identifying trends in accident data. They do have limitations, for example, many accidents result from complex chains of events that cannot be adequately described by existing classification systems. Further, these data bases rely for their data on individuals who are often not trained in accident investigation, and the coding of the basic data is sometimes inaccurate and incomplete. In addition, the data tend to focus on the injured person rather than on the person(s) causing the accident. Even with these limitations, large-scale accident data bases are valuable sources of information. Without them, we would effectively be at a loss in our accident reduction efforts.

Critical Incident Technique   As applied to accident research, this procedure provides for those who have experienced or have observed unsafe acts or actual accidents to describe those events in detail. Such reports usually provide clues that reflect patterns of behaviors and events that can be useful in developing preventive measures.

Although the critical incident technique probably is used most commonly in reporting unsafe acts (rather than actual accidents), there is evidence from various sources that observed unsafe acts and conditions are definitely related to accidents and injuries. Such confirmation, for example, comes from a survey by Edwards and Hahn (1980) of over 4000 workers in 19 plants. They used a variation of the critical incident technique for obtaining data on observations by workers of unsafe acts and conditions correlated, on an across-the-board basis, .61 with accidents and .55 with disabling injuries. Certain conditions were most highly related to accidents:

Failure to ground (electric) material or equipment
Handling dangerous material unsafely
Worker did not understand how to do job
Climbing on or over moving equipment
Loading, feeding material too fast

On the basis of their data, Edwards and Hahn suggested that accidents "happen" where they have a chance to happen—where people report the existence of unsafe acts and conditions.

Fault-Tree Analysis   *Fault-tree analysis* (FTA) is based on deductive logic in the analysis of a system or operation that depicts graphically the interrelationship of combinations of events that can result in accidents or injuries. The development of a fault tree can be a rather complicated process, as discussed by Vesely et al. (1981); and although the details are not described here, an example is given. In the diagraming of fault trees certain standard geometric symbols represent certain classes of "events." Some of these symbols are shown in Figure 21-9.

Given that an illustration cannot be fully comprehended without more detailed description, Figure 21-9 illustrates a fault-tree analysis of the event of a chip in the eye occurring. As shown, this event could occur if (1) an operator fails to wear safety glasses or (2) a person without safety glasses (other than the operator) gets too close to the operation. The second possibility, in turn, would exist only if other events

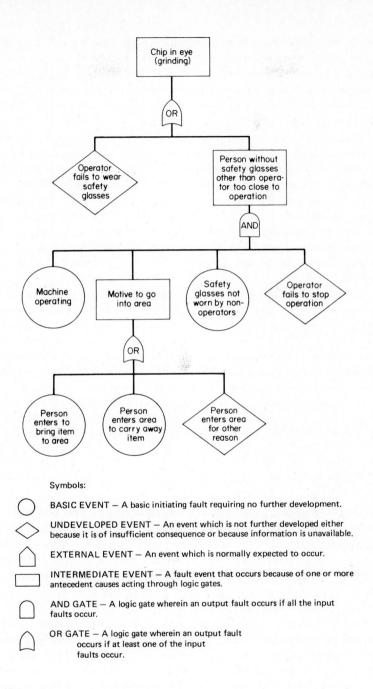

Symbols:

◯ BASIC EVENT — A basic initiating fault requiring no further development.

◇ UNDEVELOPED EVENT — An event which is not further developed either because it is of insufficient consequence or because information is unavailable.

⌂ EXTERNAL EVENT — An event which is normally expected to occur.

▭ INTERMEDIATE EVENT — A fault event that occurs because of one or more antecedent causes acting through logic gates.

⌂ AND GATE — A logic gate wherein an output fault occurs if all the input faults occur.

⌂ OR GATE — A logic gate wherein an output fault occurs if at least one of the input faults occur.

FIGURE 21-9
Example of a fault-tree analysis for use in analyzing the interrelationship between events associated with accidents and injuries. This example is for the event of a chip to the eye from a grinding operation. (*Source: Brown, 1973; fig. 6, p. 77.*)

(those under the AND symbol) were present. (In the case of one of those, there is another OR gate.)

Although fault-tree analysis has a place in accident analysis procedures, Christensen (1980) points out that there is an obvious danger with it (as with many techniques): its form and content may blind the user to alternative relationships between events that are not made explicit.

Discussion   There are various other methods for collecting and analyzing data relevant to accidents and injuries, including various forms of task analysis and the use of accident review teams (as in industry and aviation). Some of these and other techniques are discussed by Ramsey (1973) and Christensen (1980) and are not described here. In general terms, however, the various techniques provide for obtaining and analyzing data relating to the behaviors, to the physical conditions, or to both that presumably contribute to accidents or injuries. Such data, when appropriately developed, can serve as the basis for taking remedial action to reduce the incidence of such events.

Reduction of Accidents and Injuries

As mentioned earlier, the application of relevant human factors principles and data to the design of things people use in their jobs and everyday lives can reduce the likelihood of accidents and injuries. Some of the techniques for collection and analysis of data described above can point up some of the features of the things people use that could be redesigned toward this end. Aside from strictly design matters, however, certain other actions can be taken to reduce accident and injury risks, particularly those more directly related to operational procedures or to the behavior of people. Although this is not the place for a how-to-do-it or what-to-do manual on safety, a few examples may illustrate some such approaches.

Procedural Checklists   One technique that has become standard practice in aircraft operation, certain military operations, and certain other situations is the use of checklists that are to be followed in executing some specified routine. The individual follows a list of steps or functions that are to be executed, checking off each one as it is completed. This is, in effect, a substitute for memory in carrying out the stated activities.

Training   Training is, of course, one of the standard procedures used by safety directors to aid people in acquiring safe behavior practices. Cohen, Smith, and Anger (1979), on the basis of a review of approaches for fostering self-protective measures against workplace hazards, concluded that training remains the fundamental method for effecting such self-protection. However, they point out that the success of such training depends on (1) positive approaches that stress the learning of safe behaviors (not the avoidance of unsafe acts); (2) suitable conditions for practice that ensure the transferability of these learned behaviors to the real settings and their resistance to stress or other interferences; and (3) the inclusion of means for evaluating their effectiveness in reaching specified protection goals with frequent feedback to mark progress.

Feedback   Closely related to training is the matter of feedback, specifically feedback following desirable work behaviors on the part of workers. Cohen, Smith, and Anger (1979), for example, report the effects of feedback following training in practices that would reduce exposure to styrene (a suspected carcinogen) in a work situation. After initial training, the instructor visited each worker once or twice each day to provide encouragement or feedback in observing the work procedures in question. Observers, from remote vantage points, logged the number of times that the behaviors in question occurred both before and after training and found reductions of 36 to 57 percent in exposure levels to styrene.

In another situation, in two departments of a large bakery, a brief safety training program was instituted. This was followed by an "intervention" period in which there was an updated posting of the percentage of behavioral incidents that were performed safely (based on the observations of an observer). As another form of feedback the supervisors took occasion to recognize workers when they performed certain selected incidents safely.

The results are summarized in Figure 21-10, which shows the percentage of incidents performed safely for "baseline" observation sessions, for "intervention" sessions, and a "reversal" period at the end of the 25-week study period, the four

FIGURE 21-10
Percentage of behavioral incidents performed safely by bakery employees during a baseline period, an intervention period (during which feedback was provided), and a reversal period (after removal of feedback). (*Source: Komaki, Barwick, and Scott, 1978, fig. 1, p. 439. Copyright 1978 by the American Psychological Association. Reprinted by permission.*)

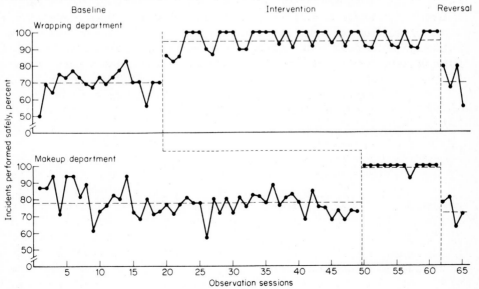

reversal observation sessions being those after the feedback of the intervention sessions were terminated. The immediate drop following the intervention sessions was so pronounced that within 2 weeks the management assigned and trained an employee to reinstitute the posting of data on a weekly basis. Within a year the injury frequency rate had stabilized at less than 10 lost-time accidents per million hours worked, as contrasted with the previous rates of 35 and above. In connection with feedback regarding safety practices, clearly this must be a continuing process and not a one-shot proposition.

Contingency Reinforcement Strategies Contingency reinforcement has its roots in the early work of Skinner, whose extensive research demonstrated that the probabilty of repetition of specific behaviors may be increased by providing rewards (i.e., reinforcement) when the specific behaviors occur. In discussing such reinforcement in the safety context, Cohen, Smith, and Anger (1979) suggest that the reinforcement rewards for safe behaviors might include bonuses, promotions, informal feedback (as discussed in the previous section), social recognition, and special privileges (time off, preferred parking locations, etc.).

As an example of the effectiveness of such a strategy, M.J. Smith, Anger, and Uslan (1978) report the results of their study with a shipyard employing 20,000 workers. They concentrated on the use of reinforcement (what they referred to as *behavioral modification*) in the reduction of eye injuries, which accounted for over 60 percent of all injuries. The program consisted of training five supervisors in the fundamentals of behavior modification: (1) observing worker behaviors (specifically the use of safety equipment), (2) recording worker behaviors; and (3) giving praise for wearing safety equipment. The work groups of these five supervisors had had among the highest eye accident rates. A before and after comparison of the eye accident rates for the subordinates of these five supervisors and for a control group of 39 supervisors is given in Table 21-4.

Cohen, Smith, and Anger (1979) report the use of reinforcement strategies in a couple of organizations aimed toward encouraging workers to use ear protection devices more consistently, with good results. In one instance the reinforcements consisted of posting of the percentage of workers wearing earplugs, social praise, coffee,

TABLE 21-4
EYE ACCIDENT RATES OF SHIPYARD
WORKERS BEFORE AND AFTER BEHAVIOR
MODIFICATION TRAINING

Work group	Eye accident rates		Change
	Before	After	
Experimental (5)	11.8	4.3	−7.5%
Control (39)	5.8	4.7	−1.2%

Source: M.J. Smith, Anger, and Usan, 1978.

doughnuts, money, and market goods; in the other instance it consisted of providing employees with the results of hearing tests. As Cohen, Smith, and Anger (1979) caution us, the potential value of reinforcement strategies depends very much on the careful selection of target behaviors and rewards.

Incentive Programs   Incentive safety programs consist of offering some type of incentive to individuals, groups, or supervisors of groups for achieving certain safety records. Among the incentives cited by Cohen, Smith, and Anger (1979) in certain organizations are the use of group safety records as a part of supervisors' performance evaluations; tokens redeemable for catalog merchandise; trading stamps given to individual employees with bonus stamps to work groups with no accidents; and direct payment of money to individuals and groups with good safety records. Cohen, Smith, and Anger (1979) cite the experiences of certain organizations that have been very successful in reducing accident rates. However, they caution against excessive preoccupation with or continued use of such programs, on the grounds that a succession of such efforts, each requiring a bigger prize than the one before, would appear unwise. They therefore recommend limited use of this plan.

Propaganda   It is evident that the distinctions among training, feedback, reinforcement, and incentives as related to safety are intricately intertwined, and do not exist in independent airtight compartments. Propaganda in the form of posters and other forms of safety communications also gets mixed up in this conglomeration. Recognizing these interrelationships, Sell (1977) sets forth the following aims of safety propaganda:

- To give more knowledge of safety factors.
- To change the attitudes of work people so that they are more inclined to act safely.
- To ensure that safe behavior takes place. Obviously this is the most important aim and the one at which all propaganda must be directed.

Sell points out that in the analysis of accidents the typical approach is to trace the cause to the person on the spot and to conclude that person was careless or lacked attention. But Sell goes on to indicate that in many accidents there are contributory factors which generally are out of control of the individuals involved—factors such as improper design in terms of human factors, fatigue (which individuals cannot avoid), lack of training, and failure by management to ensure that guards and protective clothing are available. In effect, Sell focuses on the fact that safety propaganda directed toward workers can be useful only if the accident-causing factors are under the control of the workers. (The responsibility for factors that workers cannot control should, of course, fall in the lap of management.)

On the basis of his analysis of various studies dealing with safety propaganda directed toward workers, Sell emphasizes (1) the need to try to modify the attitudes of workers that, in turn, influence their behavior and (2) for any communication to have an effect, it must be perceived and understood by the individuals in question. Drawing further from his survey, he states that there is little doubt that safety posters and other propaganda can be made to produce the desired behaviors. However, he believes that propaganda, to be really effective, must

1 Be specific to a particular task and situation
2 Back-up a training program
3 Give a positive instruction
4 Be placed close to where the desired action is to take place
5 Build on existing attitudes and knowledge
6 Emphasize nonsafety aspects

The propaganda should not

1 Involve horror, because in the present state of our knowledge this appears to bring in defense mechanisms in the people at whom the propaganda is most directed.

2 Be negative, because this can show the wrong way of acting when what is required is the correct way.

3 Be general, because almost all people think they act safely. This type of propaganda is thus seen as relevant only to other people.

Relatively few studies have provided evidence of the actual modification of behavior following the use of safety posters. In one such study in a steel mill, posters were used to encourage crane slingers to hook back the chain slings onto the crane hook when they were not in use, as a safety precaution. Three types of large posters with instructions and illustrations to "hook that sling" were posted in relevant areas where gantry cranes were used. The percentage of slings hooked back before and after posting of the signs, as shown in Figure 21-11, shows a systematic difference, with the improvement actually carrying over to a follow-up period. Such data suggest that posters can be useful in changing safety behavior if they are used where workers have the opportunity to control the occurrence of accidents by their own actions.

FIGURE 21-11
Percentage of crane slings hooked back as a safety precaution by crane operators in a British steel mill before and after the posting of safety posters. (*Source: Sell, 1977, fig. 3, p. 212. This figure appeared as fig. 3 of vol. 8, no. 4, p. 212, of* Applied Ergonomics, *published by IPC Science and Technology Press Ltd., Guilford, Surrey, U. K.*)

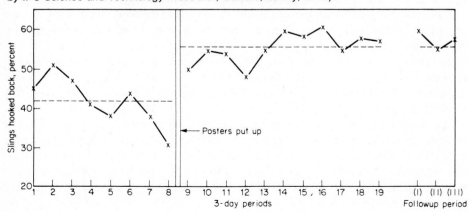

Discussion of Accidents and Safety

There is no such thing as a completely accident-free circumstance in human life. However, the human race should not assume a completely fatalistic attitude toward the inevitability of accidents, since actions can be taken to reduce the likelihood of such events. These actions involve the design of equipment and facilities in terms of human factors considerations that will contribute to safety; the development of procedures that contribute to safety and the training of personnel to follow such procedures; and the consistent use of appropriate protective devices by personnel.

SOME CLOSING REFLECTIONS

Several earlier chapters dealt with certain features of human beings relevant to the design of the things people use in their work and everyday lives, including people's sensory and perceptual skills, psychomotor abilities, and anthropometric characteristics. In addition, various chapters have dealt with some of the implications of such human characteristics to the design of some of the things people use, as displays, controls, work-space arrangement and layout, and the physical environment. Particular attention has been paid to the human factors related to the built environment, vehicles and related facilities, and (in this chapter) certain specific work-related topics such as monitoring tasks, inspection, and safety.

In reflecting about this coverage, it should be kept in mind that the material included represents only a small sample of the relevant material that could have been included. This text offers an overview of the human factors field and is not a complete compendium of relevant information. (Such a compendium would occupy more than the proverbial 5-ft bookshelf.) Given that this text represents a sample of the human factors domain, it has been an objective to help the reader become aware of, or become more sensitive to, the human factors implications of the things people use.

In connection with human factors research and applications to date, we could develop quite a list of areas that it has not been feasible to include in this text or that are touched on very briefly. Such a list could include such areas as consumer products, health services, recreation, the expanding use of computers, law enforcement, exploration of outer space, communication systems, mining, agriculture, (in the words of the king of Siam in *The King and I*) etcetera, etcetera, etcetera. (It should be added that the human factors inroads into some of these, and other, areas have to date still been fairly limited.)

But now, what of the future? There are many items of unfinished business relating to the human factors implications of the many facets of present-day work and living that have not yet benefited from the possible application of human factors data and principles. Beyond such concerns, however, there are at least a couple of broad areas of attention that should serve as challenges to the human factors clan.

One of these deals with what (in present-day expressions) is called the *quality of working life* or, more broadly, the *quality of life*. There is some truth to the charges that human factors efforts have tended toward making work easier for people to do and perhaps toward greater work specialization. In recent years, however, we have heard much about job enrichment programs that are directed toward opposite objectives of broadening work responsibilities, increasing the decision-making aspects of jobs, etc. These efforts have as their aim the enhancement of job satisfaction and the improvement of the quality of working life. Although job enrichment probably is not for everybody, its general goals are on the positive side of the ledger. In recent years several people prominent in human factors affairs have expressed concern about the quality of working life and have urged the human factors community to work toward such a goal. However, there seems to be a basic need for efforts to blend the human factors and the job enrichment approaches into an integrated effort which would offer reasonable possibilities for people to engage in work activities that are efficient and safe and that also would enhance the quality of working life.

The second area of challenge to the human factors discipline deals with the shape of things to come and actually is related to the previous discussion of the quality of working life. The future is always somewhat shrouded, but we certainly can expect increased automation in the production of goods and services with increased use of computer-controlled processes, significant changes in technology and in the products and services that such changes may bring, and the development of new products and services that we simply cannot envision. The "brave new world" of the future should indeed be developed with people—all of us—in mind. Thus, the human factors discipline must be at the cutting edge of future developments to ensure that such developments will, in reality, contribute to the improvement of the quality of working life and of life in general.

REFERENCES

Ainsworth, L. (1982). An RSM investigation of defect rate and other variables which influence inspection. *Proceedings of the Human Factors Society 26th Annual Meeting*. Santa Monica, CA: Human Factors Society, pp. 868–872.

Bergum, B. O., and Klein, I. C. (1961, November). *A survey and analysis of vigilance research*, Res. Rept. 8. Alexandria, VA: Human Resources Research Office.

Brown, D. B. (1973). Cost/benefit of safety investments using fault tree analysis. *Journal of Safety Research*, 5(3), 73–79.

Chaney, F. B., and Teel, K. S. (1967). Improving inspector performance through training and visual aids. *Journal of Applied Psychology*, 51(4), 311–315.

Christensen, J. M. (1980). Human factors in hazard/risk evaluation. In *Proceedings of the Symposium—Human Factors and Industrial Design in Consumer Products*. Medford, MA: Tufts University, pp. 442–447.

Cohen, A., Smith, M. J., and Anger, W. K. (1979). Self-protective measures against workplace hazards. *Journal of Safety Research*, 11(3), 121–131.

Colquhoun, W. P. (1959). The effect of a short rest pause on inspection efficiency. *Ergonomics*, 2, 367–372.

Conway, E., Muckler, F., and Peay, J. (1980). Human error injuries and accidents—A review and analysis. *Proceedings of the Human Factors Society 24th Annual Meeting*. Santa Monica, CA: Human Factors Society.

Davies, D., and Parasuraman, R. (1982). *The psychology of vigilance*. New York: Academic.

DeGreen, K. (1972). *Systems psychology*. New York: McGraw-Hill.

Drury, C. G. (1975). Inspection of sheet materials: Model and data. *Human Factors*, 17(3), 257–265.

Drury, C. G., and Brill, M. (1980). New methods of consumer product accident investigation. In *Proceedings of the Symposium—Human Factors and Industrial Design in Consumer Products*. Medford, MA: Tufts University, pp. 196–211.

Drury, C., and Sinclair, M. (1983). Human and machine performance in an inspection task. *Human Factors*, 25(4), 391–399.

Edwards, D. S., and Hahn, C. P. (1980). A chance to happen. *Journal of Safety Research*, 12(2), 59–67.

Harris, D. H., and Chaney, F. B. (1969). *Human factors in quality assurance*. New York: Wiley.

Heinrich, H. (1959). *Industrial accident prevention* (4th ed). New York: McGraw-Hill.

Kelley, M. L. (1955). A study of industrial inspection by the method of compared comparison. *Psychological Monographs*, 69(9), no. 394.

Komaki, J., Barwick, K. D., and Scott, L. R. (1978). A behavioral approach to occupational safety: Pinpointing and reinforcing safe performance in a food manufacturing plant. *Journal of Applied Psychology*, 63(4), 434–445.

Lawrence, A. (1974). Human error as a cause of accidents in gold mining. *Journal of Safety Research*, 6, 78–88.

Lion, J. S., Richardson, E., Weightman, D., and Browne, B. C. (1975). The influence of the visual arrangement of material, and of working singly or in pairs, upon performance at simulated industrial inspection. *Ergonomics*, 18(2), 195–204.

Loeb, M., and Alluisi, E. (1980). Theories of vigilance: A modern perspective. *Proceedings of the Human Factors Society 24th Annual Meeting*. Santa Monica, CA: Human Factors Society, pp. 600–603.

Mackie, R. (1984). Research relevance and the information glut. In F. A. Muckler (ed.), *Human factors review 1984*. Santa Monica, CA: Human Factors Society.

Meister, D. (1966). Human factors in reliability. In W. G. Ireson (ed.), *Reliability handbook* (sec. 12). New York: McGraw-Hill.

Purswell, J. L., Greenhaw, L. N., and Oats, C. (1972). An inspection task experiment. In *Proceedings of the Human Factors Society*. Santa Monica, CA: Human Factors Society, pp. 297–300.

Purswell, J. L., and Hoag, L. L. (1974). Strategies for improving visual inspection performance. *Proceedings of the Human Factors Society*. Santa Monica, CA: Human Factors Society, pp. 397–403.

Ramsey, J. D. (1973). Identification of contributory factors in occupational injury. *Journal of Safety Research*, 5(4), 260–267.

Ramsey, J. D. (1978, May). Ergonomic support of consumer product safety. Paper presented at the American Industrial Association Conference.

Rook, L. W., Jr. (1962). *Reduction of human error in industrial production* [SCTM 93-62(14)]. Albuquerque, NM: Sandia Corporation.

Sell, R. G. (1977). What does safety propaganda do for safety? A review. *Applied Ergonomics*, 8(4), 203–214.

Smith, G. L., Jr., (1972). Signal detection theory and industrial inspection. *Proceedings of the Human Factors Society*. Santa Monica, CA: Human Factors Society, pp. 284–290.

Smith, M. J., Anger, W. K., and Uslan, S. S. (1978). Behavioral modification applied to occupational safety. *Journal of Safety Research*, 10(2), 87–88.

Swain, A., and Guttmann, H. (1983). *Handbook of human reliability analysis with emphasis on nuclear power plant applications* (NUREG/CR-1278). Washington: Nuclear Regulatory Commission.

Teichner, W. H. (1974). The detection of a simple vision signal as a function of time of watch. *Human Factors*, 16(4), 339–353.

Vesely, W., Goldberg, F., Roberts, N., and Haasl, D. (1981). *Fault tree handbook*, NUREG-0492. Washington: Nuclear Regulatory Commission.

Waikar, A. (1973). Quality improvement using multiple inspector systems. Unpublished master's thesis, University of Oklahoma.

Wiener, E. L. (1977). Stimulus presentation rate in vigilance. *Human Factors*, 19(3), 301–303.

Williges, R. C., and Streeter, H. (1972). Influence of static and dynamic displays on inspection performance. *Proceedings of the Human Factors Society*. Santa Monica, CA: Human Factors Society, pp. 291–296.

APPENDIXES

LIST OF ABBREVIATIONS

A	ampere
ACGIH	American Conference of Governmental Industrial Hygienists
AFB	Air Force Base
AFHRL	Air Force Human Resources Laboratory
AFSC	Air Force Systems Command
AI	articulation index
AL	action limit
AMD	Aerospace Medical Division, Air Force Systems Command
AMRL (of USAF)	Aerospace Medical Research Laboratory, Aerospace Medical Division, Air Force Systems Command
ANSI	American National Standards Institute
APA	American Psychological Association
ASD	Aeronautical Systems Division, Air Force Systems Command
ASHA	American Speech and Hearing Association
ASHRAE	American Society of Heating, Refrigerating, and Air-Conditioning Engineers
ASHVE	American Society of Heating and Ventilating Engineers (now ASHRAE)
BB	Botsball index
BCD	borderline between comfort and discomfort
°C	Celsius
C	convection (when used as a method of heat exchange)
C	contrast (when used regarding differences in reflectance of visual stimuli)
CAFES	computer-aided function-allocation evaluation system
CAR	crew station assessment of reach
cd	candela
C/D	control-display
CFF	critical flicker fusion

CHESS	crew human engineering software system
CIE	Commission International de l'Eclairage
clo	clo unit (a measure of insulation of clothing)
cm	centimeter
CNEL	community noise equivalent level
C/R	control-response
CRI	color rendering index
CRT	cathode ray tube
d	day
D	dose (when used in reference to noise exposure)
dB	decibel
DGR	discomfort glare rating
DHHS	Department of Health and Human Services
DVA	dynamic visual acuity
E	evaporation
ECP	evoked cortical potential
ed.	edition
ed(s).	editor(s)
EEG	electroencephalogram
EMG	electromyogram
EPA	Environmental Protection Agency
EPRI	Electric Power Research Institute
ESI	equivalent sphere illumination
ET	original effective-temperature scale
ET*	current (new) effective-temperature scale
°F	Fahrenheit
FAM	function allocation model
fc	footcandle
FDP	fatigue-decreasing proficiency
FFF	flicker fusion frequency
fL	foot-lambert
ft	foot
FTA	fault-tree analysis
g	gram
G	gravity (referring to acceleration)
g	gravity (referring to vibration)
gal	gallon
GSR	galvanic skin response
h	hour
H	amount of information in bits
HECAD	human engineering computer-aided design
HFS	Human Factors Society
HRV	heart rate variability (same as SA)
HST	hand-skin temperature
Hz	hertz
IEEE	Institute of Electrical and Electronics Engineers
IES	Illuminating Engineering Society
in	inch
IRA	instrument readability analysis
IRE	Institute of Radio Engineers

ISO	International Organization for Standardization
J	joule
JND	just-noticeable difference
JSI	job severity index
K	kelvin
kc	kilocycle
kcal	kilocalorie
kg	kilogram
km	kilometer
lb	pound
lm	lumen
lx	lux
m	meter
M	metabolism
MCE	modular comfort envelope
mi	mile
mi/h	miles per hour
min	minute
mL	millilambert
mm	millimeter
MMH	manual materials handling
MPL	maximum permissible limit
ms	millisecond
N	newton
NADC	Naval Air Development Center
NAS	National Academy of Sciences
NASA	National Aeronautics and Space Administration
NC	noise criteria
NIOSH	National Institute for Occupational Safety and Health
NIPTS	noise-induced permanent threshold shift
nm	nanometer
NRC	National Research Council
NRR	noise reduction rating
OSD	operational-sequence diagram
OSHA	Occupational Safety and Health Administration
PLdB	perceived level of noise, decibel
PNC	preferred noise criteria
PNdB	perceived noise, decibel
PNL	perceived noise level
PSD	power spectral density
PSIL	preferred-octave speech interference level
PTS	permanent threshold shift (of hearing)
R	radiation (when used as method of heat exchange)
R	reliability (when used with regard to performance)
RH	relative humidity
rms	root mean square
s	second
SA	sinus arrhythmia (same as HRV)
SAE	Society of Automotive Engineers
SCRT	serial choice reaction time

SDT	signal detection theory
SI	International System of units
SIL	speech interference level
SOS	spatial operational sequence
SPL	sound-pressure level
sr	steradian
SRP	seat reference point
SWAM	statistical workload assessment model
SWAT	subjective workload assessment technique
THERP	technique for human error rate prediction
TOT	time on target
TR	technical report (term used by various organizations)
TTS	temporary threshold shift
TVSS	tactile vision substitution system
TWA	time-weighted average
UMTA	Urban Mass Transportation Administration
USA	United States of America; United States Army
USAF	United States Air Force
USASI	United States of America Standards Institute (now ANSI)
USN	United States Navy
USPHS	United States Public Health Service
UV	ultraviolet
VCP	visual comfort probability
VDT	visual display terminal
VDU	visual display unit
VL	visibility level
VL_{eff}	effective VL
VTE	visual task evaluator
VWF	vibration-induced white finger
W	watt
WADC	Wright Air Development Center, USAF (see AMRL and AFHRL)
WADD	Wright Air Development Division, USAF (see AMRL and AFHRL)
WAM	workload assessment model
WBGT	wet-bulb globe temperature
WCI	wind chill index
μ	micrometer (10^{-6})
π	pi

CONTROL DEVICES

Table B-1 presents a brief evaluation of the operational characteristics of certain types of control devices.[1] Table B-2 summarizes recommendations regarding certain features of these types of control devices.[2] In the use of these and other recommendations, it should be kept in mind that the unique situation in which a control device is to be used and the purposes for which it is to be used can affect materially the appropriateness of a given type of control and can justify (or virtually require) variations from a set of general recommendations or from general practice based on research or experience. For further information regarding these, refer to the original sources given in the reports from which these are drawn.

COMMENTS REGARDING CONTROLS[3]

- *Hand pushbutton*: Surface should be concave or provide friction. Preferably there should be an audible click when activated. Use elastic resistance plus slight sliding friction, starting low, building up rapidly, to a sudden drop. Minimize viscous damping and inertial resistance.
- *Foot pushbutton*: Use elastic resistance, aided by static friction, to support foot. Resistance to start low, build up rapidly, drop suddenly. Minimize viscous damping and inertial resistance.
- *Toggle switch*: Use elastic resistance which builds up and then decreases as position is approached. Minimize frictional and inertial resistance.
- *Rotary selector switch*: Provide detent for each control position (setting). Use elastic resistance which builds up and then decreases as detent is approached. Minimize friction and inertial resistance. Separation of detents should be at least ¼ in (6 mm).

[1]Adapted largely from A. Chapanis. "Design of Controls," chap. 8 in H. P. Van Cott and R. G. Kinkade, *Human engineering guide to equipment design* (rev. ed.). Washington: Government Printing Office, 1972.
[2]Adapted from ibid.
[3]Adapted from ibid.

- *Knob*: Preferably code by shade if knob is used without vision. Type of desirable resistance depends on performance requirements.
- *Crank*: Use when task involves two rotations or more. Friction [2 to 5 lb (0.9 to 2.3 kg)] reduces effects of jolting but degrades constant-speed rotation at slow or moderate speeds. Inertial resistance aids performance for small cranks and low rates. Grip handle should rotate.
- *Lever*: Provide elbow support for large adjustment, forearm support for small hand movements, wrist support for finger movements. Limit movement to 90°.
- *Handwheel*: For small movements, minimize inertia. Indentations in grip rim aid holding. Displacement usually should not exceed ±60° from normal. For displacements less than 120°, only two sections need to be provided, each of which is at least 6 in (15 cm) long. Rim should have frictional resistance.
- *Pedal*: Pedal should return to null position when force is removed; hence, elastic resistance should be provided. Pedals operated by entire leg should have 2- to 4-in (5- to 10-cm) displacement, except for automobile-brake type, for which 2 to 3 in (5 to 7 cm) of travel may be added. Displacement of 3 to 4 in (7 to 10 cm) or more should have resistance of 10 lb (4.5 kg) or more. Pedals operated by ankle action should have maximum travel of 2½ in (5.1 cm).

TABLE B-1
COMPARISON OF COMMON CONTROL CHARACTERISTICS

Characteristic	Hand push-button	Foot push-button	Toggle switch	Rotary switch	Knob	Crank	Lever	Hand wheel	Pedal
Space required	Small	Large	Small	Medium	Small–medium	Medium–large	Medium–large	Large	Large
Effectiveness of coding	Fair–good	Poor	Fair	Good	Good	Fair	Good	Fair	Poor
Ease of visual identification of control position	Poor*	Poor	Fair–good	Fair–good	Fair–good†	Poor‡	Fair–good	Poor–fair	Poor
Ease of nonvisual identification of control position	Fair	Poor	Good	Fair–good	Poor–good	Poor‡	Poor–fair	Poor–fair	Poor–fair
Ease of check reading in array of like controls	Poor*	Poor	Good	Good	Good†	Poor‡	Good	Poor	Poor
Ease of operation in array of like controls	Good	Poor	Good	Poor	Poor	Poor	Good	Poor	Poor
Effectiveness in combined control	Good	Poor	Good	Fair	Good§	Poor	Good	Good	Poor

*Except when control is backlighted and light comes on when control is activated.
†Applicable only when control makes less than one rotation and when round knobs have pointer attached.
‡Assumes control makes more than one rotation.
§Effective primarily when mounted concentrically on one axis with other knobs.

TABLE B-2
SELECTED DATA REGARDING DESIGN RECOMMENDATIONS FOR CONTROL DEVICES

Device	Size, in		Displacement		Resistanced	
	Minimum	Maximum	Minimum	Maximum	Minimum	Maximum
Hand pushbutton						
Fingertip operation	½	None	⅛ in	15 in	10 oz	40 oz
Foot pushbutton	½	None				
Normal operation			½ in			
Wearing boots			1 in	2½ in		
Ankle flexion only				4 in		
Leg movement						
Will *not* rest on control					4 lb	20 lb
May rest on control					10 lb	20 lb
Toggle switch			30°	120°	10 oz	40 oz
Control tip diameter	⅛	1				
Lever arm length	½	2				
Rotary selector switch					10 oz	40 oz
Length	1	3				
Width	½	1				
Depth	½					
Visual positioning			15°	40°*		
Nonvisual positioning			30°	40°*		
Knob, continuous adjustment†						
Finger-thumb						4½–6 in/oz
Depth	½	1				
Diameter	⅜	4				
Hand/palm diameter	1½	3				
Crank†						
For light loads, radius	½	4½				
For heavy loads, radius	½	20				
Rapid, steady turning						
3–5 in radius					2 lb	5 lb

(*Continued*)

TABLE B-2 (continued)
SUMMARY OF SELECTED DATA REGARDING DESIGN RECOMMENDATIONS FOR CONTROL DEVICES

Device	Size, in		Displacement		Resistance	
	Minimum	Maximum	Minimum	Maximum	Minimum	Maximum
5–8 in radius					5 lb	10 lb
8-in radius					?	?
For precise settings					2½ lb	8 lb
Lever§						
Fore-aft (one-hand)				14 in		
Lateral (one hand)				38 in		
Finger grasp, diameter	½	3			12 oz	32 oz
Hand grasp, diameter	1½	3			2 lb	20–100 lb
Handwheel†						
Diameter	7	21		90°–120°	5 lb	30 lb‡
Rim thickness	¾	2				
Pedal						
Length	3½					
Width	1					
Normal use			½ in			
Heavy boots			1 in			
Ankle flexion				2½ in		
Leg movement				7 in		
Will *not* rest on control					4 lb	10 lb
May rest on control					10 lb	180 lb

*When special requirements demand large separations, maximum should be 90°.
†Displacement of knobs, cranks, and handwheels should be determined by desired control-display ratio.
‡For two-handed operation, maximum resistance of handwheel can be up to 50 lb.
§Length depends on situation, including mechanical advantage required. For long movements, longer levers are desirable (so movement is more linear).

NIOSH RECOMMENDED ACTION LIMIT FORMULA FOR LIFTING TASKS

Factor	*U.S. Customary System units*	*Metric*
Horizontal location	6/H	15/H
Vertical location	$1 - [0.01 \times (V - 30)]$	$1 - (0.004 \times |V - 75|)$
Distance traveled (vertical)	$0.7 + 3/D$	$0.7 + 7.5/D$
Frequency of lift	$1 - F/F_{max}$	$1 - F/F_{max}$
Constant	90	40

Basic equations:

$$AL = (\text{constant}) \times (\text{horizontal location}) \times$$
$$(\text{vertical location}) \times (\text{distance traveled}) \times$$
$$(\text{frequency of lift})$$
$$MPL = 3 \times AL$$

where AL = action limit (lbs or kg) = absolute value of quantity
  H = horizontal location (cm or in) forward of midpoint between ankles at origin of lift [6 to 32 in (15 to 81 cm)]
  V = vertical location (cm or in) at origin of lift [0 to 70 in (0 to 178 cm)]
  D = vertical travel distance (cm or in) between origin and destination of lift [minimum 10 in (25 cm); if less than minimum, set at minimum]
  F = average frequency of lift (lifts/min); 0.2 or 1 lift per 5 min to F_{max}); if less than 0.2, set $F = 0$
  MPL = maximum permissible limit
  F_{max} = maximum frequency which can be sustained, taken from following chart:

Average vertical
location, in (cm)
>30 (75)   ≤30 (75)

		>30 (75)	≤30 (75)
	1 h or less	18	15
Period of performance	More or less continuous during shift	15	12

Source: Adapted from NIOSH, 1981.

SELECTED REFERENCES

This appendix includes a list of selected books and journals that deal with human factors. A number of the books listed are general references, while others deal with specific aspects of human factors. For additional references that deal with topics specific to the content of the individual chapters, see the references at the ends of the individual chapters.

BOOKS

Alluisi, E., and Fleishman, E. (eds.) (1982). *Human performance and productivity*, vol. 3: *Stress and performance effectiveness*. Hillsdale, NJ: Lawrence Erlbaum Associates.

Anderson, D., Istance, H., and Spencer, J. (eds.). (1977). *Human factors in the design and operation of ships*. Stockholm, Sweden: Ergonomilaboratoriet AB.

Anthropometric source book, vol. 1: *Anthropometry for designers* (NASA RP 1024). (1978). Houston, TX: National Aeronautics and Space Administration.

Bailey, R. (1982). *Human performance engineering*. Englewood Cliffs, NJ: Prentice-Hall.

Boyce, P. (1981). *Human factors in lighting*. New York: Macmillan.

Cakir, A., Hart, D., and Stewart, T. (1979). *Visual display terminals*. New York: Wiley.

Card, S., Moran, T., and Newell, A. (1983). *The psychology of human-computer interaction*. Hillsdale, NJ: Lawrence Erlbaum Associates.

Chaffin, D., and Andersson, G. (1984). *Occupational biomechanics*. New York: Wiley.

Chapanis, A. (ed.) (1975). *Ethnic variables in human factors engineering*. Baltimore, MD: Johns Hopkins.

Damon, A., Stoudt, H. W., and McFarland, R. A. (1966). *The human body in equipment design*. Cambridge, MA: Harvard.

Davies, D., and Parsumaman, R. (1982) *The psychology of vigilance*. New York: Academic.

DeGreene, K. B. (1972). *Systems psychology*. New York: McGraw-Hill.

Drury, C., and Fox, J. (eds.) (1975). *Human reliability in quality control*. London: Taylor & Francis.

Eastman Kodak Co. (1983). *Ergonomic design for people at work* (vol. 1). Belmont, CA: Lifetime Learning Publications.

Edwards, E., and Lees, F. (eds.) (1974). *The human operator in process control*. London: Taylor & Francis.

Forbes, T. W. (ed.) (1972) *Human factors in highway traffic safety research*. New York: Wiley.

Grandjean, E. (1973). *Ergonomics of the home*. London: Taylor & Francis.

Grandjean, E. (1980). *Fitting the task to the man: An ergonomic approach*. London: Taylor & Francis.

Grandjean, E. (ed.) (1984). *Ergonomics and health in modern offices*. London: Taylor & Francis.

Grandjean, E., and Vigliani, E. (eds.) (1980). *Ergonomic aspects of visual display terminals*. London: Taylor & Francis.

Harris, D. H., and Chaney, F. B. (1969). *Human factors in quality assurance*. New York: Wiley.

Helander, M. (ed.) (1981). *Human factors/ergonomics for building and construction*. New York: Wiley.

Hockey, R. (ed.) (1983). *Stress and fatigue in human performance*. New York: Wiley.

Howell, W., and Fleishman, E. (eds.) (1982). *Human performance and productivity*, vol. 2: *Information processing and decision making*. Hillsdale, NJ: Lawrence Erlbaum Associates.

Howell, W. C., and Goldstein, I. L. (1971). *Engineering psychology: Current perspectives in research*. New York: Meredith Corporation.

Kelley, C. (1968). *Manual and automatic control*. New York: Wiley.

Kinkade, R., and Anderson, J. (eds.) (1984). *Human factors guide for nuclear power plant control room development* (NP3659). Palo Alto, CA: Electric Power Research Institute.

Konz, S. (1979). *Work design*. Columbus, OH: Grid Inc.

Kraiss, K., and Moraal, J. (eds.) (1976). *Introduction to human engineering*. Bonn, Germany: Verlag TUV Rheinland GimgH.

Kryter, K. (1985). *The effects of noise on man* (2d ed.). Orlando, FL: Academic.

Kvalseth, T. (ed.) (1983). *Ergonomics of workstation design*. London: Butterworths.

Mackie, R. (ed.) (1977). *Vigilance: theory, operational performance, and physiological correlates, vol. 3: Human Factors NATO Conference Series*. New York: Plenum.

Meister, D. (1971). *Human factors: Theory and practice*. New York: Wiley.

Meister, D. (1976). *Behavioral foundations of systems development*. New York: Wiley.

Meister, D. (1984, July). *Human engineering data base for design and selection of cathode ray tube and other display systems* (AD-A145704). Washington: National Technical Information Service.

Meister, D. (1985). *Behavioral analysis and measurement methods*. New York: Wiley.

Meister, D, and Rabideau, G. F. (1965). *Human factors evaluation in system development*. New York: Wiley.

Murrell, K. (1969). *Ergonomics: man in his working environment*. London: Chapman and Hall.

Murrell, K. (1976). *Men and machines*. London: Methuen.

National Aeronautics and Space Administration. (1973). *Bioastronautics data book*, NASA SP-3006. J.F. Parker, Jr., and V. R. West (Managing eds.). Washington: Government Printing Office.

National Research Council (1983). *Video displays, work and vision*. Washington: National Academy Press.

Oborne, D. (1982). *Ergonomics at work*. New York: Wiley.

Oborne, D. and Levis, J. (eds.) (1980). *Human factors in transport research* (vols. 1 and 2). New York: Academic.

Parsons, H. M. (1972). *Man-machine system experiments*. Baltimore, MD: Johns Hopkins.

Poulton, E. (1979). *The environment at work*. Springfield, IL: Charles C. Thomas.

Roebuck, J., Kroemer, K., and Thomson, W. (1975). *Engineering anthropometry techniques*. New York: Wiley.

Roscoe, S. (ed.) (1980). *Aviation psychology*. Ames: Iowa State University Press.

Schmidtke, H. (ed.) (1984). *Ergonomic data for equipment design*. New York: Plenum.

Seminara, J., Gonzalez, W., and Parsons, S. (1977). *Human factors review of nuclear power plant control room design*. Palo Alto, CA: Electric Power Research Institute.

Shackel, B. (ed.) (1974). *Applied ergonomics handbook*. Surrey, England: IPC Science and Technology Press.

Sheridan, T., and Ferrell, W. (1974). *Man-machine systems: Information, control, and decision models of human performance*. Cambridge, MA: M.I.T. Press.

Shurtleff, D. (1980). *How to make displays legible*. La Mirada, CA: Human Interface Design.

Singleton, W. (1972). *Introduction to ergonomics*. Geneva, Switzerland: World Health Organization.

Singleton, W. (ed.) (1982). *The body at work: Biological ergonomics*. New York: Cambridge University Press.

Singleton, W., Easterby, R., and Whitfield, D. (eds.) (1971). *The human operator in complex systems*. London: Taylor & Francis.

Singleton, W., Fox, J., and Whitfield, D. (eds.) (1971). *Measurement of man at work*. London: Taylor & Francis.

Swain, A., and Guttmann, H. (1983). *Handbook of human reliability analysis with emphasis on nuclear power plant applications* (NUREG/CR-1278). Washington: Nuclear Regulatory Commission.

Van Cott, H. P., and Kinkade, R. G. (1972) *Human engineering guide to equipment design* (rev. ed.) Washington: Government Printing Office.

Weiner, J., and Maule, H. (eds.) (1977). *Case studies in ergonomic practice*, vol. 1: *Human factors in work, design, and production*. London: Taylor & Francis.

Wickens, C. (1984). *Engineering psychology and human performance*. Columbus, OH: Merrill.

Winter, D. (1979). *Biomechanics of human movement*. c21 New York: Wiley.

Woodson, W. (1981). *Human factors design handbook*. New York: McGraw-Hill.

PERIODICALS

Applied Ergonomics. Guildford, Surrey, England: IPC House.

Behaviour and Information Technology, London: Taylor & Francis.

Ergonomics, London: Taylor & Francis.

Human Factors, Santa Monica, CA: Human Factors Society.

Proceedings of the Human Factors Society Annual Meetings. Santa Monica, CA: Human Factors Society.

INDEXES

NAME INDEX

651

SUBJECT INDEX